Energy Demand Studies: Major Consuming Countries

Analyses of 1972 Demand and Projections of 1985 Demand

First Technical Report of the Workshop on Alternative Energy Strategies (WAES)

Paul S. Basile
Editor

Carroll L. Wilson
WAES Project Director

The MIT Press
Cambridge, Massachusetts, and London, England

This material is based upon research supported in part by the National Science Foundation under Grant No. OEP74-13902 AO1. Any opinions, findings, and conclusions or recommendations expressed in this publication are those of the authors and do not necessarily reflect the views of the National Science Foundation.

This book was set in IBM Composer Century by Eastern Composition, Inc. and printed and bound by The Colonial Press Inc. in the United States of America.

Second printing, 1977

Library of Congress Cataloging in Publication Data

Workshop on Alternative Energy Strategies.
Energy demand studies, major consuming countries.

1. Power resources. 2. Energy consumption.
3. Economic forecasting. I. Title.
HD9502.A2W66 1976 333.7 76-26933
ISBN 0-262-23076-3

Contents

Preface

In October 1974, leaders of business, industry, government, and academia from a dozen countries convened the first session of the Workshop on Alternative Energy Strategies (WAES). Since that time, the Workshop has carried forward studies of energy demand and supply, of resource availability, and of the critical factors in alternative energy futures, both nationally and globally. This book represents the first published technical report of the Workshop.

The Workshop is an ad hoc, international project, involving over seventy-five people from fifteen countries.[1] It has set itself the goal of assembling a set of coherent global energy pictures of the future. To this end, it has developed methods for estimating energy supply and demand through the year 2000 and, using these projections as a basis, for identifying alternative strategies that individual nations might employ. These strategies will be both global and national in scope.

This goal is set against a background which recognizes that, according to the best estimates, world crude oil production will peak and begin to decline within the next twenty-five years. The transition from oil to other forms of energy is coming more rapidly than most people realize. WAES evolved out of the common concern of a number of influential people who believed that this transition needed to be widely understood and effectively managed in order to avoid major national and international dislocations.

Yet much hinges on the future demand for energy. Will demand growth revert to historical rates—or will it be moderated in the future by lowered economic growth, or higher energy prices, or greater conservation efforts? What is the likely range of estimated levels of energy demand in the future? What is the potential for reduced demand, and for greater efficiency of energy use?

1. Members of the Workshop are from nations outside COMECON and China. These countries use a large fraction of the world's energy. COMECON is the Council for Mutual Economic Assistance, an international economic organization consisting primarily of the U.S.S.R. and most East European countries.

Although the U.S.S.R. and China are major world energy producers and consumers, their trade in fuels with the countries in the world outside COMECON and China (WOCC) has to date been relatively small. The future role of COMECON and China as potential energy exporters will be assessed by WAES as part of its analysis.

This first technical report of the Workshop seeks to treat these issues, country by country, in a consistent manner for the year 1985. It contains detailed 1972 energy demand data and projections of energy demand to 1985 for thirteen major consuming countries under five different "scenarios" based on a combination of variables representing economic growth, energy price, and national energy policy. Other issues, relating to supply potentials, the availability of substitute fuels, and technological and lead-time constraints, will be taken up in later reports of the Workshop.

A final report, to be published in the spring of 1977, will identify major long-term energy problems and describe alternative strategies for dealing with them. The report will focus on feasible global and national energy strategies that reflect cooperation between consumer and producer countries, including countries of the Third World.

The Workshop on Alternative Energy Strategies is an independent organization, with no official ties to any governments or private firms.[2] It is organized into three groups:

○ The *Participants* are 35 senior decision makers from around the world. They are experienced in dealing with problems involving the interplay of technology, economics, finance, and government policy. Members include officials from organizations that are suppliers and users of fuels or that produce electricity, government officials with energy-related responsibilities, and others whose insights enrich the Workshop. The Participants are taking part in this study as individuals, not as official representatives of their organization or country. They provide or arrange for financial support for WAES activities in their own country and pay for the expenses involved in attending and sponsoring meetings.

○ The *Associates*, chosen by the Participants from their own or cooperating organizations, are scientists, engineers, economists, and managers with expertise in many energy-related fields. They conduct the technical studies that are used by the Participants in formulating energy strategies.

2. Further information about the Workshop can be found in "A Global Energy Assessment 1985–2000," a brochure available from any of the Workshop Participants or from the Workshop Secretariat.

In order to expand their capabilities and influence, in some countries Workshop members have recruited Reference Groups—people from government and industry who serve as advisors, provide financial support, and do some of the analytical work.

○ The *Secretariat* includes the Project Director and a small staff at the Massachusetts Institute of Technology. The Director provides executive leadership, and the Secretariat provides project coordination. Secretariat expenses are supported by foundation grants.

This distinctive character allows the Workshop to have two important effects. Through its technical analyses it provides the necessary quantitative basis for policy decisions. More directly, involvement in such an in-depth study influences the members themselves, who will, it is hoped, begin the process of turning the options identified by the Workshop into realities.

The Workshop objectives are to establish a quantitative basis for analyzing probable future energy developments through the year 2000 and then to identify feasible alternative global and national energy strategies to deal with these developments. The analyses are designed to provide information that will raise the level of public understanding and assist governments and others in making the choices and formulating the strategies necessary to assure continuity of energy supply during the period 1985 to 2000 and beyond.

An energy strategy is a set of sustained actions taken over a long enough time to achieve a feasible energy supply-demand balance for the future. The principal energy strategy decisions are made by individual nations. Each nation has its goals, priorities, and policies that will influence the way in which it evaluates alternative energy strategies. Yet national strategies must be related to what is happening in the rest of the world if they are to be feasible.

Few nations, acting alone, have the resources to ensure their own energy supplies. Therefore, it is in the common interest to avoid placing insupportable demands on specific energy sources. For example, it makes no sense for one country to build power stations that will depend on coal from another country if the latter does not develop the needed capacity for coal production, transport, and export. The world energy system may be moving from a demand-driven to a supply- and infrastructure-constrained one, at least in the medium term. The international nature of WAES enables it to approach such issues on a global basis.

In looking at the period to the end of the century, it is important to determine what can be achieved in the time available. Economic growth and energy "savings" will be the main determinants of energy demand up to 1985, with oil remaining the principal fuel. Beyond 1985, alternative sources will be needed on an increasing scale. Because of the long lead times required, however, their contributions will be determined by actions taken now or in the near future.

The national studies of energy demand to 1985 in this volume are a result of many months' effort by Workshop members. A brief description of the overall methodology developed by the Workshop is given in Chapter 1, together with a detailed description of the energy demand studies methodology. This unique approach to energy strategy analysis draws partially on many traditional analytical techniques and tools, which have been modified and adapted by the Workshop Associates with the guidance and assistance of the Participants. The WAES methodology is endorsed and supported by all Workshop members. Each national study is, however, attributable solely to the Workshop members from a particular country, and although all studies were done in a common and consistent manner, each has its own distinctive aspects.

As with any study of the future, the projections in this volume should be taken with care. The Workshop claims no special ability to see into the future. It does claim, however, to have developed reasonable expectations for energy demand in its member countries—based on careful and explicit assumptions about several alternative states of the world.

There are differences among national approaches. Many of these variations in national style and decision making are revealed by the studies in this book. These differences add richness to the studies and illustrate some of the potentials for change in present-day patterns of energy use. Such potentials should be carefully assessed by the many analysts, scientists, and research groups outside the Workshop who have the ability and time to do so.

Carroll L. Wilson
Project Director

Workshop Participants, Associates, and Staff

PROJECT DIRECTOR
Carroll L. Wilson
Massachusetts Institute of Technology

PARTICIPANTS

Canada
Mr. Marshall A. Crowe
Chairman, National Energy Board

Mr. Maurice F. Strong
Chairman, Petro-Canada

Denmark
Professor Bent Elbek
Niels Bohr Institute

Finland
Professor Jorma Routti
Helsinki University of Technology

France
M. Jean Couture
Conseiller de Président
Société Générale

M. Jean-Marie Martin
Director, Institut Economique et Juridique de l'Energie
Centre National de la Recherche Scientifique

Germany
Director Dr. Hans Detzer
Head of Central Planning
Badische Anilin & Soda Fabrik AG

Professor Heinrich Mandel
Member of the Board of Directors
Rheinisch-Westfälisches Elektrizitätswerk AG

Professor Hans K. Schneider
Director, Institute of Energy Economics
University of Cologne

Iran
Dr. Khodadad Farmanfarmaian
Chairman, Bank Sanaye Iran

Italy
Professor Umberto Colombo
Director, Research & Development Division
Montedison

Professor Sergio Vaccà
Director, Istituto di Economia delle Fonti de Energia
Università L. Bocconi, Milano

Japan
Mr. Toshio Doko
President, Japan Federation of Economic Organizations (Keidanren)

Mr. Shuzo Inaba
President, Japan Computer Usage Development Institute

Mr. Soichi Matsune
Chairman, Committee on Energy
Japan Federation of Economic Organizations (Keidanren)

Mr. Tatsuzo Mizukami
President, Japan Foreign Trade Council, Inc.

Dr. Saburo Okita
President, The Overseas Economic Cooperation Fund

Mr. Masao Sakisaka
President, National Institute for Research Advancement

Mr. Shigefumi Tamiya
Advisor to the Chairman
Enrichment and Reprocessing Group

Mexico
Ing. Juan Eibenschutz
Executive Secretary
Mexican National Energy Commission

Dr. Victor Urquidi
President, El Colégio de México

The Netherlands
Dr. A. A. T. van Rhijn
Deputy Director General for Energy
Ministry of Economic Affairs

Netherlands/U.K.
Mr. J. C. Davidson
Director, Shell International Petroleum Company, Ltd.
Royal Dutch/Shell Group

Norway
Mr. Christian Sommerfelt
Chairman
Elkem-Spigerverket A/S

Sweden
Mr. Erland Waldenström
Chairman
Gränges AB

United Kingdom
Mr. Robert Belgrave
Policy Advisor
British Petroleum Company, Ltd.

Professor Sir William Hawthorne
Master, Churchill College
University of Cambridge

United States
Mr. Thornton F. Bradshaw
President, Atlantic Richfield Company

Mr. Walker L. Cisler
Retired Chairman of the Board
Detroit Edison Company

Mr. John T. Connor
Chairman of the Board
Allied Chemical Corporation

Mr. Richard C. Gerstenberg
Director, General Motors Corporation

Dr. H. Guyford Stever
Director, National Science Foundation

Venezuela
Dr. José A. Mayobre
Banco Central de Venezuela

Ing. Ulises Ramírez
Executive Secretary
National Energy Council

ASSOCIATES

Canada
Mr. Marc LeClerc
Director General, Special Projects
National Energy Board

Finland
Mr. Seppo Hannus
Ministry of Trade and Industry
Energy Department

France
M. Bertrand Chateau
Institut Economique et Juridique de l'Energie
Centre National de la Recherche Scientifique

Dr. Maxime Kleinpeter
Eléctricité de France
Service Etudes Economique Générales

Germany
Dr. Georg Klotmann
Strategic Planning Department
Badische Anilin & Soda Fabrik AG

Dr. Dieter Schmitt
Head, Institute of Energy Economics
University of Cologne

Mr. Paul H. Suding
Economist, Institute of Energy Economics
University of Cologne

Italy
Dr. Oliviero Bernardini
Department of Technology Assessment
Research & Development Division
Montedison

Dr. Riccardo Galli
Director, Department of Technology Assessment
Research & Development Division
Montedison

Mr. William Mebane
Department of Technology Assessment
Research & Development Division
Montedison

Japan
Mr. Shinichiro Aoyama
Senior Staff Researcher, National Institute for Research Advancement

Mr. Kenichi Matsui
Senior Staff Economist
The Institute of Energy Economics

Mr. Yasuhiro Murota
Staff Economist
The Japan Economic Research Center

Mr. Mitsuo Takei
Director, Research Affairs
The Institute of Energy Economics

Mr. Hisashi Watanabe
Executive Director, Japan Alternative Energy Strategies Organization

Mexico
Ing. Gerardo Bazán
Assessor
Mexican National Energy Commission

Ing. Alberto Escofet
Assistant to the Executive Secretary
Mexican National Energy Commission

The Netherlands
Dr. André C. Sjoerdsma
Director, Future Shape of Technology Foundation

Netherlands/U.K.
Drs. Hans DuMoulin
Head of Energy & Oil Economics Div.
Shell International Petroleum Co., Ltd.
Royal Dutch/Shell Group

Dr. Gareth Price
Group Planning, Long Term Future
Shell International Petroleum Co., Ltd.
Royal Dutch/Shell Group

Norway
Mr. Henrik Ager-Hanssen
Deputy Managing Director
Statoil, Den Norske Stats Oljeselskap A/S

Mr. Kai Killerud
Manager, Power Plant Engineering
Scandpower A/S

Sweden
Mr. Harry Albinsson
Energy Secretary
Federation of Swedish Industries

Mr. Bertil Eneroth
Special Assignments
Skandinaviska Enskilda Banken

United Kingdom
Mr. Michael Clegg
Manager, Systems Group, Corporate Planning
British Petroleum Company, Ltd.

Dr. Richard J. Eden
Head, Energy Research Group
Cavendish Laboratory
University of Cambridge

Mr. Andrew R. Flower
Assistant Policy Analyst
Policy Review Unit
British Petroleum Company, Ltd.

United States
Mr. Walter F. Allaire
Director, Energy Resources
Allied Chemical Corporation

Dr. Paul P. Craig
Director, Energy & Resources Council
University of California

Dr. Henry L. Duncombe, Jr.
Chief Economist
General Motors Corporation

Mr. Edward D. Griffith
Senior Planning Consultant
Atlantic Richfield Company

Dr. H. Paul Root
Director of Economic Studies
General Motors Corporation

Dr. David Sternlight
Chief Economist
Atlantic Richfield Company

STAFF

MIT Program Staff

Mr. Robert P. Greene
Program/Administrative Officer

Mr. Paul S. Basile
Program Officer

Mr. William F. Martin
Program Officer

WAES Consultants

Mr. Steven C. Carhart
Assistant Scientist
National Center for Analysis of Energy Systems
Brookhaven National Laboratory

Dr. Kenneth C. Hoffman
Associate Chairman for Energy Programs
Head, National Center for Analysis of Energy Systems
Brookhaven National Laboratory

MIT Support Staff

Ms. Elaine R. Goldberg
Ms. Susan M. Leland

European Support Staff

Ms. Karin Berntsen
Ms. Dalia Jackbo

Acknowledgments

Many organizations and institutions have provided direct financial support and other services for WAES. The Workshop gratefully acknowledges the following sponsors for their generous support of the Secretariat and related activities:

The Ford Foundation

The German Marshal Fund of the United States

The Andrew W. Mellon Foundation

The National Science Foundation

The Rockefeller Brothers Fund

The Rockefeller Foundation

The Alfred P. Sloan Foundation

The Workshop also gratefully acknowledges the many individuals and institutions in each country that are contributing financial and/or professional support for the Workshop and/or the WAES national studies, including:

Canada
National Energy Board
Petro-Canada

Denmark
Danish Research Council for Science
Niels Bohr Institute

Finland
Academy of Finland
Helsinki University of Technology
Ministry of Trade and Industries

France
Centre National de la Recherche Scientifique
Eléctricité de France
Société Générale

Germany
Badische Anilin & Soda Fabrik AG
Energiewirtschaftliches Institut, University of Cologne
Rheinisch-Westfälisches Elektrizitätswerk AG

Iran
Bank Sanaye Iran

Italy
CISE
Confindustria
ENEL
ENI
FIAT
Istituto Economia Fonti de Energia (IEFE), Università L. Bocconi, Milano
IRI
ISPE
Montedison
Istituto di Economia e Politica Industriale, Università di Bologna
Istituto di Fisica, Università di Napoli
Zanussi

Japan
Electric Machinery Industry Federation
Electric Power Industry Federation
Industrial Research Institute
Institute of Energy Economics
Institute for Future Technology
Japan Computer Usage Development Institution
Japan Economic Research Center
Japan EXPO Fund
Japan Foreign Trade Council, Inc.
Japan Iron and Steel Federation
Japan Techno-Economic Society
Mitsubishi Research Institute
National Institute for Research Advancement
Nomura Research Institute

Mexico
Mexican National Energy Commission

The Netherlands
Akzo Noutchemie
Centraal Planbureau
Estel Hoesch-Hoogovens
Ministerie van Economische Zaken
Ogem Holding
Paktank
Stichting Toekomstbeeld Der Techniek
Verenigde Machinefabrieken

Netherlands/U.K.
Shell International Petroleum Company, Ltd.
Royal Dutch/Shell Group

Norway
Den norske Creditbank
Elkem-Spigerverket A/S
Federation of Norwegian Industries
Institute of Atomic Energy
Ministry of Finance
Ministry of Foreign Affairs
Ministry of Industry
Royal Norwegian Council for Scientific and Industrial Research
Scandpower A/S

Sweden
Gränges AB
Skandinaviska Enskilda Banken
Federation of Swedish Industries
Kraangede AB
Ministry of Industry
National Swedish Industrial Board
Secretariat for Future Studies
Swedish Academy of Engineering Sciences
Swedish State Power Board
The International Federation of Institutes for Advanced Study

United Kingdom
British Petroleum Company, Ltd.
Department of Energy
University of Cambridge:
—Churchill College
—Cavendish Laboratory
—Engineering Department

United States
Allied Chemical Corporation
Atlantic Richfield Company
Brookhaven National Laboratory
General Motors Corporation
National Science Foundation
Overseas Advisory Associates, Inc.

Venezuela
National Energy Board

Part I
Introduction and Overview

1 The Workshop Methodology

1.1 THE CONCEPTUAL FRAMEWORK

No widely accepted methodology exists for projecting energy supply and energy demand for periods of ten to twenty-five years into the future. Therefore, the first major tasks of the Workshop were to identify and agree on the major determinants of future energy supply and demand, to select a range of likely values for these determinants, and to develop a conceptual framework for synthesizing the various national and global studies in a way that is internally consistent and comprehensible.

The period under study is divided into two time frames: from 1972 to 1985, when the contribution of each energy source is largely determined by existing infrastructure and projects already under development; and from 1985 to 2000, when pressures on oil and natural gas supplies increase, alternative energy supply systems can make major contributions, and energy conservation measures now only beginning can reduce energy demand—provided that their development on a large enough scale is started shortly and that their implementation is sustained.

The future is so uncertain that the objective of making forecasts of a specific future world environment is probably unrealistic. The Workshop has therefore identified a range of different but plausible futures; that is, a "scenario" approach has been adopted. "Scenario" is the term used to represent a plausible energy future. Several scenarios are defined and tested by the Workshop. For each scenario, values of key variables are specified and the consequent energy picture evaluated. Scenarios are not forecasts; they are chosen to span a wide range of possible futures and lead to different estimates of energy demand and supply. When significant findings remain invariant under several different scenarios, confidence in those findings is increased.

The scenarios for the first period, 1972 to 1985, are defined in terms of world economic growth, which assumes two numerical values (low and high); world oil price, with three values (oil price constant, rising, or falling); and government energy policy actions, with two values (designated as "vigorous" or "restrained"). The world economic growth variable is retained for the second period, 1985 to 2000. A more general energy price variable is substituted for the oil price variable, and a principal replacement fuel (coal or nuclear) is considered as the variable under control at the national level. A vigorous energy policy is assumed in all 1985 to 2000 scenario cases. Consistent scenarios for the entire 1972 to 2000 period are developed by coupling appropriate 1972 to 1985 cases with several 1985 to 2000 scenarios.

In addition to national studies of energy demand for 1972 and for 1985, the Workshop is studying demand to year 2000 and energy supply possibilities from 1972 to 1985 and year 2000, using the coupled scenarios for the entire period.

National supply studies by WAES countries for each scenario, 1972 to 2000, include estimates of fuel reserves, future discovery rates, and potential production rates, together with the expected contribution of renewable energy sources such as solar, geothermal, wind, biomass, etc. Similar estimates will be made for potential fuel supplies from non-WAES countries aggregated by regions. Much of the total reserves of oil, natural gas, and coal in the world outside COMECON and China (WOCC) are located in these countries, so that the assessment of their availability for export forms an important part of the analysis.

The supply studies will also include estimates of the expected scale, timing, costs, and yields of conversion processes including oil-, coal-, and gas-fired electric plants, hydroelectric generators, nuclear power stations, and coal conversion plants.

The process of balancing available energy supplies with expected energy demands is called "supply-demand integration." In the global studies this analysis takes account of international energy flows and regional fuel mix alternatives, production constraints, consumer market preferences, and infrastructure and expansion costs. National studies make use of this information about the global setting to develop national supply-demand integrations, leading to estimates of net imports or exports by fuel type for each scenario.

Such calculations can be greatly assisted by the use of computer models. A linear programming model has been developed for use in the global studies—the Global Energy Mini-Model (GEMM). In addition, balancings of individual national expectations of demand by fuel types is done manually as an input to WAES by the Energy Research Group at Cambridge University. For supply-demand integration at the national level, some countries are considering using the Brookhaven Energy System Optimization Model (BESOM), developed by the Brookhaven National Laboratory in the United States, while others rely on manual approaches.

If, in the integration analysis, it is found that desired energy demand exceeds expected supply, a reassessment is necessary to consider alteration of energy growth rates through price, conservation, or economic effects. After such feedback influences have been considered, the supply-demand integration process will assist in identifying the scale and timing of the potential energy actions that need to be taken. WAES will then analyze the possible choices of strategies to balance energy supply and demand.

Environmental factors are considered in the analysis in several ways. Supply studies take account of current environmental standards in determining resource availability, including the effects of proposed environmental standards if standards have not yet been established. Some examples may help to illustrate this process. In the development of United States western coal reserves, costs are included for the restoration of land that is strip-mined; the Japanese study assumes in its cost calculation that sulphur dioxide scrubbers will be installed in all new coal- or oil-fired electric power plants; to preserve the rivers in Northern Sweden, the Swedish study assumes that only about 70 percent of the potential hydroelectric power will be developed.

The demand and supply studies and supply-demand integration analyses include present and projected environmental standards and the capital costs, energy efficiencies, and constraints associated with such standards. In the final analysis, some conservation and environmental considerations may be in conflict. Such conflicts will be taken into account when significant.

The Workshop methodology was carefully and extensively developed during the first year of operation of WAES. This book focuses on 1985 energy demand studies, rather than on methodology. A more comprehensive description of the Workshop methodology for studies to 1985 and year 2000 will be included in a later technical report. A summary of the methodology used for projecting energy demand to 1985 follows.

1.2 DEMAND STUDIES METHODOLOGY

A common methodology has been developed for the analysis of 1972 energy use and for projections of energy demand to 1985. The approach classifies national energy use into as many as 70 (or, in one case, 230) economic sectors. These represent the principal energy-consuming activities in national economies and, taken together, account for the total energy demand.

Sectoral energy use is projected as the product of an economic activity level (e.g., number of vehicle kilometers traveled) and an energy-use efficiency (e.g., kilometers per liter). The approach of combining economic and energy efficiency factors is adapted from the methodology used in the Ford Foundation Energy Policy Project (EPP) in the United States[1] and in other studies.

The year before the oil crisis, 1972, was selected as the base year. In each Workshop country, base year energy-use data have been compiled by the country's Associates from a variety of sources. This collection of base year studies from 13 major energy-consuming countries represents a unique body of information.

Using the detailed 1972 energy-use data as a base, projections of national energy demand to 1985 under five different scenarios are made by Associates in each of the Workshop countries. In order to allow comparison of results among the national research teams, three common parameters are used to construct the scenarios, as described in general terms in Section 1.1 above. These parameters or scenario variables—world economic growth, oil price, and national energy policy—are detailed in Table 1.1.

1.2.1 World Economic Growth

World economic growth has been defined by the

1. *A Time To Choose: America's Energy Future*, Final Report by the Energy Policy Project of the Ford Foundation. Cambridge, Mass.: Ballinger Publishing Company.

Table 1.1
WAES scenario variables to 1985

Economic Growth	World Oil Price	National Policy Response
Two alternatives are considered: 1. low global GWP (3.5% p.a. global target rate)[1] 2. high global GWP (6% p.a. global target rate)[1]	Three alternatives are considered: 1. constant price ($11.50) 2. falling price (from $11.50 in 1975 to $7.66 in 1985) 3. rising price (from $11.50 in 1975 to $17.25 in 1985)	Two alternatives are considered: 1. restrained 2. vigorous
To reflect the recession since 1973 on a global scale it was assumed that the world will have recovered to 1973 levels by the end of 1976. National scenarios reflect the actual situation in the country concerned. The low global growth scenario assumes continuing international economic, political, and financial problems that limit trade growth, so that countries and regions increasingly seek internal solutions to energy and trade difficulties. The high global growth scenario assumes resolution of current international political and financial problems resulting in high world trade. The national response to the global scenarios includes national judgments of sensitivity to outside factors and of how individual countries might be expected to perform under different world trends.	All prices are in constant 1975 U.S. dollars per barrel of Arabian light crude oil FOB Persian Gulf. The price chosen for the constant price case is the January 1976 price (in 1975 dollars), which is assumed constant in real terms to 1985. The lower 1985 price in the falling price case has been chosen to be equivalent to the OECD study value.[2] The higher 1985 price in the rising price case has been chosen to lie above the levels at which some particular alternatives might become economically feasible. The particular value of $17.25 was selected so that the three 1985 prices would reflect an increase of 50% between any level and the next higher level.	A "restrained" national policy response on conservation and supply is the least that can be expected under the oil price and economic growth assumptions of the case. Market forces are assumed to act appropriately with the price and growth assumptions. Global stability is assumed. Only changes in social and individual requirements induced by market forces are included. A "vigorous" national policy response on conservation and supply includes the strongest responses that can be expected under the oil price and economic growth assumptions of the case. Market forces are assumed to act appropriately with the price and growth assumptions. Global stability is assumed. Changes in social and individual requirements may be induced by national policy responses, but these should not extend beyond what is considered politically likely in each country.

1. These figures refer to the world outside COMECON and China.

2. *Energy Prospects to 1985*, 2 vols. Paris: Organisation for Economic Cooperation and Development, 1974.

Workshop in terms of real growth in global GWP.[2] Two alternatives have been defined for the 1977 to 1985 period: a *high growth* case with global GWP growing at an average rate of 6 percent per annum, and a *low growth* case with world GNP growing at 3.5 percent per annum. The low world economic growth scenario reflects a world in which continuing international economic, political, and financial problems limit trade growth, so that countries and regions increasingly seek internal solutions to energy and trade difficulties. The high world economic growth scenario assumes a reasonable resolution of international political and financial problems, resulting in a high level of world trade.

The specifications of 6 and 3.5 percent per annum for world real economic growth were considered as "targets" for the estimation of GNP growth in each country.

Regional and national growth rate assumptions for this period, based on the target world economic growth rates, were made by Associates and national teams with advice from experts within their countries. Each national team endeavored to have its national estimate of GNP growth consistent with the world economic growth rate targets. National economic rates given in individual reports should be viewed in this context.[3] The process of establishing explicit target world economic growth rates ensures a degree of consistency among the national reports that is unusual in international studies of future energy demand.

It should be noted that the world economic growth rate assumptions are for the period from the end of 1976 to 1985. Actual historical growth rates are used to project energy demand from the

2. "World" here excludes the COMECON countries and China.

3. As part of the ongoing WAES world energy studies, the Workshop has been continually evaluating the impact of the national studies on the global studies. It would appear from our initial analysis that the world economic growth rates resulting from the national growth rates assumed in the demand studies, together with ongoing WAES energy demand studies for the rest of WOCC, imply a world economic growth rate of perhaps 5.1 percent for the high world economic growth case rather than 6 percent.

base year 1972 through the end of 1975, and it is assumed that the world GNP will have recovered to its 1973 level by the end of 1976.

1.2.2 World Oil Price

The second scenario variable selected for the 1972 to 1985 period is world oil price. For consistency, all world oil prices are stated in constant 1975 U.S. dollars per barrel of Arabian light crude oil FOB Persian Gulf. Three possible price situations are considered.

1. a constant price of \$11.50 per barrel for the period;[4]
2. a falling price, going from \$11.50 per barrel in 1976 to \$7.66 per barrel by 1985; and
3. a rising price, going from \$11.50 per barrel in 1976 to \$17.25 per barrel by 1985.

4. The \$11.50 per barrel price in October of 1975 represents the actual world price at that time. The \$11.50 per barrel price in 1975 U.S. dollars is roughly equivalent to a \$9 per barrel price in 1972 U.S. dollars, and the \$7.66 per barrel price in 1975 U.S. dollars is roughly equivalent to a \$6 per barrel price in 1972 U.S. dollars. These WAES price assumptions are therefore quite similar to the price assumptions used in the OECD study, *Energy Prospects to 1985*, and in several national energy studies such as the U.S. Federal Energy Administration's *1976 National Energy Outlook.*

In both the rising and falling price cases, a linear change in prices is assumed for simplicity; the real price schedule might more realistically be a step function with increases (or decreases) every one or two years. The rising price case assumes that costs reach a level at which alternatives to oil become economically feasible.

The 1985 prices for the rising and falling cases were determined from the constant price (\$11.50) by arbitrarily specifying a 50 percent difference between adjacent price levels. That is,

\$7.66 + (0.5 x \$7.66) = \$11.50

\$11.50 + (0.5 x \$11.50) = \$17.25

The year 1972 is the base year for WAES analyses. Oil prices are taken at their historical, actual values for the period from 1972 to the end of 1975. Price excursions from 1976 to 1985 can be represented as in Figure 1.1.

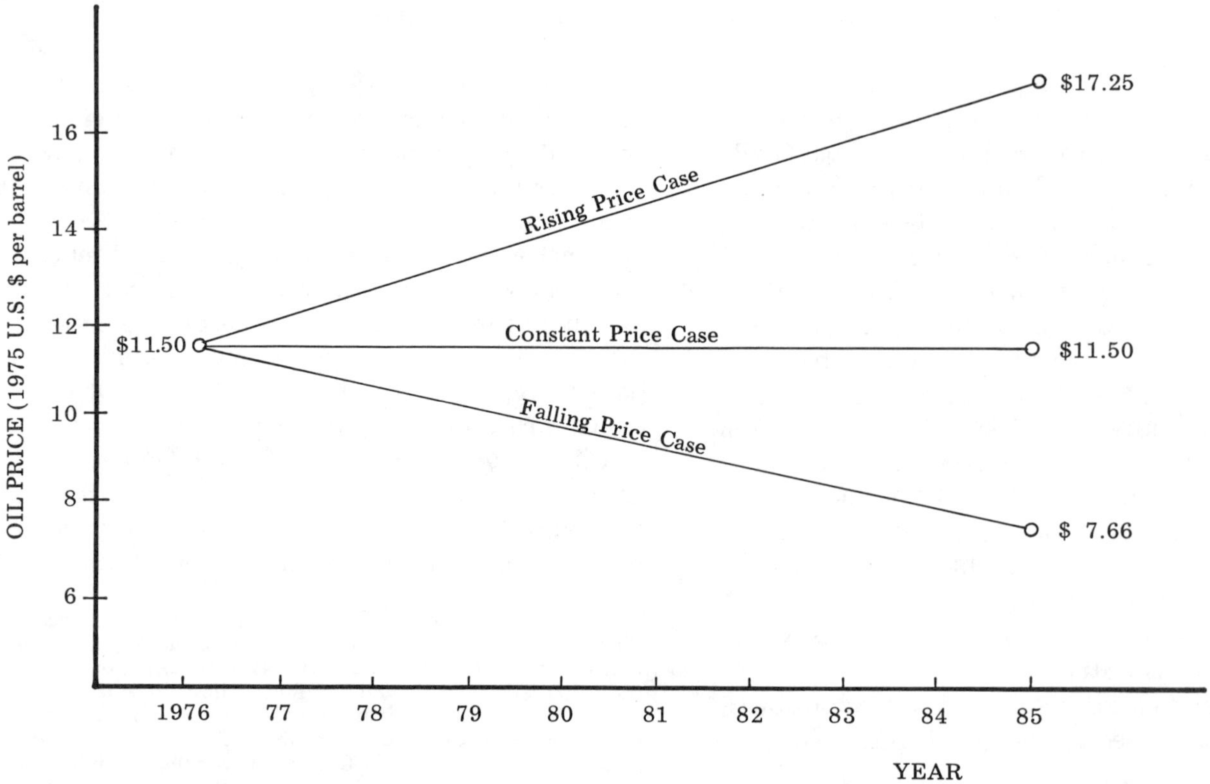

Figure 1.1 Three alternative oil price assumptions to 1985.

1.2.3 National Policy Response

Two types of national energy policy are hypothesized for the third scenario variable for 1972 to 1985: a "restrained" and a "vigorous" policy. With a restrained policy, government action on conservation and supply is the least that can be expected under the oil price and economic growth assumptions of the case. Market forces are assumed to act appropriately with the price and economic growth assumptions. Global stability is assumed. Only changes in social and individual requirements induced by market forces are to be included.

Under the vigorous policy assumption, strong government action on conservation and supply is taken. The responses are the strongest that can be expected under the oil price and economic growth assumptions of the case. Market forces are assumed to act appropriately with the price and economic growth assumptions. Global stability is assumed. Changes in social and individual requirements may be induced by national policy responses, but these should not extend beyond what is felt to be politically likely.

Within the definitions of "vigorous" and "restrained" policy actions there is enough flexibility to allow for different individual national approaches. The WAES methodology distinguishes, to the extent possible, effects due to price changes from effects due to nonprice policies alone. While recognizing that price and policy effects are related, the national studies attempt to consider the particular influence of each independently. In this way, sensitivities of energy demand to the separate variables can be determined. Explicit policies can be considered which have an effect above and beyond the results of price changes. Such policies are designed to improve the efficiency of energy use in specific ways without altering the economic activities already established for any case.

"Conservation" in these studies refers primarily to improvements in the efficiency of energy use, rather than to reductions of energy-using activities. In certain cases, conservation procedures are considered which involve only slight modifications of user practices (e.g., lowering of lighting levels where clearly acceptable). It remains true that, in all cases, economic assumptions are substantially unaltered by "conservation" in these studies.

Environmental considerations also enter the analyses. The potential environmental undesirability of an energy policy is weighed in the decision about adopting or rejecting that policy.

1.2.4 Scenario Cases to 1985

The scenario variables for the period to 1985 and the parameters assigned to them allow for twelve possible scenario cases to 1985. Five cases were selected for analysis; they were chosen to span a broad range of future energy demand and to enable an analysis of the sensitivity of energy demand to the three scenario variables. Table 1.2 shows these five cases, labeled A, B, C, D, and E. In the national demand studies, each national team reports its analysis of energy demand for the WAES base year, 1972, and projects energy demand to 1985 for each of these cases.

1.2.5 Energy-Using Sectors

As already noted, energy demand projections are

Table 1.2
WAES scenario cases to 1985

WAES Case	World Economic Growth Targets (Real) 1977–1985[1]	World Oil Price[2]	National Policy Response
A	High (6%)	$11.50 Rising to $17.25	Vigorous
B	Low (3.5%)	$11.50 Rising to $17.25	Vigorous
C	High (6%)	$11.50 Constant	Vigorous
D	Low (3.5%)	$11.50 Constant	Restrained
E	High (6%)	$11.50 Falling to $7.66	Restrained

1. The period from 1973 to the end of 1975 is assumed to correspond to actual world economic growth, with recovery to 1973 levels by the end of 1976 postulated.

2. All prices are in terms of constant 1975 U.S. dollars per barrel of Arabian light crude oil FOB Persian Gulf.

made by estimating economic activity levels and energy efficiencies for each energy-using category for each case. While not every country has used precisely the same categories in their projections, nearly all used the set developed by WAES or some variation of that set. These categories are listed in Table 1.3, together with the appropriate measure of economic activity. Energy efficiencies for each category are measured as energy used per unit of economic activity.

1.3 INTERPRETATION OF NATIONAL DEMAND STUDIES DATA

Every attempt has been made to develop national demand data based on as consistent a form as possible. Nevertheless, the careful analyst of these studies should observe a few important cautions.

First it should be remembered that the energy demands calculated in these studies are end-use energy demands—energy used by the final consumer. Energy used in the energy industries—that is, in the transport, refining, processing, and conversion of fuels—is not given in detail. Thus, these studies consider only the variety of ways in which the final consumer may change his use of energy. Yet it must not be forgotten that final energy demand varies according to the types of fuels used to meet that demand. For example, different amounts of energy may be required to provide the same amount of room heating in a home, depending on whether the heat is provided by solar energy, oil, natural gas, electricity, or district heating from power plants. Some interplay of this kind is recognized and incorporated in the demand projections to 1985. However, most of the complex questions involving the processing of energy (including losses), and the effects of varying the mix of fuels supplied, are separately considered in the Workshop's supply-demand integration procedure.

Second, the studies do not reflect any feedback from the global setting to the national demand projections. That is, they have been done, for the most part, without any consideration of the global (or national) availability of fuels to meet the projected demands. In the real world, the demand for energy is affected by the supply available. The Workshop has chosen to start by considering the energy demand resulting from the definitions of particular cases described above as "desired" energy demand, or energy demand at the case assumptions for economic growth, oil price, and energy policy, without constraints imposed by physical or political limitation of any fuel. Then, in the supply-demand integration process, "potentially available" supplies will be matched to "desired" demands. If sufficient supplies are available, the desired demands will be met. If available supplies are insufficient to meet desired demands, then the actual amount of energy consumed will necessarily be less than the amount desired.

In both cases, an examination will be made of the resolution of the supply-demand imbalance that is likely to result from some combination of market price equilibrations and producing and consuming country policy actions. This examination is part of the WAES supply-demand integration procedure.

Table 1.3
WAES energy-using categories with their measures of economic activity

Sector	Economic Activity Level
Transportation	
Urban Auto	total vehicle kilometers[1]
Rural Auto	total vehicle kilometers
Urban Bus	passenger-kilometers
Intercity Bus	passenger-kilometers
Rail (Passenger)	passenger-kilometers
Air (Passenger, Domestic)	passenger-kilometers
Air (Passenger, International)	passenger-kilometers
Ship (Passenger)	passenger-kilometers
Rail (Freight)	ton-kilometers
Air (Freight, Domestic)	ton-kilometers
Air (Freight, International)	ton-kilometers
Truck (Local)	ton-kilometers
Truck (Intercity)	ton-kilometers
Ship (Freight, Domestic)	ton-kilometers
Ship (Freight, International)[2]	ton-kilometers
Miscellaneous	
Agriculture, Forestry, and Fishing (ISIC Major Division 1)	
Farm Machinery	value added or area harvested
Greenhouses	value added
Other Agricultural Uses	value added
Forestry (ISIC 12)	value added
Fishing Boats (ISIC 13)	value added
Other Fishing Use (ISIC 13)	value added
Mining	
Handling and Internal Transportation	value added or physical output
Other Energy Use	value added or physical output
Construction	
Handling and Internal Transportation	value added or physical output
Other Energy Use	value added or physical output
Commercial[3]	
Space Heat (Fossil Fuel)[4]	floor area
Space Heat (Electricity)[4]	floor area
Air Conditioning (Fossil Fuel)[4]	floor area
Air Conditioning (Electricity)[4]	floor area
Lighting	floor area
Miscellaneous	floor area
Presently Unknown	floor area
Public[3]	
Space Heat (Fossil Fuel)[4]	floor area
Space Heat (Electricity)[4]	floor area
Air Conditioning (Fossil Fuel)[4]	floor area
Air Conditioning (Electricity)[4]	floor area
Lighting	floor area

Table 1.3 continued

Sector	Economic Activity Level
Miscellaneous	floor area
Presently Unknown	floor area
Industrial[5]	
Iron and Steel (ISIC 371)	value added or physical output
Nonferrous Metals (ISIC 372)	value added or physical output
Paper and Allied (ISIC 34)	value added or physical output
Chemicals and Allied (ISIC 35 minus ISIC 353)	value added or physical output
Mineral Products (ISIC 36)	value added or physical output
Food and Related (ISIC 31)	value added or physical output
Transport Eqpt. (ISIC 384)	value added or physical output
Key Industry (Not Elsewhere Classified)	value added or physical output
Key Industry (Not Elsewhere Classified)	value added or physical output
Misc. Manufacturing	value added or physical output
Residential[6]	
Central Space Heating:	
Single Family (Fossil Fuels)[7]	number of dwellings
Single Family (Electricity)[7]	number of dwellings
Town or Row Houses (Fossil Fuels)[7]	number of dwellings
Town or Row Houses (Electricity)[7]	number of dwellings
Apartments (Fossil Fuels)[7]	number of dwellings
Apartments (Electricity)[7]	number of dwellings
Room Heating	% saturation by fossil fuels and by electricity
Air Conditioning (Window)	% saturation by fossil fuels and by electricity
Air Conditioning (Central)	% saturation by fossil fuels and by electricity
Water Heating	% saturation by fossil fuels and by electricity
Refrigerators	% saturation by fossil fuels and by electricity
Lighting	% saturation by fossil fuels and by electricity
Cooking	% saturation by fossil fuels and by electricity
Dishwashers	% saturation by fossil fuels and by electricity
Freezers	% saturation by fossil fuels and by electricity
Television	% saturation by fossil fuels and by electricity
Other Key Appliances	% saturation by fossil fuels and by electricity
Misc. Appliances	% saturation by fossil fuels and by electricity
Potential Appliances	% saturation by fossil fuels and by electricity

1. Alternatively, the number of autos multiplied by the distance per auto per year.

2. The energy used in the international transport of fuels is calculated on a global basis and is not counted here.

3. Floor areas in these sectors is divided into "old" (pre-1972 construction) and "new" (post-1972 construction).

4. For these space heating/cooling categories, two energy efficiencies are calculated: one measuring the heat gain or loss due to building insulation, and another measuring the efficiency of the heating/cooling system itself (e.g., oil burner, electric resistance heater, fuel pump).

5. Energy used in petroleum refining, in utilities, and in coal mining is not counted here but is calculated in the supply-demand integration procedure.

6. The number of dwellings is divided into "old" (pre-1972 construction) and "new" (post-1972 construction).

7. For these central space heating categories, two efficiencies are calculated: one measuring the heat gain or loss due to building insulation, and another the efficiency of the heating system itself.

2 Summary of the National Demand Studies

2.1 INTRODUCTION

This chapter contains brief summaries of each of the national energy demand studies. All data in this chapter are given in joules, the recognized standard international energy unit. In the individual national reports, data are given in the appropriate units for the particular country. The Workshop uses a standard set of conversions from volumetric to energy units. These conversions are summarized in Table 2.1 They should be used when comparing data in different units among countries.

Verbal summaries of the national demand studies are collected in Section 2.3. In Section 2.4 may be found summary tables and bar charts for the thirteen contributors, in the following order: Canada, Denmark, Finland, France, German Federal Republic, Italy, Japan, Mexico, The Netherlands, Norway, Sweden, United Kingdom, United States.

Part II of this book contains the thirteen complete national reports. Each report consists of an overview section, a description of the specific methodology used to derive the energy demand projections, a discussion of the results, and, most importantly, a collection of detailed worksheets describing energy demand by sectors for the base year, 1972, and for the 1985 scenarios.

2.2 BASE YEAR (1972) ENERGY USED

In 1972, the nations represented in the Workshop consumed approximately 80 percent of the total energy consumed in the world outside the COMECON countries and China. This is illustrated in Figure 2.1, which also shows the approximate distribution of the 80 percent among Workshop countries. There are many similarities among the ways in which these largely industrialized countries use energy. This fact has made possible the completion of these national energy demand studies in a manner that allows easy comparison between countries.

Figure 2.2 illustrates the pattern of energy use in most WAES countries in 1972, the base year chosen for WAES analysis.

Workshop countries use similar fuels to meet their energy demands. Figure 2.3 shows the division of final energy demand in each country according to the fuels used to meet that demand. The heavy reliance of these nations on oil and gas is immediately apparent: these fuels provide from 54 to 75 percent of the energy used in each country.

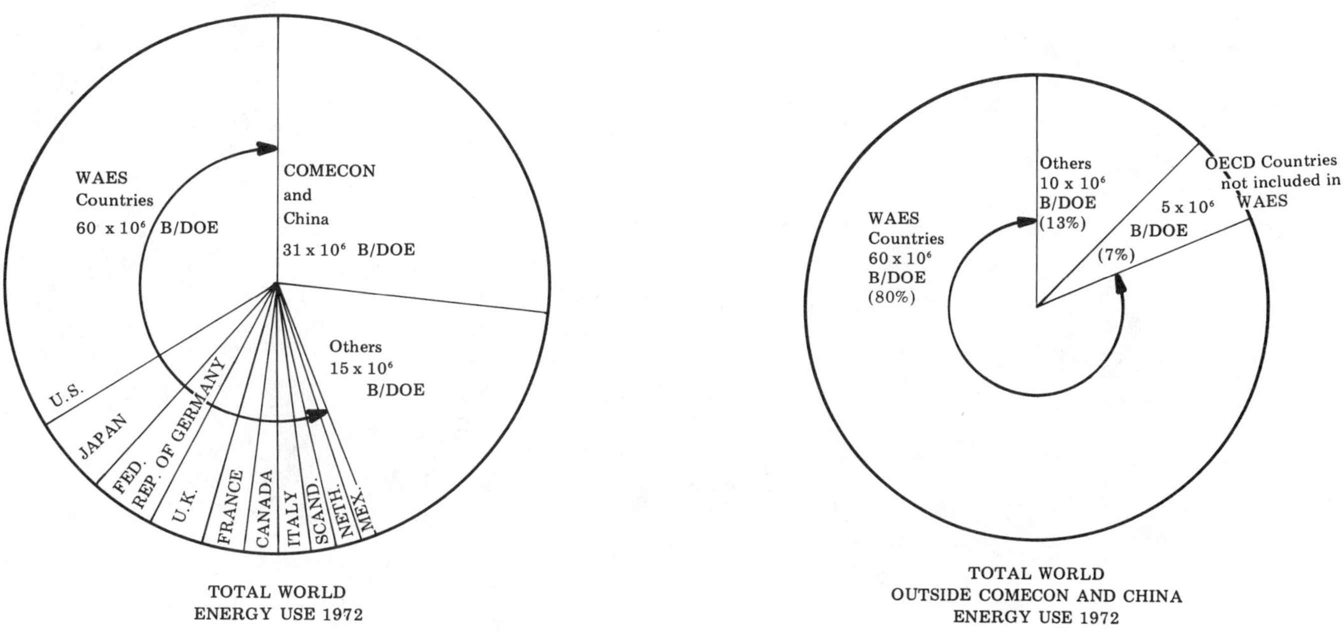

Figure 2.1 Global energy use 1972 (figures are in barrels per day of oil equivalent).

Table 2.1
Standard energy conversions (equivalent values lie in vertical columns)

Unit									
Barrels per Day of Oil Equivalent[1]	--	--	--	--	--	--	--	--	--
Tons of Oil Equivalent[2]	--	--	--	--	--	--	--	0.022	0.023
Metric Tons of Coal Equivalent[3]	--	--	--	--	--	--	--	0.034	0.036
Barrels of Oil Equivalent[4]	--	--	--	--	--	0.0064	0.02	0.16	0.17
Cubic Meters of "Average" Natural Gas[5]	--	--	--	0.027	0.09	1	2.7	25	27
Kilowatthours	--	--	--	0.3	1	11	29	280	293
Cubic Feet of "Average" Natural Gas[5]	--	--	--	1	3.4	37.3	100	950	1000
Kilocalories	0.24	0.25	1	252	860	9400	25,200	0.24 million	0.25 million
British Thermal Units	0.95	1	4.0	1000	3400	37,300	1 therm	0.95 million	1 million
Kilojoules	1	1.06	4.2	1055	3600	39,400	105,500	1 million	1.06 million

Calorific values are measured gross. Rounded equivalents only are given.

1. Equivalents in other units are shown on a per annum basis.

2. of 43 million Btu (≈ 10,000 kcal/kg net cal. val.).

3. of 12,000 Btu/lb = 7000 kcal/kg.

4. of 5.8 million Btu.

5. of 1000 Btu/ft^3 or 9400 kcal/m^3.

Source: Energy Conversion Equivalents Table, Shell International Corporation.

								Abbreviation
--	0.003	0.013	0.02	1	2.7	18,000	470 million	B/d.O.E.
0.09	0.13	0.65	1	50	135	0.9 million	23 x 10^9	TOE
0.14	0.21	1	1.5	76	209	1.3 million	36 x 10^9	MTCE
0.68	1	4.8	7.4	365	1 TBOE	6.4 million	170 x 10^9	BOE
106	155	745	1150	57,000	0.155 million	1 x 10^9	27 x 10^{12}	Nm^3 NG
1160	1700	8140	12,600	0.62 million	1.7 million	11 x 10^9	290 x 10^{12}	kWh
4000	5800	27,800	43,000	2.1 million	5.8 million	37.3 x 10^9	1 x 10^{15}	ft^3 NG
1 million	1.5 million	7 million	10.8 million	530 million	1.5 x 10^9	9.4 x 10^{12}	250 x 10^{15}	kcal
4 million	5.8 million	27.8 million	43 million	2.1 x 10^9	5.8 x 10^9	37.3 x 10^{12}	1 Q	Btu
4.2 million	6.1 million	29.3 million	45.4 million	2.2 x 10^9	6.1 x 10^9	39.4 x 10^{12}	1.06 x 10^{18}	kJ

-- = insignificant
1 therm = 100,000 Btu
1 TBOE = 1000 BOE
1 Q = 10^{18} Btu
1 Quad = 10^{15} Btu

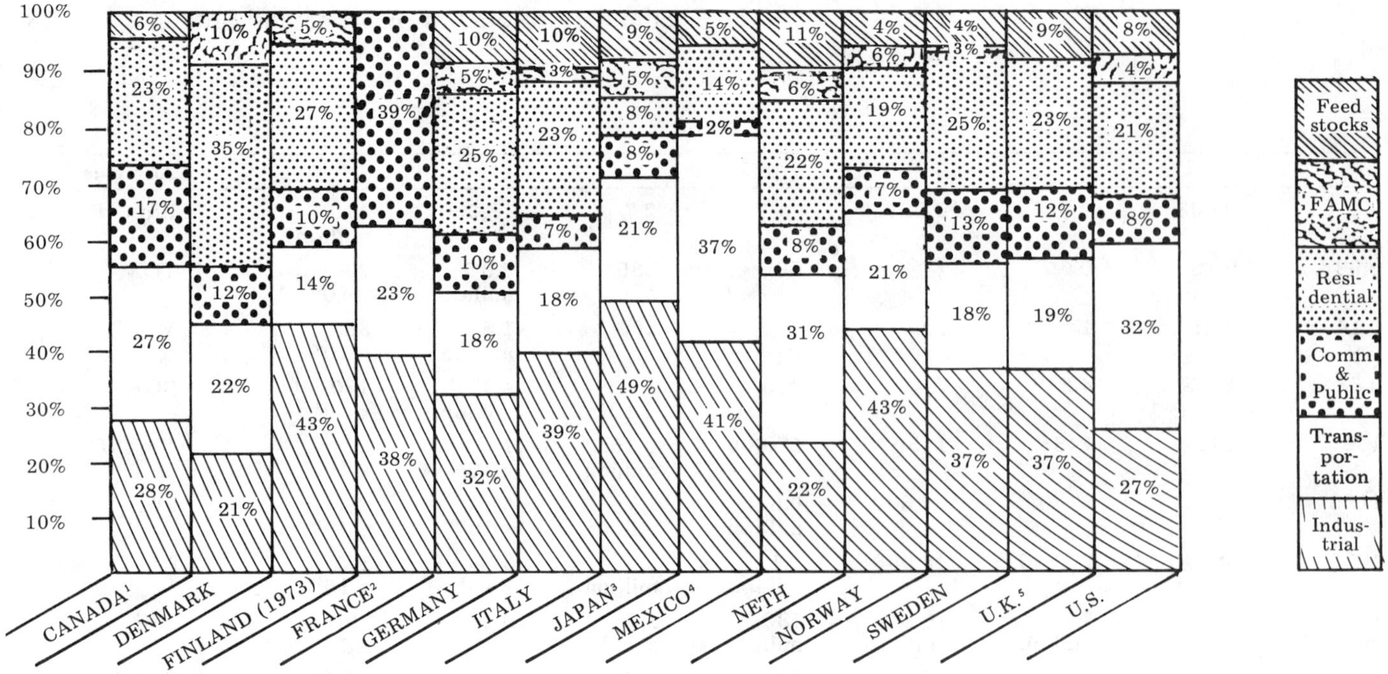

[1] Fishing, agriculture, mining and construction are aggregated into other sectors.
[2] Only three sectors are shown: domestic (including residential and commercial), transportation and industry.
[3] Transportation includes bunkers.
[4] FAMC is 1%
[5] Commercial and public also includes fishing and agriculture.
All data in this chart are submitted in the national reports to WAES contained in this book.
FAMC: fishing, agriculture, mining and construction.

Figure 2.2 Base Year (1972) energy end-use data by demand sector (from the national demand studies).

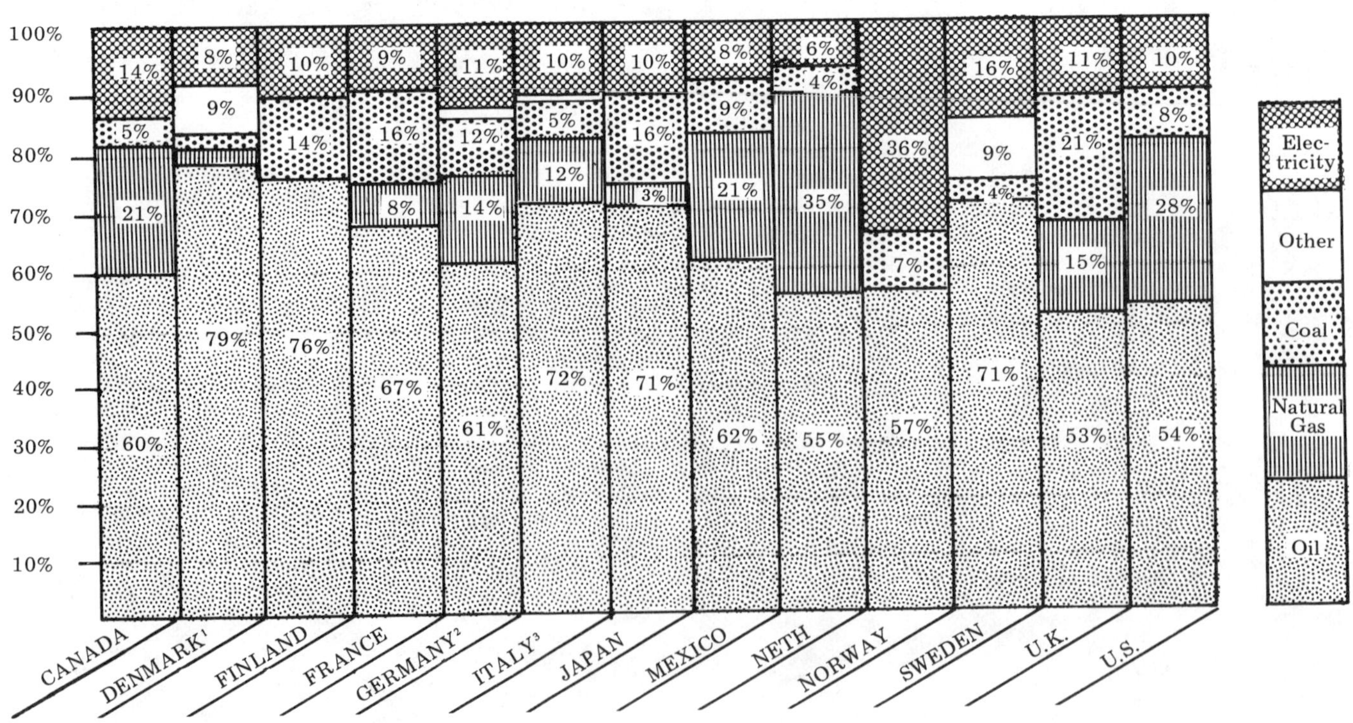

[1] Natural gas and coal are each 2%
[2] Other is 2%
[3] Other is 1%
All data in this chart are submitted in the national reports to WAES contained in this book.

Figure 2.3 Base Year (1972) energy end-use data by fuel type.

2.3 SUMMARIES OF INDIVIDUAL NATIONAL DEMAND STUDIES

2.3.1 Canada

The 1972 and projected 1985 energy demand estimates for the five WAES scenario cases are presented for Canada. The cases were determined so that the resulting estimates represent the range of reasonable possibilities for Canadian energy consumption in 1985.

The projections reveal a substantial growth in fossil fuel requirements in the feedstocks sector (i.e., nonenergy use) to 1985. A large growth of electricity use in the residential and commercial sectors from 1972 to 1985 is also projected, while direct use of fossil fuels in these sectors actually declines in one case.

The distribution by end-use sector is relatively stable across all scenarios. The share of the residential and industrial sectors is projected to decline over time as the relative importance of the transportation and feedstocks sectors increases. Part of the decline in the residential sector is attributable to the relative increase in multiple dwelling units, many of which purchase energy at commercial rates and are accordingly classified in the commercial sector. In addition, a significant decrease in the residential sector's share of fossil fuels reflects the large growth of electricity use in this sector as noted above.

The potentials for energy savings through changes in economic growth, oil price, and national energy policy response are observed in this report. In general, the results obtained from the projections indicate that the improvement of energy efficiencies in response to a vigorous national energy policy would have more of an impact on reducing energy demand than would lower economic growth.

2.3.2 Denmark

The present study concerns the energy demand in Denmark until 1985. It is based on the common WAES assumptions about the future global economic growth rate, the price of energy, and the degree to which the government acts on the energy scene. The resulting five different scenarios for the energy demand are not necessarily realistic and may differ from official energy plans. They are, however, designed to span a range of possibilities which illuminate some of the basic questions concerning the future demand for energy in Denmark.

The study assumes that there will be *economic growth* in the future as there has been in the past. The growth rates are, however, expected to be somewhat lower than in the 1960s, and the relative contributions to the growth from the various sectors are different.

It is furthermore assumed that *energy prices* in Denmark will follow those on the world market. The 1973 price increases are in the study considered as causing a three-year delay in the development of demand. In scenarios where the relative price of energy is increasing, a significant additional reduction of the energy growth rate is envisaged.

The government's *energy policy* is expected to have a considerable influence on the future demand. The study considers the penetration of energy-saving measures which are introduced partly as a result of legislative or regulatory actions and partly as a result of price.

The study has paid special attention to a number of *inherent trends* in the energy economy which might have considerable influence on the future demand. The most important trend is a continued shift from direct consumption of fossil fuel to consumption of electricity. Other inherent trends are associated with an emerging saturation for several energy-consuming appliances or activities. In the study such trends are expected to continue for the period considered.

The study is based on a quantification of the assumptions on economic growth, energy prices, energy policy, and inherent trends which have been combined to yield the five different scenarios for the energy demand. The scenarios contain a considerable amount of detail, but some of the main findings can be summarized as follows:

- In all scenarios the demand for energy has a lower growth rate than the average of 4.7 percent per annum for the period 1950 to 1973. The projected growth rates range from 0.8 percent p.a. in the case of low economic growth, increasing energy prices, and vigorous policy to 3.3 percent p.a. in the case of high economic growth, decreasing energy prices, and restrained policy.

- In most scenarios the energy demand grows more slowly than the economy. Thus the energy coefficient, which is the ratio of the growth rate of energy demand to the growth rate of the economy, ranges from 0.3 to 1.05. This can be compared to an energy coefficient around 1.5 in the 1960s.
- The growth rate for electricity is in all scenarios consistently higher than that for other forms of energy. Electric growth rates range from 4.2 percent p.a. to 7.2 percent p.a. This is significantly below the historic growth rates of 10–11 percent p.a.
- There is a considerable potential for energy conservation, especially in the heating of buildings, which accounts for approximately 40 percent of the gross energy consumption in Denmark. It seems technically and economically feasible to reduce this fraction to about 30 percent by 1985.
- The large savings in the heating sectors tend to make transportation and industry the dominant energy-consuming sectors. Thus by 1985 the Danish energy demand is distributed among sectors in a way which more closely resembles that of other industrialized nations than is the case today.

2.3.3 Finland

The Finnish energy demand study includes the five WAES scenarios from 1972 to 1985. The data are based on forecasts and studies that have been made at the Energy Department of the Ministry of Trade and Industry independently of the WAES project.

The original forecasts include two alternatives: high and low economic growth. They can be considered as compatible with the WAES high and low global growth cases, since the Finnish economy is very open and dependent on other countries. These forecasts have been done in a manner similar to that developed by WAES. On the lowest disaggregation level, the number of economic line items is close to 100.

The WAES cases C and D have been prepared rather directly on the basis of the forecasts mentioned above. The other scenarios have been prepared using incremental assumptions.

The government response is assumed to be constant in all scenarios for two reasons. First, the present response is rather vigorous, and many of the conservation measures have already been taken. Second, the effect of new measure would be very difficult to analyze, and the figures could not be considered reliable.

The effect of higher or lower oil price is assumed to be similar to that in other countries. The average price elasticity is expected to be about -0.2.

As a result of the analysis, the GNP elasticity of energy demand is expected to decrease from the pre-energy crisis value of 1.3 down to 1.0. Many of the historical trends are expected to continue until 1985, although at a slower rate. For example, the share of indigenous energy supply is still decreasing.

2.3.4 France

The French demand studies have been conducted in part in relation to official French government forecasts (Cases C, D, and E) and in part in relation to the long-term forecasts of the Institut Economique et Juridique de l'Energie, Grenoble (Cases A and B). The French economic growth rates, derived from the general assumptions of WAES, are taken as 5.5 percent per annum in the high cases (A, C, E) and 3.5 percent per annum in the low cases (B, D).

The scenarios yield a range of primary energy consumption in 1985 from 190 million TOE (Case B) to 285 million TOE (Case E), with electricity consumption ranging from 260 TWh (Case B) to 360 TWh (Case E). The official projection (Case C) yields a primary energy input of 240 million TOE and an electricity consumption of 330 TWh. The structure of the final energy demand, disaggregated by consuming sectors, does not change in a significant way from 1972 to 1985 in cases with low or medium oil price (C, D, E); in the other cases (A, B) there is a switch in consumption from the industrial to the residential, commercial, and public sectors that is explained by the start of profound changes in industrial structuring. (In 1972 the industrial sector consumed 60 percent of the total final energy demand; in 1985 this sector will consume only 33 percent in Cases A and B.)

The supply-demand integration was manual and guided mainly by the following assumptions:

- The installed nuclear power capacity in 1985 will be 35 GWe at the minimum and 45 GWe at the maximum.
- Oil imports are constrained by the need for light

products for transportation and by the existing refinery capacity and structure.

- Natural gas imports are constrained by trade agreements and by the possible development of a transportation and distribution infrastructure (with a maximum capacity of about 37 million TOE).
- About 15 million TOE of coal will be available.

The two main priorities of French energy policy are assumed, in all cases, to be energy conservation and a reduction in the ratio of energy imports to total energy consumption through the development of nuclear and other new forms of energy.

2.3.5 German Federal Republic

Final 1985 energy demand in the Federal Republic of Germany, given the WAES scenario assumptions, is estimated to range from 9.3 to 11.1 x 10^{18} joules. The lowest demand occurs in Case B (low growth, high oil price, vigorous policy) and the highest in Case E (high growth, low oil price, restrained policy).

The dominating factor for final energy demand is economic growth. Crude oil price changes in the assumed order are estimated to cause an almost 10 percent overall difference in final consumption (from Case A to Case E). A vigorous policy does not bring considerably more conservation than what is already achieved by a restrained policy, which is assumed to represent the actual policy.

In contrast to the fossil fuel final energy demand, electricity demand is higher in the case of higher oil prices due to cross-elasticity effects, since the price of electricity is not seen to develop proportionally to the crude oil price.

Due to saturation effects, the share of residential and transportation energy demand decreases in all cases. The industrial demand share grows in the case of high economic growth and decreases with low economic growth. Total residential and transportation energy demand is estimated to be not very sensitive to economic growth, though residential electricity demand supposedly reacts on economic growth, which is correlative to private income.

In the summary bar chart, it should be noted that processing losses are smaller in the lower oil price cases because of the smaller share of electricity in these cases.

2.3.6 Italy

The dynamics of energy growth in Italy over the period to 1985 are susceptible to considerable changes relative to the past, as a consequence of a number of basic factors: increases in the price of oil, variation in the type and rate of economic development, the impact of energy policies, and general changes in the structure of energy consumption.

Total demand for primary energy, projected to 1985 in the different WAES cases, ranges from a low of 5393x10^{15} joules in scenario B to a high of 9432x10^{15} joules in scenario E, corresponding to average energy growth rates of 0.7 and 5.1 percent, respectively, over the period 1972 to 1985.

The very low energy growth rate of scenario B is a consequence of the multiplicative effect of three depressing circumstances: low economic growth, high oil price, and vigorous energy policy.

Major shifts are projected to occur between the basic energy-use categories in response both to partial saturation effects within the system and to a vigorous energy policy acting with different emphasis in the various categories.

The basic overall results can be summarized in the following terms.

The income elasticity of energy demand, which increased from 1.21 for 1953 to 1960 to 1.55 for 1960 to 1966 and 1.83 for 1966 to 1972, is projected to drop to 1.45 quite independently of the price of oil and energy policy.

The oil price elasticity of energy demand is calculated as -0.07, -0.16, and -0.29 for oil price increases in the ranges \$2.90 to \$7.66, \$7.66 to \$11.50, and \$11.50 to \$17.25, respectively.

A vigorous energy policy acting over and above increases in the price of oil and leaving life style essentially unaltered can save as much as 16 percent in delivered energy demand.

Electricity is projected to grow 1.69 times faster than gross national product, representing a continuation of the historic trend of 1.24 for the period 1953 to 1963, 1.48 for 1960 to 1966, and 1.59 for 1966 to 1972.

Thanks to existing international contracts for natural gas supply to 1985, petroleum is projected to play a somewhat smaller role in the Italian energy system, dropping from almost 74 to as low as 63 percent of total primary energy (from 67 to as little as 58 percent of total delivered energy).

2.3.7 Japan

For the WAES high world growth case, a GNP growth rate of 8 percent is assumed for Japan, and for the WAES low case of 3.5 percent, 4.5 percent is assumed. This implies that the Japanese economy, which had grown with the elasticity of 1.9 to the world GNP growth rate in the last fifteen years, will grow with the lower elasticity of about 1.3 in the coming decade. This elasticity is derived from the forecast of the world economy of the Japanese Economic Research Center.

Based on these assumed economic growth rates, we have estimated energy demand for the five WAES cases, taking into account the effect of energy conservation.

In cases of high economic growth (A, C, E), the energy demand is estimated at around 2940 x 10^{16} joules (700 x 10^{13} kcal), and in cases of low economic growth (B, D), around 2310 x 10^{16} joules (550 x 10^{13} kcal).

The elasticity of the GNP growth rate to the rate of primary energy demand growth between 1972 and 1985 is estimated to range between 0.82 and 0.94.

By way of reference, a recent study of the Institute of Energy Economics estimates an energy demand of around 2690 x 10^{16} joules (640 x 10^{13} kcal) with energy conservation for 1985 based on an assumed economic growth rate of 6.2 percent between 1975 and 1985 and an estimate of about 0.80 for energy-GNP elasticity in this time period.

2.3.8 Mexico

Mexico's energy demand has been satisfied by indigenous reserves in the past, and this situation is expected to continue in the future.

The availability of oil, gas, and coal to satisfy the national demand has made it possible for the internal energy market to remain undisturbed by international oil prices.

In the analysis of the five cases presented, it has been assumed that a rise in the oil prices will stimulate conservation, which will allow Mexico to export the fuel saved, thus improving the balance of payments and contributing to a larger extent to the development of the country. This is shown in Cases A and B where a vigorous national policy is estimated to yield a decrease of 10 to 15 percent in the energy demand of the industrial and transportation sectors.

No significant modifications in consumption patterns are expected before 1985. Modifications will be encouraged in the 1976 to 2000 period in order to allow a diversification of energy supply sources, which are presently composed almost entirely (90 percent) of oil and gas. The residential and commercial sectors are very low energy consumers; therefore, the shift to a larger share for electricity (as end use) should be addressed to the transportation and industrial sectors and will take more than ten years.

The energy supply sector is nationalized and managed by the government, through several agencies. However, the application of national policies directed to energy conservation will require special legislation in order to accomplish the expected demand reductions.

2.3.9 The Netherlands

A study has been made of energy demand in The Netherlands to 1985 using two WAES scenario cases. Case C assumes a high rate of economic growth in The Netherlands (3.4 percent per year during the period 1972 to 1985), an oil price of \$11.50, and vigorous national policy response. Case D assumes a low rate of growth (2.2 percent), an oil price of \$11.50, and restrained national policy response.

The remaining WAES scenario cases—A, B, and E—have been worked out for the industrial sector only. A rough calculation has been made for the total energy demand in other sectors for Cases A, B, and E.

It becomes clear that the industrial sector—where also the main body of feedstocks is consumed—is by far the most sensitive to changes in economic growth and energy price. This is to a large extent caused by the structure of The Netherlands' industrial sector, which leans heavily toward energy-intensive process industry. The petrochemical, metallurgical, and electrotechnical industries in particular are very sensitive to world trade growth.

Agriculture is much less influenced by the world economy.

In the transportation sector, the rates of growth of private automobile and aircraft use are independent of the rate of economic growth, whereas trucking (mainly international) is very sensitive to trade growth.

Energy use in the commercial and public sector will probably grow slowly because of energy conservation measures.

Residential energy use is growing not only because the population is growing in number, but also because of the decreasing number of persons per dwelling. There is considerable room for energy saving in this sector.

2.3.10 Norway

All of the scenarios analyzed for Norway assume quite high economic growth. The analysis shows that under such conditions the growth of energy demand will be of the same character as the economic growth; that is, if the rate of growth for the economy (percent per annum) is constant, then the rate of growth of the energy demand will be fairly constant also. The analysis also shows, however, that the rate of growth of the energy demand will be substantially lower than that of the economy in the period to 1985. During the period 1950 to 1972, the ratio between the rate of energy growth and the rate of GNP growth averaged approximately 1.2, and in the period 1960 to 1972 it was even higher. However, the scenarios analyzed for the period to 1985 all have ratios in the range 0.7–0.8.

The results of the projections for the five WAES scenarios indicate that the growth rate for energy demand will be in the range of 2.5 percent per annum (for scenario B) to 3.2 percent per annum (for scenario E). The difference in demand between these extreme cases is relatively small—only about 8 percent—for the following reasons:

1. Economic activity related to the exploration for and production of oil and gas represents a significant part of the total economic activity in Norway. It has been assumed that this situation will make it possible to use the activity in the oil sector as a lever to control overall economic activity. For this reason, the economic growth rates for the high and low growth scenarios are only slightly different.

2. In Norway, electric energy has a share of the total demand that is much higher than in most other countries. Virtually all of this electric energy will be produced by hydro power even in 1985, and the price of electricity is therefore not directly influenced by the price of oil. A large and increasing share of the total amount of fossil fuels consumed will be used as feedstocks. For this reason, the effect of oil prices on the total energy demand will be relatively small.

Comparison of the results of the various scenarios makes it possible to estimate the influence of the three scenario parameters. The summary bar chart provides a basis for the following conclusions:

1. The lower economic growth rate of scenario B leads to a reduction of total energy demand by about 4 percent as compared to scenario A.

2. The lower oil price of scenario C leads to an increase in total energy demand by about 2 percent as compared to scenario A.

3. The restrained policy response and lower oil price of scenario D lead to an increase in total energy demand by about 6 percent as compared to scenario B.

4. The restrained policy response and decreasing oil price of scenario E lead to an increase in total energy demand by about 8 percent as compared to scenario A.

2.3.11 Sweden

Our report is based mainly on the extensive energy studies made by Swedish Government Commissions in 1973 and 1974 and, in particular, on the findings of the Energy Forecast Commission (EPU), which used a methodology quite similar to that of WAES.

Of the factors studied, our findings show that the economic growth rate has the most influence on energy demand; indeed, its effect in the 1975 to 1985 period is two to three times that of oil price changes. Oil price and national policy can, however, have a significant effect in shifting demand in some sectors from fossil fuels to electricity.

During the 1960s, total energy input to Sweden increased by about 5 percent per year. The WAES projections indicate a much slower rate for the period to 1985, ranging from 2.5 (Case B) to 4.6 (Case E) percent per year. Projected total energy demand in 1985 ranges from 27.6 x 10^{17} joules in Case B to 35.8 x 10^{17} joules in Case E.

Regardless of variations in the WAES parameters, the various sectors of the Swedish economy will have about the same percentage of the total energy demand in 1985, although the emphasis in each sector may shift between fossil fuels and electricity. However, if one compares the 1985 with the 1972 situation, one finds that the transportation, commercial, public, and especially the residential sector will in 1985 have lower percentages of total primary energy input than in 1972 due to savings and increases in processing losses resulting from the projected increase in electricity use. The growth rates for processing losses are, however, slightly lower than those for electricity demand as a whole due to the projected increased use of district heating. The possibility of utilizing nuclear power plants more efficiently by district heating is presently being discussed in Sweden.

2.3.12 United Kingdom

Since energy consumption in the UK has been, and will be, strongly influenced by the availability of indigenous fuels, the demand study necessarily includes an analysis of available energy supplies. The projections imply that the UK will move from a position, in 1972, where it imported more than half the energy it consumed, to become a net exporter in 1985. Thus in the medium-term future the UK energy prospects are favourable compared to those of many industrialised countries.

However, the prospects may become far less favourable in the 1985 to 2000 period. Therefore, in order to analyse the policy options and the influence of policy on demand in the period to 1985 it is necessary to project demand beyond 1985. Hence the UK demand study includes summary projections of supply and demand in the year 2000 to supplement those for 1985.

The principal conclusion of the 1985 study is that UK energy policy to 1985 should be largely concerned with the 1990s. Policy makers should consider how the benefits of a period of comparative plenty could be used to prepare for the scarcity to follow.

For example, at a time when the UK is exporting oil, and coal is considered an unpopular, 'old-fashioned', fuel it may not appear to be desirable to invest heavily in developing the coal mining industry. However, the lead time for developing a coalfield being about a decade, the investment is necessary in the eighties to decrease the cost of petroleum imports in the nineties and beyond when the price may rise far above the long-term cost of substitutes *because* of their long lead times. Clearly the development of more efficient and convenient techniques for using coal and the expansion of coal-fired electricity generating capacity must accompany the development of the coal fields.

The study also shows a need for as much nuclear power capacity in the year 2000 as is possible, but suggests that, without determined effort and foresight, the installed nuclear capacity is unlikely to exceed about 60 GWe. To exceed this figure would require investment in the electricity generating industry at a time when capacity is not short and in technology which may not have cured all its teething troubles.

Energy policy before 1985 can help the 1990s in ways other than foresighted investment. The demand for North Sea gas has been developed rapidly in the early seventies, and the projections indicate a continuation of this trend. As supplies contract in the 1990s it will become necessary to replace indigenous natural gas with either imported LNG or syngas. In both cases there will be a substantial increase in costs and considerable investment requirements. These requirements can be reduced and delayed by restraining the build up of demand for gas in the eighties, before supplies decline. Such a policy could complement the development of coal demand through favourable relative pricing and incentives.

The lead times for the achievement of effective results from energy conservation measures (other than 'good housekeeping') are also such that active programs should be maintained in the seventies and eighties in order to reduce imports later in the century, though of course, benefits from energy conservation could also be considerable even when the country is a net exporter of energy.

2.3.13 United States

Total energy demand in the United States in 1985

given the set of explicit Workshop assumptions, is estimated to range from 85.2 to 112.7 x 10^{18} joules. Within this 25 percent variation in energy demand for the WAES cases a few substantial shifts in the division of energy use among the major economic sectors can be seen. Industrial energy use, as a percentage of the total, increases from some 27 percent in 1972 to about 30–32 percent in the 1985 cases—due to relatively greater efficiency improvements in other sectors as well as to higher growth rates in the industrial sector. Transportation energy use, on the other hand, declines from 32 percent of the total in 1972 to about 28 percent in the 1985 cases—reflecting the projected large improvements in energy efficiency in this sector in all 1985 cases.

Energy use in some subsectors is projected to grow at significantly higher or lower rates than the total energy demand. For example, chemicals, iron and steel, petrochemical feedstocks, and residential energy use are expected to be high energy growth sectors. In contrast, energy use by automobiles will decline from 1972 to 1985 because of the substantial improvements projected in mileage efficiency.

This report provides estimates of energy demand based on specific economic growth, oil price, and national energy policy assumptions. The estimates show both the economic activity and the energy efficiency associated with each of nearly 70 energy-use categories.

Sectoral totals for final energy demand in joules are shown in the accompanying summary table and bar graph. Case B—low economic growth coupled with a rising oil price and vigorous national policy measures—results in the lowest final energy demand, while Case E—high economic growth coupled with a falling oil price and restrained national policy—results in the largest final energy demand. There is a difference of almost 28 x 10^{18} joules between the highest and lowest 1985 energy demand estimates.

2.4 Summary Tables and Bar Charts

On the following pages will be found summary tables and bar charts for the thirteen countries contributing to this book. Units of joules have been used throughout to facilitate comparisons between countries.

WAES NATIONAL ENERGY DEMAND STUDIES, 1972-1985
ENERGY USE SUMMARY

Country Canada

Scenario Assumptions

WAES Case	A		B		C		D		E			
GNP 1976-1985	High		Low		High		Low		High			
Oil Price*	\$11.50 → \$17.25		\$11.50 → \$17.25		\$11.50		\$11.50		\$11.50 → \$7.66			
Natl. Policy Response	Vigorous		Vigorous		Vigorous		Restrained		Restrained			
Natl. GNP Growth Rate 1976-1985 (%/yr.) **	5.4 5.2		5.4 3.6		5.4 5.2		5.4 3.6		5.4 5.2		1972 Base Year	
Fuel Type	Fossil Fuel	Elec.	Fossil Fuel	Elec.	Fossil Fuel	Elec.	Fossil Fuel	Elec.	Fossil Fuel	Elec.	Fossil Fuel	Elec.
Units	10^{15} J		10^{15} J		10^{15} J		10^{15} J		10^{15} J		10^{15} J	
Energy-Using Sectors												
Industrial (excluding feedstocks)	1432.5	589.0	1422.6	586.6	1521.3	626.4	1601.9	661.3	1758.1	738.1	1129.8	374.6
Transportation	2081.2	0.0	1990.7	0.0	2136.8	0.0	2317.7	0.0	2548.8	0.0	1413.0	0.0
Commercial and Public	972.6	449.9	901.3	417.7	974.5	449.9	904.1	418.1	1044.2	479.1	671.3	203.0
Residential	1017.4	351.4	984.1	340.9	1030.5	356.0	1051.3	364.1	1143.3	394.9	1013.1	180.7
Fishing, Agriculture, Mining, Construction												
Feedstocks	896.9	0.0	888.5	0.0	896.9	0.0	935.3	0.0	944.1	0.0	318.2	0.0
Final Energy Demand (excluding processing losses)	6400.6	1390.3	6187.2	1345.2	6560.0	1432.3	6810.3	1443.5	7438.4	1612.0	4545.4	748.3
Processing Losses	3861.2		3731.1		3977.5		3845.9		4282.3		2223.2	
Primary Energy Input	11652.1		11263.5		11969.8		12099.7		13332.7		7526.9	

*Oil price 1975 US\$ per barrel Arabian light crude oil FOB Persian Gulf.
**Upper figure applies to 1976-80; lower figure applies to 1981-85.

CANADA

All numbers are in 10^{15} Joules

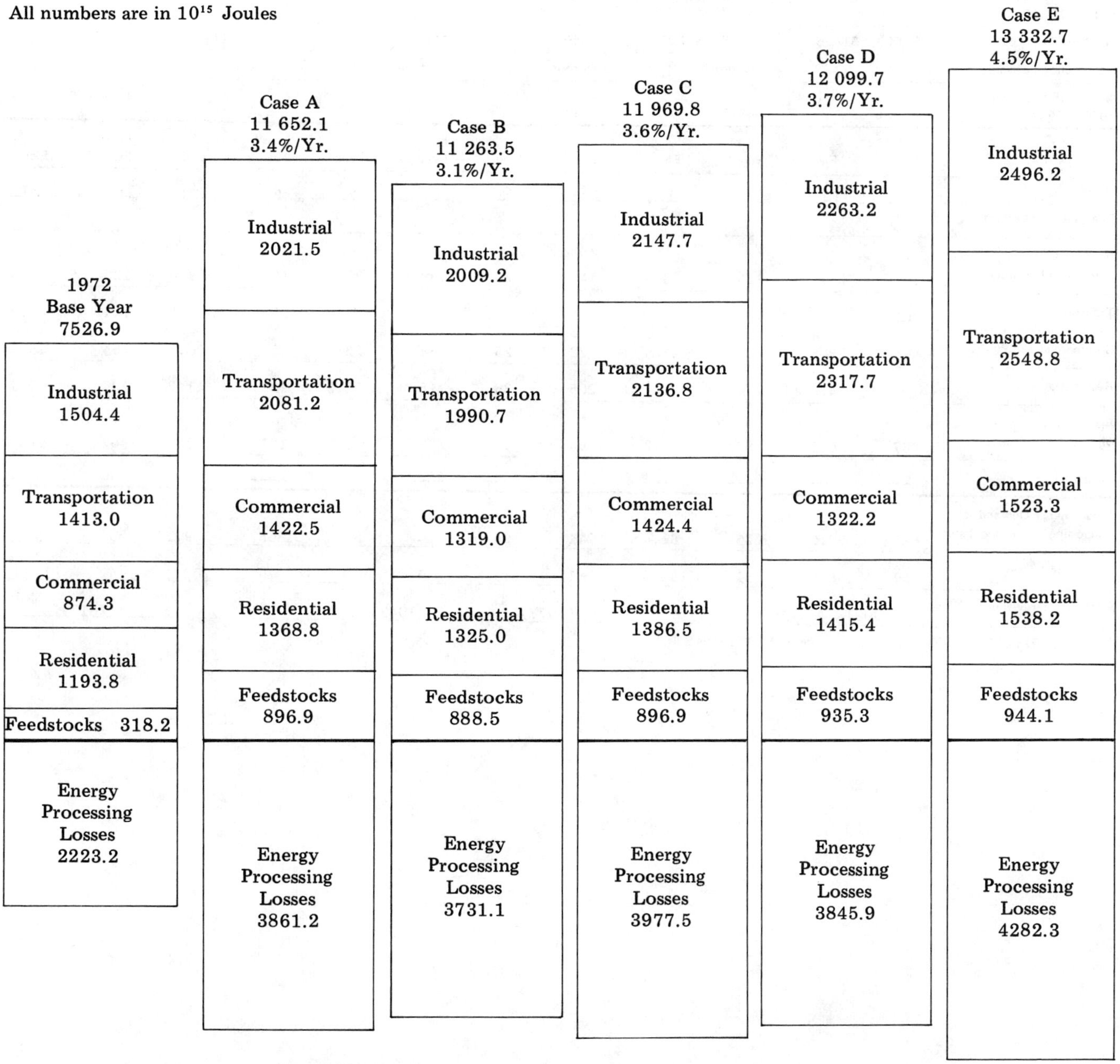

Energy growth rate per year from 1972 to 1985
Commercial includes the public sector

WAES NATIONAL ENERGY DEMAND STUDIES, 1972-1985
ENERGY USE SUMMARY

Country Denmark

1 PJ = 10^{15} J

Scenario Assumptions												
WAES Case	A		B		C		D		E			
GNP 1976-1985	High		Low		High		Low		High			
Oil Price*	\$11.50 → \$17.25		\$11.50 → \$17.25		\$11.50		\$11.50		\$11.50 → \$7.66			
Natl. Policy Response	Vigorous		Vigorous		Vigorous		Restrained		Restrained		1972 Base Year	
Natl. GNP Growth Rate 1976-1985 (%/yr.)	4.1		3.1		4.1		3.1		4.1			
Fuel Type	Fossil Fuel	Elec.	Fossil Fuel	Elec.	Fossil Fuel	Elec.	Fossil Fuel	Elec.	Fossil Fuel	Elec.	Fossil Fuel	Elec.
Units	PJ	PJ	PJ	PJ	PJ	PJ	PJ	PJ	PJ	PJ	PJ	PJ
Energy-Using Sectors												
Industrial (excluding feedstocks)	142	25.5	126	23.0	149	27	139	25	164	29.4	120	14.8
Transportation	203	.4	184	.4	223	.4	209	.4	253	.5	141	.3
Commercial and Public	65	23.8	59	21.6	70	25.8	68	25.3	82	30.2	65	15.2
Residential	191	34.0	174	31.2	205	36.3	215	35.5	252	41.2	207	20.1
Fishing, Agriculture, Mining, Construction	87	6.5	82	6.2	94	7.0	93	6.9	107	7.9	62	4.3
Feedstocks	----	----	----	----	----	----	----	----	----	----	----	----
Final Energy Demand (excluding processing losses)	688	90.2	625	82.4	741	96.5	724	93.1	858	109.2	595	54.7
Processing Losses	57	168	51	153.0	61	179	59	173	70	203	49	118
Primary Energy Input	(1003)		(911)		(1077)		(1049)		(1240)		(817)	

*Oil price 1975 US\$ per barrel Arabian light crude oil FOB Persian Gulf.

DENMARK

All numbers are in 10^{15} Joules (PJ)

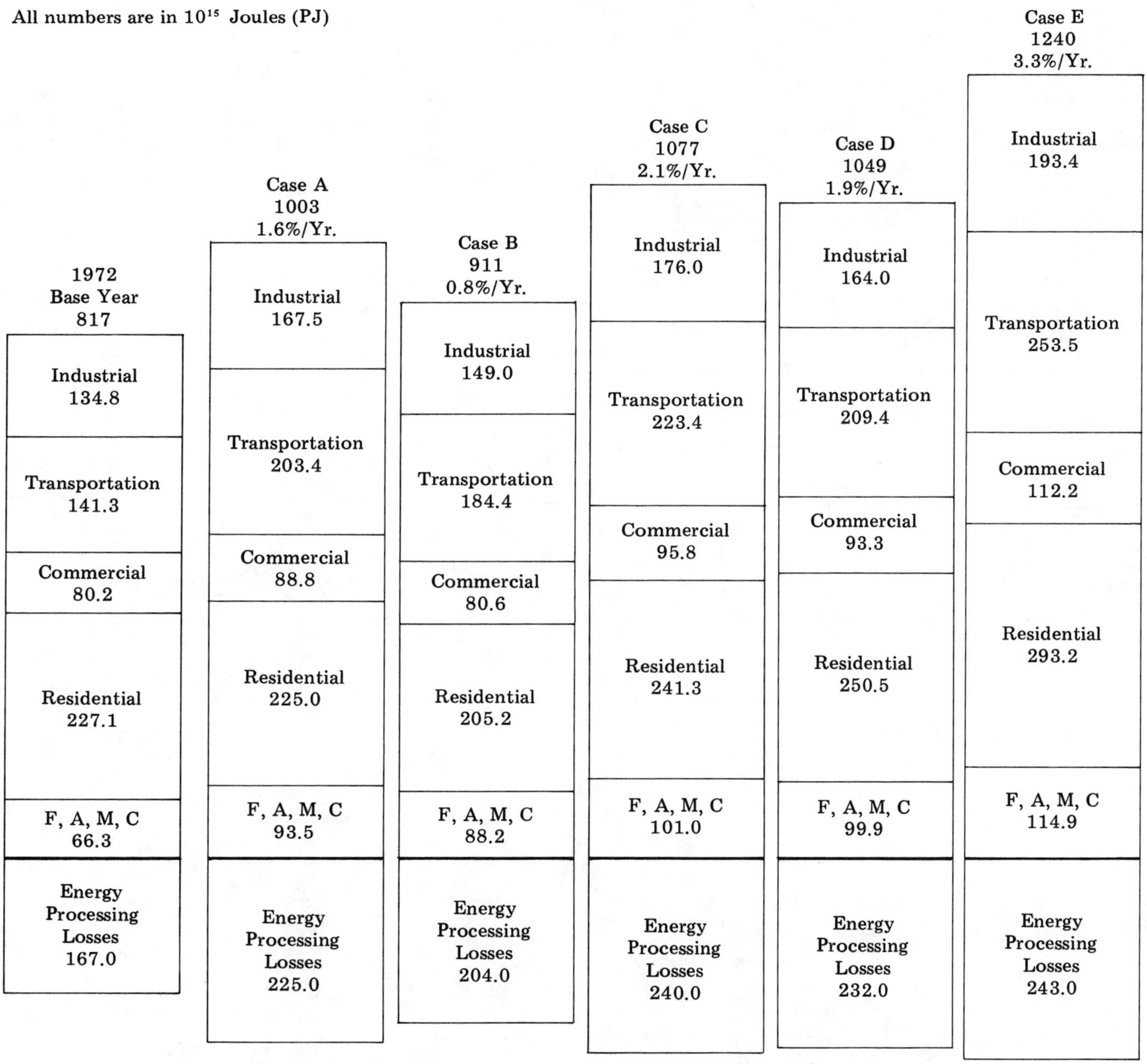

Energy growth rate per year from 1972 to 1985
Commercial includes the public sector
FAMC: fishing, agriculture, mining and construction

WAES NATIONAL ENERGY DEMAND STUDIES, 1972-1985
ENERGY USE SUMMARY

Country Finland

Scenario Assumptions												
WAES Case	A		B		C		D		E			
GNP 1976-1985	High		Low		High		Low		High			
Oil Price*	\$11.50 → \$17.25		\$11.50 → \$17.25		\$11.50		\$11.50		\$11.50 → \$7.66			
Natl. Policy Response [1]	~~Vigorous~~		~~Vigorous~~		~~Vigorous~~		~~Relaxed~~		~~Restrained~~			
Natl. GNP Growth Rate 1976-1985 (%/yr.)	4.4		3.25		4.4		3.25		4.4		1972 Base Year	
Fuel Type	Fossil Fuel [2]	Elec.	Fossil Fuel	Elec.	Fossil Fuel	Elec.	Fossil Fuel	Elec.	Fossil Fuel	Elec.	Fossil Fuel	Elec.
Units 10^{15} J												
Energy-Using Sectors												
Industrial (excluding feedstocks)	423	64	406	61	448	73	415	66	500	82	262	39
Transportation	167	2	154	2	181	2	167	2	190	3	98	0
Commercial and Public	66	30	57	28	70	35	60	31	73	36	60	10
Residential	173	44	172	43	181	52	180	48	190	55	176	15
Fishing, Agriculture, Mining, Construction	32	6	32	6	34	7	34	6	37	7	31	3
Feedstocks	4		4		4		4		4		2	
Final Energy Demand (excluding processing losses)	865	146	825	140	918	169	860	153	994	183	629	68
Processing Losses	94	141	96	133	103	175	101	147	109	198	41	51
Primary Energy Input	959	287	921	273	1021	344	961	300	1103	381	670	118

*Oil price 1975 US\$ per barrel Arabian light crude oil FOB Persian Gulf.

1. National policy options not analyzed

2. "Fossil fuel" includes domestic fuels (wood, peat, bark, industrial wastes etc.)

FINLAND

All numbers are in 10^{15} Joules

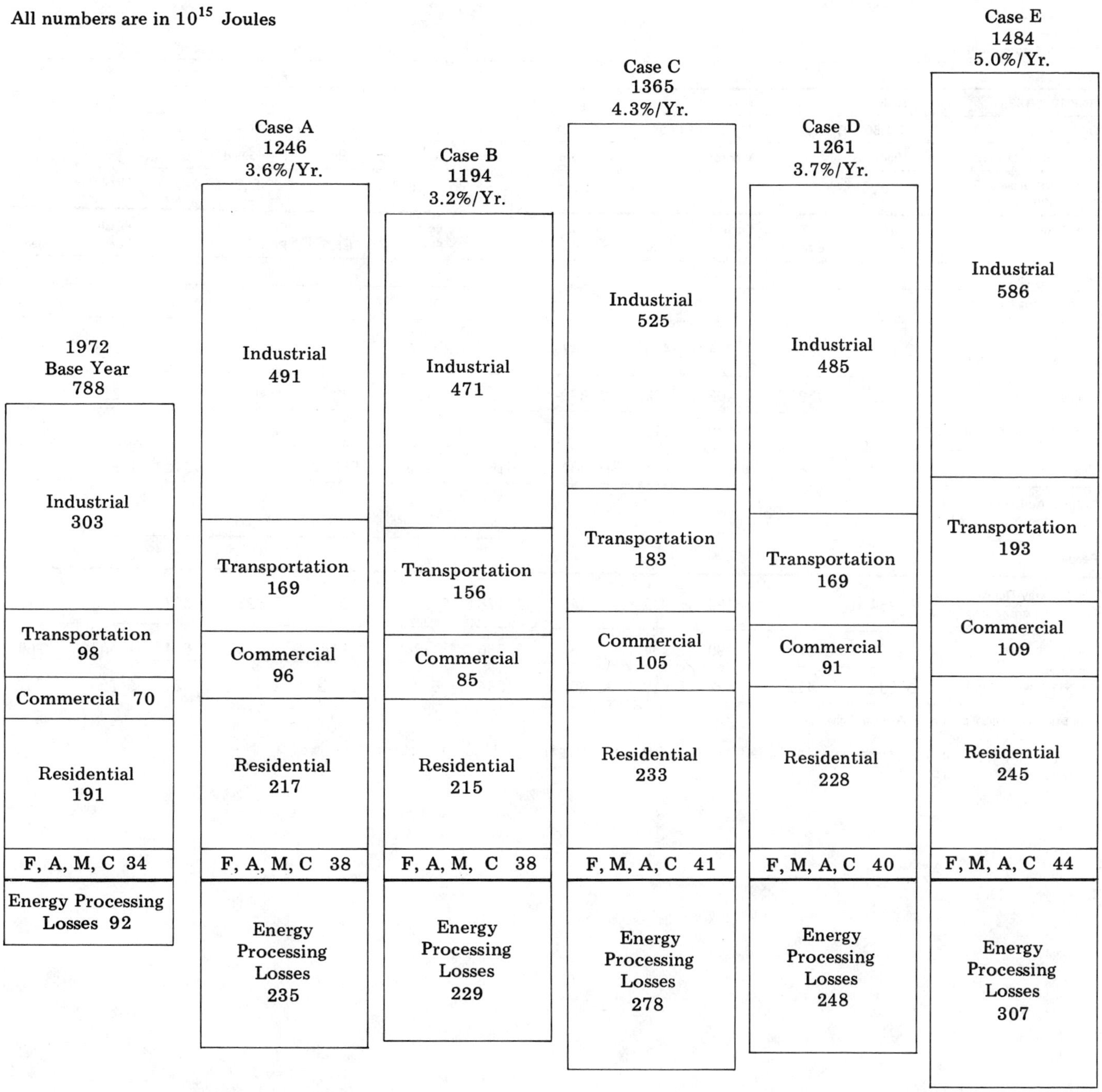

Energy growth rate per year from 1972 to 1985
Industrial includes feedstocks
Commercial includes the public sector
FAMC: fishing, agriculture, mining and construction

WAES NATIONAL ENERGY DEMAND STUDIES, 1972-1985
ENERGY USE SUMMARY

Country FRANCE

Scenario Assumptions												
WAES Case	A		B		C		D		E			
GNP 1976-1985	High		Low		High		Low		High			
Oil Price*	$11.50 → $17.25		$11.50 → $17.25		$11.50		$11.50		$11.50 → $7.66			
Natl. Policy Response	Vigorous		Vigorous		Vigorous		Restrained		Restrained			
Natl. GNP Growth Rate 1976-1985 (%/yr.)	5,5 %		3,5 %		5,5%		3,5 %		5,5 %		1972 Base Year	
Fuel Type	Fossil Fuel	Elec.	Fossil Fuel	Elec.	Fossil Fuel	Elec.	Fossil Fuel	Elec.	Fossil Fuel	Elec.	Fossil Fuel	Elec.
Units 10^{15} J.												
Energy-Using Sectors												
Industrial (excluding feedstocks)	1 672	526	1 364	425	2 347	522	2 126	472	2 699	580	1 763	274
Transportation	1 848	29	1 408	32	1 898	40	1 770	32	2 405	43	1 217	25
Commercial and Public	484	227	528	212								
Residential	1 760	378	1 892	266	2 200	616	2 109	568	2 787	684	1 913	178
Fishing, Agriculture, Mining, Construction 1)												
Feedstocks												
Final Energy Demand (excluding processing losses)	5 764	1 160	5 192	935	6 445	1 178	6 005	1 072	7 891	1 307	4 893	477
Processing Losses	332	2 276	290	1 784	364	2 356	344	2 144	516	2 614	402	960
Primary Energy Input	6 096	3 436	5 482	2 719	6 809	3 534	6 349	3 216	8 407	3 921	5 295	1 337

*Oil price 1975 US$ per barrel Arabian light crude oil FOB Persian Gulf.

1) Fishing and Agriculture are included in Residential. Mining and Construction are included in Industrial.

FRANCE

All numbers are in 10^{15} Joules

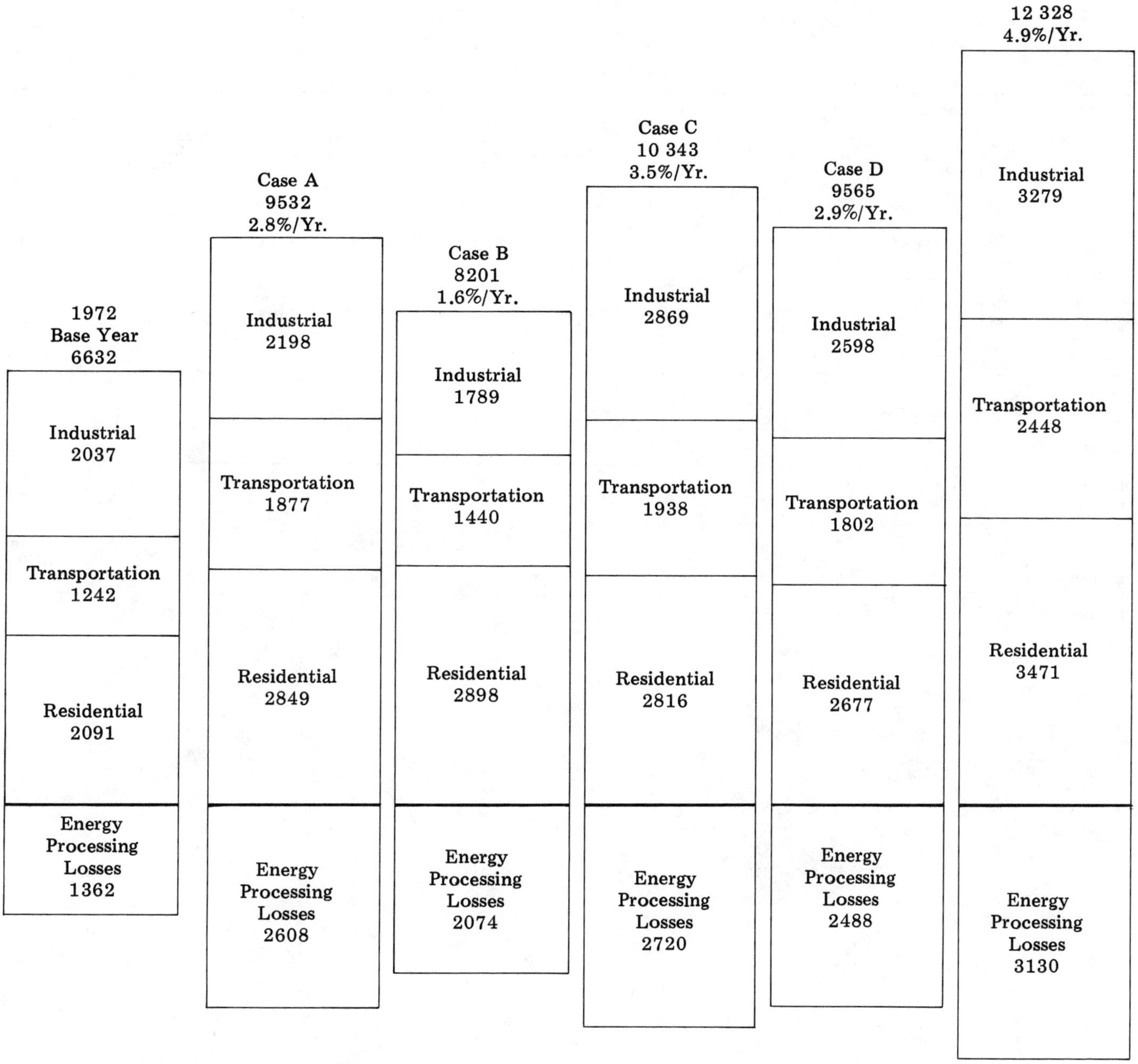

Energy growth rate per year from 1972 to 1985
Residential includes the commercial and public sectors

WAES NATIONAL ENERGY DEMAND STUDIES, 1972-1985
ENERGY USE SUMMARY

Country Federal Republic of Germany

Scenario Assumptions

WAES Case	A		B		C		D		E		1972 Base Year	
GNP 1976-1985	High		Low		High		Low		High			
Oil Price*	\$11.50 → \$17.25		\$11.50 → \$17.25		\$11.50		\$11.50		\$11.50 → \$7.66			
Natl. Policy Response	Vigorous		Vigorous		Vigorous		Restrained		Restrained			
Natl. GNP Growth Rate 1976-1985 (%/yr.)	4.5%		2.5%		4.5%		2.5%		4.5%			
Fuel Type	Fossil Fuel	Elec.	Fossil Fuel	Elec.	Fossil Fuel	Elec.	Fossil Fuel	Elec.	Fossil Fuel	Elec.	Fossil Fuel	Elec.
Units	10^{15}J	10^{15}J	10^{15}J	10^{15}J	10^{15}J	10^{15}J	10^{15}J	10^{15}J	10^{15}J	10^{15}J	10^{15}J	10^{15}J
Energy-Using Sectors												
Industrial (excluding feedstocks)	2585	823	2276	659	2665	827	2363	664	2846	838	2158	435
Transportation	1727	45	1662	45	1802	46	1742	46	1839	45	1428	31
Commercial and Public	793	293	752	275	811	293	783	275	852	293	639	122
Residential	1798	451	1775	389	1974	424	1981	372	2151	397	1816	202
Fishing, Agriculture, Mining, Construction	403	58	385	57	428	59	410	58	430	58	352	36
Feedstocks	1152		1079		1232		1129		1322		765	
Final Energy Demand (excluding processing losses)	8458	1670	7929	1425	8912	1649	8408	1415	9440	1631	7158	826
Processing Losses	512	3758	477	3228	523	3710	523	3205	566	3666	937	1695
Primary Energy Input	8970	5428	8406	4653	9435	5359	8931	4620	10006	5297	8095	2521

*Oil price 1975 US\$ per barrel Arabian light crude oil FOB Persian Gulf.

FEDERAL REPUBLIC OF GERMANY

All numbers are in 10^{15} Joules

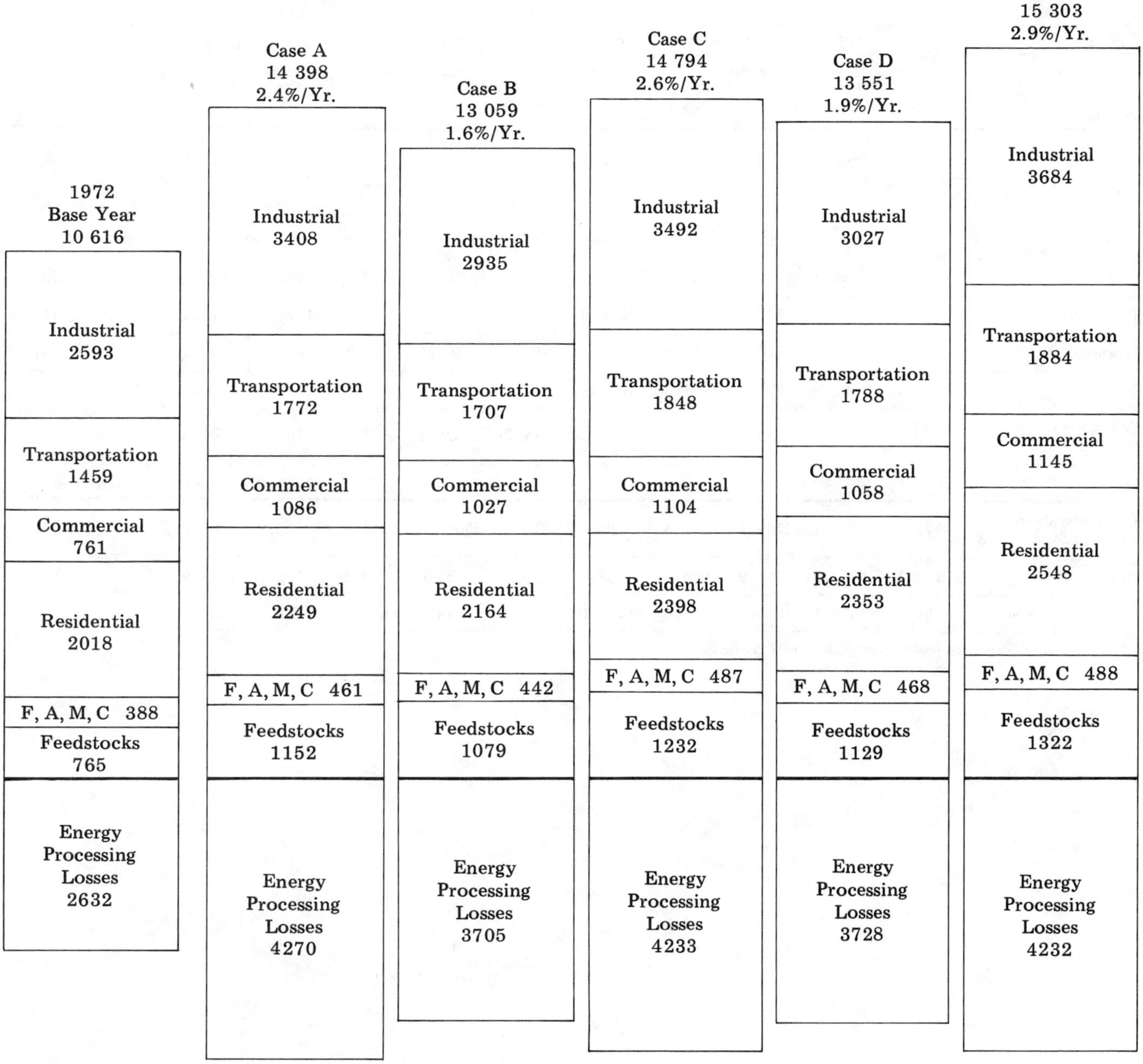

Energy growth rate per year from 1972 to 1985
Commercial includes the public sector
FAMC: fishing, agriculture, mining and construction

WAES NATIONAL ENERGY DEMAND STUDIES, 1972-1985
ENERGY USE SUMMARY

Country Italy

Scenario Assumptions												
WAES Case	A		B		C		D		E			
GNP 1976-1985	High		Low		High		Low		High			
Oil Price*	$11.50 → $17.25		$11.50 → $17.25		$11.50		$11.50		$11.50 → $7.66			
Natl. Policy Response	Vigorous		Vigorous		Vigorous		Restrained		Restrained			
Natl. GNP Growth Rate 1976-1985 (%/yr.)	4.5		2.5		4.5		2.5		4.5		1972 Base Year	
Fuel Type	Fossil Fuel	Elec.	Fossil Fuel	Elec.	Fossil Fuel	Elec.	Fossil Fuel	Elec.	Fossil Fuel	Elec.	Fossil Fuel	Elec.
Units 10^{15} Joules												
Energy-Using Sectors												
Industrial (excluding feedstocks)	1818.8	448.3	1415.3	336.6	2002.6	471.8	1651.4	367.1	2242.4	500.6	1313.1	253.3
Transportation	658.5	23.4	549.6	16.9	672.7	26.1	891.6	18.0	1224.0	18.5	712.9	13.8
Commercial and Public	231.1	97.5	218.9	78.3	301.4	98.4	367.5	81.6	430.3	105.1	240.3	44.4
Residential	843.1	192.6	807.1	137.3	1121.8	197.2	1372.6	161.2	1610.4	234.8	829.7	90.8
Fishing, Agriculture, Mining, Construction	203.0	24.3	193.4	22.2	204.3	24.3	194.2	22.6	204.7	24.3	89.6	10.9
Feedstocks	817.1		524.9		897.9		619.9		1031.0		393.5	
Final Energy Demand (excluding processing losses)	4571.6	786.1	3709.2	591.3	5200.7	817.8	5097.2	650.5	6742.8	883.2	3579.1	413.2
Processing Losses	223.1	1242.0	185.9	906.3	258.3	1286.4	256.6	1015.1	326.9	1478.9	234.3	691.1
Primary Energy Input	4794.7	2028.1	3895.1	1497.6	5459.0	2104.2	5353.8	1665.6	7069.7	2362.1	3813.4	1104.3

*Oil price 1975 US$ per barrel Arabian light crude oil FOB Persian Gulf.

ITALY

All numbers are in 10^{15} Joules

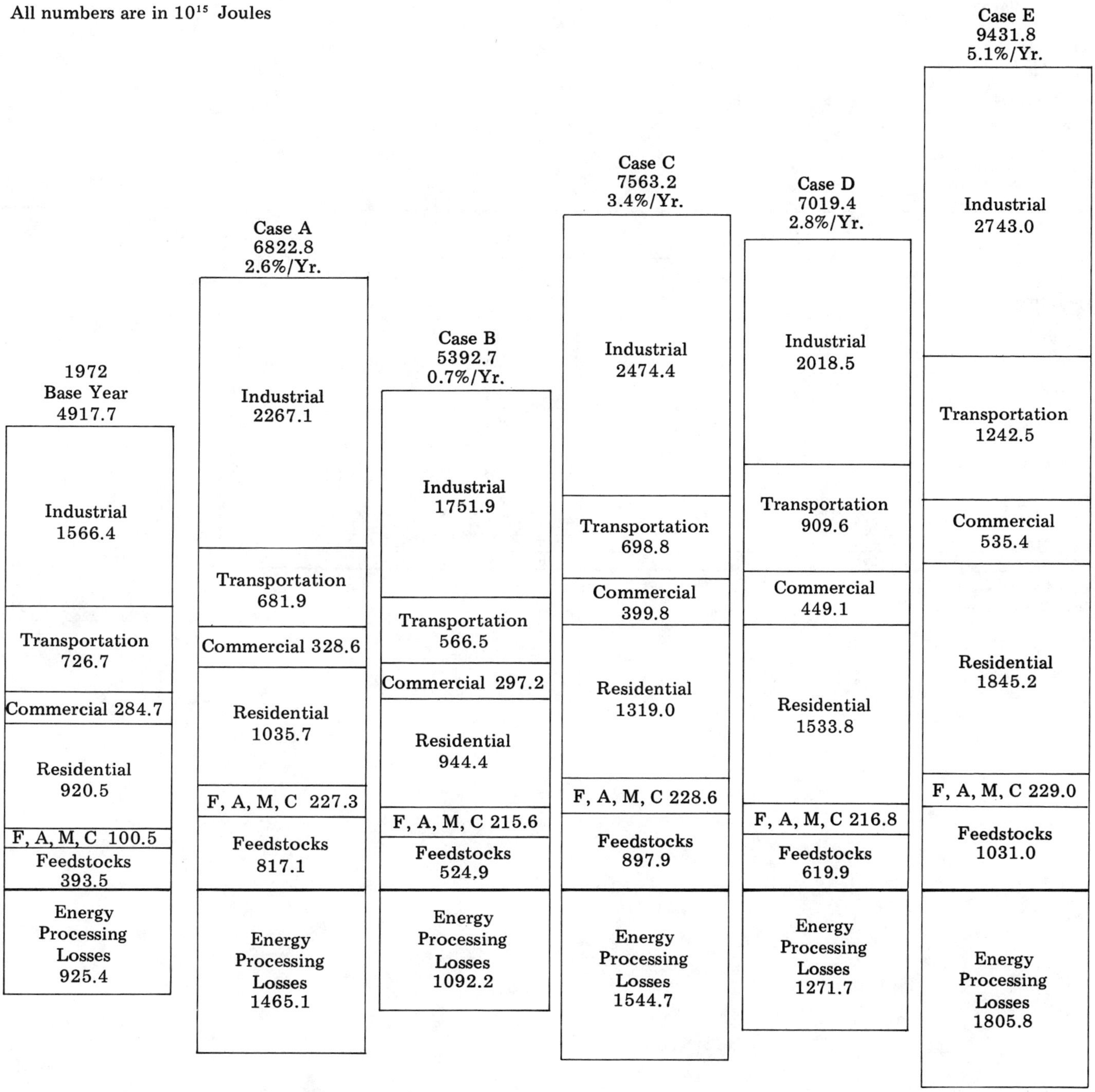

Energy growth rate per year from 1972 to 1985
Commercial includes the public sector
F, A, M, C: fishing, agriculture, mining and construction
Figures do not include bunkerage for non-domestic use, change in stocks, imports and exports of refinery products

WAES NATIONAL ENERGY DEMAND STUDIES, 1972-1985
ENERGY USE SUMMARY

Country JAPAN

Scenario Assumptions

WAES Case	A		B		C		D		E			
GNP 1976-1985	High		Low		High		Low		High			
Oil Price*	\$11.50 → \$17.25		\$11.50 → \$17.25		\$11.50		\$11.50		\$11.50 → \$7.66			
Natl. Policy Response	Vigorous		Vigorous		Vigorous		Restrained		Restrained			
Natl. GNP Growth Rate (1977-1985) (%/yr.)	8.0		4.5		8.0		4.5		8.0		1972 Base Year	
Fuel Type	Fossil Fuel	Elec.	Fossil Fuel	Elec.	Fossil Fuel	Elec.	Fossil Fuel	Elec.	Fossil Fuel	Elec.	Fossil Fuel	Elec.
Units	10^{16}J		10^{16}J		10^{16}J		10^{16}J		10^{16}J		10^{16}J	
Energy-Using Sectors												
Industrial (excluding feedstocks)	994.1	146.6	736.3	107.5	1020.6	149.9	765.2	110.9	1070.6	156.7	504.4	70.6
Transportation	340.6	9.7	286.4	8.0	353.2	9.2	312.5	7.1	387.7	8.4	176.4	4.2
Commercial and Public	177.2	30.7	168.0	23.9	183.1	31.9	177.7	25.2	193.2	33.6	78.1	21.0
Residential	147.4	38.6	127.7	33.6	152.5	39.9	134.8	35.3	164.2	42.8	81.9	16.4
Fishing, Agriculture, Mining, Construction	71.4	4.2	62.2	3.4	72.7	4.2	63.8	3.8	74.8	4.2	51.2	2.9
Feedstocks	160.4		128.5		160.4		128.5		160.4		100.0	
Others Incld. Bunkers	221.8		164.2		221.8		164.2		229.3		65.5	
Final Energy Demand (excluding processing losses)	2113.0	229.7	1673.3	176.4	2164.3	235.2	1746.8	182.3	2280.2	245.7	1057.6	115.1
Processing Losses	560.3		439.7		570.4		450.2		591.3		282.2	
Primary Energy Input	2587.2	315.8	1976.5	312.9	2654.0	315.8	2066.4	312.9	2801.4	315.8	1355.8	99.1

*Oil price 1975 US\$ per barrel Arabian light crude oil FOB Persian Gulf.

JAPAN

All numbers in 10^{16} Joules

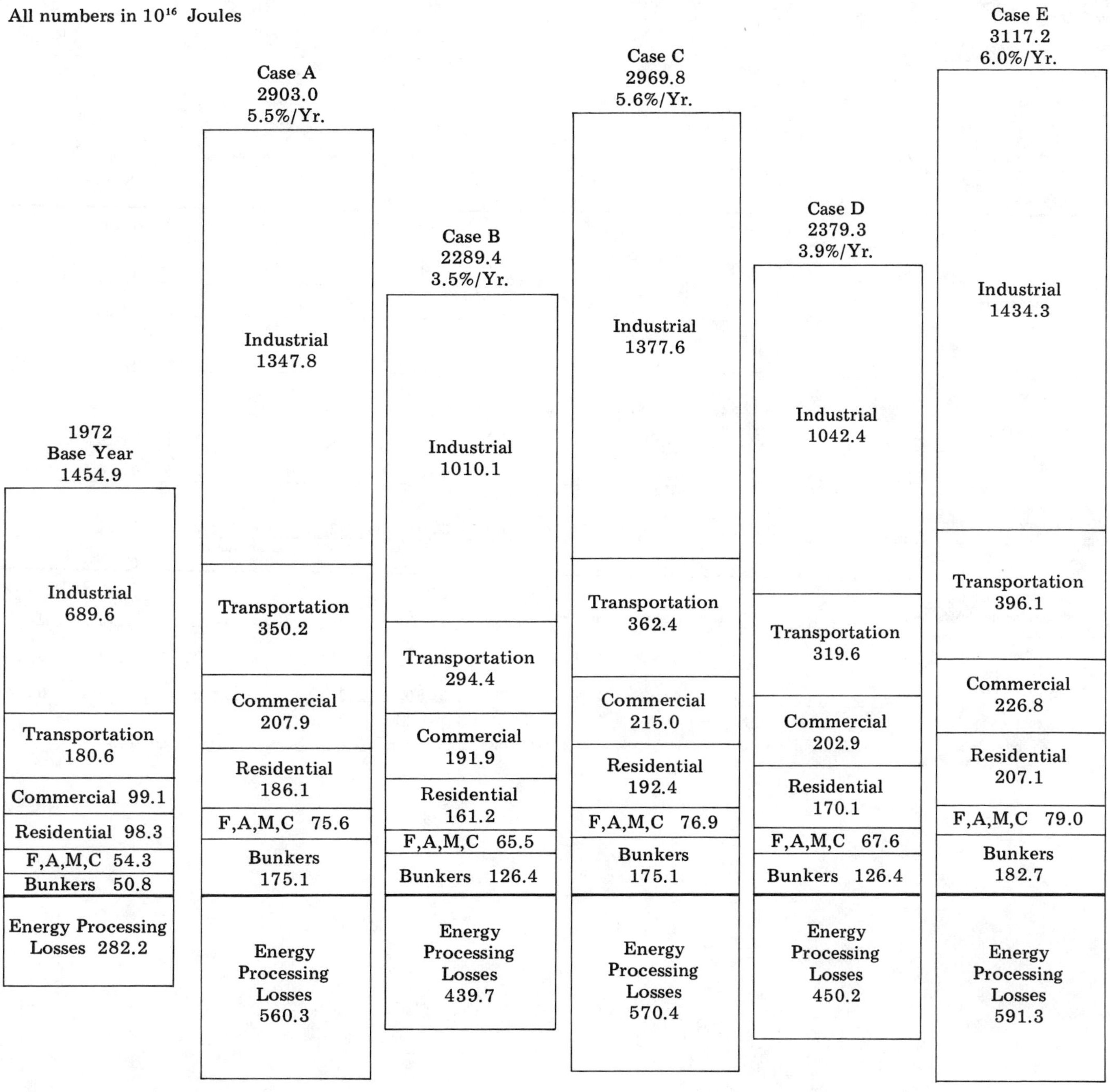

Energy growth rate per year from 1972 to 1985
Industrial includes non-energy uses
Commercial includes the public sector
F,A,M,C: fishing, agriculture, mining and construction
Bunkers: energy used in international transport of goods

WAES NATIONAL ENERGY DEMAND STUDIES, 1972-1985
ENERGY USE SUMMARY

Country MEXICO

Scenario Assumptions

WAES Case	A		B		C		D		E**			
GNP 1976-1985	High		Low		High		Low		High			
Oil Price*	\$11.50 → \$17.25		\$11.50 → \$17.25		\$11.50		\$11.50		\$11.50 → \$7.66			
Natl. Policy Response	Vigorous		Vigorous		Vigorous		Restrained		Restrained			
Natl. GNP Growth Rate 1976-1985 (%/yr.)	7.5%		6%		7.5%		6%		6%		1972 Base Year	
Fuel Type	Fossil Fuel	Elec.	Fossil Fuel	Elec.	Fossil Fuel	Elec.	Fossil Fuel	Elec.	Fossil Fuel	Elec.	Fossil Fuel	Elec.
Units	10^{18} Joules		10^{18} Joules		10^{18} Joules		10^{18} Joules		10^{18} Joules		10^{18} Joules	
Energy-Using Sectors												
Industrial (excluding feedstocks)	1.50	0.19	1.14	0.16	1.36	0.18	1.27	0.16	1.26	0.16	0.56	0.05
Transportation	1.49		1.21		1.36		1.32		1.28		0.55	
Commercial and Public		0.14		0.11		0.13		0.11		0.11		0.03
Residential	0.54	0.06	0.42	0.05	0.53	0.05	0.43	0.05	0.43	0.05	0.18	0.02
Fishing, Agriculture, Mining, Construction	0.01	0.03	0.01	0.02	0.01	0.03	0.01	0.02	0.01	0.02	0.01	0.01
Feedstocks	0.22		0.17		0.21		0.17		0.17		0.07	
Final Energy Demand (excluding processing losses)	3.76	0.42	2.95	0.34	3.47	0.39	3.20	0.34	3.15	0.34	1.37	0.11
Processing Losses	0.75	0.77	0.54	0.64	0.67	0.64	0.60	0.58	0.59	0.58	0.29	0.23
Primary Energy Input	4.51	1.19	3.49	0.98	4.14	1.03	3.80	0.92	3.74	0.92	1.66	0.34

*Oil price 1975 US\$ per barrel Arabian light crude oil FOB Persian Gulf.

**Low energy demand in Case E is a consequence of low GNP growth due to elimination of potential oil exports.

MEXICO

All numbers are in 10^{18} Joules

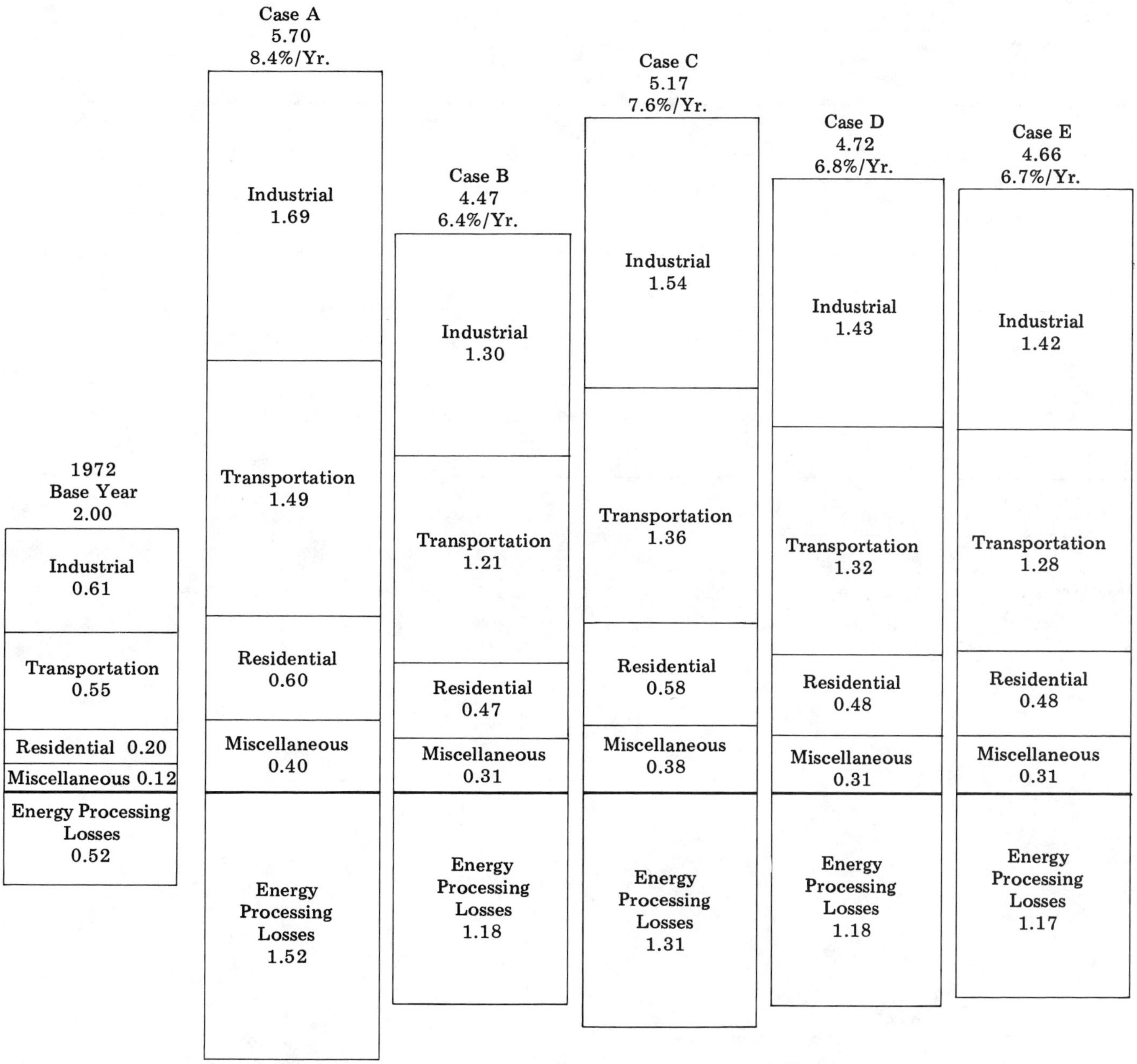

Energy growth rate per year from 1972 to 1985
Miscellaneous includes feedstocks, commercial & public, fishing, agriculture, mining & construction

WAES NATIONAL ENERGY DEMAND STUDIES, 1972-1985
ENERGY USE SUMMARY

Country THE NETHERLANDS

Scenario Assumptions

WAES Case	A		B		C		D		E			
GNP 1976-1985	High		Low		High		Low		High			
Oil Price*	\$11.50 → \$17.25		\$11.50 → \$17.25		\$11.50		\$11.50		\$11.50 → \$7.66			
Natl. Policy Response	Vigorous		Vigorous		Vigorous		Restrained		Restrained			
Natl. GNP Growth Rate 1976-1985 (%/yr.) **	3.4		2.2		3.4		2.2		3.4		1972 Base Year	
Fuel Type	Fossil Fuel	Elec.	Fossil Fuel	Elec.	Fossil Fuel	Elec.	Fossil Fuel	Elec.	Fossil Fuel	Elec.	Fossil Fuel	Elec.
Units $(10^{15} J)$												
Energy-Using Sectors	soft		soft						soft			
Industrial (excluding feedstocks)	979.8	127.4	647.3	83.2	1,028.4	133.6	683.6	88.0	1,038.3	135.4	498.3	63.9
Transportation					967.9	4.9	968.1	4.9			795.5	3.3
Commercial and Public	165.8	50.2	152.9	46.3	186.5	51.2	172.0	47.3	207.2	52.3	155.1	39.1
Residential					744.7	73.7	722.7	73.7			524.7	36.0
Fishing, Agriculture, Mining, Construction					231.9	11.9	225.4	11.4			152.9	9.8
Feedstocks	759.3	-	444.6	-	759.3	-	444.6	-	759.3	-	276.3	-
Final Energy Demand (excluding processing losses)	3,946.5		3,106.2		3,918.7	275.3	3,216.4	225.3	4,656.4		2,402.8	152.1
Processing Losses	486.3		404.9		173.8	342.8	168.2	281.3	597.5		170.9	304.0
Primary Energy Input	4,432.8		3,511.1		4,710.6		3,891.2		5,253.9		3,029.8	

*Oil price 1975 US\$ per barrel Arabian light crude oil FOB Persian Gulf.

** 1972-1985

THE NETHERLANDS

All numbers are in 10^{15} Joules

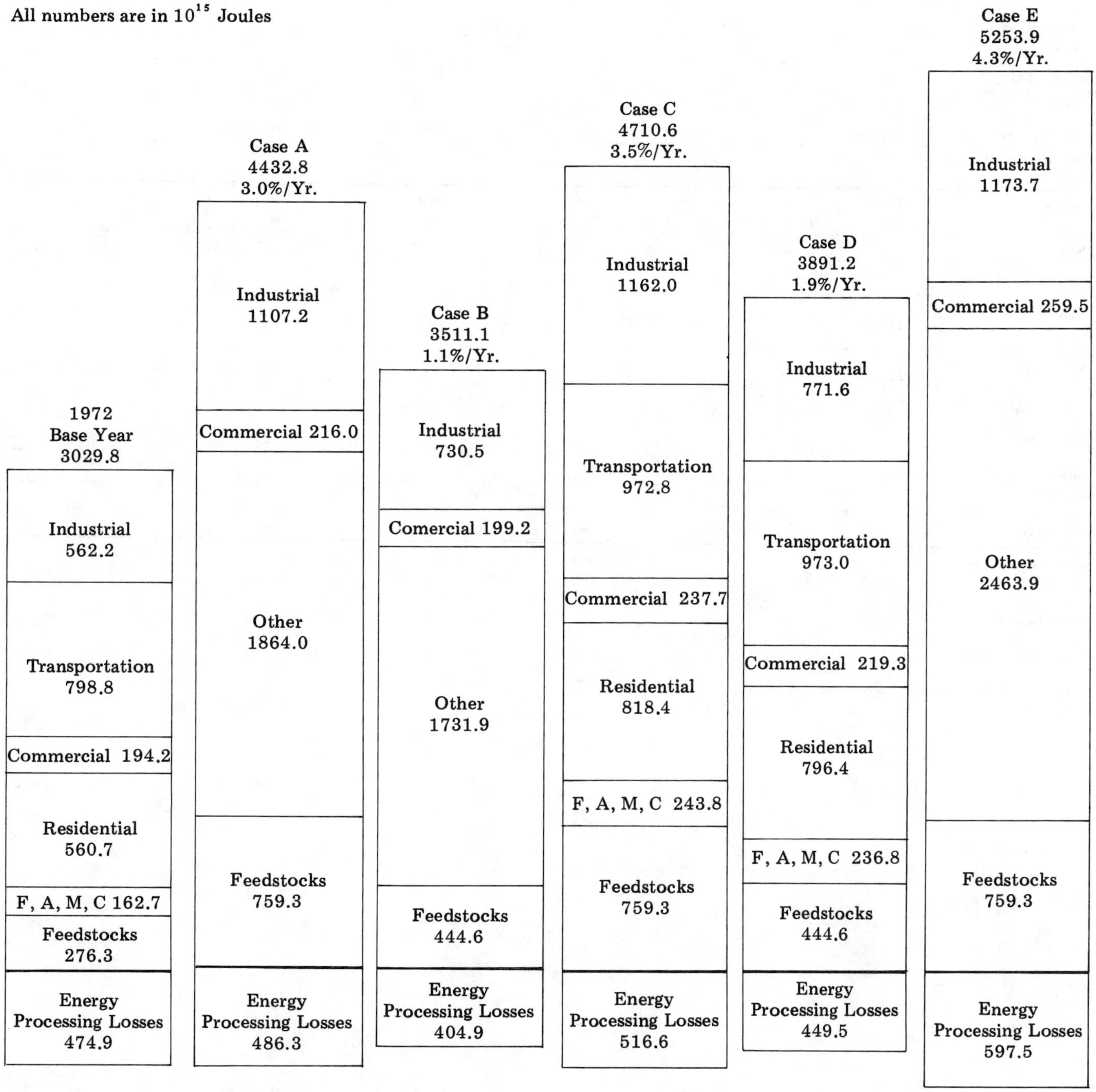

Energy growth rate per year from 1972 to 1985
Commercial includes the public sector
F, A, M, C: fishing, agriculture, mining and construction
Cases A, B, and E energy totals are "soft" data
Other, in cases A, B, and E includes Transportation, Residential and FAMC

WAES NATIONAL ENERGY DEMAND STUDIES, 1972-1985
ENERGY USE SUMMARY

Country NORWAY

Scenario Assumptions												
WAES Case	A		B		C		D		E			
GNP 1976-1985	High		Low		High		Low		High			
Oil Price*	$11.50 → $17.25		$11.50 → $17.25		$11.50		$11.50		$11.50 → $7.66			
Natl. Policy Response	Vigorous		Vigorous		Vigorous		Restrained		Restrained			
Natl. GNP Growth Rate 1976-1985 (%/yr.)	4.4		4.2		4.4		4.2		4.4		1972 Base Year	
Fuel Type	Fossil Fuel	Elec.	Fossil Fuel	Elec.	Fossil Fuel	Elec.	Fossil Fuel	Elec.	Fossil Fuel	Elec.	Fossil Fuel	Elec.
Units	10^{15} J/year		10^{15} J/year		10^{15} J/year		10^{15} J/year		10^{15} J/year		10^{15} J/year	
Energy-Using Sectors												
Industrial (excluding feedstocks)	175.2	184.3	171.8	180.3	180.9	184.3	180.7	182.9	190.2	187.0	116.8	127.8
Transportation	221.4	2.9	204.6	2.8	221.4	2.9	218.2	2.8	235.0	2.9	115.3	2.0
Commercial and Public	37.2	24.4	36.8	24.1	37.2	24.4	37.6	24.7	38.0	25.0	23.0	15.0
Residential	37.2	71.6	36.1	69.5	48.0	73.0	50.5	80.5	66.3	77.7	58.0	51.2
Fishing, Agriculture, Mining, Construction	35.3	10.5	35.4	10.6	35.3	10.5	35.4	10.6	35.3	10.5	28.1	8.1
Feedstocks	41.5		38.6		41.5		38.6		41.5		20.0	
Final Energy Demand (excluding processing losses)	547.8	293.7	523.3	287.4	564.3	295.1	561.0	301.5	606.3	303.0	361.2	204.1
Processing Losses	61.5	625.9	58.6	612.3	63.8	628.9	63.5	642.4	69.0	645.8	45.4	445.7
Primary Energy Input	609.3	919.6	581.9	899.7	628.1	924.0	624.5	943.9	675.3	948.8	406.6	649.8

*Oil price 1975 US$ per barrel Arabian light crude oil FOB Persian Gulf.

NORWAY

All numbers are in 10^{15} Joules

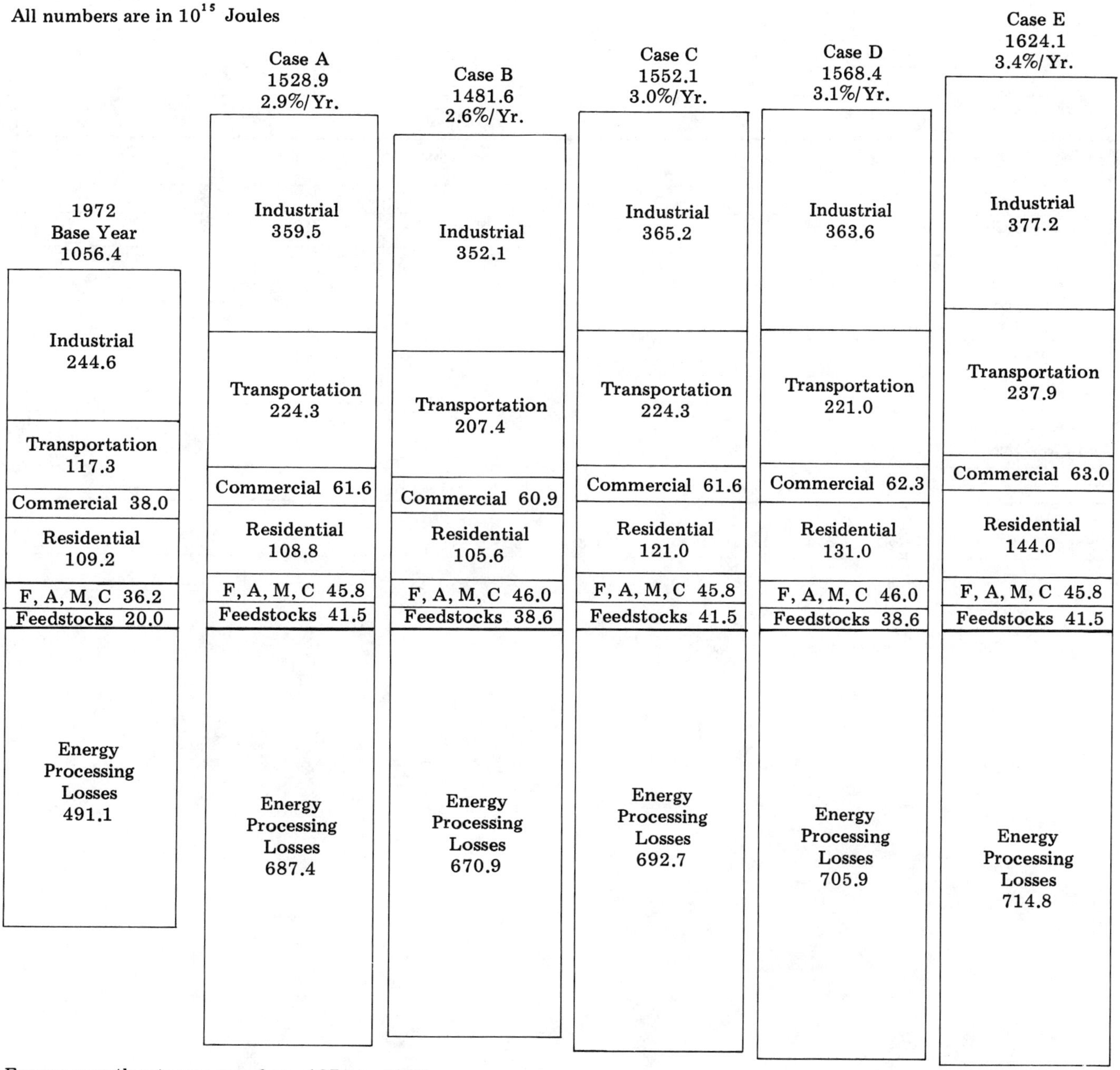

Energy growth rate per year from 1972 to 1985
Commercial includes the public sector
F, A, M, C: fishing, agriculture, mining and construction

WAES NATIONAL ENERGY DEMAND STUDIES, 1972-1985
ENERGY USE SUMMARY

Country Sweden

Scenario Assumptions

WAES Case	A		B		C		D		E			
GNP 1976-1985	High		Low		High		Low		High			
Oil Price*	$11.50 → $17.25		$11.50 → $17.25		$11.50		$11.50		$11.50 → $7.66			
Natl. Policy Response	Vigorous		Vigorous		Vigorous		Restrained		Restrained			
Natl. GNP Growth Rate 1976-1985 (%/yr.)	4		2.5		4		2.5		4		1972 Base Year	
Fuel Type	Fossil Fuel	Elec.	Fossil Fuel	Elec.	Fossil Fuel	Elec.	Fossil Fuel	Elec.	Fossil Fuel	Elec.	Fossil Fuel	Elec.
Units	10^{17}J		10^{17}J		10^{17}J		10^{17}J		10^{17}J		10^{17}J	
Energy-Using Sectors												
Industrial (excluding feedstocks)	5.8	2.3	5.2	2.1	6.1	2.5	5.7	2.2	6.9	2.7	4.3	1.2
Transportation	3.2	.1	2.8	.1	3.5	.1	3.3	.1	4.1	.2	2.5	.1
Commercial and Public	1.0	1.0	1.3	.8	1.4	.8	1.6	.8	1.6	1.0	1.6	.3
Residential	1.4	1.6	1.5	1.5	2.5	1.1	2.6	1.2	3.2	1.2	3.2	.6
Fishing, Agriculture, Mining, Construction	.6	.4	.5	.3	.7	.4	.5	.3	.7	.5	.3	.1
Feedstocks	1.4		1.3		1.6		1.4		1.8		.5	
Final Energy Demand (excluding processing losses)	13.5	5.4	12.5	4.7	15.8	5.0	15.1	4.5	18.3	5.5	12.3	2.3
Processing Losses	.8	10.3	.8	9.6	.9	10.0	.9	9.2	1.0	11.0	.6	4.8
Primary Energy Input	14.3	15.6	13.3	14.3	16.7	15.0	16.0	13.7	19.3	16.5	12.9	7.1

*Oil price 1975 US$ per barrel Arabian light crude oil FOB Persian Gulf.

SWEDEN

All numbers are in 10^{17} Joules

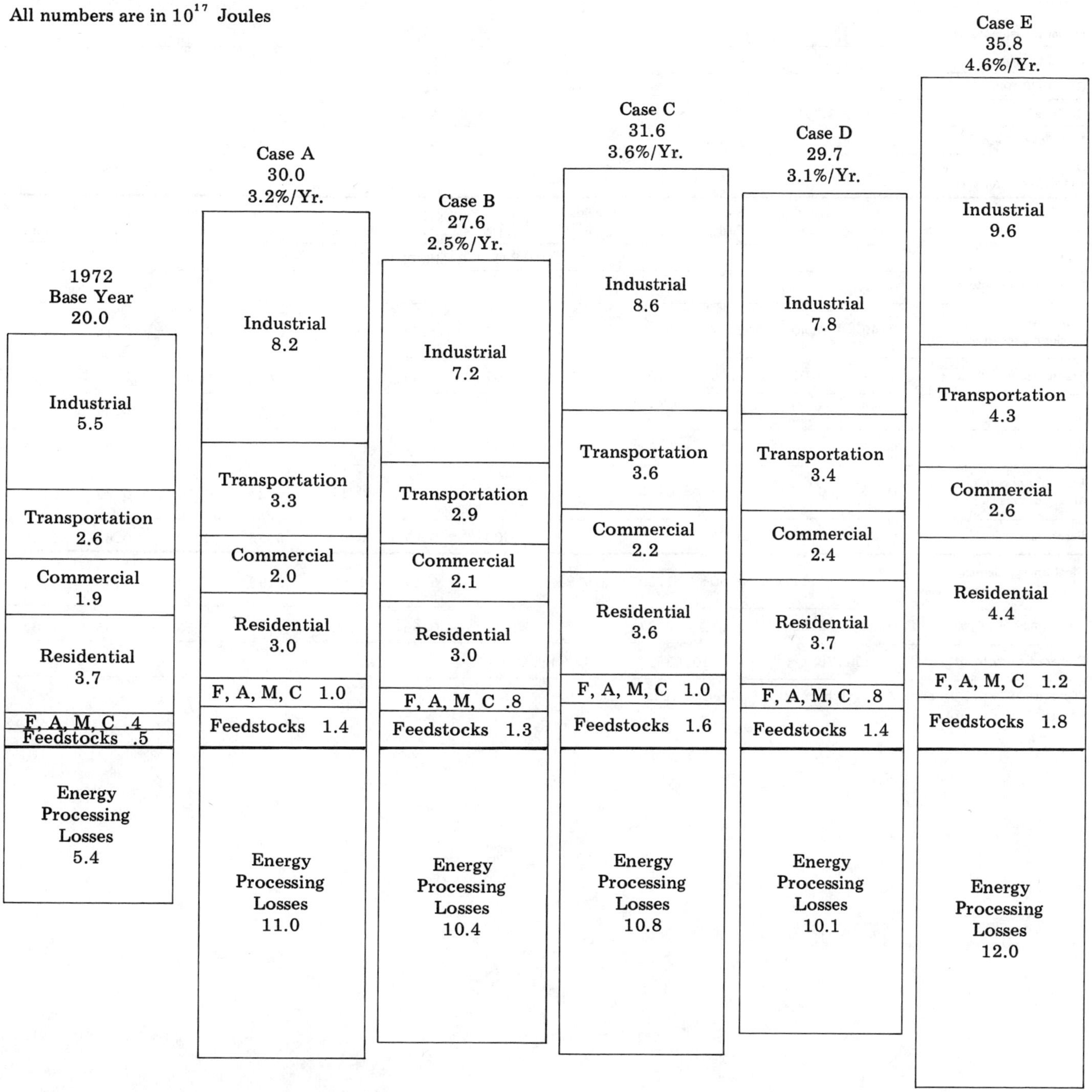

Energy growth rate per year from 1972 to 1985
Commercial includes the public sector
F, A, M, C: fishing, agriculture, mining and construction

WAES NATIONAL ENERGY DEMAND STUDIES, 1972-1985
ENERGY USE SUMMARY

Country United Kingdom

Scenario Assumptions

WAES Case	A		B		C		D		E			
GNP 1976-1985	High		Low		High		Low		High			
Oil Price*	$11.50 → $17.25		$11.50 → $17.25		$11.50		$11.50		$11.50 → $7.66			
Natl. Policy Response	Expected		Expected		Expected		Expected		Expected			
Natl. GDP Growth Rate 1977-1985 (%/yr.) **	3.1		1.8		3.1		1.8		3.1		1972 Base Year	
Fuel Type	Fossil Fuel	Elec.	Fossil Fuel	Elec.	Fossil Fuel	Elec.	Fossil Fuel	Elec.	Fossil Fuel	Elec.	Fossil Fuel	Elec.
Units	10^{18} J PER ANNUM											
Energy-Using Sectors												
Industrial (excluding feedstocks)	2.40	0.40	2.37	0.37	2.44	0.42	2.36	0.38	2.47	0.43	2.16	0.26
Transportation	1.57	0.01	1.46	0.01	1.59	0.01	1.49	0.01	1.61	0.01	1.24	0.01
Public	0.33	0.07	0.30	0.06	0.32	0.08	0.31	0.06	0.32	0.08	0.32	0.05
Residential	1.31	0.46	1.24	0.43	1.33	0.48	1.24	0.44	1.35	0.49	1.21	0.31
Fishing, Agriculture, and Commercial	0.25	0.17	0.28	0.14	0.24	0.18	0.28	0.14	0.25	0.18	0.29	0.11
Feedstocks	0.97	0.00	0.84	0.00	1.00	0.00	0.87	0.00	0.98	0.00	0.62	0.00
Final Energy Demand (excluding processing losses)	6.83	1.11	6.50	1.00	6.93	1.16	6.57	1.02	6.98	1.19	5.85	0.74
Processing Losses	0.71	2.63	0.65	2.37	0.73	2.73	0.66	2.42	0.74	2.80	0.65	2.07
Primary Energy Input	11.28		10.53		11.55		10.67		11.71		9.30	

*Oil price 1975 US$ per barrel Arabian light crude oil FOB Persian Gulf.
** excluding oil

UNITED KINGDOM

All numbers are in 10^{18} Joules

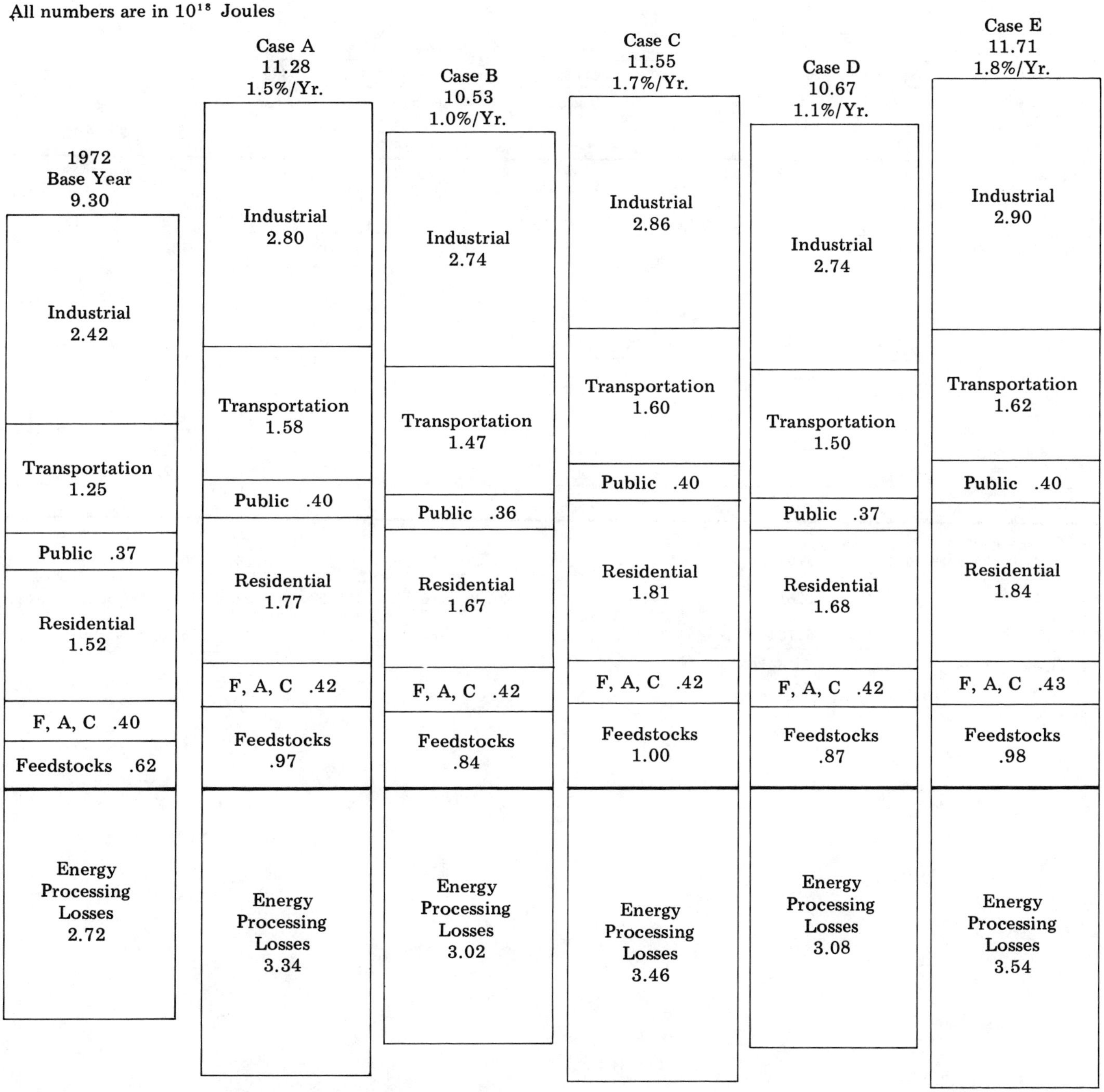

Energy growth rate per year from 1972 to 1985
F, A, C: fishing, agriculture and commercial

WAES NATIONAL ENERGY DEMAND STUDIES, 1972-1985
ENERGY USE SUMMARY

Country USA

Scenario Assumptions												
WAES Case	A		B		C		D		E			
GNP 1976-1985	High		Low		High		Low		High			
Oil Price*	$11.50 → $17.25		$11.50 → $17.25		$11.50		$11.50		$11.50 → $7.66			
Natl. Policy Response	Vigorous		Vigorous		Vigorous		Restrained		Restrained			
Natl. GNP Growth Rate 1976-1985 (%/yr.)	4.4		3.2		4.4		3.2		4.4		1972 Base Year	
Fuel Type	Fossil Fuel	Elec.	Fossil Fuel	Elec.	Fossil Fuel	Elec.	Fossil Fuel	Elec.	Fossil Fuel	Elec.	Fossil Fuel	Elec.
Units	$10^{18}J$		$10^{18}J$		$10^{18}J$		$10^{18}J$		$10^{18}J$		$10^{18}J$	
Energy-Using Sectors												
Industrial (excluding feedstocks)	19.13	2.82	16.37	2.45	19.79	2.95	17.89	2.67	21.67	3.20	13.13	1.91
Transportation	19.49	.02	18.63	.02	19.86	.02	20.87	.02	22.34	.02	17.91	.01
Commercial and Public	3.10	2.09	3.05	1.98	3.44	2.17	3.85	2.58	4.10	2.90	2.90	1.50
Residential	7.73	2.86	7.78	2.71	8.68	3.15	10.52	3.67	10.75	4.26	9.99	1.98
Fishing, Agriculture, Mining, Construction	2.57	.18	2.03	.15	2.64	.19	2.07	.17	2.64	.20	2.00	.14
Feedstocks	8.74		7.28		8.65		7.49		8.67		4.42	
Final Energy Demand (excluding processing losses)	60.76	7.97	55.13	7.31	63.06	8.49	62.69	9.12	70.18	10.58	50.37	5.54
Processing Losses	6.08	18.81	5.51	17.26	6.31	20.06	6.27	21.52	7.02	24.98	5.23	14.21
Primary Energy Input	66.84	26.78	60.64	24.57	69.37	28.55	68.96	30.64	77.20	35.56	55.60	19.75

*Oil price 1975 US$ per barrel Arabian light crude oil FOB Persian Gulf.

UNITED STATES

All numbers are in 10^{18} Joules

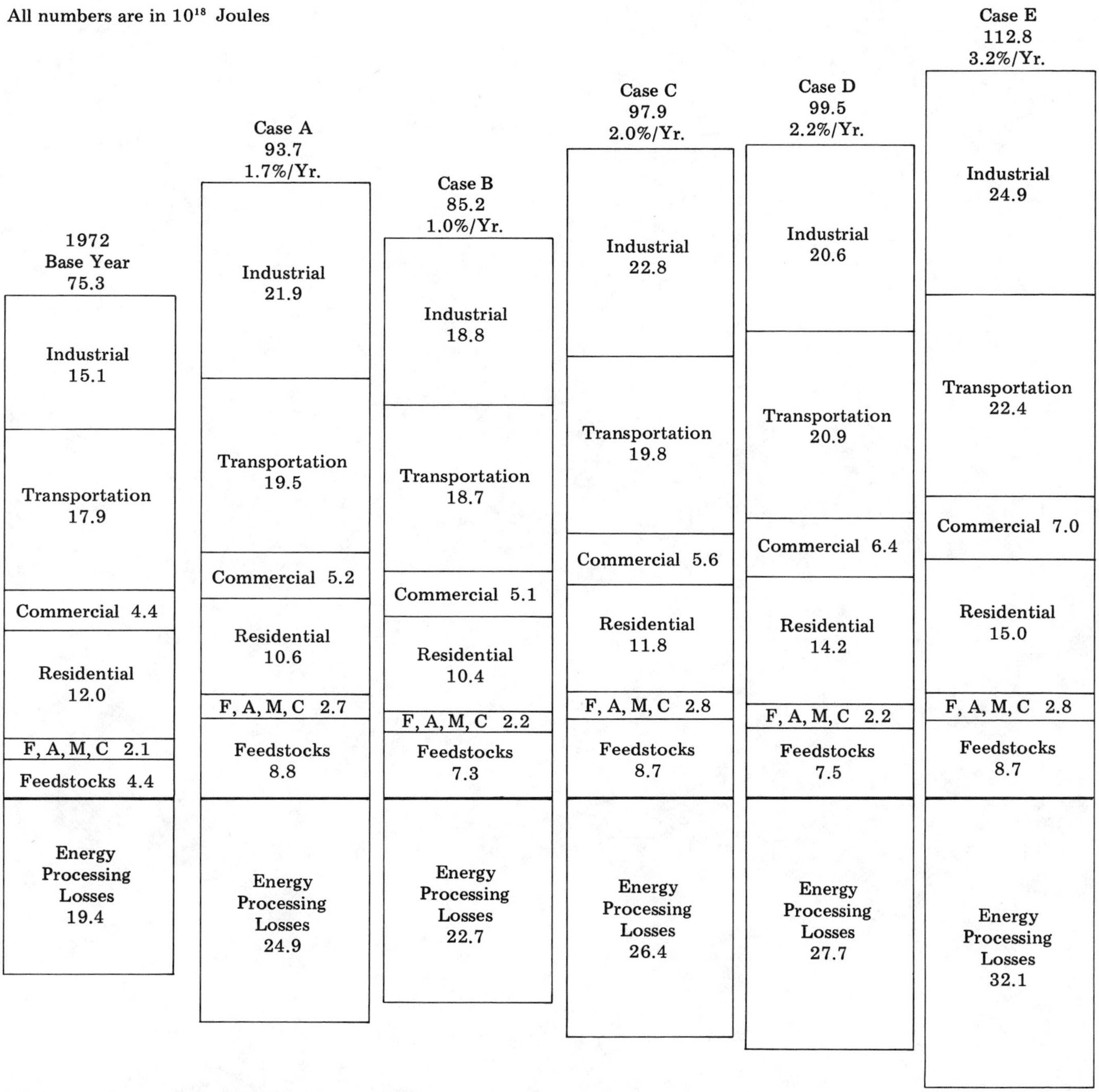

Energy growth rate per year from 1972 to 1985
Commercial includes the public sector
F, A, M, C: fishing, agriculture, mining and construction

Part II
National Energy Demand Studies

HOW TO USE PART II

Each chapter of Part II contains the energy demand study of a Workshop country. Each chapter was written according to a common general format, but differences, reflecting the different energy situations in the several countries, appear from chapter to chapter. Nearly all chapters contain the following elements:

- An overview section, which either describes the overall observations and conclusions or in some way introduces the report.
- A methodology section, which describes the adaptation of the overall WAES methodology to individual national circumstances and the specific procedures used in projecting energy demand.
- An analysis section, which includes a discussion of the economic activity level and energy efficiency assumptions used in estimating energy use in the major market sectors (transportation, industry, commercial, public, and residential). The impacts of changes in the scenario variables (economic growth, oil price, and national energy policy) are also discussed.
- A section containing summary tables and detailed worksheets presenting economic, efficiency, and energy-use data for all sectors for the cases studied. While different countries show different levels of detail on the worksheets, in each case the worksheets represent the basic data about which the report has been written.

Canada

3.1 OVERVIEW

The intent of this report is to focus attention on specific changes in energy use that result from the explicit assumptions outlined in each scenario case.

The range of projected energy use for the five cases developed is about 1450 x 10^{12} Btu or about 17 percent of the total requirements. The range represents the total spread of energy use arising from detailed assumptions of high and low economic growth rates, three levels of oil prices, and the estimated extent of government policy initiatives in the energy field. Thus, the potentials for energy savings through changes in economic growth, oil price, and government policy initiatives may be observed in this report. It must be borne in mind, of course, that none of the scenarios presented in this report will necessarily occur. The scenario approach adopted by WAES attempts merely to indicate the direction in which energy requirements in Canada may develop under the assumptions outlined.

Through the utilization of the national policy response variable, WAES has attempted to provide a preliminary indication of the potential for conservation in the Canadian economy. To predict future energy requirements without this important variable would result in the exclusion of a major determining factor of energy requirements.

In general, the scenarios indicate that the conservation possibilities through price and nonprice effects could be significant in the next decade. Based on the estimates provided in this report, which have been developed from background studies done elsewhere, it would appear that while the price effects on energy demand—for example, from a doubling of price—are considerable, a government-induced nonprice conservation program could further reduce the growth in energy requirements in the next decade.

As would be expected, the impacts of the price and nonprice effects vary according to the sector. Bearing in mind the difficulties of accurate forecasting, the scenarios indicate, for example, that the price effects in the commercial sector could account for virtually all of the energy savings relative to the nonprice effects. In the residential sector, energy savings resulting from price and nonprice effects would be approximately equal. In the transportation sector, energy savings from government-induced nonprice conservation efforts would have more than twice the impact of price effects.

3.2 INTRODUCTION

3.2.1 Basis of the Canadian Submission

This study provides estimates of Canadian energy requirements for 1972 and for 1985 under five different scenarios based on the criteria established by WAES. In the preparation of this study, advice and background information were sought from various representatives of Energy, Mines and Resources Canada, Statistics Canada, and the National Energy Board of Canada.

The basic tool used to prepare the Canadian submission to WAES was an integrated macroeconomic energy demand forecasting model developed by Energy, Mines and Resources Canada. Extensive use was also made of their published report, *An Energy Strategy for Canada: Policies for Self-Reliance* of April 1976. Use was made of the recent reports dealing with natural gas and oil issued by the National Energy Board: *Canadian Natural Gas—Supply and Requirements* of April 1975, and *Canadian Oil—Supply and Requirements* of September 1975. In addition, a background study, *Energy Conservation*, prepared for The Science Council of Canada and published in July 1975, was used as a basis for quantifying the effects on energy consumption of government-induced conservation measures. The 1972 base year estimates were derived from a Statistics Canada publication, *Detailed Energy Supply and Demand in Canada*.

It must be stressed that these scenarios were derived solely to meet WAES requirements and are in no sense official forecasts of the National Energy Board or the Government of Canada.

3.2.2 Statement on Data

The estimates in this study are of final end-use energy demands by sectors, which are then summed to provide the line item "final energy demand" as required on the WAES worksheets.

These estimates should be considered as "final" according to the WAES definition, although global and national supply-demand integrations may result in minor revisions to the delivered demand data. The final two line items on the worksheets (i.e., processing losses and primary energy input) may also be subject to minor revisions following the global supply-demand integration process.

3.2.3 Recent Government Initiatives

In April 1976, the Government of Canada, in support of the objective of energy self-reliance within ten years, proposed nine major policy thrusts. These policy elements provide a coordinated framework for the development of specific programs and measures. They include (1) appropriate energy pricing; (2) energy conservation; (3) increased exploration and development; (4) increased resource information; (5) interfuel substitution; (6) new delivery systems; (7) emergency preparedness; (8) increased research and development; and (9) greater Canadian content and participation in resource development.

The Government of Canada has adopted the following targets:

i. To move domestic oil prices toward international levels; and to move domestic prices for natural gas to an appropriate competitive relationship with oil over the next two to four years. (Thus, domestic prices must continue to increase in order to reinforce efficiency and restraint in energy use and to encourage the development of additional Canadian supplies.)

ii. To reduce Canadian net dependence on imported oil in 1985 to one-third of Canadian total oil demands. (Canada was roughly in balance in terms of oil exports and imports in 1975.)

iii. To maintain self-reliance in natural gas until such time as northern resources can be brought to market under acceptable conditions.

iv. To double, at a minimum, exploration and development activity in the frontier regions of Canada over the next three years, under acceptable social and environmental conditions. (An active role is anticipated for Petro-Canada, and more vigorous land and lease obligations have been announced.)

v. To reduce the average rate of growth of energy use in Canada, over the next ten years, to less than 3.5 percent per year. (This represents a reduction of the historically observed growth rate by over one-third.)

The federal government's energy conservation program is designed to encourage efficiencies in the use of energy and, as noted, to reduce the rate at which Canadian energy requirements will grow in the future. Over the past year the Canadian government's program has consisted mainly of public information, consultation, and in-house changes.

In February 1976, the Government of Canada announced a number of new federal government initiatives directed at strengthening the energy conservation program introduced one year earlier. The new initiatives are briefly summarized below.

Automobiles
New mileage standards were introduced for automobiles sold in Canada. Each automobile manufacturer and importer with significant sales volume in Canada will be required to meet sales-weighted fleet-average targets of at least 24 miles per gallon by 1980 and 33 miles per gallon by 1985. In addition, beginning in 1976 a listing of all new automobiles, ranked in order of fuel performance, will be compiled and published by the Ministry of Transport.

Buildings
The Government of Canada has announced that new guidelines for the design, construction, and operation of energy-efficient buildings of all sizes in Canada will be completed before the end of 1976. These guidelines will either be adopted as part of the National Building Code or be embodied in new federal standards. In addition, financial encouragement and assistance is being given for the improvement of existing residences, through the revision of existing loan and grant programs, to emphasize the purchase of insulation material and energy-saving equipment.

Appliances
Minimum energy-efficiency standards are being reviewed for furnaces and prepared for major home appliances and office equipment. In addition, an appliance-labeling program is being introduced. It should promote both the design and the purchase of more efficient appliances.

Industry
Existing industrial assistance programs are now being administered in a manner that supports energy conservation, including the recovery of waste heat.

Federal Government
Specific elements of the Government's in-house conservation program were announced in early 1975. The initiatives recently announced will strengthen this program by instituting energy conservation targets and reporting systems. Beginning in the 1976-1977 fiscal year, all government departments and agencies have been asked to decrease their energy consumption by 10 percent and target to hold energy use at that level for the next ten years. A monitoring and reporting system is being set up, and the Minister of Energy, Mines and Resources will report to the public on the results of this program. In addition it has been decided that all federally funded programs will follow the energy conservation guidelines established for federal buildings and will contain provision for energy auditing to ensure that energy conservation approaches are adopted.

3.3 METHODOLOGY USED IN PROJECTING ENERGY REQUIREMENTS

3.3.1 The Forecasting Model

An integrated macroeconomic energy demand forecasting model developed by Energy, Mines and Resources Canada was used initially for providing the various scenarios requested by WAES. This analytical framework projects energy demands on the basis of historical relationships between energy use and demographic and economic activity. However, it does allow substantial scope for the adjustment of these historical relationships on the basis of anticipated reactions to changing energy prices and technology. With regard to the effect of increases in future Canadian energy price levels, it is recognized that only limited historical information is available, so that the quantification of these price effects is a difficult task. Future pricing relationships among competing energy sources are difficult to assess, as is the potential for interfuel substitutions. These critical areas where uncertainties are most pronounced are the subjects of further investigation.

The projections of future Canadian energy requirements presented in this report are internally consistent, in the sense that once total energy requirements are estimated for each end-use sector of the economy, the demands for all energy sources are determined on a "market share" basis, which ensures that the sum of the energy sources equals total energy demands.

The relationships used in projecting energy demands are more detailed than simple energy/GNP or energy/population ratios. In the residential sector, for example, energy demand depends upon such factors as personal disposable income, the number of households, the split between single and multiple housing, and the relative price of energy. Demand in the commercial sector also depends on the nature of the housing stock since energy used by large multiple dwellings is priced at commercial rates and included as commercial demand. Commercial sector energy requirements also depend on the level of retail trade and the energy price. Demand for energy in the industrial sector depends not only on price and industrial production, but also on the capital/labour mix employed in the productive process. In general, the relationships used reflect the structural characteristics of the sector examined to the greatest extent possible.

It should be noted that the forecasting model does not take into account explicit energy conservation measures, although the projections do reflect a more efficient use of energy, as might reasonably be expected to occur in response to higher energy prices. The method chosen to estimate the effect of government-induced conservation measures (i.e., nonprice measures) on the demand projections is outlined in Section 3.3.4.

3.3.2 Economic Growth Rates

Two global economic growth parameters were established by WAES in an attempt to span the range of most probable futures. The high growth rate—6 percent per annum—assumes resolution of current international political and financial problems, resulting in high world trade. The low global growth scenario—3.5 percent per annum—reflects continuing international economic, political, and financial problems which limit trade growth, so that countries and regions increasingly seek internal solutions to energy and trade difficulties.

The Canadian economic growth rates corresponding to the WAES high growth case were assumed to be 5.4 percent for the 1976 to 1980 period and 5.2 percent for 1981 to 1985. Thus, the upper end of the range for Canada is determined by assuming that, on average, demographic and economic trends through the 1980s will be similar to those observed in the 1960s. For the WAES low global growth rate case, the Canadian economic growth rates were assumed to be 5.4 percent for 1976 to 1980 and 3.6 percent for 1981 to 1985. The Canadian lower end of the range assumes that potential economic growth in the 1980s slows markedly from the historical experience, reflecting a reduced rate of growth in the Canadian population, labour force, and labour productivity. The major characteristics of these assumptions are displayed in Table 3.1.

All differences between the two economic profiles occur from 1981. It is felt that the alternative growth paths used for each scenario bracket fairly accurately the feasible growth potential of the Canadian economy through this period.

3.3.3 Oil Prices

The WAES scenario oil price criteria have been referred to as the low price case, the mid-1975 price case, and the high price case.

For the purpose of determining Canadian energy requirements subject to the WAES low oil price criteria, it is assumed that the price of oil in Canada remains at end-1975 levels in constant 1975 dollars. This price was established at $8.00 per barrel at the wellhead in Alberta (about $8.80 in Toronto) and was chosen to approximate the WAES low oil price case of US$7.66 per barrel, FOB Persian Gulf. The Canadian oil price increases only at the general rate of inflation, and does not increase relative to the prices of other goods and services. It is further assumed that the price of natural gas increases to "commodity-equivalent" value with crude oil ($1.52/Mcf at the Toronto city-gate in 1975 dollars) in the latter part of the 1970s, and that the prices for electricity and coal also remain constant, in real terms, at end-1975 levels.

In order to determine Canadian energy requirements subject to the WAES mid-1975 oil price case (US$11.50 per barrel, FOB Persian Gulf), it is assumed that oil prices increase relatively faster than the prices of other goods and services until about 1978, when they reach a level that is roughly equivalent to the current international price (about $13.00 per barrel landed in Montreal in 1975 dollars). Prices for electricity and coal are assumed to increase at the same rate as oil prices, and the price of natural gas is again assumed to adjust to "commodity-equivalent" value with crude oil ($2.25/Mcf at the Toronto city-gate in 1975 dollars) by the late 1970s. After 1978, all energy prices are assumed to increase at only the general rate of inflation. It is recognized that future pricing

Table 3.1
Major assumptions underlying energy demand projections

	1974 Level	Average Annual Growth Rate (%)			1990 Level
		1961–1960	1976–1980	1981–1990[2]	
Gross National Product ($1971 billions)	108.8	5.2	5.4	3.6	203.0
Population (millions)	22.5	1.8	1.3	1.2	27.5
Households (millions)	6.7	2.9	2.7	2.1	9.8
Unemployment Rate (%)[1]	5.4	5.0	6.3	5.4	5.9
Employment (millions)	9.1	2.8	2.8	1.7	12.7
Ratio of Multiple Dwellings to Total Housing Stock[1]	0.417	0.363	0.432	0.458	0.472
Personal Disposable Income per Capita ($1971 thousands)	3.2	2.7	2.8	1.9	4.5
Industrial Real Domestic Product (Index, 1971 = 100)	117.9	5.7	5.9	3.7	226.1

1. Average annual level shown, rather than growth rate.
2. These are the growth rates assumed for the "low" economic growth case; for the "high" case the average annual growth rates for the period 1961 to 1970 were assumed to apply in the 1980s.

relationships among competing energy sources are difficult to assess.

In an attempt to estimate Canadian energy requirements based on the WAES high oil price case (US$17.25 per barrel, FOB Persian Gulf), it is assumed that energy prices increase to 1978 as outlined in the immediately preceding price case. For the 1979 to 1985 period, Canadian oil prices are assumed to continue to increase relatively faster than the prices of other goods and services, until in 1985 they reach a level that is roughly equivalent to the WAES high oil price case (about $18.35 per barrel landed in Montreal in 1975 dollars). Prices for electricity and coal are assumed to increase at a slightly lower rate than oil prices. The price of natural gas is assumed to continue to adjust to "commodity-equivalent" value with crude oil on a year by year basis for the 1979 to 1985 period. This would result in a natural gas price at the Toronto city-gate of approximately $3.16/Mcf, in 1975 dollars, by 1985. Once again, it should be pointed out that future pricing relationships among competing energy sources are difficult to assess and that several simplifying assumptions were required.

3.3.4 National Policy Response

The energy demand forecast model does not take into account explicit nonprice conservation measures, although it does reflect a more efficient use of energy, as might reasonably be expected to occur in response to higher energy prices.

The methodology adopted on the demand side to take into consideration the effect of a restrained national policy response, as requested by WAES, is to assume that the estimates forecast by the model would meet the criteria of a restrained response. Thus, the scenarios represent likely developments in the absence of new government initiatives.

The approach taken in determining the effects of a vigorous government policy response is to project the nonconservation case first, and to apply to it estimated savings similar to those outlined in *Energy Conservation*. This methodology was adopted because it was felt that the conservation efforts employed in the study seemed intuitively to be reasonable and that the savings resulting from those efforts appeared to be generally in line with what one could expect if vigorous measures were adopted. In researching the effects of conservation in Canada, it was found that a very limited number of empirical studies were available as reference material. However, it should be noted that effects of energy conservation in Canada are the subject of further study at the National Energy Board and Energy, Mines and Resources Canada.

Thus, the effects of conservation outlined in this report should be treated as first approximations and are subject to further refinement as more research in this area is undertaken.

The Science Council of Canada publication, *Energy Conservation*, recognizes that there are a number of rationales for a Canadian energy conservation policy. The publication recommends a program under which planned measures minimize social costs, social inequities, and economic disruptions. To meet these requirements, the publication has proposed a single form of conservation policy: demand reduction through efficiency measures which do not decrease the effective consumption and therefore do not decrease the level of economic activity. Thus, if an energy efficiency equation is formulated such that

INPUT = OUTPUT + WASTE

it should be possible to achieve any particular output goal at a lower input by decreasing the waste factor through higher efficiency. This approach has been referred to in some studies as a "technical fix" scenario.

The various conservation measures recommended in the study straddle a broad range of implementation techniques which fall into three general areas: (1) voluntary programs, (2) price or market mechanisms, and (3) proscriptive or regulatory (i.e., mandatory) measures, involving time use restrictions, consumption quantity restrictions, establishment of legal standards, etc. In some instances, multiple techniques may be necessary.

The direct application of the above methodology to the scenario projections would result in an overstatement of the effects of conservation efforts, since one of the techniques employed in *Energy Conservation* is an increase in the prices of energy supplies relative to the prices experienced since the end of 1972. Energy price increases are, of course, an integral part of the energy forecasting model.

In an effort to reduce the problem of double-counting the price effects, estimates were made by sector of the effect on energy requirements of moving from the WAES low price case to the WAES mid-1975 (constant) price case and of moving from the WAES low price case to the WAES high price case. These estimates were derived by sector from the projections developed by the model. The estimated savings resulting from national conservation efforts (which include price effects) were then reduced according to the estimates of the price effects, as determined above, for movements from the WAES low price case to the WAES mid-1975 price case. These residual estimated nonprice savings were then applied to the energy projections of the model for Case C. The same estimated nonprice savings were also applied to the model's energy projections for Cases A and B.

Thus, for Cases A and B, a slightly higher conservation effect would result than for Case C inasmuch as higher prices are assumed for these two cases. The effects of government-induced conservation measures (i.e., nonprice measures) were not needed for Cases D and E.

Assumed savings by sector resulting from national conservation efforts (*including* price effects) are as follows: transportation sector, 16.7 percent; nonenergy use, 5.0 percent; commercial sector, 6.5 percent; industrial sector, 14.5 percent; residential sector, 10.1 percent.

Many of the conservation efforts employed in arriving at the above estimates are outlined in *Energy Conservation.* The approach taken is to assume a number of specific initiatives and various degrees of response. These are outlined in Table 3.2.

Table 3.2
Components of national policy response

Sector	Vigorous Policy
Transportation	Make 50% of new cars average a 30% improvement in fuel economy.
	Improve engine maintenance and use radial tires on 50% of all cars.
	Lower maximum highway speeds by 10 mph.
	Persuade 10% of urban commuters to carpool.
	Shift 20% of urban city-centre passenger-miles to bus.
	Shift 20% of intercity auto passenger-miles to bus and rail.
	Shift 30% of short-haul domestic civil air-passenger-miles to intercity bus.
	Persuade 10% of people who now drive urban distances of 2 miles or less to walk.
	Shift 20% of intercity trucking to rail freight.
	Shift 10% of domestic civil air freight to rail.
	Improvements in engine and car-body design, in 30% of new cars bought after 1980, giving average fuel economy of 25%.
Commercial	Overall improvements in efficiency:
	• 35% improvement in efficiency, applied to 55% of new population.
	• 15% improvement in efficiency, applied to 30% of old population.
Industrial	Overall improvements allowing 20% reduction in energy consumption by 40% of population.
Residential	Improve efficiency in space heating and cooling by 30% by enhanced thermal insulation applied to 70% of new dwellings and 20% of old dwellings.
	Reduce air infiltration and increase furnace efficiency, leading to 15% reduction in energy use for space heating and cooling, applied to 30% of total population.
	Lower (winter) or raise (summer) thermostat by 3°C (4°F), 24 hours daily; apply to 40% of total population; assume 9% reduction in fuel use for space heating/cooling.
	Improve efficiency of hot water heating (including maintenance, insulation, lower water temperature, etc.). Assume 30% increased efficiency for new population, 20% for old; apply to 70% of new population and 20% of old population.
	Decrease illumination levels, leading to 20% increase in fuel economy; apply to 80% of new dwellings and 20% of old dwellings.
	Improve efficiency of window-type air conditioning equipment by 20%; apply to 70% of new population and 15% of old population.
	Improve efficiency of other appliances by 10%; apply to 70% of new population and 15% of old population.

3.4 ENERGY PROJECTIONS

Generalized observations from the projections of Canadian energy demand for the five WAES 1985 scenario cases are presented in this section.

The sectoral totals for final energy requirements show that Case B (low economic growth coupled with high oil price and vigorous national policy response) results in the lowest final energy demand, while Case E (high economic growth coupled with low oil price and restrained national policy response) results in the largest final energy demand.

Cases C and D, the medium oil price cases, are approximately midway in the range of energy use from Case B to Case E. The results obtained from the projections indicate that the improvement of energy efficiencies through a vigorous national policy response had more of an impact on reducing energy demand than did lower economic growth. This probably resulted from our assumption that all differences between the two economic profiles occur from 1981, whereas the vigorous national policy response begins to be effective almost immediately.

The "processing loss" numbers in the energy-use summary worksheet result from national supply-demand balances (not included in this report) and may be further modified following the WAES global supply-demand integration process. However, it is thought that these figures represent a close-to-final estimate for this category.

The energy growth rates for 1972 to 1985 show some general trends. The most striking is the large growth in fossil fuel requirements in the feedstocks sector (i.e., nonenergy use). Another noticeable trend is the large growth of electricity use in the residential and commercial sectors from 1972 to 1985, while direct use of fossil fuels in these sectors actually declines in one case. For Case C, electricity use is projected to comprise 26 percent of the total energy demand in the residential sector and nearly 32 percent of the total energy demand in the commercial sector in 1985. In 1972, electricity comprised 15 percent of the total energy demand in the residential sector and 23 percent of the total energy demand in the commercial sector.

The distribution of energy demand by end-use sector is relatively stable across all scenarios and somewhat stable over time. The share of the residential and industrial sectors is projected to decline over time as the relative importance of the transportation and feedstocks sectors increases. Part of the decline in the residential sector is attributable to the relative increase in multiple dwelling units, many of which purchase energy at commercial rates and are accordingly classified in the commercial sector. In addition, the significant decrease in the residential sector's share of fossil fuels reflects the large growth of electricity use in this sector.

3.5 DEMAND STUDY WORKSHEETS

In this section may be found the worksheets for the cases presented in this report. The worksheets appear in the following sequence:

Energy Use Summary (in 10^{12} Btu)

Energy Growth Rates Summary

Sectoral Percentages of Total Final Energy Demand

Base Year (1972) Worksheets

Base Year (1972) Summary

Case A Summary

Case B Summary

Case C Summary

Case D Summary

Case E Summary

WAES NATIONAL ENERGY DEMAND STUDIES, 1972-1985
ENERGY USE SUMMARY

Country Canada

Scenario Assumptions

WAES Case	A		B		C		D		E			
GNP 1976-1985	High		Low		High		Low		High			
Oil Price*	$11.50 → $17.25		$11.50 → $17.25		$11.50		$11.50		$11.50 → $7.66			
Natl. Policy Response	Vigorous		Vigorous		Vigorous		Restrained		Restrained			
Natl. GNP Growth Rate 1976-1985 (%/yr.) **	5.4 5.2		5.4 3.6		5.4 5.2		5.4 3.6		5.4 5.2		1972 Base Year	
Fuel Type	Fossil Fuel	Elec.	Fossil Fuel	Elec.	Fossil Fuel	Elec.	Fossil Fuel	Elec.	Fossil Fuel	Elec.	Fossil Fuel	Elec.
Units	10^{12} Btu		10^{12} Btu		10^{12} Btu		10^{12} Btu		10^{12} Btu		10^{12} Btu	
Energy-Using Sectors												
Industrial (excluding feedstocks)	1357.8	558.3	1348.4	556.0	1442.0	593.7	1518.4	626.8	1666.4	699.6	1070.9	355.1
Transportation	1972.7	0.0	1886.9	0.0	2025.4	0.0	2196.9	0.0	2415.9	0.0	1339.3	0.0
Commercial and Public	921.9	426.4	854.3	395.9	923.7	426.4	857.0	396.3	989.8	454.1	636.3	192.4
Residential	964.4	333.1	932.8	323.1	976.8	337.4	996.5	345.1	1083.7	374.3	960.3	171.3
Fishing, Agriculture, Mining, Construction												
Feedstocks	850.1	0.0	842.2	0.0	850.1	0.0	886.5	0.0	894.8	0.0	301.6	0.0
Final Energy Demand (excluding processing losses)	6066.9	1317.8	5864.6	1275.0	6218.0	1357.5	6455.3	1368.2	7050.6	1528.0	4308.4	718.8
Processing Losses	3659.9		3536.6		3770.1		3645.4		4059.0		2107.3	
Primary Energy Input	11044.6		10676.2		11345.6		11468.9		12637.6		7134.5	

*Oil price 1975 US$ per barrel Arabian light crude oil FOB Persian Gulf.

**Upper figure applies to 1976-80; lower figure applies to 1981-85.

WAES NATIONAL ENERGY DEMAND STUDIES, 1972-1985
ENERGY GROWTH RATES SUMMARY

Country Canada

Scenario Assumptions

WAES Case	A		B		C		D		E	
GNP 1976-1985	High		Low		High		Low		High	
Oil Price*	\$11.50 → \$17.25		\$11.50 → \$17.25		\$11.50		\$11.50		\$11.50 → \$7.66	
Natl. Policy Response	Vigorous		Vigorous		Vigorous		Restrained		Restrained	
Natl. GNP Growth Rate 1976-1985 (%/yr.) **	5.4 5.2		5.4 3.6		5.4 5.2		5.4 3.6		5.4 5.2	
Fuel Type	Fossil Fuel	Elec.	Fossil Fuel	Elec.	Fossil Fuel	Elec.	Fossil Fuel	Elec.	Fossil Fuel	Elec.
Units (%/yr. 1972-1985)	%/yr.		%/yr.		%/yr.		%/yr.		%/yr.	
Energy-Using Sectors										
Industrial (excluding feedstocks)	1.8	3.5	1.8	3.5	2.3	4.0	2.7	4.5	3.5	5.4
Transportation	3.0	0.0	2.6	0.0	3.2	0.0	3.9	0.0	4.6	0.0
Commercial and Public	2.9	6.3	2.3	5.7	2.9	6.3	2.3	5.7	3.5	6.8
Residential	.1	5.2	-.2	5.0	.1	5.4	.3	5.5	.9	6.2
Fishing, Agriculture, Mining, Construction										
Feedstocks	8.3	0.0	8.2	0.0	8.3	0.0	8.6	0.0	8.7	0.0
Final Energy Demand (excluding processing losses)	2.7	4.8	2.6	4.5	2.9	5.0	3.2	5.1	3.9	6.0
Processing Losses	4.3		4.1		4.6		4.3		5.2	
Primary Energy Input	3.4		3.1		3.6		3.7		4.5	

*Oil price 1975 US\$ per barrel Arabian light crude oil FOB Persian Gulf.

**Upper figure applies to 1976-80; lower figure applies to 1981-85.

WAES NATIONAL ENERGY DEMAND STUDIES, 1972-1985
SECTORAL PERCENTAGES OF TOTAL

Country Canada

Scenario Assumptions

WAES Case	A		B		C		D		E			
GNP 1976-1985	High		Low		High		Low		High			
Oil Price*	$11.50 → $17.25		$11.50 → $17.25		$11.50		$11.50		$11.50 → $7.66			
Natl. Policy Response	Vigorous		Vigorous		Vigorous		Restrained		Restrained			
Natl. GNP Growth Rate 1976-1985 (%/yr.) **	5.4 5.2		5.4 3.6		5.4 5.2		5.4 3.6		5.4 5.2		1972 Base Year	
Fuel Type	Fossil Fuel	Elec.	Fossil Fuel	Elec.	Fossil Fuel	Elec.	Fossil Fuel	Elec.	Fossil Fuel	Elec.	Fossil Fuel	Elec.
Units	%		%		%		%		%		%	
Energy-Using Sectors												
Industrial (excluding feedstocks)	22.4	42.4	23.0	43.6	23.2	43.7	23.5	45.8	23.6	45.8	24.8	49.4
Transportation	32.5	0.0	32.2	0.0	32.6	0.0	34.0	0.0	34.3	0.0	31.1	0.0
Commercial and Public	15.2	32.3	14.6	31.1	14.8	31.4	13.3	29.0	14.0	29.7	14.8	26.8
Residential	15.9	25.3	15.9	25.3	15.7	24.9	15.5	25.2	15.4	24.5	22.3	23.8
Fishing, Agriculture, Mining, Construction												
Feedstocks	14.0	0.0	14.3	0.0	13.7	0.0	13.7	0.0	12.7	0.0	7.0	0.0
Final Energy Demand (excluding processing losses)	100.	100.	100.	100.	100.	100.	100.	100.	100.	100.	100.	100.
Processing Losses												
Primary Energy Input												

*Oil price 1975 US$ per barrel Arabian light crude oil FOB Persian Gulf.

** Upper figure applies to 1976-80; lower figure applies to 1981-85.

NATIONAL ENERGY DEMAND PROJECTIONS

Country Canada

Year 1972

WAES Case Base Year

Transportation

	(A)	(B)	(C)	(D)	(E)	(F)(1)	(G)	
	Population/Auto	Distance/Auto/Yr.	Load Factor	Fraction Urban or Rural	Energy/Distance	Fossil Fuel	Electricity	
Urban Auto	2.95	10,000 (2)		50	1.9 x 10^3 (8)	641.6		(1)
Rural Auto				50				(2)
	Activity Level	% Electrified	Load Factor	Fossil Fuel/ Activity Level	Electricity/Activity			
Urban Bus	4.7 x 10^9 (3)			1.32 x 10^3 (4)		6.2		(3)
Intercity Bus	2.8 x 10^9 (3)			.96 x 10^3 (4)		2.7		(4)
Rail (Passenger)	2.0 x 10^9 (3)			4.95 x 10^3 (4)		9.9		(5)
Air (Passenger, Domestic)	6.2 x 10^9 (3)					105.4		(6)
Air (Passenger, International)	7.3 x 10^9 (3)							(7)
Ship (Passenger)								(8)
Rail (Freight)	123.6 x 10^9 (5)			.52 x 10^3 (6)		64.8		(9)
Air (Freight, Domestic)	.21 x 10^9 (5)							(10)
Air (Freight, International)	.23 x 10^9 (5)							(11)
Truck (Local)								(12)
Truck (Intercity)	20.6 x 10^9 (5)			5.69 x 10^3 (6)		117.2		(13)
Ship (Freight, Domestic)	122 x 10^6 (7)							(14)
Ship (Freight, International)	176 x 10^6 (7)							(15)
Miscellaneous								(16)
					Total	1339.3		

(1) 10^{12} Btu
(2) Miles
(3) Passenger-miles
(4) Btu ber passenger mile
(5) Ton(2000 lb)-miles
(6) Btu per ton-mile
(7) Tons handled
(8) Btu per mile

NATIONAL ENERGY DEMAND PROJECTIONS

Country Canada

Year 1972

WAES Case Base Year

Commercial and Public

	(A)	(B)	(C)	(D)	(E)	
Commercial	Floor Area	Heat Gain or Loss/Area	System Efficiency	Fossil Fuel	Electricity	
Space Heat (Fossil Fuel)						(27)
Space Heat (Electricity)						(28)
Air Conditioning (Fossil Fuel)						(29)
Air Conditioning (Electricity)						(30)
		Energy/Area	Electricity/Area			
Lighting						(31)
Miscellaneous						(32)
Presently Unknown						(33)
			Total	636.3	192.4	

Public	Floor Area	Heat Gain or Loss/Area	System Efficiency	Fossil Fuel	Electricity	
Space Heat (Fossil Fuel)						(34)
Space Heat (Electricity)						(35)
Air Conditioning (Fossil Fuel)						(36)
Air Conditioning (Electricity)						(37)
		Energy/Area	Electricity/Area			
Lighting						(38)
Miscellaneous						(39)
Presently Unknown						(40)
			Total			
			Commercial and Public Total			

NATIONAL ENERGY DEMAND PROJECTIONS

Country Canada

Year 1972

WAES Case Base Year

Industrial

	(A) (1) Value Added	(B) (2) Energy/Value Added	(C) Physical Output	(D) Energy/Output	(E) % Exported	(F)(3) Fossil Fuel	(G)(3) Electricity	(H) Feedstocks	
Iron and Steel (ISIC 371)	909.4×10^6	89.3×10^3			14.8	60.4	20.4		(41)
Nonferrous Metals (ISIC 372)	1007.9×10^6	120.3×10^3			33.3	90.6	30.6		(42)
Paper and Allied (ISIC 34)	1928.4×10^6	162.9×10^3			50.8	243.0	71.2		(43)
Chemicals and Allied (ISIC 35 minus ISIC 353)	1524.1×10^6	91.1×10^3			18.3	102.2	36.6		(44)
Mineral Products (ISIC 36)	973.2×10^6	122.9×10^3			12.3	107.1	12.5		(45)
Food and Related (ISIC 31)	3470.4×10^6	28.9×10^3			21.2	89.4	10.8		(46)
Transport Eqpt. (ISIC 384)	2631.5×10^6	15.4×10^3			67.9	31.7	8.7		(47)
Key Industry (N.E.C.)									(48)
Key Industry (N.E.C.) Non-Energy						301.6			(49)
Misc. Manufacturing						346.5	164.3		(50)
N.E.C. = Not Elsewhere Classified					**Total**	1372.5	355.1		

(1) 1972 Canadian dollars
(2) Btu per Canadian dollar
(3) 10^{12} Btu

NATIONAL ENERGY DEMAND PROJECTIONS

Country Canada

Year 1972

WAES Case Base Year

Residential

Population 21.9 x 10^6 Persons/Dwelling 3.6 Dwellings 6.1 x 10^6

		(A)	(B)	(C)	(D)	(E) (1)	(F) (1)	
Central Space Heating		Dwellings	Heat Loss	System Efficiency		Fossil Fuel	Electricity	
Single Family	Fossil Fuels	3,621,000		0.7		553.2		(51)
	Electricity	181,000		1.0			12.7	(52)
Town or Row Houses	Fossil Fuels	425,000		0.7		64.9		(53)
	Electricity	21,000		1.0			1.5	(54)
Apartments	Fossil Fuels	1,673,000		0.7		151.3		(55)
	Electricity	188,000		1.0			7.5	(56)
		% Saturation (Fossil Fuels)	% Saturation (Electricity)	Energy/Unit (Fossil Fuels)	Energy/Unit (Electricity)			
Room Heating								(57)
Air Conditioning (Window)			5.8		400 kwh		.5	(58)
Air Conditioning (Central)								(59)
Water Heating		45.1	48.5	27.2 x 10^6	4400 kwh	74.8	44.4	(60)
Refrigerators			98.7		800 kwh		16.4	(61)
Lighting			100		700 kwh		14.6	(62)
Cooking		19.3	80.7	10.6 x 10^6	1150 kwh	12.5	19.3	(63)
Dishwashers			9.2		200 kwh		.4	(64)
Freezers			35.6		900 kwh		6.7	(65)
Television			95.8		380 kwh		7.6	(66)
Other Key Appliances (2)		3.5	42.1	9 x 10^6	600 kwh	1.9	5.3	(67)
Misc. Appliances						101.7	34.4	(68)
Potential Appliances								(69)
					Total	960.3	171.3	

(1) 1972 Canadian dollars
(2) Btu per Canadian dollar
(3) 10^{12} Btu

NATIONAL ENERGY DEMAND PROJECTIONS

Country Canada

Year 1972

WAES Case BASE YEAR

UNITS = 10^{12} Btu

Summary

Category	Oil	Gas	Coal	Electricity	Other	Total by Category (Excluding Processing Losses)
Transportation	1334.9	0.0	4.4	0.0		1339.3
Agriculture, Mining, and Construction						
Commercial ↗ Public	365.1	266.2	5.0	192.4		828.7
Industrial	667.5	489.0	216.0	355.1		1727.6
Residential	667.4	281.2	11.7	171.3		1131.6
Total Delivered Energy by Fuel	3034.9	1036.4	237.1	718.8		
Processing Losses by Fuel	216.7	300.9	0.5	1589.2		
Gross Input by Fuel	3251.6	1337.3	237.6	2308.0		7134.5

Gross Energy Input (Add across bottom row)

NATIONAL ENERGY DEMAND PROJECTIONS

Country Canada

Year 1985

WAES Case A

UNITS = 10^{12} Btu

Summary

Category	Oil	Gas	Coal	Electricity	Other	Total by Category (Excluding Processing Losses)
Transportation	1971.0	0.0	1.7	0.0		1972.7
Agriculture, Mining, and Construction						
Commercial ↗ Public	459.2	459.0	3.7	426.4		1348.3
Industrial	992.4	919.8	295.7	558.3		2766.2
Residential	504.4	455.4	4.6	333.1		1297.5
Total Delivered Energy by Fuel	3927.0	1834.2	305.7	1317.8		
Processing Losses by Fuel	421.6	232.5	1.0	3004.8		
Gross Input by Fuel	4348.6	2066.7	306.7	4322.6		11044.6

Gross Energy Input (Add across bottom row)

NATIONAL ENERGY DEMAND PROJECTIONS

Country Canada

Year 1985

WAES Case B

UNITS = 10^{12} Btu

Summary

Category	Oil	Gas	Coal	Electricity	Other	Total by Category (Excluding Processing Losses)
Transportation	1885.3	0.0	1.6	0.0		1886.9
Agriculture, Mining, and Construction						
Commercial / Public	428.3	422.6	3.4	395.9		1250.2
Industrial	982.9	917.1	290.6	556.0		2746.6
Residential	488.7	439.7	4.4	323.1		1255.9
Total Delivered Energy by Fuel	3785.2	1799.4	300.0	1275.0		
Processing Losses by Fuel	405.9	225.3	1.0	2904.4		
Gross Input by Fuel	4191.1	2004.7	301.0	4179.4		10676.2

Gross Energy Input (Add across bottom row)

NATIONAL ENERGY DEMAND PROJECTIONS

Country Canada

Year 1985

WAES Case C

UNITS = 10^{12} Btu

Summary

Category	Oil	Gas	Coal	Electricity	Other	Total by Category (Excluding Processing Losses)
Transportation	2023.6	0.0	1.8	0.0		2025.4
Agriculture, Mining, and Construction						
Commercial / Public	458.2	461.8	3.7	426.4		1350.1
Industrial	1011.1	983.9	297.1	593.7		2885.8
Residential	512.6	459.5	4.7	337.4		1314.2
Total Delivered Energy by Fuel	4005.5	1905.2	307.3	1357.5		
Processing Losses by Fuel	430.3	241.8	1.0	3097.0		
Gross Input by Fuel	4435.8	2147.0	308.3	4454.5		11345.6

Gross Energy Input (Add across bottom row)

NATIONAL ENERGY DEMAND PROJECTIONS

Country Canada

Year 1985

WAES Case D

Summary

UNITS = 10^{12} Btu

Category	Oil	Gas	Coal	Electricity	Other	Total by Category (Excluding Processing Losses)
Transportation	2195.0	0.0	1.9	0.0		2196.9
Agriculture, Mining, and Construction						
Commercial / Public	427.7	425.9	3.4	396.3		1253.3
Industrial	1057.3	1038.0	309.6	626.8		3031.7
Residential	523.8	468.0	4.7	345.1		1341.6
Total Delivered Energy by Fuel	4203.8	1931.9	319.6	1368.2		
Processing Losses by Fuel	414.4	234.5	1.0	2995.5		
Gross Input by Fuel	4618.2	2166.4	320.6	4363.7		11468.9

Gross Energy Input (Add across bottom row)

NATIONAL ENERGY DEMAND PROJECTIONS

Country Canada

Year 1985

WAES Case E

Summary

UNITS = 10^{12} Btu

Category	Oil	Gas	Coal	Electricity	Other	Total by Category (Excluding Processing Losses)
Transportation	2413.8	0.0	2.1	0.0		2415.9
Agriculture, Mining, and Construction						
Commercial / Public	490.7	495.2	3.9	454.1		1443.9
Industrial	1114.3	1129.7	317.2	699.6		3260.8
Residential	568.7	509.8	5.2	374.3		1458.0
Total Delivered Energy by Fuel	4587.5	2134.7	328.4	1528.0		
Processing Losses by Fuel	452.7	259.4	1.1	3345.8		
Gross Input by Fuel	5040.2	2394.1	329.5	4873.8		12637.6

Gross Energy Input (Add across bottom row)

4.1 INTRODUCTION

At the time when WAES was being organized, a Danish national project was being set up along lines similar to those envisaged for WAES. The Danish project was initiated in December 1973 in connection with an international panel on energy problems being held in Copenhagen under the auspices of the International Federation of Institutes of Advanced Study (IFIAS). The panel recommended that studies of "Alternative Organizations of Society in Terms of Energy" be initiated and that Denmark should serve as a test case for this type of study. The study was to be located at the Niels Bohr Institute, University of Copenhagen, which is one of the IFIAS member institutes.

In view of the similarities of the WAES studies and the Danish IFIAS study, it was natural to establish a close collaboration between the group of WAES countries and the Danish study. This connection was formally established by the connection of Prof. Bent Elbek of the Niels Bohr Institute as advisor to WAES. Later Prof. Elbek was invited to become a Danish Participant in the WAES group.

As the work progressed, it became clear that the WAES project and the Danish IFIAS project, in spite of similarities, also differed in essential aspects. It therefore became necessary to perform a number of separate studies of the Danish energy situation according to the methodology developed by WAES. These studies were carried out with the assistance of Civ. Ing. Dorthe Klitgaard and Civ. Ing. Benny Petersen, who were already engaged in studies within the field of energy analysis under the direction of Prof. Elbek. Further assistance in the numerous calculations was obtained from Mrs. Eja Elbek, who joined the group as a voluntary worker.

The study has profited from information from a large number of sources. It has especially drawn freely on the results of the IFIAS project in the field of potential energy savings. Other important sources have been Handelsministeriet (the Ministry of Trade), Danmarks Statistik, and Dansk Shell.

The data for the year 1972, which form the basis for the demand projections, have been compiled from a number of sources including those mentioned above. Information taken directly from the energy analysis project has also been used to a considerable extent. Some of the data are thus new and not necessarily in agreement with other sources. In order to keep the confusion in the data field as low as possible, it has been attempted to adhere to the more aggregated numbers contained in "Danmarks Energibalance 1972" issued by "Dansk Kedelforening."

The study appears at a time when the official Danish energy policy is being formulated. The government has recently issued a report, "Dansk Energipolitik 1976," which presents a plan for the energy supply extending to 1995.[1] This plan also considers in some detail the energy demand in 1985 and can in some aspects be directly compared with the scenarios presented here. The government plan most closely resembles the scenario C, which assumes high economic growth, high but constant energy prices, and vigorous policy. This scenario yields a growth rate for delivered energy (other than for electric generation) of 1.7 percent p.a., 1972 to 1985, whereas the government plan from 1975 projects as little as 0.2 percent p.a. For electric energy the scenario yields 4.5 percent p.a. and the government plan 4.9 percent p.a. Finally, the total demand for primary energy in the scenario C grows by 2.1 percent p.a. while the government plan envisages 1.2 percent p.a.

Although there is general agreement as to the lowering of the energy growth rates, differences are thus apparent. The government plan envisages larger energy savings in the space heating sectors and essentially assumes a complete reinsulation of buildings. The scenario takes a more pessimistic view on the possibility of accomplishing this within the ten-year period considered. Furthermore, the scenario as a basis for projection assumes a 1975 consumption equal to that of the base year 1972. The government plan assumes permanence of the energy savings that have reduced consumption by 10 percent from 1972 to 1975.

The significance of these differences should not be exaggerated although they do have a large effect on the final demand in 1985, for which the scenario projects 1077 PJ (257 x 10^6 Gcal) and the

1. "Dansk Energipolitik 1976." Copenhagen: Handelsministeriet, 1976. ISBN 87-503-1969-8.

energy plan 800 PJ (191 x 10^6 Gcal). These differences diminish, however, if the projections for 1995 are considered.

This study has in part been supported by Statens Naturvidenskabelige Forskningsråd. (Danish Research Council for Science.)

4.1.1 Note on Units

All quantities of energy in this report are given in joules (J) or multiples thereof:

1 PJ (Peta-joule) = 10^{15} J

1 TJ (Tera-joule) = 10^{12} J

1 GJ (Giga-joule) = 10^9 J

1 MJ (Mega-joule) = 10^6 J

Conversion to other units:

1 PJ = 0.239x10^6	Gcal	1 MJ	= 239 kcal
= 0.278	TWh		= 0.278 kWh
= 0.947x10^{12}	Btu		= 947 Btu
= 0.239x10^6	TOE		
= 0.16x10^6	BOE		

1 PJ/yr. = 0.0317 GW = 450 B/d.O.E.

Monetary units are Danish kroner (D.kr.). Conversion: 1 D.kr. 1972 = US$0.143 (1972).

4.2 METHODOLOGY

4.2.1 Scenarios

The five WAES scenarios are applied to the Danish case in this report. To correspond to the WAES high and low world economic growth cases of 6 and 3.5 percent p.a., respectively, national real economic growth rates of 4.1 and 3.1 percent p.a., respectively, have been used.

For further studies the scenarios C and D have been extended to year 2000 in a preliminary fashion. These extended scenarios are not included here, but occasionally they will be referred to in order to underline some long-term developments in the energy demand.

The quantification of the parameters for the Danish energy case is discussed in the following sections.

4.2.2 Total Growth Rate of the Economy

The Danish economy is in general expected to grow a little slower than the world average. This is the basis for the selection of national growth rates until 1985. The high growth rate of 4.1 percent is in accordance with the official plan PPII issued in 1973.[2] The low growth rate of 3.1 percent is rather arbitrarily chosen as 75 percent of the rate envisaged in PPII.

In accordance with the general WAES methodology, the economic growth rates are considered to be independent of energy price. This might not be a very good assumption for an energy-importing country like Denmark, where balance of payments problems resulting from the price of energy could affect the sustainable economic growth rate.

In order to obtain a detailed projection of the energy demand it is necessary to estimate the economic growth rates for the major sectors of the economy. Table 4.1 shows the growth rates used for the projections until 1985. They are by and large in agreement with the PPII plan for the case of high growth.

Low growth rates are taken as 75 percent of the high growth rates for all sectors. Some differences in the behavior of the sectors might be expected, but it has not been possible to get reliable estimates of such effects. For some of the industrial sectors the projections have been based on separate runs of a multisector economic growth model.[3] These numbers are given in Table 4.2.

Table 4.1
GNP growth rates for individual sectors, 1976 to 1985

Sector	High Growth (% p.a.)	Low Growth (% p.a.)
Transportation	4.0	3.0
Industrial	4.9	3.7
Agriculture	3.0	2.3
Construction	2.0	1.5
Commercial	4.1	3.1
Public	5.0	3.8
Residential	3.7	2.8
Total	4.1	3.1

2. "Perspektivplanredegörelse 1972–1987" (PPII). Copenhagen, 1973.

3. These runs were performed by Cand. Polit. Asger Hansen.

Table 4.2
GNP growth rates for industrial sectors

Sector	ISIC	High Growth (% p.a.)	Low Growth (% p.a.)
Iron and Steel	371	6.7	5.0
Nonferrous Metals	372	6.7	5.0
Paper	34	1.6	1.2
Chemicals	35	4.7	3.5
Mineral Products	36	4.6	3.4
Food and Related	31	3.1	2.3
Transport Eqpt.	384	6.7	5.0
Machinery	38-384	6.7	5.0
Textiles	32	1.6	1.2
Wood	33, 39	6.7	5.0
All Industry		4.8	3.6

The growth rates in Tables 4.1 and 4.2 are interpreted as averages for the period 1976 to 1985. In all scenarios it has thus been assumed that the present recession ceases by the end of 1975. For the projections it has furthermore been assumed that the energy situation in the base year 1972 is reestablished in 1976. In actual fact the gross energy consumption in 1975 was 10 percent lower than in 1972 whereas the consumption of electricity was about the same in the two years.

4.2.3 Influence of Oil Prices on Demand

Very little is known about the price elasticities of energy. In general these elasticities are low, perhaps of the order of -0.15 for the total economy. (That is, demand is reduced by 0.15 percent for each 1 percent increase in price.) The most sensitive sectors are residential heating and private transportation. Based upon the survey of price elasticities presented by Fletcher,[4] the long-term elasticities in Table 4.3 have been chosen.

With an overall elasticity of -0.15, the differences among the scenarios relative to the constant price cases are *ceteris paribus* assumed to be as listed in Table 4.4.

It should be noted that the price increases already experienced could have a long-term influence on demand. In all the scenarios it has been assumed that the oil crisis will have some lasting effects. This is quantified in the assumption that the 1972

4. "Energi 1985–2000." SOU 1974:64 and SOU 1974:65. Stockholm, 1975.

Table 4.3
Price elasticities for individual sectors

Sector	Price Elasticity	Change in Energy Intensity until 1985 (% p.a.)[1]
Transportation	-0.2	-1.0
Industry	-0.1	-0.5
Agriculture	-0.2	-1.0
Construction	-0.0	-0.0
Commercial	-0.2	-1.0
Public	-0.1	-0.5
Residential	-0.15	-0.7
Average	-0.15	-0.7

1. Price increase of 50 percent as in scenarios A and C. Signs reversed for scenario E.

Table 4.4
The effect of oil price on demand in the five WAES scenarios

Case	Price Effect on Demand
A	-8%
B	-8%
C	0
D	0
E	+8%

consumption can be taken as a basis for projections starting January 1, 1976. It is thus assumed that the savings, which are mostly of psychological origin, will disappear, but that growth in energy consumption is set back by three years. If one assumes a "normal" growth of 3 percent p.a., this corresponds to a 9 percent reduction in total demand. Since the actual 1975 consumption was 10 percent lower than in 1972, the scenarios presuppose reestablishment of former consumer habits within a foreseeable future. This type of behavior is well-known for such highly taxed goods as cigarettes and liquor. That it might be found for energy as well is perhaps indicated by the most recent developments in energy demand which indicate a return to former habits.

4.2.4 National Policy Response

The parameter "policy response" is hard to quantify, especially under prevailing political conditions in Denmark.

There does, however, seem to be wide support for action in the field of energy conservation, which is also an important component in the government's energy plan, "Dansk Energipolitik 1976."

The plan envisages total energy savings of 140 PJ/year (34 x 10^6 Gcal/year) by 1985. This amounts to 20 percent of the 1975 consumption. In Section 4.3 the specific suggestions are discussed in greater detail.

It should be realized, however, that quite vigorous political action would be necessary to bring the conservation plan into reality. In all probability in 1985 there will still be houses with imperfect insulation and waste heat from industrial processes. Therefore, referring to the general WAES qualifier that the policy should be "vigorous, but plausible," the savings which in the scenarios are ascribed to government intervention amount to only about 60 percent of those assumed in the government plan. The savings proposed by the government but not necessarily ascribed to policy actions are in some instances close to the technical potential for energy conservation given reasonable economic constraints. It is not clear to what extent the technical conservation potential can be approached, but *if* approached, it will be through the combined action of energy price, government action, and the long-term trends to be discussed in Section 4.2.5. Therefore the technical potential establishes a reasonable upper limit for the combined effect.

With these considerations in mind, the *extra* savings obtainable through government intervention were estimated for the major sectors of the economy. These estimates are listed in Table 4.5 as absolute savings and as percentages of the 1972 demand. In order to facilitate the calculations the savings have also been expressed as equivalent reductions in the percent p.a. growth in the energy intensity (energy per unit output) for the sectors. In some cases conservation estimates have also been made for more specific energy demands (e.g., within the transportation sector). Further details about the energy conservation estimates are given in Section 4.3.

The specific policy measures that might be taken to ensure energy savings of a certain magnitude have not been considered in great detail in either the government plan or the present scenarios. The policy tools will probably have to be tested in practice in order to judge their efficiency. Section 4.3 reviews some scattered information about impact of policy in the period 1973 to 1975.

A restrained policy, in the sense that almost no political attempt is made to check the growth of energy demand, is not a very likely development in Denmark. Nevertheless it is considered here as a limiting case which by comparison allows an estimate of the impact of government policy. Restrained policy is quantified as a policy where the savings listed in Table 4.5 are all set equal to zero.

4.2.5 Inherent Trends in Energy Intensity

Much of the development in energy demand can be related to the general economic development as described by the gross national product. It is obvious, however, that an assumption of constant energy intensity (constant energy/GNP ratio) would result in grossly misleading projections. In fact, all of the WAES demand studies can be viewed as attempts to estimate how much and how fast the energy intensities can be lowered in the

Table 4.5
Energy conservation from vigorous policy response

Sector	Gross Energy, 1972 (PJ/yr.)	Savings by 1985		Change in Energy Intensity, 1972–1985 (% p.a.)
		PJ/yr.	% of 1972	
Transportation	139	6	4	-0.3
Industrial	162	11	7	-0.5
Agriculture	56	4	7	-0.5
Construction	26	2	7	-0.5
Commercial	61	9	14	-1.0
Public	45	4	9	-0.7
Residential	283	47	17	-1.4
Total	772	83	11	-0.9

individual countries without adverse influences on the economic development.

As discussed in Sections 4.2.3 and 4.2.4, some of the changes in energy intensity can be ascribed to price changes for energy or to government policy. There remain, however, some changes that occur for no obvious reason and that are best described as *inherent trends* in the energy system. These trends can cover a variety of technical changes such as the mechanization of agriculture or the substitution of diesel trains for locomotives. They can also reflect the introduction of new consumer goods, changes in the structure or output of an industry, or saturation in the demand for energy-consuming products.

In order to quantify these inherent trends a special study was made of the main sectors of the economy for the period 1948 to 1973, during which the parameters of price and policy can be assumed to have had very little significance. Unfortunately, the data for the distribution of energy demand in this period are of dubious quality, and some information can only be obtained by indirect methods. Nevertheless, it seems possible to isolate a number of interesting developments.

Figure 4.1 shows the development in the energy intensity for *other energy*, that is, all energy used in the particular sector except electricity. Similarly, Figure 4.2 shows the development for the *electric energy* intensity. The figures also show likely extrapolations until 1980 which have been used to estimate average inherent intensity changes until 1985. These have in Table 4.6 been expressed as percent p.a. changes for the major sectors. The values in Table 4.6 have been used in all the scenario calculations. The intensity changes are thus assumed to be independent of those induced by price and policy.

The inherent trends significantly affect the distribution of energy among sectors and reflect the shift from other forms of energy to electricity. It therefore seems important to incorporate the inherent trends in the projections. On the other hand, it must be realized that it is difficult to determine the trends accurately. The gross energy is, however, only slightly affected by the trends because the increase in electric energy more or less compensates the decrease in other energy. For

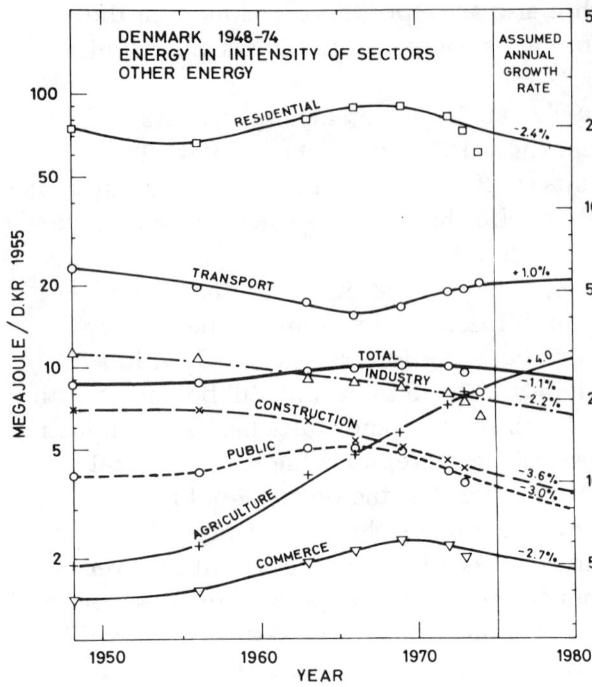

Figure 4.1 Historic development in energy intensity for all energy except electricity. Note the decreasing intensity for most sectors in recent years. Percentages to the right are projected trends. 1 D.kr. 1955 = 2.27 D.kr. 1972 = US $0.33 (1972).

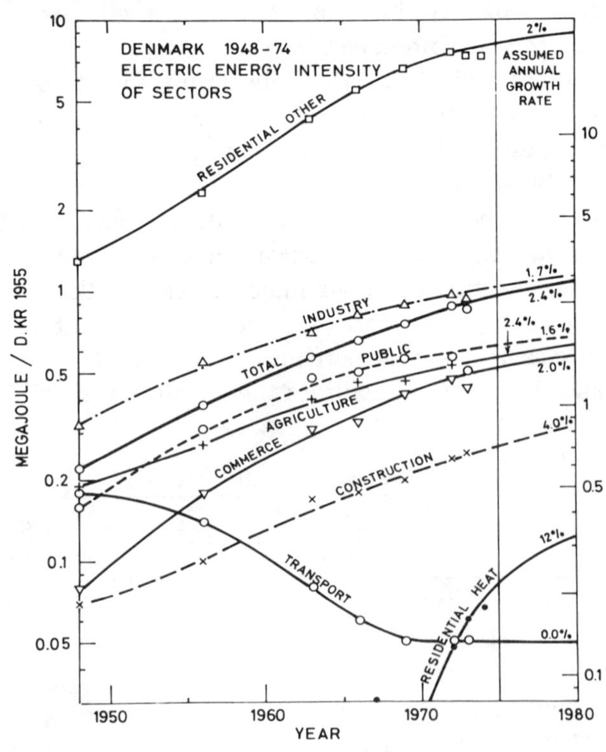

Figure 4.2 Historic trends for electric energy intensity. Most sectors have shown strongly increasing intensities, but with a tendency to level off in recent years. Percentages to the right are projected trends. 1 D.kr. 1955 = 2.27 D.kr. 1972 = US$0.33 (1972).

Table 4.6
Inherent energy trends

Sector	Inherent Intensity Change	
	Electricity (% p.a.)	Other Energy (% p.a.)
Transportation	0.0	1.0
Industrial	1.7	-2.2
Agriculture	2.4	4.0
Construction	4.0	-3.6
Commercial	2.0	-2.7
Public	1.6	-3.0
Residential, heat	12.0	-2.4
Residential, other	2.0	-
Total Intensity	2.1	-1.0
Gross Energy	-0.3	

the same reason the gross energy projection is rather insensitive to the exact values chosen for the trends.

4.2.6 Details of Demand Calculations

In order to generate the energy demand in the scenarios in a consistent manner a simple mathematical procedure was followed as described below. Apart from its simplicity the method has the advantage that it shows directly how the growth rates are affected by the scenario assumptions.

The basic assumption behind the calculation of the energy demand E corresponding to a certain level of economic activity A is

$$E = A \times I, \tag{4.1}$$

where I is the energy intensity of the activity measured, for example, in joules per constant dollar. This equation is valid for each of the sectors considered.

The time development of A is given by the economic growth rate α. Likewise the energy intensity is supposed to be time-dependent. The parameters affecting I are expressed as exponents β giving the price elasticity of demand, γ giving the influence of the government policy, and δ giving the inherent trends in the energy intensity. Thus

$$\begin{aligned} E(t) &= A_0(1+\alpha)^t I_0(1+\beta+\gamma+\delta)^t \\ &\approx E_0(1+\alpha+\beta+\gamma+\delta)^t. \end{aligned} \tag{4.2}$$

In the WAES approach the economic growth rate α is exogenous. The coefficients β, γ, δ have been estimated in the following manner:

1. Price elasticities (see Section 4.2.3) are interpreted as long-term elasticities applicable to the 10–15 year periods considered in the project. Thus demand D is affected by price P according to

$$D \propto P^q, \tag{4.3}$$

q being the long-term elasticity coefficient. In the scenarios considered here, the price is expected to develop exponentially:

$$P = P_0(1+r)^t, \tag{4.4}$$

where r is the rate of price increase. Thus

$$D \propto (1+r)^{qt} \approx (1+qr)^t. \tag{4.5}$$

The coefficient β in Equation 4.2 is thus

$$\beta \approx qr. \tag{4.6}$$

This coefficient is, for the increasing price scenarios, on the order of -0.5 percent p.a.

2. The policy parameter γ is estimated from the policy-induced energy conservation judged possible over the period considered (see Section 4.2.4). If demand in the period t is reduced from D_0 to D by conservation measures, then

$$\gamma \approx \frac{1}{t} \ln \frac{D}{D_0}. \tag{4.7}$$

In the scenarios, D/D_0 for the individual sectors ranges from 0.95 to 0.85 over a ten-year period. Thus γ ranges from -0.5 percent p.a. to -1.5 percent p.a.

3. The inherent trend parameter δ is estimated from the survey (see Section 4.2.5) of energy intensities in the period 1950 to 1970, when there was no particular government response. The relative price of energy was declining slightly, but no correction of the trends has been made for a possible price effect. Such a correction would make the trend rates slightly smaller (change in the negative direction). The trend estimates tend, if anything, to overestimate the demand.

If the energy intensity has changed from I_0 to I over t years, then

$$\delta \approx \frac{1}{t} \ln \frac{I}{I_0}. \tag{4.8}$$

The trends estimated from the period 1965 to 1973 are on the order of +2.0 percent p.a. for electricity and -1.0 percent p.a. for other energy.

For the scenarios C and D, detailed calculations

were performed for all 69 demand groups defined by WAES. These calculations are, for most sectors, based on the method outlined above, but in some cases additional estimates were employed to illuminate fine structure, which is lost in the more aggregated studies performed using the 8 sectors listed in Table 4.6. In general there is good agreement between the 69-sector studies and the 8-sector studies.

Table 4.7 shows an example of the build-up of an aggregated demand projection for 8 sectors, and the energy-use summary in Section 4.5 shows the results for all five scenarios.

4.3 ANALYSIS AND RESULTS

4.3.1 Comparisons of Scenarios

Many of the differences among the five scenarios have already been discussed in Section 4.2 because such differences result from built-in assumptions. Nevertheless, it is of interest to underline some of the main conclusions that can be drawn from the scenarios, which are summarized in Figure 4.3.

Perhaps the first observation to be made is that final energy demand does not vary greatly among the scenarios. Expressed as gross energy relative to 1972, the 1985 demand varies from 111 percent (scenario B) to 152 percent (scenario E). The relatively small difference between these extremes illustrates that the energy system does not react strongly even to considerable changes in the external conditions.

The annual growth rates for energy in the scenarios range from 1.1 percent p.a. to 4.3 percent p.a., the extremes again being scenario B and scenario E. All energy growth rates are thus considerably reduced compared to the 4.7 percent p.a. experienced in the period 1950 to 1973. Even scenario E with declining relative energy prices leads to a lower energy growth. The reasons for the reduced growth rates are, depending on the scenario considered, some combination of increased energy prices, government policy, lower economic growth, and inherent trends. The relative importance of these contributions is shown in Table 4.8 and, for a particular case, in Figure 4.4. The trends appear to be of minor importance. It should be remembered, however, that in the 1960s the trends alone implied an energy growth of 2–2.5 percent p.a. in excess of the economic growth, whereas the trends in recent years and as projected here imply an energy growth rate slightly less than the economic growth rate. These trends reflect changes in the economic system toward lighter production and services, improved technology, and a certain saturation for many energy-intensive consumer products such as automobiles and appliances (Figure 4.5). These trends have been studied elsewhere in more detail.[5]

The relationship between economic growth and

5. Bent Elbek and Uffe Korsbech, "Saturation Scenarios for Denmark." Copenhagen: Niels Bohr Institute, 1975.

Table 4.7
Example of scenario generation, Denmark

	1972		Growth Rates (% p.a. from 1975 to 1985)									1985	
				Intensity, Electricity				Intensity, Other Energy					
Sector	Electricity (PJ)	Other Energy (PJ)	Economy	Price	Gov.	Inh.	Electricity (total)	Price	Gov.	Inh.	Other Energy (total)	Electricity (PJ)	Other Energy (PJ)
Transportation	0.3	141	4.0	−1.0	−0.3	0.0	2.7	−1.0	−0.3	+1.0	3.7	0.4	203
Industrial	14.8	120	4.9	−0.5	−0.5	1.7	5.6	−0.5	−0.5	−2.2	1.7	25.5	142
Agriculture	3.1	41	3.0	−1.0	−0.5	2.4	3.9	−1.0	−0.5	+4.0	5.5	4.5	70
Construction	1.2	21	2.0	0.0	−0.5	4.0	5.5	0.0	−0.5	−3.6	−2.1	2.0	17
Commercial	9.4	35	4.1	−1.0	−1.0	2.0	4.1	−1.0	−1.0	−2.7	−0.6	14.0	33
Public	5.8	30	5.0	−0.5	−0.7	1.6	5.4	−0.5	−0.7	−3.0	0.8	9.8	32
Residential, Heat	1.4	207	3.7	−0.7	−1.4	12.0	13.6	−0.7	−1.4	−2.8	−0.8	5.0	191
Residential, Electricity	18.7		3.7	−0.7	−0.5	2.0	4.5					29.0	
Net Energy	54.7	595	4.1				5.1				1.4	90.2	688

These figures are for Case A (4.1% p.a. economic growth in Denmark, rising oil price, vigorous national policy response). "Price," "Gov.," and "Inh." refer to price, government policy, and inherent trend components of intensity, respectively.

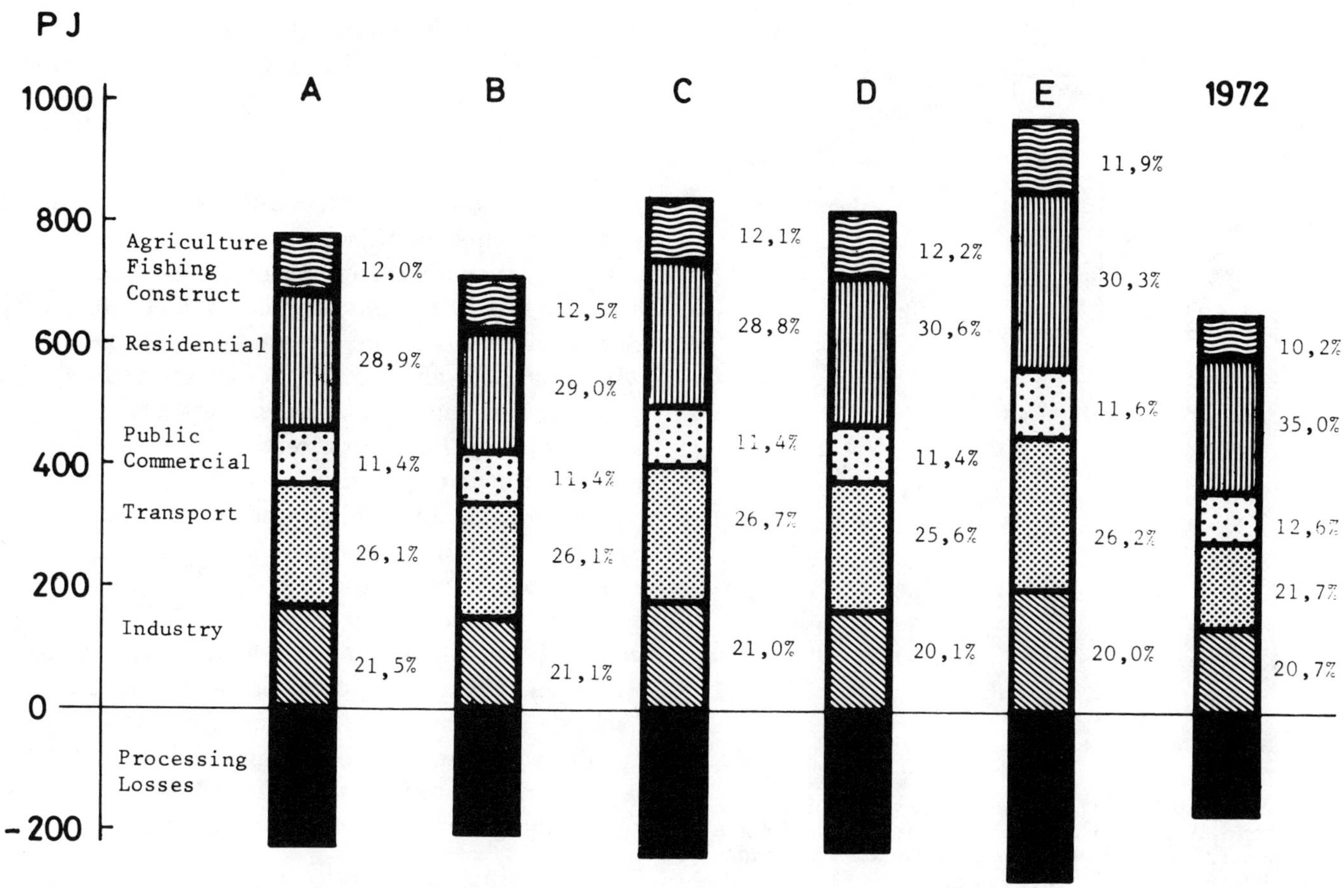

Figure 4.3 Comparison of the five WAES cases to 1985.

energy growth is often expressed as an *energy coefficient*, which is the ratio of the growth rate of the energy demand to the growth rate of the economy. These energy coefficients are given in Table 4.8 for the five scenarios. Except for scenario E, the energy coefficients are well below unity in contrast to an energy coefficient of 1.5 in the 1960s.

The very low energy coefficients might make some scenarios appear less creditable. The low energy growth rates are, however, natural consequences of the scenario assumptions, all of which seem reasonably realistic. As discussed in Section 4.3.2, the implied reductions in energy demand all appear to be well within the limits of present-day technical and economic capability.

The WAES scenarios are, as repeatedly stressed, *not* forecasts. Neither are they supposed to represent grossly unrealistic borderline cases. Nevertheless, some scenarios seem more probable than others. In the absence of major external disturbances, the energy future of Denmark for the next

Table 4.8
Scenario comparisons

Scenario	Gross Energy (PJ/yr.)	Energy Coefficient	1975–1985 Growth Rate (% p.a.)	Composition of Energy Growth			
				Economy (% p.a.)	Price (% p.a.)	Government (% p.a.)	Trends (% p.a.)
A	1002	0.51	2.1	4.1	-0.7	-0.9	-0.3
B	908	0.35	1.1	3.1	-0.7	-0.9	-0.3
C	1076	0.68	2.8	4.1	0.0	-0.9	-0.3
D	1049	0.81	2.5	3.1	0.0	0.0	-0.3
E	1240	1.05	4.3	4.1	+0.7	0.0	-0.3
1972	817						

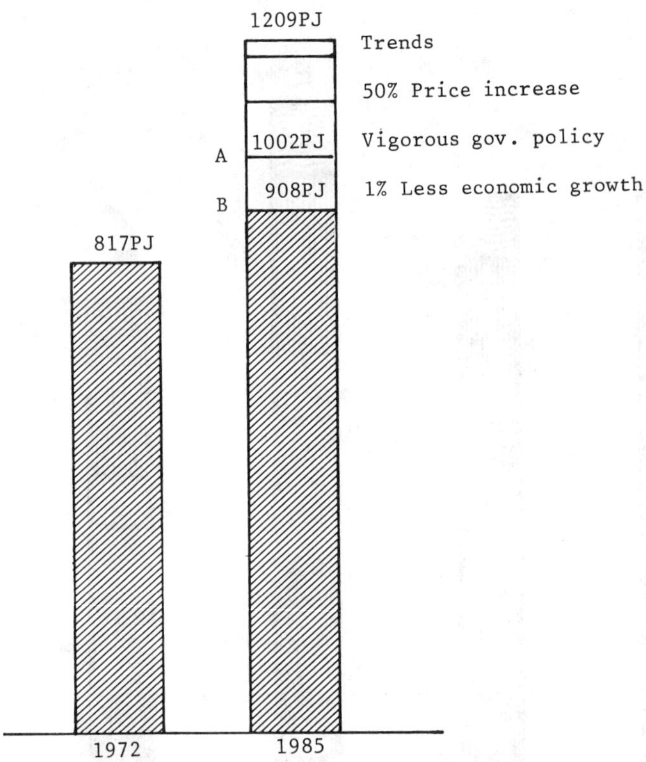

Figure 4.4 Contributions to lower energy growth illustrated for scenarios A and B. The full 1209 PJ of the right column corresponds to an economic growth rate of 4.1% p.a. and energy intensities at their 1972 values.

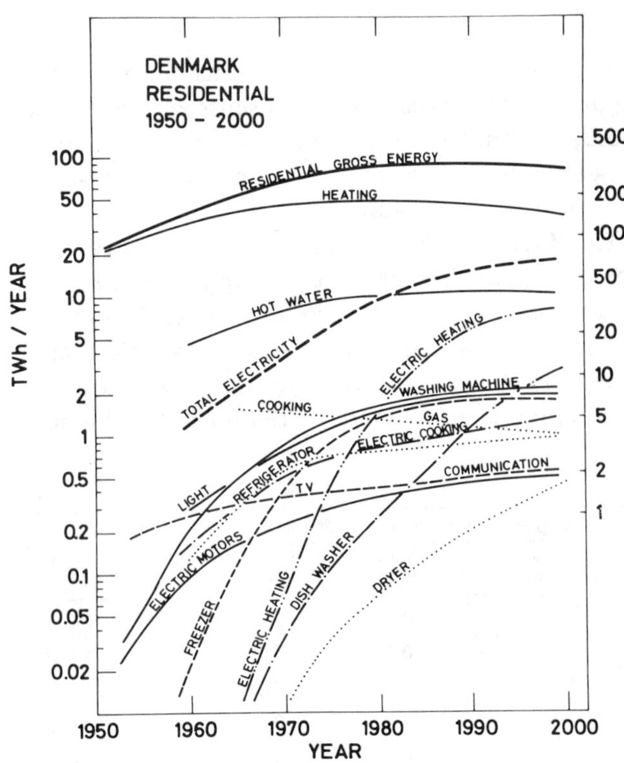

Figure 4.5 Saturation scenario for the residential sector to the year 2000.

ten years is likely to resemble scenarios B, C, or D more than scenarios A and E. Scenario C seems close to the government's energy plan, which, however, envisages energy growth rates as low as those of scenario B.

The demand scenarios to 1985 are here presented without explicit reference to the projections extending scenarios C and D to year 2000. Neither have the supply options been considered except for casual references to the gross energy demand, which can be evaluated only for definite supply systems. Although year 2000 scenarios and the supply options have not directly influenced the demand scenarios presented here, they have provided a valuable background for the testing of assumptions and methods.

In connection with the year 2000 projections, some interesting questions arise which perhaps cannot be answered in a very satisfactory manner. These questions bear closely on the much-discussed "society in a state of energy equilibrium." Does continued economic growth imply continued energy growth, or is it possible to imagine a course of economic development with constant energy input?

Such questions may not need to be answered in the 1985 analysis, but they do become pressing over the longer time span in which severe energy shortages might well occur. In this connection it is worth noting the considerably reduced 1985 energy coefficients discussed above. Much of the reduction in these coefficients can be ascribed to a mending of former habits. Energy savings from insulation of old buildings and plugging of energy leaks do not come again. Nevertheless, real changes in the energy system also point to lower energy growth. It is doubtful, however, that such changes can occur fast enough, and reduced economic growth might be the inevitable consequence of stagnant energy supplies toward the end of the century. These questions deserve full attention in the later phases of the WAES project.

4.3.2 Potential for Energy Conservation

This section briefly summarizes some estimates of the technical potential for energy conservation in Denmark. The estimates are limited to measures which are judged to be economical at present energy prices and real interest rates. This in general means that the investment should have a payback period of not more than 10–15 years. A favorable economy, however, does not ensure full penetration of a given conservation measure. The estimates should therefore be considered as showing the maximum potentials rather than the conservation likely to be achieved. In the scenarios it has been a guideline that the combined effect of energy price, government policy, and inherent trends should not exceed the technical potential.

Although it is possible to consider very fine details (e.g., improved efficiencies of refrigerator compressors), the discussion here concentrates upon the major sectors of the economy which form the basis of the aggregated projections (see Table 4.1). For these sectors Table 4.9 summarizes the estimates of the conservation potentials.

Transportation

In the transportation sector the technical possibilities are limited. The sector is dominated by the private automobile, which is responsible for 45 percent of the energy used for transport or 8 percent of the gross energy demand in the country. The energy intensity of 3.1 MJ/km corresponds to an average performance of 10 km per liter of gasoline. Although some improvements in the performance of engines and tires can be expected, such improvements might be more or less compensated for by a general tendency toward slightly larger automobiles. This tendency is probably the reason for the increasing energy intensity of transport shown in Figure 4.1. The greatest improvements in energy efficiency within the transportation sector are associated with moderately lower speeds (automobiles, airplanes), removal of the least efficient automobiles (large and heavy vehicles), and taxation of trucks on a distance basis (fuel tax instead of yearly vehicle tax).

Industrial

Danish industry is mostly light industry and only a few products are energy-intensive. Within the food industry, increased heat recovery from heating and

Table 4.9
Estimates of energy conservation potentials

Sector	Type of Policy	Policy Actions	Savings by 1985 DE75[1] (PJ/yr.)	IFIAS[2] (PJ/yr.)	DE76[3] (PJ/yr.)	Demand in 1985 (PJ/yr.)[4]	Savings/ Demand (%)[5]
Transportation	Mode shift Capacity utilization	Enforced speed limits Tax large cars Tax diesel oil	7	19	(8)	206	6
Industrial	Heat recovery Insulation Process improvement	Cheap loans Subsidy R&D, information	18	23	20	261	8
Agriculture	Insulation of stables Insulation of greenhouses Reduced tilling	Subsidy Cheap loans R&D, information	4	18	(8)	75	14
Construction	Planning	R&D, information	2			32	6
Commercial	Insulation Heat recovery from cooling and ventilation Automatic controls	Subsidy Cheap loans Energy norms	11	12		91	13
Public	Insulation Heat recovery from ventilation Automatic controls	Gov. decision Energy norms	7	9	100	73	11
Residential	Insulation Improved appliances Automatic controls	Subsidy Energy norms Appliances, furnaces	55	90		297	23
Total			104	171	146	1145	

(The DE76 value 100 is bracketed across Commercial, Public, and Residential.)

1. B. Elbek and U. Korsbech, Dansk Energipolitik (1975).
2. IFIAS, Report No. 2 (1976).
3. Handelsministeriet, Dansk Energipolitik 1976. Parentheses indicate inferences from this report.
4. Gross energy calculated for 4.1% p.a. growth in GNP and energy intensities as of 1972.
5. Average of three estimates of savings in % of 1985 demand.

cooling processes is possible in a number of cases. A general awareness of the energy problems might strengthen the tendency toward lower energy intensity in general. Altogether the main contributions to energy conservation in industry can be expected from numerous small improvements rather than from spectacular improvements within a few industries. Increased research and development and utilization of already known technology are cheap ways to assist the process.

Agriculture and Construction

Agriculture has shown a steady increase in energy intensity for more than twenty years. This development might slow down in the coming years, but drastic changes are unlikely. Greenhouse gardening consumes approximately 40 percent of the sector's fuel and is quite sensitive to the energy price. Improvements from insulation, new cultures, and slightly shorter growing seasons have already reduced the energy intensities by about 20 percent.[6]

Construction is responsible for only a small fraction of the country's energy demand. Indirectly, through the use of energy-intensive products such as cement, steel, and plastics, construction does have a considerable impact on energy demand which should be taken into account in the planning stage for new construction.

Commercial and Public

The commercial sector is not particularly energy-intensive. A large fraction of the energy consumed is electricity for cooling, freezing, and lighting. Considerable improvements seem possible from moderate investments in heat recovery systems, automatic control of heating and ventilation, and adjustments in lighting levels. These remarks are also valid for the public sector, where a fast penetration of conservation measures can be ensured since most buildings are under government jurisdiction.

Residential

The largest single contribution to the conservation potential is found in the residential sector. Increased insulation in existing single-family houses and apartments could reduce their energy consumption by approximately 20 percent. Table 4.10 shows a proposed level of insulation in the residential sector.

From these insulation thicknesses and the corresponding thicknesses for apartment buildings, the total savings for all residential buildings can be calculated. The yearly gross energy savings and the investments necessary are given in Table 4.11.

The IFIAS study which forms the basis for the data in Table 4.11 clearly shows the conservation potential that lies in better thermal insulation of buildings.[7] If the savings listed in Table 4.11 are scaled to include public and commercial buildings and savings that can be obtained through increased use of thermostatic controls, the total conservation potential listed in Table 4.9 is approached. It should again be stressed that these estimates concern the technical conservation potential only. The extent to which this potential is exploited is uncertain but of course dependent on such parameters as government policy and energy price.

In this connection it is interesting to consider the experiences with energy conservation in 1974 and 1975. As mentioned earlier, the gross energy consumption in these two years was approximately 10 percent less than in 1972. This reduction is without doubt mostly a lingering effect of the oil crisis, but part of the savings can in fact be ascribed to technical changes in the energy system.

As seen from Table 4.9, approximately 70 percent of the potential savings is found in the space heating sectors. The 1974–1975 savings in these sectors can be estimated from a market analysis performed for one of the manufacturers of insulation materials.[8] If the data from the market analysis are combined with the heat loss estimates in Table 4.11, one obtains the savings listed in Table 4.12.

The combined effort has resulted in total savings of 5 PJ/year, which corresponds to about 5 percent of the energy used for heating houses. As shown in Table 4.11, the technical potential is about 33 PJ/year. If the yearly addition to the savings is kept at about 2 PJ/year (the later savings are harder to obtain), it is indicated that the full potential will be reached 15 to 20 years hence. This is somewhat

6. Benny Petersen and Bent Elbek, "Energi-värdier of Väksthusprodukter." Copenhagen: Niels Bohr Institute, 1975.

7. The IFIAS energy group has issued six reports on "Energy in Denmark 1973–2005" (ISBN 87-87585-01-4).

8. The analysis was done by Henrigues Marketing for A/S Roockwool.

Table 4.10
Proposed insulation for existing single-family houses

	Future Insulation (mm)[1]	Future k-value ($W/m^2 \cdot {}^\circ C$)
Roof	150	0.25
Wall	80	0.50
Floor	125	0.25
Windows	double	3.0

1. Mineral wool or equivalent.

Source: Asger Hansen and Niels Moe, "Boligens Energiforbrug 1973–2005." IFIAS Report No. 1. Copenhagen, 1975. ISBN 87-87585-02-2.

later than envisaged in the government energy plan, according to which the full potential should be reached by 1985.

The insulation campaign in 1974 and 1975 was supported by a 25 percent government subsidy in addition to the pressure from oil price and general awareness of the energy problem. Nevertheless, the total savings obtained corresponded to only 0.8 percent of the country's gross energy demand. It will take a very considerable effort to better this result in the future.

4.3.3 Economics of Energy Conservation

Saving energy by technical measures usually requires that certain investments be made. These investments can vary greatly from case to case, and it is by no means easy to see which investments are the most profitable. In general, of course, the investments that give the largest savings per unit cost are to be preferred.

Table 4.13 lists some presently available estimates of the cost of saving energy in a number of different fields and by different methods. The estimates have been arranged so that the cheapest measures come first, the most expensive last. The investments are expressed in D.kr. per GJ/year saved (which quite accurately corresponds to U.S. dollars per barrel/year saved). In addition to the cost, Table 4.13 lists the potential savings for each measure if it penetrated 100 percent by 1985.

In order to give an impression of the energy savings involved, these figures can be compared to an expected energy demand of 1200 PJ in 1985 if the demand developed along traditional lines. The sum of the potential savings in Table 4.13 is 110 PJ/year, which corresponds to about 9 percent of

Table 4.11
Energy savings from insulation of existing dwellings

	Heat Loss, Average			Heat Loss, Total			
	Present[1]	Future[1]	Reduction Factor	Present (PJ/yr.)	Future (PJ/yr.)	Savings in Gross Energy (PJ/yr.)	Investment (10^6 D.kr. 1975)
Single-Family Houses[2]							
Roof	17.1	6.7	0.39	17.9	7.0	15.6	1782
Wall	18.7	12.7	0.68	19.6	13.3	9.0	672
Floor	10.5	7.7	0.74	11.0	8.1	4.2	703
Windows	17.6	15.0	0.85	18.5	15.8	3.9	750
Total	63.9	42.1	0.66	67.0	44.2	32.7	3907
Apartments[3]							
Roof	2.0	0.9	0.46	1.6	0.7	1.2	144
Wall	4.2	2.8	0.66	3.2	2.2	1.5	400
Floor	1.2	0.8	0.70	0.9	0.6	0.4	90
Windows	11.4	7.7	0.67	8.9	6.0	4.1	800
Total	18.8	12.2	0.65	14.6	9.5	7.2	1434

1. Figures for single-family houses are in GJ/house-yr. Those for apartments are in MJ/apartment-yr.
2. Based on 1.05×10^6 houses, averaging $100m^2$, and 2900°C days.
3. Based on 0.78×10^6 apartments, averaging $60m^2$.

General Note:
The heat losses given here are smaller than those derived for the base year.
The main reason for this is that this table considers transmission losses but not ventilation losses.

Table 4.12
Energy savings from insulation, Denmark, 1973 to 1975

Action	Percentage of Houses Affected	Savings in Gross Energy (PJ/yr.)[1]
Roofs Insulated	19	3.0
Walls Insulated	10	0.9
Windows Replaced[2]	10	0.4
Furnaces Replaced[2]	10	0.7
Total		5.0

1. Based on two years' effort (1974 and 1975).
2. Estimates.

the projected demand. The total bill for this amount of energy saved is about 20×10^9 D.kr.

Understanding of the economic consequences of the energy savings is facilitated if they are expressed in terms of the time it would take to pay the investment back if payments came only from the saved fuel costs. This payback period is also listed in Table 4.13, based on present energy prices to consumers. Obviously some of the investments are quite profitable, but those having payback periods of 10–15 years or more will probably be made only if subsidized or otherwise promoted politically.

The investment costs are displayed graphically in Figure 4.6, which clearly illustrates that the cost of saving energy increases steeply (in fact exponentially) with the amount of energy saved.

In order to judge the soundness of the analysis, it is of course important to ascertain that the list in Table 4.13 is reasonably complete in the sense that no cheap conservation measures of considerable impact have been left out. Apart from conservation resulting from improved habits under the influence of information or energy-saving campaigns, this does not seem to be the case. The list does not, however, mention any conservation measures within the electricity-consuming areas. This is partly because the efficiency there is in many cases already quite high, although without doubt household appliance standards, for example, could be improved considerably at moderate cost. Electricity is the fastest-growing component of the energy system and might in 1985 comprise as much as 30

Table 4.13
The cost of saving energy in Denmark

Object	Measure	Potential Savings (PJ/yr.)[1]	Investment per Unit Saved (D.kr/GJ/yr.)[2]	Payback Period (years)[3]
1. Greenhouses	Insulation	5.0	36	2
2. Buildings (Public and Commercial)	Thermostats	4.2	42	3
3. Industrial Processes	Heat Recovery	10.5	48	3
4. Buildings (Public and Commercial)	Heat Exchangers	2.4	64	4
5. Walls (Houses)	Insulation	10.0	66	4
6. Roofs (Houses)	Insulation	17.6	101	6
7. Houses (New)	Thermostats	2.4	105	7
8. Ceilings (Apartments)	Insulation	1.3	107	7
9. Houses (Old)	Thermostats	11.7	142	9
10. Floors (Houses)	Insulation	5.1	150	10
11. Windows	Double Glass	9.1	173	11
12. Buildings	New Building Codes	8.0	191	12
13. Floors (Apartments)	Insulation	0.4	224	15
14. Walls (Apartments)	Insulation	1.8	229	15
15. Apartments	Heat Exchangers	1.2	367	24
16. Automobiles	Diesel Engines	13.6	524	34
17. Windows	Triple Glass	4.9	574	37

Based on data from the IFIAS project.
1. In 1985.
2. 1975 D.kr.
3. At average energy prices of 15.5 D.kr./GJ (650 D.kr./TOE). This happens to equal 15.5 \$/bbl (!).

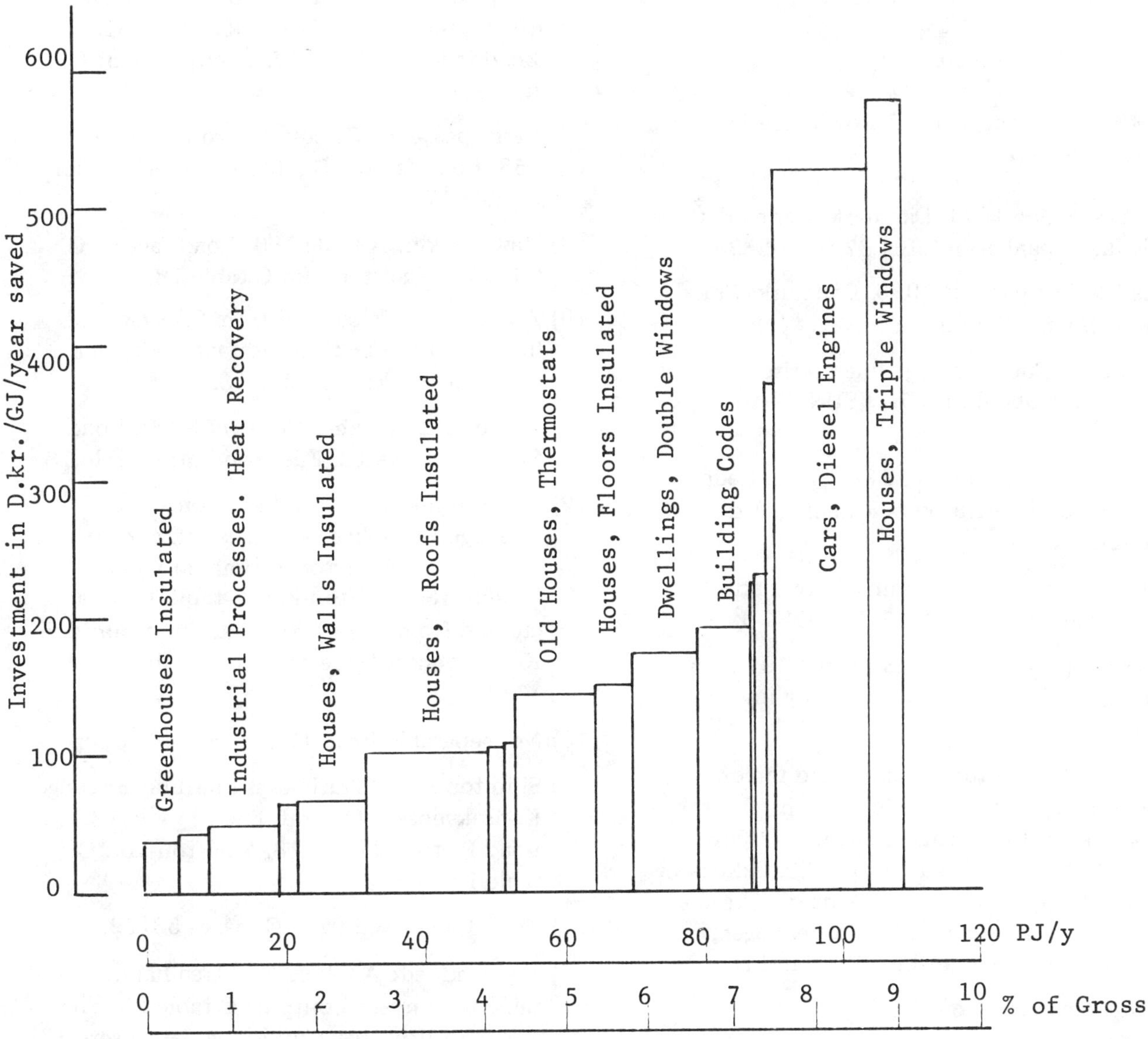

Figure 4.6 Investment necessary for the yearly saving of one unit of energy. The vertical scale can be read as U.S. dollars per barrel/year saved. The area in each block is the total investment.

percent of the primary energy demand. The significance of improvements in this field is therefore obvious.

In spite of the possible omissions, the qualitative statements of this section are probably generally true: a large number of conservation measures are profitable at present energy prices. The cost of saving energy does, however, increase steeply with the amount saved, and very drastic savings by technical measures alone would be very expensive indeed.

4.4 CONCLUSIONS

The main result of this study is that in the period to 1985 Denmark can obtain a growth rate for energy that is considerably lower than the rate for any other extended period in this century. The low energy growth rate is the combined result of a slight reduction in the expected economic growth rate, increased energy prices, the energy policy of the government, and inherent trends in the energy system.

4.5 DEMAND STUDY WORKSHEETS

4.5.1 References and Notes

For the 1972 worksheets the following basic references were used:

A: Statistisk Årbog 1974, Danmark, Danmarks Statistik, Copenhagen ISBN 87-501-0282 6

B: Statistisk tiårs oversigt 1973, Danmarks Statistik, ISBN 87-501-0241-9

C: Bent Elbek: Energi, energi, energikrise, Munksgård, Copenhagen 74, ISBN 87-16-05147-9

D: Danmarks Energibalance 1972, Dansk Kedelforening/Danske Elværkers Forenings Udredningsafd

E: Industristatistik 1970, Danmarks Statistik, Copenhagen 1973, ISBN 87-501-0224-9

Sources of specific line items for the 1972 worksheets are given in the following list (by line number):

1) Population/auto: A table 1 and table 154, Distance/auto-year: "Vejtransporten i tal og tekst", Copenhagen, DK ISSN 0083-5358 table 66. Load factor: "Trafikkens samfundsmæssige konsekvenser", Trafikforskningsgruppen, ATV, Copenhagen, page 185. Fuel/pass-km: C, table 17.

2) Not separable from 1).

3) Bus pass-km: "Vejtransporten" (see ref. line 1) table 67. Load factor: C table 17. Fuel/pass-km: C table 17.

4) Not Separable from 3).

5) Rail pass-km: Calculated from A table 166. % electric: calculated from "Person-transport", VTS, Danmarks Tekniske Højskole, page 7. Load factor: C table 17, fuel/pass km and elec/pass-km: C table 17.

6) Air pass-km: A table 171 and SAS, Årsberetning 1971–72, page 26. Load Factor: calculated from: SAS (see ref. line 6) page 21+26. Fuel/pass-km: C table 17.

7) Air pass-km: Calculated from "Statistiske efterretninger", 1974 nr. 35, page 731. Load factor: C table 17. Fuel/pass-km: C table 17.

8) Ferry pass-km: Calculated from A, table 158. Load factor: C table 17. Fuel/pass-km: C table 17.

9) Rail ton-km: A table 166. Load factor: C table 18. Fuel/ton-km: C table 18.

10) Air ton-km: Calculated from SAS (see ref. line 6) pages 21+26. Load factor: C table 18. Fuel/ton-km: C table 18.

11) Air ton-km: A table 171 (2/7 SAS). Load factor: C table 18. Fuel/ton-km: C table 18.

12) Truck ton-km: "Trafikkens samfundsmæssige Konsekvenser" (see ref. above) table 12. Load factor: estimated from ref. C table 18. Fuel/ton-km: C table 18 + calculated from "Trafikkens samfundsmæssige Konsekvenser" (see ref. line 1) tables 11+16.

13) Not separable from 12).

14) Ship ton-km: "Trafikkens samfundsmæssige Konsekvenser" (see ref. line 1) table 12. Load factor: C table 18. Fuel/ton-km: C table 18.

15) Fuel: Calculated from C, pages 58+59.

17) Value added: A table 389. Area harv.: A table 69. Physical output: C table 20. Fuel + elec.: "Vækstsamfundet og dets reserver", L.O.K.-studiebøgerne 1974, page 53.

18) Value added: A table 389. Area harvested: A table 94. Energy/area: B. Petersen, B. Elbek & M. G. Amsen: "Energiværdier af gartneriprodukter", Niels Bohr Institutet 1975, page 6.

19) Value added: A table 389. Energy/value added: estimated.

20) Value added: A table 389. Energy/value added: estimated.

21) Value added: A table 389. Energy: Estimated from data on oil consumption.

23) Value added: A table 389. Fuel + elect.: Calculated from E page 180.

25) Value added: A table 389. Energy: Estimated.

27) Floor area: Estimated from C table 26. Heat loss/area: Calculated from line 51. Efficiency: as in line 51.

31) Electricity/area: From survey of selected shops. D. Klitgaard and B. Elbek, to be published.

33) Fuel + electr.: Estimated from employment.

34) Floor area: C table 26. Heat loss: Estimated from survey, see line 31.

38) Electricity/area: Estimated from survey, see line 31.

41-49) Value added: Calculated from A table 389 and E table 3.07. Fuel and electricity: Calculated from D table 3.1, E table 7.03 and E section 7.1.

51-56) Dwellings: A table 14 and B page 19. Heat loss: Estimated for summer cottages, else calculated from D table 3.3 and "Prognose for elopvarmede boliger 1970-90", Appendix 9, DEFU, Copenhagen. System Efficiency: D sektion 3.3.

57) Not separable from central space heating.

58-59) Negligible.

60) % Saturation: Estimated. Energy/unit, fuel: "Notat om energiforbrug til varmt brugsvand", IFIAS, Niels Bohr Institutet, Copenhagen, page 5. Energy/unit, elect.: C table 7.

61) % Saturation and energy/unit: C table 7.

62) % Saturation: Estimated. Electricity: Estimated as 20% of residential electricity cf. D table 3.3.

63) % Saturation: A table 45. Fuel: Calculated from D table 3.3. Energy/unit elect.: C table 7.

64-68) % Saturation and energy/unit: C table 7.

69) Balanced to D table 3.3.

The methods used to calculate the specific line items in the scenario D worksheets are given in the following list (by line number). Because of the different level of aggregation and slightly different assumptions, the 69-line study presented here does not give exactly the same results as the 8-line study summarized in Table 4.7. All monetary units are 1972 D.kr. = 1972 US$0.144.

1) Population 1985, 5.3 mill. Autos + 3.0% p.a. from economy, distance/auto-year 16600→17000 km. Unchanged fuel economy.

3) Bus + 2.4% p.a. from economy. Unchanged load factor and fuel economy.

5) Rail passenger + 20% from improved service (inter-city trains) + 1.0% p.a. from historic trends. Electric fraction 19%→25%, because of suburban growth. Unchanged fuel economy.

6)-7) Air travel, domestic and international, + 8.0% p.a. (historic).

8) Ship travel, mostly ferry + 3% p.a. from economy.

9) Rail freight, 2.4% p.a. from economy less historic trend.

10)-11) Air freight + 9.1% p.a. (historic).

12) Truck freight, + 2.4% p.a. from economy less historic trend.

14) Freight, ship, domestic. Zero growth.

15) Bunker + 2.4% p.a. from economy.

17) Farm machinery 2.5% p.a. from trends + 2.3% p.a. from economy. Electric + 3.9% p.a. from trends + 2.3% p.a. from economy.

18) Greenhouses, 2.3% p.a. from economy, 0% p.a. from intensity. Electric + 3.9% p.a. from trends + 2.3% p.a. from economy.

19)-20) Other agriculture as 17).

21) Fishing + 2.5% p.a. from intensity, 0% p.a. from economy.

23) Mining + 1.5% p.a. from growth in construction. Electricity in addition 1.5% p.a. from intensity.

25) Construction + 1.5% p.a. from economy, -4% p.a. from trends. Electric 0% intensity change.

27)-33) Commercial area + 3.1% p.a. from economy, heat loss -1.8% p.a. from trend. System efficiency 0.75→0.80. Electric intensity + 3.1% p.a. from trend.

34)-40) Public area + 3.8% p.a. from economy, heat loss -1.8% p.a. from trend. System efficiency 0.75→0.80. Electric intensity + 5.8% p.a. from trend.

41)-50) Value added calculated from growth rates from econometric model (cf. table 3). Fossil fuel -2.1% p.a. from trends. Electric intensities unchanged.

51) Summer cottages + 5000 pr. year. Doubling of heating standard. Single family house + 20000 pr year (net in 13 years). Average heat loss reduced 11%. System efficiency 0.7→0.75.

52) Electric heating + 12% p.a. from trend.

53)-54) Row houses + 2500 pr. year. Else as lines 51 and 52.

54) Apartments + 10000 pr. year net. Else as lines 51 and 52.

60)-69) % Saturation estimated from logistic growth curves. (See fig. 5 and Elbek and Korsbech, Dansk Energipolitik ref. 8.) Energy/unit essentially as base year except for TV.

4.5.2 Worksheets

In this section summary tables and detailed worksheets are presented in the following order:

Energy Use Summary

Energy Growth Rates Summary

Sectoral Percentages of Total Final Energy Demand

1972 Base Year Data (summary plus five sheets)

Case D (five sheets).

WAES NATIONAL ENERGY DEMAND STUDIES, 1972-1985
ENERGY USE SUMMARY

Country Denmark

Scenario Assumptions

WAES Case	A		B		C		D		E			
GNP 1976-1985	High		Low		High		Low		High			
Oil Price*	$11.50 → $17.25		$11.50 → $17.25		$11.50		$11.50		$11.50 → $7.66			
Natl. Policy Response	Vigorous		Vigorous		Vigorous		Restrained		Restrained			
Natl. GNP Growth Rate 1976-1985 (%/yr.)	4.1		3.1		4.1		3.1		4.1		1972 Base Year	
Fuel Type	Fossil Fuel	Elec.	Fossil Fuel	Elec.	Fossil Fuel	Elec.	Fossil Fuel	Elec.	Fossil Fuel	Elec.	Fossil Fuel	Elec.
Units	PJ	PJ	PJ	PJ	PJ	PJ	PJ	PJ	PJ	PJ	PJ	PJ
Energy-Using Sectors												
Industrial (excluding feedstocks)	142	25.5	126	23.0	149	27	139	25	164	29.4	120	14.8
Transportation	203	.4	184	.4	223	.4	209	.4	253	.5	141	.3
Commercial and Public	65	23.8	59	21.6	70	25.8	68	25.3	82	30.2	65	15.2
Residential	191	34.0	174	31.2	205	36.3	215	35.5	252	41.2	207	20.1
Fishing, Agriculture, Mining, Construction	87	6.5	82	6.2	94	7.0	93	6.9	107	7.9	62	4.3
Feedstocks	----	----	----	----	----	----	----	----	----	----	----	----
Final Energy Demand (excluding processing losses)	688	90.2	625	82.4	741	96.5	724	93.1	858	109.2	595	54.7
Processing Losses	57	168	51	153.0	61	179	59	173	70	203	49	118
Primary Energy Input	(1003)		(911)		(1077)		(1049)		(1240)		(817)	

*Oil price 1975 US$ per barrel Arabian light crude oil FOB Persian Gulf.

WAES NATIONAL ENERGY DEMAND STUDIES, 1972-1985
ENERGY GROWTH RATES SUMMARY

Country Denmark

Scenario Assumptions

WAES Case	A		B		C		D		E	
GNP 1976-1985	High		Low		High		Low		High	
Oil Price*	$11.50 → $17.25		$11.50 → $17.25		$11.50		$11.50		$11.50 → $7.66	
Natl. Policy Response	Vigorous		Vigorous		Vigorous		Restrained		Restrained	
Natl. GNP Growth Rate 1976-1985 (%/yr.)	4.1		3.1		4.1		3.1		4.1	
Fuel Type	Fossil Fuel	Elec.	Fossil Fuel	Elec.	Fossil Fuel	Elec.	Fossil Fuel	Elec.	Fossil Fuel	Elec.
Units (%/yr. 1972-1985)	%	%	%	%	%	%	%	%	%	%
Energy-Using Sectors										
Industrial (excluding feedstocks)	1.3	4.3	.4	3.4	1.7	4.7	1.1	4.1	4.1	5.4
Transportation	2.8	2.2	2.1	2.2	3.6	2.2	3.1	2.2	4.6	4.0
Commercial and Public	0.0	3.5	- .7	2.7	.6	4.1	.3	4.0	1.8	5.4
Residential	- .6	4.1	-1.3	3.4	- .1	4.7	.3	4.5	1.5	5.7
Fishing, Agriculture, Mining, Construction	2.6	3.2	2.2	2.9	3.3	3.8	3.2	3.7	4.3	4.8
Feedstocks	----	----	----	----	----	----	----	----	----	----
Final Energy Demand (excluding processing losses)	1.1	3.9	.4	3.2	1.7	4.5	1.5	4.2	2.9	5.5
Processing Losses	1.1	3.9	.4	3.2	1.7	4.5	1.5	4.2	2.9	5.5
Primary Energy Input	(1.6)		(.8)		(2.1)		(1.9)		(3.3)	

*Oil price 1975 US$ per barrel Arabian light crude oil FOB Persian Gulf.

WAES NATIONAL ENERGY DEMAND STUDIES, 1972-1985
SECTORAL PERCENTAGES OF TOTAL

Country Denmark

Scenario Assumptions

WAES Case	A		B		C		D		E			
GNP 1976-1985	High		Low		High		Low		High			
Oil Price*	$11.50 → $17.25		$11.50 → $17.25		$11.50		$11.50		$11.50 → $7.66			
Natl. Policy Response	Vigorous		Vigorous		Vigorous		Restrained		Restrained			
Natl. GNP Growth Rate 1976-1985 (%/yr.)	4.1		3.1		4.1		3.1		4.1		1972 Base Year	
Fuel Type	Fossil Fuel	Elec.	Fossil Fuel	Elec.	Fossil Fuel	Elec.	Fossil Fuel	Elec.	Fossil Fuel	Elec.	Fossil Fuel	Elec.
Units	%	%	%	%	%	%	%	%	%	%	%	%
Energy-Using Sectors												
Industrial (excluding feedstocks)	20.6	28.3	20.2	27.9	20.1	28.0	19.2	26.9	19.1	26.9	20.2	27.1
Transportation	29.5	.4	29.4	.5	30.1	.4	28.9	.4	29.5	.4	23.7	.5
Commercial and Public	9.4	26.4	9.4	26.2	9.4	26.7	9.4	27.2	9.6	27.7	10.9	27.8
Residential	27.8	37.7	27.8	37.9	27.7	37.6	29.7	38.1	29.4	37.7	34.8	36.7
Fishing, Agriculture, Mining, Construction	12.6	7.2	13.1	7.5	12.7	7.3	12.8	7.4	12.5	7.2	10.4	7.9
Feedstocks												
Final Energy Demand (excluding processing losses)	100.0	100.0	100.0	100.0	100.0	100.0	100.0	100.0	100.0	100.0	100.0	100.0
Processing Losses												
Primary Energy Input												

*Oil price 1975 US$ per barrel Arabian light crude oil FOB Persian Gulf.

NATIONAL ENERGY DEMAND PROJECTIONS

Country Denmark

Year 1972

WAES Case BASE YEAR

Summary

Category	PJ Oil	PJ Gas	PJ Coal	PJ Electricity	PJ Other	Total by Category PJ (Excluding Processing Losses)
Transportation	138.7	1.4	.4	.3	----	140.8
Agriculture, Mining, and Construction	61.6	.6	----	4.3	----	66.5
Commercial	27.6	.7	.5	9.4	6.1	44.3
Public	22.4	.9	.6	5.8	6.1	35.8
Industrial	108.6	2.9	7.2	14.8	1.8	135.3
Residential	156.2	7.2	2.5	20.1	41.4	227.4
Total Delivered Energy by Fuel	515.1	13.7	11.2	54.7	55.4	650.1
Processing Losses by Fuel	36.4	1.4	----	118.1	10.5	166.5
Gross Input by Fuel	551.5	15.1	11.2	172.8	65.9	816.5

Gross Energy Input (Add across bottom row)

NATIONAL ENERGY DEMAND PROJECTIONS

Country Denmark

Year 1972

WAES Case BASE YEAR

Transportation

	(A)	(B) km	(C)	(D)	(E) MJ/km	(F) TJ	(G) TJ	
	Population/Auto	Distance/Auto/Yr.	Load Factor 2)	Fraction Urban or Rural	Energy/Distance	Fossil Fuel	Electricity	
Urban Auto	4.2	16600	33		3.13	62612		(1)
Rural Auto								(2)
	Activity Level 1)	% Electrified	Load Factor 2)	Fossil Fuel/ Activity Level 3)	MJ/Pkm Electricity/Activity			
Urban Bus	7400		30	.80		5920		(3)
Intercity Bus								(4)
Rail (Passenger)	2986	19.5	31	1.42	.56	3413	328	(5)
Air (Passenger, Domestic) 4)	329		58	2.68		882		(6)
Air (Passenger, International) 5)	3873		61	2.68		10380		(7)
Ship (Passenger) 6)	1327		14	5.57		7391		(8)
Rail (Freight)	1907		32	.54		1038		(9)
Air (Freight, Domestic)	1.4		54	18.02		25		(10)
Air (Freight, International) 7)	82.4		54	18.02		1485		(11)
Truck (Local)	13400		50	2.39		32026		(12)
Truck (Intercity)								(13)
Ship (Freight, Domestic)	1600		50	0.63		1008		(14)
Ship (Freight, International) 8)						14400		(15)
Miscellaneous (Balance)						- 148		(16)
					Total	140432	328	

1) A 3-8 10^6 pass-km/yr
   A 9-15 10^6 ton-km/yr

2) C 3-8 % capacity utilized (person-km basis)
   C 9-15 % capacity utilized (ton-km basis)

3) D 3-8 MJ/pass-km
   D 9-15 MJ/ton-km

4) Including Greenland

5) To first destination. Aviation fuel sold in Denmark is 31320 TJ

6) Ferry

7) 2/7 of SAS freight & post international

8) Bunkers

NATIONAL ENERGY DEMAND PROJECTIONS

Country Denmark

Year 1972

WAES Case BASE YEAR

Agriculture, Mining, and Construction

Agriculture, Forestry, and Fishing (ISIC Major Division 1)	(A) 10^6D.kr. Value Added	(B) MJ/D.kr. Energy/Value Added	(C) 10^3 ha Area Harvested	(D) GJ/ha Energy/Area	(E) TJ Physical Output	(F) J/J Energy/Output	(G) TJ Fossil Fuel	(H) TJ Electricity	
Farm Machinery	8807	2.3	2921	7.0	$318.9x10^3$	.064	17560	2880	(17)
Greenhouses	671	25.0	.580	28960			16580	216	(18)
Other Agricultural Uses	(furs) 278	.5					149		(19)
Forestry (ISIC 12)	118	4.0					472		(20)
Fishing Boats (ISIC 13)	850	7.9				.92	6692		(21)
Other Fishing Use (ISIC 13)									(22)
Total Fertilizer (tons)		N 308300	P 58100	K 158400		Total	41453	3096	

Mining	(A) 10^6D.kr. Value Added	(B) MJ/D.kr. Energy/Value Added	(C) Physical Output	(D) Energy/Physical Output	(E) TJ Fossil Fuel	(F) TJ Electricity	
Handling and Internal Transportation	183	15.7			2710	160	(23)
Other Energy Use							(24)
				Total	2710	160	

Construction	(A) 10^6D.kr. Value Added	(B) MJ/D.kr. Energy/Value Added	(C) Physical Output	(D) Energy/Physical Output	(E) TJ Fossil Fuel	(F) TJ Electricity	
Handling and Internal Transportation	13250	1.4			18000	1080	(25)
Other Energy Use							(26)
				Total	18000	1080	
				Agriculture, Mining, and Construction Total	62163	4336	

NATIONAL ENERGY DEMAND PROJECTIONS

Country Denmark

Year 1972

WAES Case BASE YEAR

Commercial and Public

Commercial	(A) 10^6 m^2 Floor Area	(B) MJ/m^2 year Heat Gain or Loss/Area	(C) System Efficiency	(D) TJ Fossil Fuel	(E) TJ Electricity	
Space Heat (Fossil Fuel)	17	640	.75	14500		(27)
Space Heat (Electricity)						(28)
Air Conditioning (Fossil Fuel)						(29)
Air Conditioning (Electricity)						(30)
		Energy/Area	MJ/m^2/year Electricity/Area			
Lighting	17		388		6600	(31)
Miscellaneous						(32)
Presently Unknown small crafts				20400	2800	(33)
			Total	34900	9400	

Public	10^6 m^2 Floor Area	MJ/m^2/year Heat Gain or Loss/Area	System Efficiency	TJ Fossil Fuel	TJ Electricity	
Space Heat (Fossil Fuel)	30	750	.75	30000		(34)
Space Heat (Electricity)						(35)
Air Conditioning (Fossil Fuel)						(36)
Air Conditioning (Electricity)						(37)
		Energy/Area	MJ/m^2/year Electricity/Area			
Lighting	30		144		4320	(38)
Miscellaneous					1440	(39)
Presently Unknown						(40)
			Total	30000	5760	
			Commercial and Public Total	64900	15160	

NATIONAL ENERGY DEMAND PROJECTIONS

Country Denmark

Year 1972

WAES Case BASE YEAR

Industrial 9)

	(A) 10^6 D.kr. 10) Value Added	(B) MJ/D.kr. Energy/Value Added	(C) Physical Output	(D) Energy/Output	(E) % Exported	(F) TJ 11) Fossil Fuel	(G) TJ Electricity	(H) Feedstocks	
Iron and Steel (ISIC 371)	392	21.8				7780	780		(41)
Nonferrous Metals (ISIC 372)	186	13.0				2200	220		(42)
Paper and Allied (ISIC 34)	3003	3.1				7980	1350		(43)
Chemicals and Allied (ISIC 35 minus ISIC 353)	3016	4.6				11130	2710	small	(44)
Mineral Products (ISIC 36)	2207	21.3				44730	2240		(45)
Food and Related (ISIC 31)	5478	4.6				22850	2320		(46)
Transport Eqpt. (ISIC 384)	1549	2.7				3240	890		(47)
Key Industry (N.E.C.) 12)	7040	1.8				10060	2400		(48)
Key Industry (N.E.C.) 13)	2299	2.2				4190	840		(49)
Misc. Manufacturing 14)	3065	2.4				6340	1050		(50)
N.E.C. = Not Elsewhere Classified					**Total**	120500	14800		

9) More than 6 employees

10) Scaled from 1970 data
Industry GNP(1972)/ GNP(1970)= 1.183
Value added is in 1972 kr.

11) Scaled from 1970 data. Energy for industry
E(1972)/E(1970) = 1.39

12) Machinery (JSIC 38-384)

13) Textile (JSIC 32)

14) Wood (JSIC 39+33)

NATIONAL ENERGY DEMAND PROJECTIONS

Country Denmark

Year 1972

WAES Case BASE YEAR

Residential

Population $5.0x10^6$ Persons/Dwelling 2.7 Dwellings $1.86x10^6$

		(A) 10^3	(B) MJ/yr	(C)	(D)	(E) TJ	(F) TJ	
Central Space Heating		Dwellings	Heat Loss	System Efficiency		Fossil Fuel	Electricity	
Single Family	Fossil Fuels 15)	130 911	1000 78470	1.0 .7		102123	130	(51)
	Electricity	12	81875	1.0			982	(52)
Town or Row Houses	Fossil Fuels	122	78470	.7	$137x10^6 m^2$	13676		(53)
	Electricity	2	81875	1.0			164	(54)
Apartments	Fossil Fuels	811	48830	.75		52802		(55)
	Electricity	2	81875	1.0			164	(56)
		% Saturation (Fossil Fuels)	% Saturation (Electricity)	MJ/year Energy/Unit (Fossil Fuels)	MJ/year Energy/Unit (Electricity)			
Room Heating								(57)
Air Conditioning (Window)								(58)
Air Conditioning (Central)								(59)
Water Heating		75	2	24171	13140	33719	489	(60)
Refrigerators			95		1440		2544	(61)
Lighting			100		2950		5500	(62)
Cooking		53	46	5112	2880	5040	2464	(63)
Dishwashers			5		2430		226	(64)
Freezers			45		2628		2200	(65)
Television			95		590		1043	(66)
Other Key Appliances 16)			40		4932		3669	(67)
Misc. Appliances 17)			2		1980		74	(68)
Potential Appliances 18)							451	(69)
					Total	207360	20100	

15) Summer cottages (upper line)

16) Clothes washer

17) Dryer

18) Balance

NATIONAL ENERGY DEMAND PROJECTIONS

Country Denmark

Year 1985

WAES Case D, DK 3.1%

Transportation

	(A)	(B) km	(C) %	(D)	(E) MJ/km	(F) TJ	(G) TJ	
Autos 1617000	Population/Auto	Distance/Auto/Yr.	Load Factor 2)	Fraction Urban or Rural	Energy/Distance	Fossil Fuel	Electricity	
Urban Auto	3.3	17000	33	-	3.1	85200		(1)
Rural Auto								(2)
	10^6 Activity Level 1)	% Electrified	Load Factor 2)	Fossil Fuel/ Activity Level 3)	MJ/Pkm Electricity/Activity			
Urban Bus	9400		30	0.8		7520		(3)
Intercity Bus								(4)
Rail (Passenger)	3880	25	31	1.42	0.56	4130	543	(5)
Air (Passenger, Domestic) 4)	710		58	2.7	-	1915		(6)
Air (Passenger, International) 5)	8360		61	2.7		22570		(7)
Ship (Passenger) 6)	1780		14	5.6		9970		(8)
Rail (Freight)	2410		32	0.54		1300		(9)
Air (Freight, Domestic)	3		54	18		54		(10)
Air (Freight, International)	197		54	18		3546		(11)
Truck (Local)	16990		50	2.4		40768		(12)
Truck (Intercity)								(13)
Ship (Freight, Domestic)	1600		50	0.6		960		(14)
Ship (Freight, International) 7)						18254		(15)
Miscellaneous								(16)
					Total	196187	543	

1) A 3-8 10^6 pass-km/yr
A 9-15 10^6 ton-km/yr

2) C 3-8 % capacity utilized (person-km basis)
C 9-15 % " " (ton-km basis)

3) D 3-8 MJ/pass-km
D 9-15 MJ/ton-km

4) Including Greenland

5) To first destination

6) Ferry

7) Bunker

NATIONAL ENERGY DEMAND PROJECTIONS

Country Denmark

Year 1985

WAES Case D, DK 3.1%

Agriculture, Mining, and Construction

Agriculture, Forestry, and Fishing (ISIC Major Division 1)	(A) 10^6 D.kr. Value Added	(B) MJ/D.kr. Energy/Value Added	(C) 10^3 ha Area Harvested	(D) GJ/ha Energy/Area	(E) TJ Physical Output	(F) J/J Energy/Output	(G) TJ Fossil Fuel	(H) TJ Electricity	
Farm Machinery	11055	3.0	2800×10^3	11.9	330×10^3	0.101	28063	5255	(17)
Greenhouses	842	25.2	0.73×10^3	29050			20813	394	(18)
Other Agricultural Uses	348	0.68					238		(19)
Forestry (ISIC 12) (furs)	148	5.1					754		(20)
Fishing Boats (ISIC 13)	850	10.1					8566		(21)
Other Fishing Use (ISIC 13)									(22)
Total Fertilizer (tons)		N 350000	P 60000	K 160000		Total	58434	5649	

Mining	(A) 10^6 D.kr. Value Added	(B) MJ/D.kr. Energy/Value Added	(C) Physical Output	(D) Energy/Physical Output	(E) TJ Fossil Fuel	(F) TJ Electricity	
Handling and Internal Transportation	212	15.8			3145	215	(23)
Other Energy Use							(24)
				Total	3145	215	

Construction	(A) 10^6 D.kr. Value Added	(B) MJ/D.kr. Energy/Value Added	(C) Physical Output	(D) Energy/Physical Output	(E) TJ Fossil Fuel	(F) TJ Electricity	
Handling and Internal Transportation	15377	0.99			13973	1253	(25)
Other Energy Use							(26)
				Total	13973	1253	
				Agriculture, Mining, and Construction Total	75552	7117	

NATIONAL ENERGY DEMAND PROJECTIONS

Country Denmark

Year 1985

WAES Case D, DK 3.1%

Commercial and Public

Commercial	(A) 10^6 m^2 Floor Area	(B) MJ/m^2 y Heat Gain or Loss/Area	(C) System Efficiency	(D) TJ Fossil Fuel	(E) TJ Electricity	
Space Heat (Fossil Fuel)	23	535	0.80	15400		(27)
Space Heat (Electricity)						(28)
Air Conditioning (Fossil Fuel)						(29)
Air Conditioning (Electricity)						(30)
		Energy/Area	$MJ/m^2/yr$ Electricity/Area			
Lighting	23		525		12044	(31)
Miscellaneous						(32)
Presently Unknown, small crafts				21660	5110	(33)
			Total	37060	17154	

Public	10^6 m^2 Floor Area	MJ/m^2/year Heat Gain or Loss/Area	System Efficiency	TJ Fossil Fuel	TJ Electricity	
Space Heat (Fossil Fuel)	43	635	0.80	34136		(34)
Space Heat (Electricity)						(35)
Air Conditioning (Fossil Fuel)						(36)
Air Conditioning (Electricity)						(37)
		Energy/Area	$MJ/m^2/yr$ Electricity/Area			
Lighting	43		250		10804	(38)
Miscellaneous (street light)					3601	(39)
Presently Unknown						(40)
			Total	34136	14405	
			Commercial and Public Total			

NATIONAL ENERGY DEMAND PROJECTIONS

Country Denmark

Year 1985

WAES Case D, DK 3.1%

Industrial (More than 6 employees)

	(A) 10^6 D.kr.72 Value Added	(B) MJ/D.kr. Energy/Value Added	(C) Physical Output	(D) Energy/Output	(E) % Exported	(F) TJ Fossil Fuel	(G) TJ Electricity	(H) Feedstocks	
Iron and Steel (ISIC 371)	638	18.2				10354	1270		(41)
Nonferrous Metals (ISIC 372)	302	10.9				2928	358		(42)
Paper and Allied (ISIC 34)	3519	2.5				7290	1521		(43)
Chemicals and Allied (ISIC 35 minus ISIC 353)	4254	3.9				12790	3823		(44)
Mineral Products (ISIC 36)	3083	17.5				50897	3129		(45)
Food and Related (ISIC 31)	6876	3.8				23311	2912		(46)
Transport Eqpt. (ISIC 384)	2523	2.3				4312	1450		(47)
Key Industry (N.E.C.) 32 Textile	2590	1.8				3827	946		(48)
Key Industry (N.E.C.) 38-384 Machinery	11467	1.5				13389	3909		(49)
Misc. Manufacturing 39,33 Wood	4992	2.0				8438	1710		(50)
					Total	137536	21028		

N.E.C. = Not Elsewhere Classified

NATIONAL ENERGY DEMAND PROJECTIONS

Country Denmark

Year 1985

WAES Case D, DK 3.1%

Residential

Population 5.3×10^6 Persons/Dwelling 2.35 Dwellings 2.25×10^6

Central Space Heating		(A) 10^3 Dwellings	(B) MJ/unit-year Heat Loss	(C) System Efficiency	(D)	(E) TJ Fossil Fuel	(F) TJ Electricity	
Single Family	Fossil Fuels	200	2000	1.0	Cottages		400	(51)
	Electricity	1090 60	70000 70000	0.75 1.0		101700	4200	(52)
Town or Row Houses	Fossil Fuels	150	70000	0.75		14000		(53)
	Electricity	7.5	70000	1.0			525	(54)
Apartments	Fossil Fuels	940	40000	0.8		47000		(55)
	Electricity	7	70000	1.0			490	(56)
		% Saturation (Fossil Fuels)	% Saturation (Electricity)	MJ/year Energy/Unit (Fossil Fuels)	MJ/year Energy/Unit (Electricity)	162700	5615	
Room Heating								(57)
Air Conditioning (Window)								(58)
Air Conditioning (Central)								(59)
Water Heating		85	10	25000	13000	47800	2925	(60)
Refrigerators			100		1500		3375	(61)
Lighting			100		3000		6750	(62)
Cooking		30	70		3000	4000	4725	(63)
Dishwashers			18		2500		1013	(64)
Freezers			60		2700		3645	(65)
Television			98		800		1764	(66)
Other Key Appliances	Washer		55		4500		5569	(67)
Misc. Appliances	Dryer		11		2000		495	(68)
Potential Appliances								(69)
					Total	214500	35876	

5 Finland

5.1 ABSTRACT

This report includes energy demand scenarios for Finland for the period 1972 to 1985. The five scenarios are varied according to energy price and GNP growth rate. The ratio of energy growth rate to GNP growth rate, which was about 1.3 before the energy crisis, is expected to level down to about 1.0. The present time is very inconvenient for energy demand forecasting since the last couple of years have been a period of energy price increases, economic recession, and warm winters. Thus, the figures given cannot be considered very reliable.

5.2 INTRODUCTION

The Finnish team joined WAES only in October 1975, when Professor Jorma Routti participated in the Shimoda meeting in Japan. Thus, the Finnish demand study is rather incomplete and is based mainly on information obtained independently of the WAES project.

Although there is a great deal of uncertainty concerning the scenario projections, they can be considered "hard" if measured only against the information and data base available. It is expected, however, that within a year or two there will be a definite need to make a thorough revision of the forecasts, when new information has been obtained.

Although the present energy situation is roughly the same for all industrialized, energy-importing countries, the local characteristics of each country will be the main determinants for the future energy situation. The Finnish situation seems to be especially unfavorable for the following reasons:

1. Finland is located in a cold climate, so that much energy is needed for space heating.

2. The population is spread out, and much energy is needed for transportation.

3. Finnish industries are energy-intensive.

4. There are no high-grade fossil or nuclear fuel reserves in Finland.

5. The supply of indigenous energy sources (hydro power, wood, peat, industrial wastes, etc.) cannot be substantially increased.

6. Many new energy sources are difficult to utilize in Finland, including solar, geothermal, tidal, and wave and wind energy, ocean thermal gradients, etc.

At the present time the share of indigenous energy sources is about 30 percent of the supply, but it is expected to decrease to 25 percent and below by 1985. The availability of imported energy has been secured, however, since several long-term import contracts have been made, mostly with the Soviet Union and Poland. These contracts include oil, natural gas, coal, and electricity. Thus, the problems concern mostly the costs of energy. The present situation is rather serious, since the balance of payments is already very negative for other reasons. The only policy option for the national government is thus to implement vigorous energy conservation measures.

5.3 METHODOLOGY

The method used to obtain the energy demand figures is not quite the same as that defined by WAES. The figures are based on studies made at the Energy Department of the Ministry of Trade and Industry independently of the WAES project. Nevertheless, for the most part the assumptions on which the results are based can be considered compatible with those used by WAES.

The scenarios C and D are based on the high and low growth assumptions made by the Energy Department. The corresponding GNP growth rates are, on the average, 4.4 and 3.25 percent. These are estimated to change as shown in Table 5.1. These growth rates are compatible with the WAES global high and low alternatives since the Finnish economy is very open and thus dependent on other countries, particularly those in OECD.

The scenarios C and D are based on the assumption that there are no major changes in oil price. The scenarios A, B, and E have been derived from

Table 5.1
High and low economic growth rate assumptions

	1975–1980	1981–1985
high growth	4.0%	4.8%
low growth	3.0%	3.5%

them under incremental assumptions. As is the case in other countries, there exists no hard information about long-range price elasticities. For this reason the method we have used is rather intuitive. The elasticity for total energy consumption is assumed to be about -0.2, although this of course varies in different demand sectors.

The energy demand is disaggregated into six sectors. The original disaggregation used at the Energy Department is much more detailed and includes close to 100 line items. The grouping of these items is, however, different from that of WAES, and it would be difficult to transform these figures into the WAES format. Nor would such a transformation serve the national energy policy-making work. The method used is, however, the same in principle as that of WAES. The starting point is the GNP growth rate. This is transformed into sectoral and subsectoral growth rates. The technical energy consumption efficiencies are then determined for each subsector.

The GNP elasticity of energy demand is expected to decrease from 1.3 to about 1, which means that the energy efficiency of the whole economy will stay approximately constant. The efficiency of space heating is expected to increase because of the larger share of district heating, better insulation norms, and other conservation measures. On the other hand, the demand for electricity is still expected to grow much faster than the GNP.

In this report, the government response has not been separately analyzed. First, the government response can be predicted to be rather vigorous regardless of the scenario variables. Second, although there seems to be some room left for a still more vigorous policy, the consequences of such a policy would be very difficult to predict at this moment.

5.4 PRESENTATION AND ANALYSIS OF RESULTS

The traditional pattern of energy demand growth is changing because of both the lower growth rate and the achievement of better energy efficiencies. In the following paragraphs, the main development features of the various demand sectors are summarized.

5.4.1 Industrial

The growth rate of industrial production volume is expected to be slightly greater than that of the total GNP. However, the growth rate of certain energy-intensive industries such as pulp and paper production is expected to be rather low. Conservation measures will further moderate the growth of energy demand.

5.4.2 Transportation

Major saturation effects are expected in private automobile traffic, although the number of cars per 1000 inhabitants is now not much more than 200. In the 1960s the growth rate was about 10 percent per annum. In the future it is expected to be less than 5 percent, depending on the economic growth rate.

5.4.3 Commercial and Public

The energy demand of commercial and public sectors is highly dependent on the space or floor area of the buildings. Fuel demand is expected to grow slowly because of energy conservation measures. Electricity demand, however, seems to be growing very rapidly.

5.4.4 Residential

The growth rate of the volume of residential building space will be lower than in the past. In addition, conservation measures are expected to hold fuel demand approximately constant. Electricity demand is, however, expected to grow rapidly in this sector, too. This is partly due to electric heating of small houses. The numbers of many electricity-consuming household appliances such as freezers, dishwashers, electric saunas, stoves, etc., will be growing for a long time before reaching the saturation level.

5.4.5 Agriculture, Fishing, Mining, Construction

Most of the energy consumed in this residual sector is oil needed for vehicles and machines in agriculture.

5.4.6 Feedstocks

The share of energy sources needed for feedstocks will be increasing, but in absolute terms it will be very small.

5.4.7 Processing Losses

The share of primary energy absorbed by processing losses is increasing. The main reason for this is that electricity will be produced by less efficient means than hydro power.

5.5 CONCLUSIONS

The five different scenarios seem to have no significant qualitative differences. The differences seem to be mostly in timing, that is, in the point of time at which a certain consumption level is reached.

The relationship between economic growth and energy demand is rather straightforward, but the long-range impact of rising or falling oil price is not yet well known.

5.6 DEMAND STUDY WORKSHEETS

On the next two pages will be found the energy use summary and energy growth rates summary for the five WAES scenarios.

WAES NATIONAL ENERGY DEMAND STUDIES, 1972-1985
ENERGY USE SUMMARY

Country Finland

Scenario Assumptions

WAES Case	A		B		C		D		E			
GNP 1976-1985	High		Low		High		Low		High			
Oil Price*	$11.50 → $17.25		$11.50 → $17.25		$11.50		$11.50		$11.50 → $7.66			
Natl. Policy Response [1]	~~Vigorous~~		~~Vigorous~~		~~Vigorous~~		~~Relaxed~~		~~Restrained~~			
Natl. GNP Growth Rate 1976-1985 (%/yr.)	4.4		3.25		4.4		3.25		4.4		1972 Base Year	
Fuel Type	Fossil [2] Fuel	Elec.	Fossil Fuel	Elec.	Fossil Fuel	Elec.	Fossil Fuel	Elec.	Fossil Fuel	Elec.	Fossil Fuel	Elec.
Units 10^{15} J												
Energy-Using Sectors												
Industrial (excluding feedstocks)	423	64	406	61	448	73	415	66	500	82	262	39
Transportation	167	2	154	2	181	2	167	2	190	3	98	0
Commercial and Public	66	30	57	28	70	35	60	31	73	36	60	10
Residential	173	44	172	43	181	52	180	48	190	55	176	15
Fishing, Agriculture, Mining, Construction	32	6	32	6	34	7	34	6	37	7	31	3
Feedstocks	4		4		4		4		4		2	
Final Energy Demand (excluding processing losses)	865	146	825	140	918	169	860	153	994	183	629	68
Processing Losses	94	141	96	133	103	175	101	147	109	198	41	51
Primary Energy Input	959	287	921	273	1021	344	961	300	1103	381	670	118

*Oil price 1975 US$ per barrel Arabian light crude oil FOB Persian Gulf.

1. National policy options not analyzed
2. "Fossil fuel" includes domestic fuels (wood, peat, bark, industrial wastes etc.)

WAES NATIONAL ENERGY DEMAND STUDIES, 1972-1985
ENERGY GROWTH RATES SUMMARY

Country Finland

Scenario Assumptions										
WAES Case	A		B		C		D		E	
GNP 1976-1985	High		Low		High		Low		High	
Oil Price*	\$11.50 → \$17.25		\$11.50 → \$17.25		\$11.50		\$11.50		\$11.50 → \$7.66	
Natl. Policy Response[1]	~~Vigorous~~		~~Vigorous~~		~~Vigorous~~		~~Restrained~~		~~Restrained~~	
Natl. GNP Growth Rate 1976-1985 (%/yr.)	4.4		3.25		4.4		3.25		4.4	
Fuel Type	Fossil Fuel[2]	Elec.	Fossil Fuel	Elec.	Fossil Fuel	Elec.	Fossil Fuel	Elec.	Fossil Fuel	Elec.
Units (%/yr. 1972-1985)										
Energy-Using Sectors										
Industrial (excluding feedstocks)	3.8	3.9	3.4	3.5	4.2	5.0	3.6	4.1	5.1	5.9
Transportation	4.2	-	3.5	-	4.8	-	4.2	-	5.2	-
Commercial and Public	0.7	8.7	-0.4	8.1	1.2	9.9	0.0	8.9	1.5	10.2
Residential	-0.1	8.6	-0.2	8.4	0.2	10.0	0.2	9.4	0.6	10.5
Fishing, Agriculture, Mining, Construction	0.2	4.5	0.2	4.5	0.7	5.1	0.7	4.3	1.4	5.7
Feedstocks	5.5		5.5		5.5		5.5		5.5	
Final Energy Demand (excluding processing losses)	2.5	6.1	2.1	5.8	3.0	7.3	2.4	6.5	3.6	8.0
Processing Losses	6.6	8.1	6.8	7.7	7.3	9.9	7.2	8.5	7.9	11.0
Primary Energy Input	2.8	7.1	2.5	6.7	3.3	8.6	2.8	7.4	3.9	9.4

*Oil price 1975 US\$ per barrel Arabian light crude oil FOB Persian Gulf.

1. National policy options not analyzed

2. "Fossil fuel" includes domestic fuels (wood, peat, bark, industrial wastes etc.)

6.1 INTRODUCTION

6.1.1 The Contribution to WAES

The French team has been structured in accordance with the national organization for energy planning and forecasting.

Under the leadership of M. Jean Couture, Conseiller du Président, Société Générale, former General Secretary for energy, a liaison has been assured through his Associate to the central planning office. This entity collects information from different sectors (energy sectors, political management, heavy energy consumers, etc.) to establish forecasts.

M. Jean-Marie Martin, Chairman of the Institut Economique et Juridique de l'Energie, University of Grenoble, joined the WAES study team in 1975.

Due to the particular structure of French energy planning, a reference group has been organized in a manner that is probably different from that of other countries. Nevertheless, the relevance and reliability of the studies have not suffered since the structure assures an extensive and critical exchange of views covering all the relevant economic, political, and social aspects of the energy problem.

6.1.2 The Data

The 1972 data were directly extracted from publications of the Institut National de la Statistique and can be considered "hard." They were completed by adding other information that has become available.

The 1985 figures were revised after publication in 1975 of the report of the Commission Nationale de l'Energie. This transitional entity was set up by the government to study the forecast to 1985 in the energy sector, with contributions from the interested parties indicated above.

To complete the WAES scenarios that lay a little outside the official French forecasting cases (namely, C, D, and E), the University of Grenoble has helped in collecting the necessary information.

6.2 NATIONAL ENERGY SITUATION AND POLICY

6.2.1 Evolution of the Energy Situation during the Last Ten Years

The share of oil in the French energy demand increased between 1960 and 1972 from 31.4 to 67 percent as the competitive position of this fossil fuel vis-à-vis coal accelerated the decline of solid fuel extraction (coal decreased from 54.7 percent in 1960 to 17 percent in 1972).

The low price of oil, compared with other fuels, also held back the development of replacement fuels such as nuclear.

As a consequence, the dependence of France on foreign supply increased from 41 percent in 1960 to about 75 percent in 1972.

The immediate advantages of cheap fuel prevailed in the existing world and national economic situation over objections relating to the supply security that might have been achieved by promoting other sources that were not competitive in this context.

Due to the new situation created by the 1973 oil crisis and the consequent steep price increase of oil and natural gas, the present French energy policy can be characterized by a certain number of constraints and implications:

- the need to assure the security of supply by diversification of the sources of imported fuels;
- the need to develop replacement sources that allow increased self-dependence and that have now become more competitive (such as nuclear energy for electricity generation, which is now being rapidly developed);
- the need to promote foreign trade to reduce the deficit in the balance of payments;
- the need to encourage a lower degree of importation by promoting an energy conservation policy and exploitation of the indigenous energy supply (nuclear, hydraulic, new forms of energy).

6.2.2 Energy Policy to 1985

The government's interest in economic and social forecasting is manifested by the special entity created after the Second World War to develop noncompulsory but indicative five-year plans, the latest of which was recently issued for the 1976 to 1980 period. Such forecasting allows adaptive and corrective measures to be taken in order to avoid or prevent possible conflicts.

In these plans, the energy sector takes an important place due to its implications for the medium- and long-term evolution of society in terms of the requirement for investments, whose amount must be compatible with GNP growth in the global objectives of the national policy.

6.3 NATIONAL ENERGY RESOURCES

While energy demand continues to increase due to the growth rate of the economy, despite an active energy conservation policy, exploitation of the indigenous fossil fuel energy resources will remain rather limited, at least from now to 1985.

6.3.1 Coal

The recognized resources amount to 1.4 billion tons,[1] from which 0.8 billion are economically recoverable. As in other European countries, extraction of coal from new mines will in the future depend on the price evolution of competitive forms of energy but will in any case, in France, be very limited before 1985.

Roughly 20 million tons will be extracted yearly from the national mines over the next decade. This extraction rate is expected to be roughly maintained into the future unless there is some drastic change of economic conditions. The major portion of the coal used in the steel industry must be imported, the indigenous coal being mainly used as steam coal.

According to the forecast, at least 50 percent of the coal demand will have to be satisfied by imports during the next decade.

6.3.2 Oil

Because there are practically no resources of oil on the national territory (less than 1 percent of the demand in 1975), the sources of imports will be, according to the national energy policy, diversified. Important oil research activity goes on in the Channel and in the Mediterranean Sea.

1. Some sources suggest a much higher level of coal reserves.

6.3.3 Natural Gas

The indigenous secured reserves of natural gas amount to 300 billion m^3, and the annual production in 1975 reached about 10 billion m^3, a level that will be maintained up to 1983 but will probably decrease progressively thereafter.

In 1972, about 50 percent of the natural gas consumed was imported; in 1985, it is expected that the degree of dependence on foreign supply will reach at least 75 percent.

As with oil, government policy is to diversify the foreign sources of imports by encouraging long-term contracts.

From the point of view of the security of the gas supply, the geographical distribution of gas reserves in the world is apparently more favorable for a diversification of supply than in the case of oil, and the rigidity of the gas "chain" from the producer to the consumer is much stronger due to transport constraints.

On the other side, the cost of gas transport and storage is higher than that for oil.

Natural gas will in the future be used mainly in industry and for heating in the residential and commercial sectors and practically not at all in power plants.

6.3.4 Hydraulic

Some 80 to 85 percent of the economically recoverable hydraulic resources have already been exploited.

In 1972, the yearly production reached 58 TWh, and it is expected that the remaining capacity will be developed up to 63 TWh by 1985 and about 70 TWh by the year 2000.

France has in operation a tidal plant with a yearly production of 0.5 TWh.

None of these figures takes account of the possible construction of an important new tidal plant at Mont Saint Michel in western France, with a yearly production estimated at 25–27 TWh. For the time being, the financial impact of the project has postponed its realization, although it has not been abandoned.

6.3.5 Nuclear

The situation with regard to indigenous resources of uranium is, compared to that with fossil fuel reserves, more favorable for the time being; at least 100,000 tons of uranium (up to \$30 per pound of U_3O_8) can be considered as assured and probable. Due to long-term contracts with foreign suppliers (Gabon, Niger), the expectation for the security of demand satisfaction is high.

The above-mentioned figure of 100,000 tons must be considered conservative, the real efforts of exploration having been undertaken only recently.

The French nuclear program in the next decade will be based on light-water reactors (pressured water), the needed enrichment sources for the fuel being provided by long-term contracts with the United States and by the EURODIF plant now under construction in France, with the participation of Belgium, Italy, and Spain. The yearly output of this gas diffusion plant will reach about 11 million separation work units (SWUs) after 1980. A new plant (COREDIF) is being planned, with possible participation of foreign partners.

The LWR fuel reprocessing will take place in a plant that will be in operation in the near future (final capacity 800 tons/year).

The French nuclear program started with some natural uranium, gas-graphite-moderated plants, but this line has now been abandoned in favor of the light-water reactors. Also, as part of the attempt to assure the future fuel supply, the breeder reactor technique was developed, and the 250 MWe Phénix plant has been operated successfully for two years. A new commercial breeder plant (1200 MWe) will be ordered in the near future for commissioning in 1981–1982.

The safety and other problems connected with nuclear energy are being investigated very intensively.

6.3.6 Oil Shale

Oil shale deposits in France are far from negligible. For the Parisian region alone, at a depth of about 50 meters recoverable reserves are estimated to be about 1 billion tons of oil.

Nevertheless, the total financial and environment impact—confirmed by the hesitation in some more favored countries—will contribute to postponing indefinitely the possible exploitation of this resource.

6.3.7 Geothermal Resources

In the next decade, low temperature room heating by geothermy will be developed. High temperature geothermal resources are probably very limited.

Altogether, the new forms of energy (geothermal, wind, solar) represent only a few percent of the French energy forecast for 1985. A more intensive extension after 1985 will depend on financing requirements and environmental and technical constraints.

6.3.8 Waste Heat

The recovery of waste heat for district heating and industrial low temperature heat is an important possibility for the future, although in 1985 the technical and financial constraints will limit the contribution of this source to a few percent of the energy consumption.

6.4 CONSERVATION POLICY

While the government retains the goal of not reducing economic activity and trying to maintain the expected growth rate, it has issued an important program of rational energy use, including legal measures, financial incentives, and an information campaign.

Before the 1973 energy crisis, the projected primary energy demand for 1985 (approximately WAES Case E) amounted to 285 million tons of oil equivalent.

Without changing the growth rate, a new figure (Case C) of 240 million tons of oil equivalent has been projected.

The energy-saving measures must account, therefore, for about 45 million tons. The division of this saving will be one-third in industry and two-thirds in the residential and commercial sectors.

In industry, measures will include improving control equipment and eliminating factors contributing to energy waste; incentive measures will be offered for investment in energy-saving techniques, through loans and tax deductions. The time lag of these measures is evident since in many cases the new, less energy consuming equipment is installed mainly for reasons of extension or modification. In all cases the amortization of the invested money must be sufficiently short to induce the sponsored measures.

In new buildings, a law issued in 1974 requires a standard degree of thermal insulation. For existing buildings, tax incentive measures will encourage better thermal insulation. An information campaign leading to a closer survey of room heating controls will contribute to this effort.

New metering techniques (e.g., individual hot water meters) in new collective buildings will help to confront the user with the need for individual energy saving.

For industry as well as for the commercial sector, heat recuperation techniques are being investigated and brought increasingly into use.

Better use of materials (especially the recovery of old metals) will, by lessening the energy need for manufacturing primary products, have an impact on the energy demand.

In the transportation sector, efforts are being made to promote public transport, although in this field a forecast of the possibility of energy saving is very difficult because of the position of the private car in the citizen's life.

A special entity, the Committee on Energy Saving, has been created by the government to promote and control all the energy-saving measures. Where possible, voluntary efforts will have preference over compulsory requirements.

6.5 METHODOLOGY USED FOR THE DEMAND STUDIES

Two different methodologies have been used to investigate the five WAES scenarios for the French case. For scenarios C, D, and E (which are considered by public authorities to be the most probable alternatives), the relevant indications of the National Energy Commission, with the support of the Planning Office, have been used. For scenarios A and B, the investigations of the Grenoble Energy Institute have been applied. (For a brief description of the IEJE's methodology, see the appendix to this report.)

National policy response, for which WAES has adopted two alternatives—"vigorous" and "restrained"—was not considered under this aspect, since for countries with a relatively high dependence on foreign supply, such a differentiation must be approached very cautiously. This does not, however, impair the comparability of the French studies with others. Moreover, these expressions include a certain number of subjective considerations that differ from country to country.

The presentation of the results, as requested by WAES, does not yield a judgment of preference among the different forms of energy, since the final criteria are the global price (including fixed and variable costs and their expected evolution) that must be paid by the user over a given period and also subjective factors such as personal comfort.

In the tables, the processing losses should not hide the fact that for fossil fuel in final use, the efficiency coefficient is not achieved exactly and generally varies among the different uses. The electricity figures include all the processing losses from generators to the users' main connections, during which the efficiency is very favorable. This does not mean that we do not need to consider the use of electricity in certain circumstances very carefully and only with particular adjuncts (such as thermal insulation in domestic uses) to maintain this higher efficiency.

6.6 1985 DEMAND PROJECTIONS

6.6.1 Comments on the Demand Study Worksheets

Economic Assumptions

For the economic growth scenario descriptor, high and low growth rates of 5.5 and 3.5 percent per annum have been chosen. These values are the presently expected figures for the different cases and have been checked with the national planning authorities. The French population, which is presently 52 million, is expected to reach 54.5 million in 1985, according to present estimation.

Processing Losses

In Cases A and B, the processing losses include the energy used for the transportation of crude oil products, coal, and gas; the losses in the transformation processes themselves are smaller in Cases A and B than in Cases C, D, and E because of the more vigorous energy conservation.

In every case the processing losses in the electric sector include losses in the Carnot cycle (63 percent of the primary energy input) and those in the transportation and distribution network (10 percent of used electricity).

Feedstocks
Feedstocks are not included in the "total delivered energy" indicated in the worksheets (except coal for the steel industry). They are registered only in the explanatory notes to the worksheets.

Coefficients Used
1 TCE (ton of coal equivalent) = 7000 th
1 TOE (ton of oil equivalent) = 1.5 TCE
= 10,500 th = 44 x 10^9 J
1 th (therm) = 10^6 cal = 4.18 x 10^6 J = 3968 Btu
1 kWh = 0.861 th = 3.6 x 10^6 J.

6.6.5 Discussions of the Five Scenarios

Scenario A
The assumptions about the economic activity levels of the various demand sectors have been chosen to be very close to those of an "extension" scenario (which are, in turn, very close to those of Cases C and E), and the technical assumptions about specific consumption, equipment efficiencies, and technological development are very similar to those of Case B (a "contrasted" scenario), which are based on a more vigorous energy conservation policy.[2]

The main difference between the assumptions concerning economic activity levels in Case A and in an extension case such as C occurs in the transportation sector, since a large increase in motor fuel prices will involve two kinds of consequences:

- a global decrease in the need for transportation of passengers;

- more transportation of freight and passengers by rail and less transportation by road.

The primary energy demand (without feedstocks) has been evaluated at 220 million TOE (9600 x 10^{15} J). The feedstocks will amount to 15 million TOE (660 x 10^{15} J).

2. The expressions "extension" and "contrasted" are used in some French energy studies to summarize the general evolution globally. The "contrasted" scenario is the one used for developing WAES Case B.

The two main characteristics of the supply-demand integration in Case A are the assumptions of 40 GWe of nuclear generating capacity installed by 1985 and of 4 million TOE in new energy forms and heat recovery in 1985.

Scenario B
The demand study for Case B has been developed with the methodology elaborated by the University of Grenoble for long-term forecasting of energy demand. This case is very similar to one of the scenarios that has already been developed at Grenoble, at least as far as the basic assumptions (economic growth, price of oil, policy response) are concerned.

The primary energy demand will be 190 million TOE (8300 x 10^{15} J), to which must be added 12 million TOE for feedstocks (530 x 10^{15} J).

A very rough evaluation done with figures from the National Planning Commission leads to a result nearer 200 million TOE (8800 x 10^{15} J), with no difference for the feedstocks.

Given the necessity of ensuring an acceptable standard of living, greater constraints on and evolutions of social behavior are needed in Case B than in Case D (particularly in the area of transportation policy). There is also a need for more rapid changes in industrial structure toward low energy consuming industries.

The main characteristics of the supply-demand integration in Case B are assumptions of 35 GWe of nuclear generating capacity installed by 1985 and of 6 million TOE in new energy forms and heat recovery in 1985.

Scenario C
Case C is derived from the report of the National Energy Commission issued in 1975 by a team of experts representing all French energy sectors. The correlation between the figures of the report and the basic data of Case C has been checked.

The primary energy demand has been evaluated at 240 million TOE (10,400 x 10^{15} J), with 15 million TOE (660 x 10^{15} J) foreseen for feedstocks.

The main characteristic of Case C, in the French analysis, is the effort to develop nuclear energy and, consequently, its more important place (about 25 percent) in the satisfaction of the energy demand. Coal will maintain its relatively modest place, and coal extraction from national mines will not be greatly decreased. The use of oil will remain at the 1975 level but will be reduced as a percentage of the total energy supply. Finally, the new forms of energy (solar, wind, geothermal) will represent only a very small part of the supply (about 1 percent).

Scenario D
Case D differs from Case C in France by a reduction of about 10 percent in the energy demand. Basically the strategy will remain the same, underlined by a strong program of energy conservation.

The total demand will be 220 million TOE (9600 x 10^{15} J), with an additional 13.5 million TOE (600 x 10^{15} J) used for feedstocks.

Case D has been investigated by the National Planning Office as a separate alternative to Case C with a reduced growth. Despite a small reduction of dependence on outside sources due to lower oil and gas imports, there could be some greater social constraints arising from attempts to assure the highest possible employment rate despite a lower national income.

Scenario E
With the same economic growth rate as Cases A and C, Case E is characterized mainly by a decreasing oil price. The main orientation is therefore linked to higher oil demand and more restrained energy conservation policy.

The total expected energy demand of 285 million TOE (12,400 x 10^{15} J) will be satisfied in the following way:

62.7 percent by oil

6 percent by coal

12.3 percent by gas

19 percent by primary electricity (hydro and nuclear).

This last figure of 19 percent leads to a reduced electricity demand compared with Case C. In the same line, the demand for coal decreases from 30 million TOE in Case C to 17 million TOE.

With a demand for 178 million tons of oil the degree of dependence on outside sources will be higher than in the other scenarios (feedstock demand will add another 15 million tons).

The policy of promoting new forms of energy will be very restrained and will focus mainly on environmental aspects of energy problems.

Due to the time response of new solutions (especially nuclear) and the decisions already taken, Case E must be considered a hypothetical exercise for the French situation.

6.7 APPENDIX: A NONECONOMETRIC MODEL FOR LONG-TERM FORECASTING OF ENERGY DEMAND (Institut Economique et Juridique de l'Energie, Grenoble)

6.7.1 Objectives

The central objective for researching our energy forecasting model is to take into account the long-range consequences of current changes in the energy market that cannot appear in a satisfying way in traditional econometric models (because of the problem of measuring the long-range price elasticity of energy demand, the future relationship between economic growth and energy demand growth, etc.).

6.7.2 Methodology

The methodology we propose is based upon system analysis and the scenario method.

It seems indispensable, in the present context, to analyze in great detail all of the factors that determine the formation and evolution of the energy demand of a given economic system and the mechanisms of evolution of these factors. System analysis is very helpful for this task because it permits a rational and tidy approach despite the variety and complexity of the relationships established by the energy consumption, and also because it allows us to overcome the limitations of the traditional analytic methods by ensuring consistency between the evolution of the various analyzed factors and the evolution of the global economic, political, and social context.

The scenario method is called for because many futures are possible, resulting from alternative evolutions of the prices of primary energy sources, of economic growth, of industrial strategies, etc. The scenario method can be used to explore all reasonable possibilities while maintaining consistency among the assumptions used in defining and examining each possible future.

The methodology consists of four steps:

1. Set up the framework of the analysis; that is to

say, disaggregate the global socioeconomic system to obtain a set of subsystems in which the mechanisms of formation and evolution of demand can be analyzed in an aggregated way.

2. Look for and analyze in each subsystem the factors determining the level and structure of the demand, the factors determining its evolution, the mechanisms of evolution of all these factors (i.e., the relationships between them), and the relationships between the subsystems.

3. Define the scenario of evolution of the global socioeconomic system, setting assumptions both on the evolution of factors exogenous to our study (e.g., oil price) and also on the evolution of some endogenous determining factors and relationships (e.g., industrial strategies).

4. Run the model and quantify, for each scenario, the growth of energy demand by system and by type of energy.

6.7.3 Framework of the Analysis

The global system is disaggregated at three levels. At the first level the global system is broken down into three energy-consuming systems—urban, industrial, and transportation—one energy-supplying system, and three systems constituting the environment of the foregoing systems—scientific and technological, political, and ecological. Only the energy-consuming systems are analyzed in detail in the demand studies.

At the second level each consuming system is broken down into subsystems corresponding to groups of economic actors whose "energetical" behaviors are similar:

1. Urban: households (distributed in three categories of income), commercial and public sector, municipalities.

2. Industrial: heavy energy-consuming industries, distributed in nine branches (steel, aluminum, cement, glass, paper and pulp, chlorine, nitrogen, plastics, other chemical), and the rest of industry, distributed in three groups.

3. Transportation: seven groupings for freight and two for passengers.

At the third level each subsystem is again broken down into as many subsubsystems as there are different energy uses for these actors. For example, we have the following subsubsystems:

1. Urban/households: space heating, hot water, specific uses of electricity (light, refrigeration), cooking, urban transportation.

2. Industrial/one heavy energy-consuming industry: all the known technological processes of the branch and, for each process, the following uses: specific use of electricity, steam, oven, feedstocks; uses where electricity and fossil fuel are in competition.

3. Transportation/passengers: automobiles, buses, planes, traditional trains, high-speed trains.

6.7.4 Developing the Analysis

Each subsubsystem corresponding to a certain economic actor and a certain use of energy is analyzed in three steps: first a static analysis, then a dynamic analysis, and finally an analysis of the relationships.

Static Analysis

First, we put on line the "primary need" corresponding to the energy demand (e.g., need for a minimum temperature in the domestic heating case). This primary need involves a requirement for useful energy whose level depends not only on the primary need itself but also on the technological background (e.g., for a given inside temperature, the requirement for useful energy, which is to say the level of heat losses, depends on the insulation standard). In its turn, the requirement for useful energy depends on the final energy demand, whose level and kind depend also on technological constraints (e.g., the efficiency of the equipment).

Dynamic Analysis

The evolution of the energy demand results from the evolution of the primary needs and from the evolution of the technological determining factors.

The evolution of the primary needs depends on the evolution of the global system (demography, economic growth rate, etc.) and also on the evolution of the behaviors of the relevant economic actors (e.g., the relationship between the development of automobiles and the evolution of the need for transportation).

The evolution of the technological factors is determined by the choices made by the economic actors (e.g., in changing or retrofitting old equipment or installing new equipment). These choices depend on the economic behavior postulated for the actors (represented by the socioeconomic determining factors) and, sometimes, on the policies or strategies chosen by some of them (represented by the political determining factors related to the decision makers).

Relationships
It is through the relationships between the determining factors that the energy demand is quantified for a given year in the static analysis, while it is these relationships that transmit the evolution of a given factor to the other factors until the final demand is reached in the dynamic analysis.

Other kinds of relationships—economic, physical, and political—exist between subsystems and between the energy-consuming systems. These ensure the consistency of the global system in evolution. They represent the data input of a subsystem and they transmit the influence of the evolution of each system to all the others.

6.7.5 Defining the Scenario

A scenario consists of all the assumptions about factors and relationships whose evolution cannot be extrapolated from past values and which can therefore evolve in very different ways. The method of the analysis and the necessity for consistency led us to build scenarios in which the relationships of dependency among most of the determining factors are explicitly taken into account. We have thus structured the scenarios as hierarchies with four levels of assumptions, each level depending on the assumptions at the next higher level.

1. At the top level are assumptions on the inputs of the global system (oil price, demography, economic growth, etc.).

2. At the second level is the definition of the type of society that will evolve during the period of the projection and thus of the main characteristics of permissible government policy.

3. At the third level are assumptions on the relationships between the energy-consuming systems and their environment: the technological background, scientific discoveries, ecological constraints, conditions of supply of final energy (prices, taxes, levels of demand, resources), etc.

4. At the lowest level are assumptions on determining factors and the relationships between energy-consuming systems: strategies for locating the steel industry, evolution of specific gas consumption in automobiles, etc.

6.7.6 Running the Model

The model has not yet been computered, and we are now only "hand-running" it in two scenario cases for the years 1985 and 2000 (Figure 6.1). One scenario is an extension of past economic trends (relatively fast economic growth, slowly increasing energy prices, no major changes in the social fabric), while the other contrasts sharply with these trends (slow economic growth, large jumps in energy prices, profound changes in society and in government policy).

The output of the model is the final energy demand distributed in four kinds of final energy—electricity, heating fuels, steam and hot water (district heating), and motor fuels—and three systems—urban, industrial, and transportation. The final energy demand of rural households (including energy for agriculture), which represents about 10 percent of the global demand, has been very roughly studied and quantified to allow comparisons with the results of other methods.

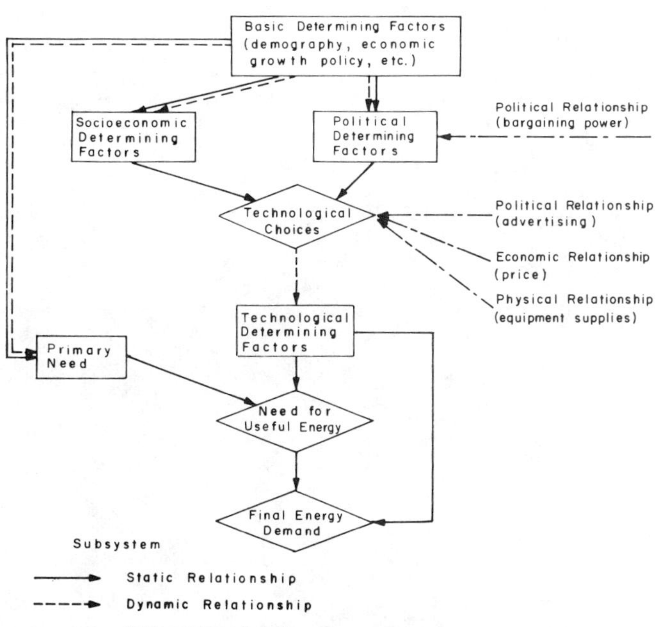

Figure 6.1 General scheme of analysis of the IEJE forecasting model.

6.8 DEMAND STUDY WORKSHEETS

In this section are included the summary tables for the French energy demand study in the following order:

Energy Use Summary

Sectoral Percentages of Total Final Energy Demand

Energy Growth Rates Summary

Summary sheets for WAES Cases A–E.

WAES NATIONAL ENERGY DEMAND STUDIES, 1972-1985
ENERGY USE SUMMARY

Country FRANCE

Scenario Assumptions

WAES Case	A		B		C		D		E			
GNP 1976-1985	High		Low		High		Low		High			
Oil Price*	\$11.50 → \$17.25		\$11.50 → \$17.25		\$11.50		\$11.50		\$11.50 → \$7.66			
Natl. Policy Response	Vigorous		Vigorous		Vigorous		Restrained		Restrained			
Natl. GNP Growth Rate 1976-1985 (%/yr.)	5,5 %		3,5 %		5,5%		3,5 %		5,5 %		1972 Base Year	
Fuel Type	Fossil Fuel	Elec.	Fossil Fuel	Elec.	Fossil Fuel	Elec.	Fossil Fuel	Elec.	Fossil Fuel	Elec.	Fossil Fuel	Elec.
Units 10^{15}J.												
Energy-Using Sectors												
Industrial (excluding feedstocks)	1 672	526	1 364	425	2 347	522	2 126	472	2 699	580	1 763	274
Transportation	1 848	29	1 408	32	1 898	40	1 770	32	2 405	43	1 217	25
Commercial and Public	484	227	528	212								
Residential	1 760	378	1 892	266	2 200	616	2 109	568	2 787	684	1 913	178
Fishing, Agriculture, Mining, Construction 1)												
Feedstocks												
Final Energy Demand (excluding processing losses)	5 764	1 160	5 192	935	6 445	1 178	6 005	1 072	7 891	1 307	4 893	477
Processing Losses	332	2 276	290	1 784	364	2 356	344	2 144	516	2 614	402	960
Primary Energy Input	6 096	3 436	5 482	2 719	6 809	3 534	6 349	3 216	8 407	3 921	5 295	1 337

*Oil price 1975 US\$ per barrel Arabian light crude oil FOB Persian Gulf.

1) Fishing and Agriculture are included in Residential. Mining and Construction are included in Industrial.

WAES NATIONAL ENERGY DEMAND STUDIES, 1972-1985

Sectoral Percentages of Total

Country FRANCE

Scenario Assumptions

WAES Case	A		B		C		D		E			
GNP 1976-1985	High		Low		High		Low		High			
Oil Price*	$11.50 → $17.25		$11.50 → $17.25		$11.50		$11.50		$11.50 → $7.66			
Natl. Policy Response	Vigorous		Vigorous		Vigorous		Restrained		Restrained			
Natl. GNP Growth Rate 1976-1985 (%/yr.)	5,5 %		3,5 %		5,5 %		3,5 %		5,5 %		1972 Base Year	
Fuel Type	Fossil Fuel	Elec.	Fossil Fuel	Elec.	Fossil Fuel	Elec.	Fossil Fuel	Elec.	Fossil Fuel	Elec.	Fossil Fuel	Elec.
Units												
Energy-Using Sectors												
Industrial (excluding feedstocks)	29	45	26	45	36	44	36	44	34	44	36	57
Transportation	32	3	27	3	29	3	29	3	30	3	25	5
Commercial and Public	8	20	10	23								
Residential	31	32	37	29	35	53	35	53	36	53	39	38
Fishing, Agriculture, Mining, Construction 1)												
Feedstocks												
Final Energy Demand (excluding processing losses)	100	100	100	100	100	100	100	100	100	100	100	100
Processing Losses												
Primary Energy Input												

*Oil price 1975 US$ per barrel Arabian light crude oil FOB Persian Gulf.

1) Fishing and Agriculture are included in Residential. Mining and Construction are included in Industrial.

WAES NATIONAL ENERGY DEMAND STUDIES, 1972-1985
ENERGY GROWTH RATES SUMMARY

Country FRANCE

Scenario Assumptions

WAES Case	A		B		C		D		E	
GNP 1976-1985	High		Low		High		Low		High	
Oil Price*	\$11.50 → \$17.25		\$11.50 → \$17.25		\$11.50		\$11.50		\$11.50 → \$7.66	
Natl. Policy Response	Vigorous		Vigorous		Vigorous		Restrained		Restrained	
Natl. GNP Growth Rate 1976-1985 (%/yr.)	5,5 %		3,5 %		5,5 %		3,5 %		5,5 %	
Fuel Type	Fossil Fuel	Elec.	Fossil Fuel	Elec.	Fossil Fuel	Elec.	Fossil Fuel	Elec.	Fossil Fuel	Elec.
Units (%/yr. 1972-1985)										
Energy-Using Sectors										
Industrial (excluding feedstocks)	- 0,4	+ 5,1	- 1,6	+ 3,4	+ 2,2	+ 5,1	+ 1,5	+ 4,3	+ 3,3	+ 5,9
Transportation	+ 3,3	+ 1,1	+ 1,1	+ 1,9	+ 3,5	+ 3,7	+ 2,9	+ 1,9	+ 5,4	+ 4,3
Commercial and Public 1)										
Residential	+ 1,2	+ 9,9	+ 1,8	+ 7,9	+ 1,1	+10,0	+ 0,7	+ 9,3	+ 2,9	+10,9
Fishing, Agriculture, Mining, Construction 2)										
Feedstocks										
Final Energy Demand (excluding processing losses)	+ 1,3	+ 7,1	+ 0,5	+ 5,3	+ 2,1	+ 7,2	+ 1,6	+ 6,4	+ 3,7	+ 8,1
Processing Losses										
Primary Energy Input										

*Oil price 1975 US\$ per barrel Arabian light crude oil FOB Persian Gulf.

1) Commercial and Public are included in Residential.
2) Fishing and Agriculture are included in Residential. Mining and Construction are included in Industrial.

NATIONAL ENERGY DEMAND PROJECTIONS

Country FRANCE

Year 1 985

WAES Case A

Summary 10^{15} J.

Category	Oil	Gas	Coal	Electricity	Other	Total by Category (Excluding Processing Losses)
Transportation	1 848			29		1 877
Agriculture, Mining, and Construction 1)						
Commercial 2)						
Public 2)						
Industrial	440	572	660	526		2 205
Residential	1 100	968		605	176	2 849
Total Delivered Energy by Fuel	3 388	1 540	660	1 160	176	6 931
Processing Losses by Fuel	272	60		2 276		
Gross Input by Fuel	3 660	1 600	660	3 436	176	9 531

1) Agriculture is included in Residential. Mining and Construction are included in Industrial.
2) Commercial and Public are included in Residential.

Gross Energy Input (Add across bottom row)

NATIONAL ENERGY DEMAND PROJECTIONS

Country FRANCE

Year 1 985

WAES Case B

Summary 10^{15} J.

Category	Oil	Gas	Coal	Electricity	Other	Total by Category (Excluding Processing Losses)
Transportation	1 408			32		1 440
Agriculture, Mining, and Construction 1)						
Commercial 2)						
Public 2)						
Industrial	396	396	572	425		1 789
Residential	1 100	1 056		479	264	2 899
Total Delivered Energy by Fuel	2 904	1 452	572	936	264	6 128
Processing Losses by Fuel	232	58		1 784		
Gross Input by Fuel	3 136	1 510	572	2 720	264	8 202

1) Agriculture is included in Residential. Mining and Construction are included in Industrial.
2) Commercial and Public are included in Residential.

Gross Energy Input (Add across bottom row)

NATIONAL ENERGY DEMAND PROJECTIONS

Country FRANCE

Year 1 985

WAES Case C

Summary 10^{15}J.

Category	Oil	Gas	Coal	Electricity	Other	Total by Category (Excluding Processing Losses)
Transportation	1 898			40		1 888
Agriculture, Mining, and Construction 1)						
Commercial 2)						
Public 2)						
Industrial	719	748	880	522		2 869
Residential	1 184	748	132	616	132	2 812
Total Delivered Energy by Fuel	3 801	1 496	1 012	1 178	132	7 619
Processing Losses by Fuel	304	60		2 356		
Gross Input by Fuel	4 105	1 556	1 012	3 534	132	10 339

1) Agriculture is included in Residential. Mining and Construction are included in Industrial.
2) Commercial and Public are included in Residential.

Gross Energy Input (Add across bottom row)

NATIONAL ENERGY DEMAND PROJECTIONS

Country FRANCE

Year 1 985

WAES Case D

Summary 10^{15}J.

Category	Oil	Gas	Coal	Electricity	Other	Total by Category (Excluding Processing Losses)
Transportation	1 770			32		1 802
Agriculture, Mining, and Construction 1)						
Commercial 2)						
Public 2)						
Industrial	704	660	762	472		2 598
Residential	1 141	704	132	568	132	2 677
Total Delivered Energy by Fuel	3 615	1 364	894	1 072	132	7 077
Processing Losses by Fuel	289	55		2 144		
Gross Input by Fuel	3 904	1 419	894	3 216	132	9 565

1) Agriculture is included in Residential. Mining and Construction are included in Industrial.
2) Commercial and Public are included in Residential.

Gross Energy Input (Add across bottom row)

NATIONAL ENERGY DEMAND PROJECTIONS

Country FRANCE

Year 1 985

WAES Case E

Summary 10^{15} J.

Category	Oil	Gas	Coal	Electricity	Other	Total by Category (Excluding Processing Losses)
Transportation	2 405			43		2 448
Agriculture, Mining, and Construction 1)						
Commercial 2)						
Public 2)						
Industrial	1 379	704	616	580		3 279
Residential	1 951	704	132	684		3 471
Total Delivered Energy by Fuel	5 735	1 408	748	1 307		9 198
Processing Losses by Fuel	460	56		2 614		
Gross Input by Fuel	6 194	1 464	748	3 921		12 327

1) Agriculture is included in Residential. Mining and Construction are included in Industrial.
2) Commercial and Public are included in Residential.

Gross Energy Input (Add across bottom row)

7.1 SUMMARY

This study contains empirical energy consumption data for the base year 1972 and five sets of data for the 1985 energy demand scenarios. It uses the WAES methodology to derive the five scenarios for the 1985 energy demand of 69 line items on the basis of 1972 energy demand. WAES has chosen three descriptors to be used in developing the scenarios: economic growth, oil price, and national policy response.

The assumptions used to determine values for the descriptors in the case of the Federal Republic of Germany are as follows:

1. High and low national economic growth rates have been set at 4.5 and 2.5 percent per annum.

2. Oil price is assumed to determine the energy price level and the price relations between energy sources. Alterations of oil prices cause alterations in oil product prices and adaptation in gas, coal, and electricity prices in this order with decreasing strength.

3. National policy response on the demand side focuses on the conservation issue. Conservation measures that are currently in parliamentary process or that have already been enacted are taken to comprise restrained policy. A vigorous policy can, under the given case assumptions, be only slightly more restrictive.

The study shows that even under the most favorable circumstances (Case E), energy demand does not grow at the same rate as in the past. This applies to the demand for electricity as well as for fossil fuels.

One reason for this is the rather high level of independence of energy demand from economic growth in several sectors (all nonindustrial sectors in the case of fossil fuel demand) and the saturation phenomenon in some sectors (transportation, residential).

Low energy prices, moreover, are not believed to provide any considerable incentive to increase energy demand. Even in the lower or constant oil price cases, the trend to lower specific consumptions that can be seen in most demand categories today is not assumed to turn around.

On the whole, it is the rather large difference between the high and low assumed economic growth rates that is the most sensitive determining factor for energy demand.

Finally, processing losses are believed to undergo an interesting development under the assumed circumstances. They are reduced considerably by the use of district heating, particularly in the high oil price case, but they are increased by the higher share of electricity.

7.2 INTRODUCTION

7.2.1 Authorship

This demand study was compiled at the Institute of Energy Economics at Cologne by Dr. Dieter Schmitt and Paul H. Suding, who are responsible for its contents. Dr. Georg Klotmann of BASF has contributed comments regarding the development of the chemicals industry.

7.2.2 References

This quantitative and empirical study has relied heavily on published data, analyses, and forecasts done by experts in the various fields touched upon during the course of the work.

Base Case

(1) Arbeitsgemeinschaft Energiebilanzen (Ed.), Energiebilanzen der Bundesrepublik Deutschland, Frankfurt.

(2) Bundesministerium Für Wirtschaft III D 3, Daten zur Entwicklung der Energiewirtschaft in der Bundesrepublik Deutschland, Bonn-Bad-Godesberg, Dezember 1974.

(3) Der Mineralölmarkt in der Bundesrepublik 1970-1974, in: Esso Magazin, 2/75, pp. 14–15.

(4) Die westdeutsche Mineralölwirtschaft 1972, in: OEL, Juni 1973, pp 175–178.

(5) Holthausen, D., Göpner, W., Lumpe, W., Die Entwicklung der Gaswirtschaft in der Bundesrepublik Deutschland im Jahre 1972, in: gwf, gas-erdgas, Heft 9, September 1973, pp. 425–454.

(6) Statistisches Bundesamt (Ed.), Statistisches Jahrbuch für die Bundesrepublik Deutschland, Mainz.

(7) Rieke, H., Die künftige Entwicklung des Straßenverkehrs in der BRD, DIW-Beiträge zur Strukturforschung, Heft 22-1972, Berlin, 1972.

(8) Kraftfahrt-Bundesamt, Bestand an Kraftfahrzeugen und Kfz-Anhängern am a. Juli 1972, Bonn-Bad-Godesberg, 1973.

(9) Frerichs, J. Sarazin, Th., PKW-Prognose 1985, in: Straßenban und Straßenverkehrstechnik, Hrsg. vom Bundesministerium für Verkehr, Abt. Straßenban, Heft 169, Bonn, 1974.

(9a) Frerichs, J., Nitsche, B., Sarazin, Th., LWK-Prognose 1985, in gleicher Reihe, Heft 168, Bonn, 1974.

(10) Wieczorek, S., Winterholler, G., Der Energieverbrauch im Verkehr, ifo-Studien zur Verkehrswirtschaft 7, München, 1975.

(11) Statistisches Bundesamt, Fachserie H, Verkehr, Reihe 5, Straßenverkehr II, Personenverkehr.

(12) dto., Reihe 3, Luftverkehr.

(13) dto., Reihe 1, Binnenschiffahrt.

(14) dto., Reihe 2, Seeschiffahrt.

(15) dto., Fachserie F, Groß- und Einzelhandel, Gastgewerbe, Reiseverkehr, Reihe 8, Reiseverkehr IV, Grenzüberschreitender Reiseverkehr.

(16) IATA (Ed.), World Air Transport Statistics 1974, Genf, 1974.

(17) Seidel, B., Untersuchung der Vorleistungsstrukturen von Produktgruppen des Bereichs Landwirtschaft, Forstwirtschaft und Fischerei in der BRD 1962 und 1966, in: DIW-Vierteljahresheft zur Wirtschaftsordnung, Heft 4/1971, pp. 301 ff.

(18) O. V. Direkter und inderekter Einsatz von Energie in der Bundesrepublik Deutschland im Jahre 1972, in: DIW-Wochenbericht 1-2/74, pp. 1 ff.

(19) Filip-Köhn, R., Input-Output-Tabelle für die BRD 1967, in: DIW-Vierteljahres-Hefte zur Wirtschaftsforschung, 1974, pp. 69-73.

(20) Batelle, Der derzeitige Stand und die künftige technisch-wirtschaftliche Entwicklung auf den Märkten für Heizung, Klimatisierung und Warmwasserversorgung, Frankfurt, 1971 (unpublished).

(21) Lüttringhaus, G., Energieverbrauch der privaten Haushalte in den Jahren 1972 und 1973, Köln, 1975 (unpublished).

(22) Information Erdgas, Daten Fakten Fälle 74/2, Essen, 1974.

(23) Esso, Esso Studie 5, So heizt der Deutsche, Hamburg, 1971.

(24) Statistisches Bundesamt (Ed.), Die Fläche der Wohnungen, in: Wirtschaft und Statistik 1/75, p. 34.

(24a) dto. Ergebnisse der 1%-Wohnungsstichprobe, in: Wirtschaft und Statistik 7/74, p. 504.

(25) Dotzenrath, W., Die elektrizitäswirtschaftliche Bedeutung von Heizstromlieferungen bei der Entwicklung des Haushaltsstromverbrauchs, Düsseldorf, 1974.

(26) HEA, HEA-Statistik, Frankfurt.

Projections

(27) Bülow, H., Eckerle, K., Entwicklungstendenzen der für die Energienachfrage relevanten Strukturelemente von Wirtschaft und Gesellschaft, Basel, 1975 (unpublished).

(28) Behrndt, E., Der Strombedarf der Haushalte in der BRD bis 1980, in: BWK 5/71, pp. 253-256.

(29) Schmitt, D., Mönig, W., Suding, P. H., Der Elektrizitätsbedarf der Bundesrepublik Deutschland bis zum Jahre 2000, Köln, 1975 (unpublished).

(30) HEA, Überlegungen zur künftigen Entwicklung der elektrischen Wohnungsbeheizung in der BRD, Frankfurt, 1974.

(31) O. V. Die voraussichtliche Entwicklung des Personenverkehrs in der BRD bis 1990, in: DIW-Wochenbericht Nr. 34/75.

(32) Deutsche Shell AG., Die Motorisierung geht weiter, Prognose des PKW-Bestandes, Hamburg, 1975.

(33) Schmitt, D., Der Energiebedarf der Bundesrepublik Deutschland, Zwischenbericht zur Studie: Technologien zur Einsparung von Energie, Köln, 1976 (unpublished).

(34) Ballensiefen, M., Zu den Absatzaussichten der deutschen Automobilindustrie bis 1985, in: RWI-Mitteilungen 1975/1, 26. Jahrgang, pp. 1-23.

(35) Aule, O., Braun, H., Heinrich, J., Munk, E., Rationalisierung aus der Sicht der Baubeteiligten, in: ifo-Schnelldienst 5, 13, Februar 1976, 29. Jahrgang, pp. 7-10.

(36) Koch, K., Energie als Kostenfaktor in der westdeutschen Wirtschaft, in: Vierteljahreshefte zur Wirtschaftsforschung des DIW, 1972, pp. 203-213.

(37) Korthaus, E., Die Ausfuhr der Bundesrepublik Deutschland nach acht Abnehmerregionen. Ökonometrische Erklärung und Modellrechnung. RWI-Mitteilungen 1975/2, Jahrgang 26, pp. 98 ff.

(38) Berden, W., Veränderungen der wirtschaftlichen und technischen Produktionsbedingungen in der Eisen- und Stahlindustrie der Bundesrepublik Deutschland und ihre Auswirkungen auf den Energieverbrauch bis 1985. Diss. Köln, 1975.

(39) Karrenberg, B., Die Anpassungsfähigkeit der Industrie an Veränderungen auf dem Energiemarkt, in: RWI-Mitteilung 1975/3, 26. Jahrgang, pp. 199-217.

(40) Birg, H., Analyse und Prognose der Bevölkerungsentwicklung in der Bundesrepublik Deutschland und ihren Regionen bis zum Jahr 1990, DIW-Beiträge zur Strukturforschung, Heft 35-1975, Berlin.

(41) Munck, E., Ruderer, I., Ifo-Bauvorausschätzung, München, Oktober 1974.

(42) Dotzenrath, W., Das EVU als Dienstleistungsunternehmen für den Verbraucher, in: Elektrizitätswirtschaft 1/1975, pp. 2-7.

(43) Krelle, W., et al., Ein Prognosesystem für die wirtschaftliche Entwicklung der Bundesrepublik Deutschland, Meisenheim am Glahn, 1969.

(44) Krelle, W., Erfahrungen mit einem ökonometrischen Prognosemodell für die Bundesrepublik Deutschland, Meisenheim am Glahn, 1974.

(45) Wirtschafts- und Sozialwissenschaftliches Institut der BGB, Vorausschätzung eines gesamtwirtschaftlichen Entwicklungsrahmens, Abschußbericht eines Teilgutachtens im Rahmen des Forschungsprogramms des Bundesministeriums für Forschung und Technologie, Künftiger Bedarf an elektrischer Energie usw. Düsseldorf, März 1975 (unpublished).

7.2.3 Data Type

This study contains both empirical data for the base case 1972 and five sets of data for the year 1985.

The degree of disaggregation required in the WAES demand study is not equaled by any other German study, particularly in the residential, public, commercial, construction, mining, and agriculture sectors. As far as these sectors are concerned, therefore, the base case data are partly the results of the authors' own calculations based mostly on input-output tables. Although the authors feel rather uncertain about them, the numbers are given to keep this study in line with the common WAES approach.

The data for 1985 are based on scenario assumptions. They must be considered in relation to base case values because some variables can be measured only by indicators, so that the relative changes in value between 1972 and 1985 are relevant.

7.3 METHODOLOGY

7.3.1 The WAES Scenarios and the Approach of This Study

This demand study uses the WAES methodology to derive five different scenario cases for the energy demand of 69 demand sectors in the year 1985. WAES has chosen three descriptors to write these scenarios and has developed five global pictures using these descriptors. Each team must consequently draw its national scenarios against the global background. It must determine values for the general scenario descriptors as well as corresponding numbers for the activity levels and the energy coefficients.

The scenario descriptors chosen by WAES are economic growth, oil price, and national energy policy. These descriptors are not mutually independent, although the scenario approach seems to deny the relationships by combining different values for one variable with the same value for another (e.g., in Case A a high oil price is combined with a high economic growth rate whereas in Case E the same high economic growth rate is assumed to evolve under conditions of low oil prices).

The same phenomenon appears in the national scenarios. As a word of clarification, it must be stressed that the relationships between economic growth, oil price, and national energy policy are realized and that the apparent neglect is caused by the scenario technique adopted and the uncertainty about the nature of the relationships. Therefore the reader must not draw any conclusion about the feasibility, for example, of the coincidence of a rising oil price with a high economic growth rate. However, there were a priori, simple plausibility considerations which suggested that these cases could not be totally excluded as not being feasible.

A variety of methods and tools have been used because a unified modeling system for all purposes arising in this study does not exist. The inquiry into the effects of oil price and the rate of growth of world product on the German economy was aided by an econometric macromodel, the "Bonner Modell"; the estimation of the activity levels is based on several studies, either on the forecasts of the PROGNOS-AG and its semieconometric approach or on other studies specializing in certain sectors. For estimations of the energy coefficients, the studies at the Energiewirtschaftliches Institut and other trend and regression analyses were used.

7.3.2 World and National Economic Growth

In determining high and low economic growth rates for the German scenarios with regard to the world and European picture, one has to take into account the fact that the economy of the Federal Republic of Germany is highly dependent on foreign trade.

In order to investigate the sensitivity of German GDP growth to world growth changes, runs were made with the Bonner Modell in its aggregated version. The assumption was made that the relationship between GWP growth and the growth of world trade volume observed in the past would persist in the future. A 5 percent growth in world trade per year for the period 1976 to 1985 results under this assumption in a 2.7 percent GDP growth rate for Germany, while a 9 percent growth in world trade results in a 3.2 percent GDP growth rate. The model suggested in the high case a steep increase of the inflation rate but only a small increase of real growth because of limitations on the production potential in the Federal Republic of Germany.

Other forecasts of the growth of the German economy have come up with higher as well as lower estimates, so that the results of the trial runs of the Bonner Modell do not represent the upper and lower limit of probable economic growth. Looking at the past relationship between world and F.R.G. economic growth, an elasticity coefficient for 1962 to 1972 of about 0.87 can be observed (GWP 5.4 percent growth per year and GDP 4.7 percent). Using this coefficient, the rate of growth of GDP from 1977 to 1985 would be 5.2 and 3.0 percent per year in the high and low growth cases, respectively. Taking into account the slight decrease in population and the expected lower labor productivity on the one hand and an economic policy that does not permit overly low growth rates on the other hand, GDP growth rates of 4.5 and 2.5 percent per year from 1977 to 1985 are sensible as upper and lower bounds for plausible growth and have therefore been chosen for the use in the German growth scenarios (see Table 7.1).

7.3.3 Oil Price, Economic Growth, and Energy Prices

It is the WAES convention to use only two values for the economic growth scenario descriptor and

Table 7.1
Actual and forecasted growth of real GDP (% p.a.)

						1976–1985[1]	
	1962–1972	1973	1974	1975	1976[1]	High	Low
World	5.4					6.0	3.5
Europe		6.0	2.0	–2.0	3.0	5.0	3.0
F.R.G.	4.7	6.0	0.4	–4.2	5.0	4.5	2.5

1. Forecasted or scenario growth rates.

not to vary the value with respect to different oil prices.

Despite the prevailing hypothesis that oil prices have a considerable impact on total economic growth, the methodology requires us to neglect those effects. Therefore, it is not the total growth but the structure of the economy or, in the terminology of this study, the "activity levels" of the demand categories that are affected by oil price.

The term "oil price" refers to crude prices. The energy price level, which is a composition of prices of end-use energy sources, does not develop proportionally to crude prices. The production cost of oil products stems partly from transport and processing costs, which do not vary as a function of crude prices. Since market forces do not allow for considerably higher profits, the oil products prices are assumed to rise more slowly in the case of higher crude prices. Gas prices as well as coal prices are assumed to undergo an adaptive development to the price leader, oil, in all substitutive markets. In the case of electricity, on the one hand the transformation inputs (oil, gas, and coal) are rising in price and pushing the production cost, thereby increasing the pressure for higher electricity prices; on the other hand, although the substitution potential for electricity is limited in certain markets (household appliances, industrial processes), the direct price elasticity must be considered as effective. Thus all cost factors (uranium, fuel fabrication, investment goods) that might try to profit from the improvement of the market situation of electricity as compared to other energy sources gain only limited success. Electricity prices are therefore the least affected prices in the case of increasing oil prices.

For this reason, the energy price level, after all, is assumed to rise in WAES Cases A and B only about half as much as oil prices.

In Case E, the price situation cannot be assumed to be symmetrical to Cases A and B around the axis Case C. Lower oil prices in real terms lead to slightly lower oil product prices in real terms compared to the stable situation in Cases C and D. Gas is assumed to be slightly cheaper than oil. Indigenous coal prices cannot be flexible downwards, and electricity prices may, under these conditions, remain about stable.

The energy price level in Case E is assumed to be considerably higher than the two-thirds of the 1975 price level that is taken as the 1985 oil price level.

7.3.4 National Policy Response

The national policy response scenario descriptor has two possible states: "restrained" and "vigorous." The content of these alternative policies must be defined by each of the national teams. The policy descriptor is not independent from the oil price and economic growth descriptors. WAES has agreed on a definition that explicitly takes this into account by referring to "policy . . . which can be expected under the price and economic assumptions of the case."

Though this literally means five different policy assumptions, since there are five different economic growth and oil price combinations assumed, only two different policies have to be defined: (1) a vigorous policy which fits in with the two high oil price cases (A and B) and the medium oil price, high economic growth case (C); and (2) a restrained policy in line with the medium oil price, low growth case (D) and the low oil price, high economic growth case (E). In fact, this means the same policies with different reasons and arguments for the different backgrounds.

The government response is supposed to include only those policy measures that find their reason in energy policy. Measures that belong to stabilization or regional policy or other nonenergy policies are, therefore, not taken into account. Also not taken into account are the results of international conservation efforts, since the individual national considerations are unknown.

The main principle in German energy policy, which will persist until 1985 in either vigorous or restrained cases, is the presumption that prices determined by market competition are the decisive means to ensure economical and therefore rational energy consumption, but that this allocation mechanism alone does not meet the needs of the economy as a whole. Market imperfections such as lack of transparency and short planning horizons in particular must be removed by information and research and development policies. For estimating a least and strongest response, this is the basic starting point.

In the context of energy demand, the main policy issue in the F.R.G. is conservation. Several laws and bills involving energy conservation have already been enacted or are in parliamentary progress. They allow for some conservation measures by decree. These policies must be taken as the legal framework for the policy scenarios in the F.R.G. They have been developed with the common understanding that conservation measures must be practicable and that in general the private discounted benefits gained by conservation must exceed the private costs. Even the rise in oil price assumed in Cases A and B seems not to be high enough to produce any considerable change in this legal framework up to 1985. The main goal in conservation policy will therefore be the establishment of long-range energy-cost rationality in the private sector. Thus the differences between restrained and vigorous policies will consist mainly of adjustments of the measures to exogenous factors such as oil prices. In addition, the interests of society may require some changes that are endogenous to national policy, such as tax allowances.

A list of considered measures for the residential, public, and transportation sectors is given in Table 7.2. The tax allowance and depreciation measures in the residential sector are extensions of the principle of private profitability; in the absence of mandatory measures, they would probably have higher individual effects.

In most parts of the transportation sector, a non-profit-maximization behavior similar to that manifested in the residential sector is assumed. Therefore the main goal will be to guide consumer behavior to a long-term least-cost orientation.

A change in energy consumption can be caused by the introduction of land-use measures and, in particular, by reorganization of the transportation system. However, the main reasons for such measures—for example, those of the "Deutsche Bundesbahn"—will not be energy conservation. Some measures might even lead to higher energy consumption. Moreover, some considered measures, such as the replacement of individual transport by public transport, are not likely to occur before 1985 under the relevant case circumstances.

Since the industrial and commercial sectors are assumed to make decisions on the basis of profit

Table 7.2
Components of restrained and vigorous energy policy responses

Policy Type	Restrained	Vigorous
Residential		
Limitation of the heat-flow coefficient for new houses	maintain present standard (DIN 4108 Beiblatt)	lower coefficient
Requirements regarding the construction and use of heating systems	apply to new houses only	apply to both old and new houses
Control and maintenance of heating systems, old and new houses	require occasional maintenance	require frequent maintenance
Modernization programs for heating systems in old homes	aim for lower effect	aim for higher effect
Income tax allowance for owner-occupied new housing (§ 76 EStG)	actual	increase with conservation effect
Depreciation allowance for owner-occupied housing (§ 82 a EStDV)	actual	increase with conservation effect
Depreciation on conservation investments in rental housing	not allowed	extraordinary
District heating	no change in legal framework	create obligation for heat suppliers and monetary incentives
Public		
Control of energy usage in public property	no additional measures	reorganization and intensification
Transportation		
Speed limits (now 130 km/h on the autobahn)	establish guidelines	establish legal limit
Insurance of passengers in private cars	leave voluntary	make obligatory
Control and maintenance of ignition and carburetor systems	no policy	introduced

maximization, they need not be guided like the residential and transportation sectors. As for possible conservation measures in these sectors, they will probably not differ very much under "restrained" or "vigorous" government policy. They will consist mainly of R&D programs and concomitant information campaigns. With rising awareness of the potential for energy saving, the profit motive will lead companies to introduce processes and organizational techniques that will, under the case circumstances, be the most profitable.

7.3.5 Major Activity Levels

The activity levels for the five scenarios are results of dynamic economic and social developments simulated under different scenario conditions. The relationships between activity levels and scenario descriptors, therefore, ought to differ somewhat from case to case due to the changing environment. To demonstrate the underlying relationships several appropriate coefficients are employed. These coefficients were not used for the calculation of the activity levels, which was done by modeling techniques and by reference to the literature, but they may be helpful in showing the underlying assumptions and the rather judgmental reaction estimates.

The sectoral growth coefficient X_{ij}, defined by

$$(A_{ij,72} / A_{ij,85}) = (GDP_{72} / GDP_{85})\, X_{ij} ,$$

where

A = activity level,

GDP = Gross Domestic Product,

i = Sector i,

j = Case j,

shows the growth of a sector or demand category in a particular case in relation to overall economic growth. If this coefficient is less than unity, the sector activity grows underproportionally; if it is greater than unity, it grows overproportionally. This coefficient does not say anything about the interdependence of sectoral and overall growth. When in different growth scenario cases (e.g., C and D), there appear in one demand category very different coefficients, this means that the interdependence of sectoral and overall growth is not very close. To demonstrate this tightness, another coefficient is calculated, the deviation coefficient, d_i:

$$d_i = X_{im} / X_{in} ,$$

where m and n are different cases.

In some cases, the growth impact of oil prices on activity levels is also considered. This can be measured by the oil price–sectoral growth coefficient β_{imn}:

$$(A_{im} / A_{in}) = (P_m / P_n)\, \beta_{imn} ,$$

where P is oil price.

Industry

Table 7.3 shows the growth of industrial subsectors in relation to overall growth. Case C represents, with small deviations, Cases A, C, and E, and Case D represents Cases D and B. The total industrial growth elasticity is close to unity in all cases. This means that the industrial fraction of the economy in terms of GDP is assumed to be the same in all cases and that the industrial growth rate equals the overall growth rate.

The large differences between the industrial subsector coefficients indicate a change in the industrial structure. The deviation coefficient can be interpreted by analogy to regression analysis as an error term showing the importance of factors other than overall growth on the growth of the sector. The closer this coefficient comes to unity, the closer is the interdependence between overall and sectoral growth.

Table 7.3
Sectoral growth coefficients and deviations, Cases C and D

	X_C	X_D	$d = X_C/X_D$
Iron and Steel	0.62	0.56	1.10
Nonferrous Metals	0.89	0.85	1.04
Paper and Allied	0.94	1.06	0.89
Chemicals	1.80	1.77	1.02
Mineral Products	0.59	0.77	0.77
Food and Related	0.86	1.23	0.70
Transport Equipment	0.25	0.26	0.96
Textiles	0.25	0	∞
Glass	0.51	0.68	0.74
Machinery	0.97	0.98	0.99
Miscellaneous	1.15	0.98	1.17
Total	1.04	1.00	1.04

In the case of the demand category food, for example, the sectoral growth coefficient X is close to unity in all cases. This means that the growth rate in this sector is close to the overall growth rate. But the deviation coefficient d shows that this does not mean a close functional relationship, since the high error term signals the importance of other determining factors.

The oil price changes between the sectors is reflected only in the chemicals, transport equipment, machinery, and miscellaneous activity levels (Table 7.4). Higher oil prices are assumed to have a small negative impact on the chemicals and transport equipment industries and a small positive impact on the machinery and other investment goods industries in Germany.

Residential

The number of dwellings and their distribution among heating systems are the major activity levels in the residential sector (Tables 7.5 to 7.8). The 1985 scenario projections of these numbers are based on:

- existing structure (1972)
- new housing construction scenario estimates
- modernization scenario estimates (changeovers to central space heating systems)
- demolition scenario estimates.

All scenario variables are assumed to have some impact on these numbers: economic growth mainly on total construction; oil price on the choice of fuel and heating system (it is assumed that electricity price reacts subproportionally on oil price changes); and energy policy on the introduction of district heating (for demonstration purposes a Case C′ is calculated to show the assumed policy impact).

Others

All nonresidential, nonindustrial activity levels are influenced by economic growth. The relation is again represented by the sectoral growth coefficient and its deviation (Table 7.9). Except for the air categories, all sectors grow subproportionally.

The deviation coefficients of the other sectors are, on average, farther from unity than those of the industrial sector. This signals sectoral developments that are rather independent of overall development.

Table 7.4
Impact of oil prices on sectoral growth, Cases A, C, and E

	β_{iAC}	β_{iCE}
Chemicals	-0.045	-0.018
Transport Equipment	-0.075	-0.049
Machinery	+0.069	+0.024
Miscellaneous	+0.011	+0.008

Table 7.5
Centrally heated dwellings by types of housing and by heating system: stock (1972) and projected demolition (1972–1985)

	Stock 1972 (thousands)	Number to Be Demolished 1972–1985 (thousands)
Fossil Fuels		
1 and 2 Family	4514	170
Others	4836	400
Total	9350	570
District Heat		
1 and 2 Family	78	0
Others	865	20
Total	943	20
Electricity		
1 and 2 Family	463	30
Others	613	230
Total	1076	260
Total		
1 and 2 Family	5055	200
Others	6314	650
Total	11369	850

7.3.6 Major Energy Coefficients

For the estimation of the scenario energy coefficients, we started by analyzing developments in the past, current trends, and other underlying factors. On this basis the coefficients for 1985 were estimated according to the scenario variables.

Industry

The analysis shows in almost all industrial categories a downward trend of the energy per value added ratio. These trends are assumed to continue in the future, but with a lesser inclination and modified by the economic growth and oil price scenario descriptors, for two reasons:

1. Investments to introduce higher efficiencies are dependent on economic growth.

Table 7.6
Modernization of housing stock by introduction of central heating, by type of housing and by heating system, 1972–1985 (thousands of dwellings)

WAES Case	A	B	C	C′	D	E
GDP	high	low	high	high	low	high
Oil Price	high	high	medium	medium	medium	low
Natl. Policy Response	vig.	vig.	vig.	res.	res.	res.
Fossil Fuels						
1 and 2 Family	500	450	650	700	650	800
Others	1000	950	1150	1300	1250	1400
Total	1500	1400	1800	2000	1900	2200
District Heat						
1 and 2 Family	150	120	150	100	100	100
Others	750	680	650	500	400	500
Total	900	800	800	600	500	600
Electricity						
1 and 2 Family	530	480	400	400	300	300
Others	1250	1120	1200	1200	1100	1100
Total	1800	1600	1600	1600	1400	1400
Total						
1 and 2 Family	1200	1050	1200	1200	1050	1200
Others	3000	2750	3000	3000	2750	3000
Total	4200	3800	4200	4200	3800	4200

Table 7.7
New housing with central space heating, by type of housing and by heating system, 1972–1985 (thousands of dwellings)

WAES Case	A	B	C	C′	D	E
GDP	high	low	high	high	low	high
Oil Price	high	high	medium	medium	medium	low
Natl. Policy Response	vig.	vig.	vig.	res.	res.	res.
Fossil Fuels						
1 and 2 Family	1480	1340	1500	1700	1550	1950
Others	1820	1560	1850	2200	1950	2350
Total	3300	2900	3400	3900	3500	4300
District Heat						
1 and 2 Family	250	230	250	200	150	200
Others	1650	1470	1650	1400	1250	1300
Total	1900	1700	1900	1600	1400	1500
Electricity						
1 and 2 Family	970	730	900	800	600	550
Others	530	470	500	400	300	350
Total	1500	1200	1400	1200	900	900
Total						
1 and 2 Family	2700	2300	2700	2700	2300	2700
Others	4000	3500	4000	4000	3500	4000
Total	6700	5800	6700	6700	5800	6700

Table 7.8
Housing stock in 1985 with central space heating, by type of housing and by heating system (thousands of dwellings)

WAES Case	A	B	C	C′	D	E
GDP	high	low	high	high	low	high
Oil Price	high	high	medium	medium	medium	low
Natl. Policy Response	vig.	vig.	vig.	res.	res.	res.
Fossil Fuels						
1 and 2 Family	6324	6134	6544	6544	6544	7094
Others	7256	6946	7436	7936	7636	8186
Total	13580	13080	13980	14680	14180	15280
District Heat						
1 and 2 Family	478	428	478	378	328	378
Others	3245	2995	3145	2745	2495	2645
Total	3723	3423	3623	3123	2823	3023
Electricity						
1 and 2 Family	1953	1643	1733	1633	1333	1283
Others	2163	1973	2083	1983	1783	1833
Total	4116	3616	3816	3616	3166	3116
Total						
1 and 2 Family	8755	8205	8755	8755	8205	8755
Others	12664	11914	12664	12664	11914	12664
Total	21419	20119	21419	21419	20119	21419

Table 7.9
Sectoral growth coefficients and deviations for transportation, commercial, public, agriculture, mining, and construction sectors

	X_C	X_D	$d=X_C/X_D$
Transportation			
Auto	0.629	0.978	0.64
Public Road Transport (Scheduled)	0.106	0.155	0.68
Public Road Transport (Unscheduled)	0.313	0.491	0.64
Air (Passenger, Domestic)	0.721	1.000	0.72
Air (Passenger, International)	1.510	1.734	0.87
Passenger (Total)	0.462	0.727	0.64
Air (Freight, Domestic)	0.221	0.348	0.64
Air (Freight, International)	2.560	3.684	0.69
Truck (Local)	0.726	0.847	0.86
Truck (Intercity)	0.727	0.850	0.86
Freight (Total Domestic)	0.673	0.556	1.21
Freight (Total)	0.549	0.639	0.86
Commercial	0.835	1.084	0.77
Public			
Floor Area	0.528	0.700	0.75
GDP Contribution	0.508	0.641	0.79
Agriculture	0.346	0.546	0.63
Mining	0.008	0.013	0.62
Construction	0.440	0.422	1.04

2. Higher oil prices lead to energy-efficient behavior and to substitution of energy sources since oil prices affect overall energy prices. The prices of nonoil energy sources are not, however, assumed to rise proportionally. Oil price is assumed to affect oil product and gas prices strongly, coal prices a little less, and electricity prices even less.

Substitution effects between energy sources are assumed to take place. To demonstrate the assumed trend effect and the impact for the scenario variables, the range of the energy coefficients in the 1985 scenarios in relation to 1972 are shown in Table 7.10. A policy impact is not considered.

Others

In the transportation sector, no major changes in the energy coefficients are assumed. For example, the energy per distance traveled coefficient in the automobile subsector is projected to vary only by about 4 percent from its present level. The coefficient is at present very low in Germany. It is true that with rising income the average size of cars will grow, but the energy efficiency will probably also increase, especially with higher oil product prices, and these effects are assumed to balance out, giving a rather small overall effect.

In the residential sector, the most important energy coefficient is heat loss (Table 7.11). In the case of fossil fuel heating systems, the coefficient will be decreased considerably, whereas in the case of electrical heating systems, the potential for improvement is quite small because the insulation standard of fully electrified dwellings was already very high in 1972.

In the other sectors the estimation procedure is similar. The coefficients are shown in Table 7.12.

7.4 PRESENTATION AND ANALYSIS OF RESULTS

The energy consumption scenarios for 1985 result from the underlying assumptions about determining factors (scenario descriptors) and their relationship to energy consumption. These assumptions were made partly explicit in the Sections 7.3.5 and 7.3.6.

It is the purpose of this section to show the overall effects of the determining factors on energy demands. It therefore contains mainly tables and figures to enable easy comparisons for consistency checks and investigations of the assumed relationships. Summary sheets may be found with the other worksheets in Section 7.6.

7.4.1 Overall Energy Demand Growth and the Impact of Economic Growth, Oil Price, and Policy

Figure 7.1 shows the development of energy consumption in Germany since 1950 and the 1985 demand scenarios drawn from 1972 consumption. Energy consumption will not continue to grow following an exponential trend. Neither fossil and electric final energy demand nor primary

Table 7.10
Range of energy coefficients in industry, 1985

	Range of Energy Coefficients in the WAES Scenarios (1972=100)
Iron and Steel	90 – 81
Nonferrous Metals	94 – 91
Paper and Allied	82 – 73
Chemicals	83 – 76
Mineral Products	90 – 82
Food and Related	98 – 83
Transport Equipment	100 – 98
Textiles	95 – 91
Glass	96 – 87
Machinery	100 – 98
Miscellaneous	98

Table 7.11
Heat Loss

	Coefficient Range 1985 (1972 = 100)	
	Fossil Fuels	Electricity
1 and 2 Family Houses	88–81	96
Others	88–81	100–98
Public Buildings	100–94	

Table 7.12
Energy coefficients for other sectors

	Range of Coefficients (1972 = 100)
Agriculture	115–107
Mining	100–95
Construction	100–95
Commercial	96–90

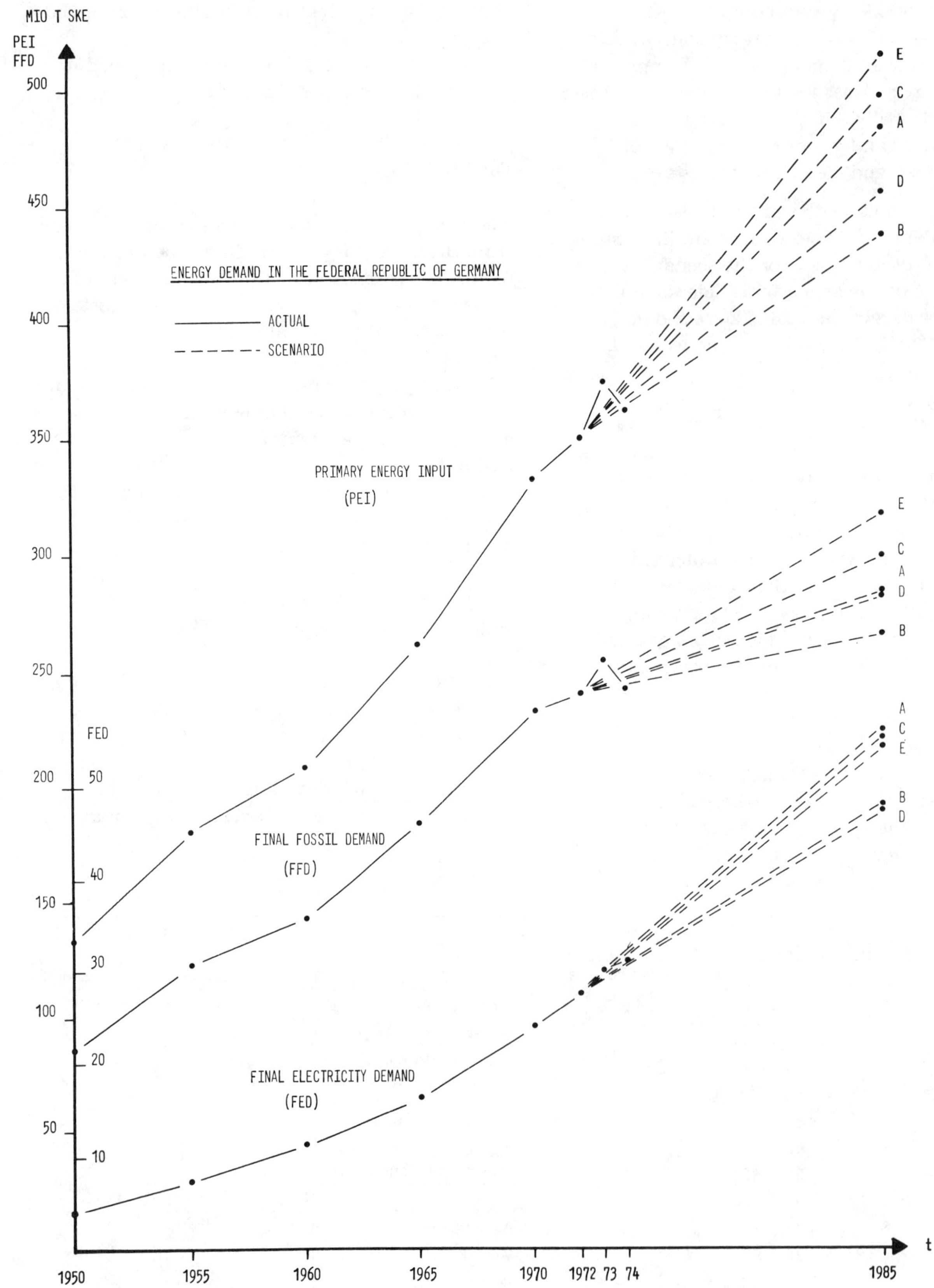

Figure 7.1 Development of energy demand in the Federal Republic of Germany 1950–1972 and projected to 1985.

energy input will have the same growth rate as in the past.

The ratio of energy demand growth to economic growth, which is often called the growth "elasticity" of energy, will probably be slightly lower in the future (Table 7.13). In this study this ratio will not be called elasticity because the *ceteris paribus* condition is not fulfilled. The "pure" elasticity—showing the impact of a single scenario descriptor—will, rather, be shown to compare scenario cases that differ in only one variable.

To express this pure elasticity several coefficients are calculated, namely, a price elasticity α defined by

$$(E_A/E_C) = (P_A/P_C)^{\alpha}$$

and an "income" elasticity γ defined by

$$(E_A/E_B) = (GDP_A/GDP_B)^{\gamma}.$$

Further, to study the significance of these coefficients, a "test" was used, namely the percentage deviation of energy demand in Case E (which is called E_E) from the functional energy demand in Case E* (called E_{E*}) calculated from the Case D demand using α and γ:

$$E_{E*} = E_D\ (P_E/P_D)^{\alpha}\ (GDP_E/GDP_D)^{\gamma}.$$

The effect of economic growth on final energy demand is much smaller in the case of fossil fuel consumption than in the case of electricity consumption. The oil price elasticity of fossil fuel demand is not regarded to be very substantial, due to the underproportional evolution of fossil fuel price levels. The oil price elasticity of electricity demand is positive because of (1) the assumed change in price relations and (2) substitution effects.

If one measures the significance of these elasticities by applying the "test" suggested above, one finds that they are quite stable, which means that the assumption of constant elasticity seems to be proved (see Table 7.14).

The impact of the considered policy measures is also rather small (Table 7.15). It can be roughly measured by calculating a Case D* or E* demand from Case C using α and γ, and comparing the results with the scenario demand for Cases D and E. The ratio shows that the difference between a vigorous and a restrained policy generates less than 1 percent savings in final energy demand, and about 3 percent in primary input. The measurement of savings is rough because the identification of savings is disturbed by income and price elasticity deviations. Given the apparently rather low savings gains, it should be considered that the restrained policy already contains a substantial amount of so-called energy-saving measures.

The bulk of the savings occurs in the transformation sector since the heat for enforced district heating is generated mostly in combined process plants, which produce heat and electric power with a high overall efficiency.

7.4.2 Sectoral Energy Demand

Sectoral percentage shifts of total energy demand are shown in the worksheets. Those shifts indicate considerable differences in the growth of energy demand between sectors. The corresponding growth rates again show an obvious distinction in the reaction of the sectoral demand to changes in the scenario descriptors.

The elasticities of energy demand with regard to oil price and economic growth vary in a very broad range among the sectors in the case of both fossil fuel and electric energy (Tables 7.16 and 7.17).

Table 7.13
Energy and economic growth rates and their ratios

			1972–1985 Scenarios				
	1950–1960	1960–1972	A	B	C	D	E
Final Energy Demand							
Fossil Fuels[1]	3.0	3.6	1.30	0.79	1.70	1.25	2.15
Electricity[1]	10.7	7.7	5.56	4.28	5.46	4.23	5.37
Primary Energy Input[1]	4.6	4.4	2.37	1.60	2.59	1.90	2.87
Economic Growth[1]	8.0	5.1	3.73	2.35	3.73	2.35	3.73
Ratio (PEI/EG)	0.57	0.86	0.64	0.68	0.69	0.81	0.77

1. %/yr.

Table 7.14
Oil price and income elasticity of total energy demand

	α	γ	"Test"
Final Energy Demand			
Fossil Fuels	−0.13	0.37	0.1%
Electricity	0.03	0.91	0.4%
Primary Energy Input	−0.07	0.56	0

The income elasticity γ, which catches all effects of different economic growth rates, shows in all cases the expected positive sign. The income elasticity of electricity demand is assumed to be higher on average than the elasticity of fossil energy demand. Also, the range of income elasticities is wider in the case of electricity.

Fossil fuel energy demand reacts in the expected manner to oil price changes. The range of elasticities is rather narrow, with the highest figure for residential and the lowest for commercial, public, and industrial demand. In the case of electric energy demand, the picture seems a little confusing because of the appearance of both positive and negative signs. The explanation is the comparatively high substitution elasticity in the residential heating demand. Since electricity prices are assumed to be the energy prices least affected by oil price changes, electric energy can substitute for other energy sources whose prices are closely connected to the price of oil. In other sectors, electricity can technically substitute for other forms of energy only in a few cases and can itself be replaced to a certain extent by capital. Therefore the "cross-elasticity" is negative in these sectors.

A very interesting phenomenon is revealed by the "test." It becomes clear that the *significance* of the elasticities is, at a sectorally more disaggregated level, considerably lower, and that, therefore, the apparent stability of the total elasticities is only an accident. The stability of the relationship on a highly aggregated level is certainly due in part to the "law of large numbers," but not completely. This can mean that the elasticity coefficient is not constant at all points of demand, so that the underlying assumption of a functional relationship of this kind is not sufficient. Since the deviations are still rather small, it will be used further as a reasonable approximation to energy-consuming behavior.

To demonstrate the effect of vigorous and restrained policy, a case comparison like Table 7.15 could be used. However, a glance at the deviation coefficients in Tables 7.16 and 7.17 and the saving percentages in Table 7.15 may disclose the worthlessness of this procedure. Considerable quantitative differences between vigorous and restrained

Table 7.16
Oil price and income elasticity of sectoral energy demand: Fossil fuels

	α	γ	"Test"
Industrial	−0.075	0.73	2.9%
Transportation	−0.1	0.22	2.5%
Commercial and Public	−0.055	0.305	0.9%
Residential	−0.23	0.074	2.4%
Agriculture, Mining, and Construction	−0.148	0.263	6.0%
Feedstocks	−0.166	0.371	2.6%
Total	−0.13	0.37	0.1%

Table 7.17
Oil price and income elasticity of sectoral energy demand: Electricity

	α	γ	"Test"
Industrial	−0.012	1.278	0.6%
Transportation	−0.054	0	4.4%
Commercial	0	0.4	0
Public	0	0.28	0
Residential	+0.1522	0.85	2.0%
Agriculture, Mining, and Construction	−0.042	0.1	3.4%
Total	+0.03	0.91	0.4%

Table 7.15
Potential savings in switching from a restrained to a vigorous policy (10^{15} J)

	Case D*[1]	Case D	Savings[3]	Case E*[2]	Case E	Savings[3]
Final Energy Demand						
Fossil Fuels	8.357	8.408	0.051 (0.6%)	9.394	9.440	0.046 (0.5%)
Electricity	1.408	1.415	0.007 (0.5%)	1.629	1.631	0.002 (0.1%)
Primary Energy Input	13.421	13.817	0.396 (3%)	15.198	15.642	0.444 (3%)

1. $E_{D*} = E_C\ (GDP_D/GDP_C)$.
2. $E_{E*} = E_C\ (P_E/P_C)$.
3. Savings = $(E_i - E_{i*})/E_i$.

policy can appear only in the residential and transport sectors. These will become apparent in the more disaggregated analysis.

7.4.3 Energy Demand in Detail

Industry
Since alternative conservation policies in industry are not considered, several possibilities for measuring the impact of oil price and economic growth on industrial energy demand are given. Cases A and B and Cases C and D differ only in the economic growth assumption; Cases A and C, B and D, and C and E only in the oil price assumption.

Again, the disaggregated analysis of Table 7.18 shows a wide range of elasticity coefficients and, in some cases, an insufficiency of the chosen demand function with constant elasticities (fossil demand of the iron and steel industry, fossil and electric energy demand of the paper and allied industries).

Table 7.19 also shows a quite divergent picture of energy demand elasticities with regard to oil price. Elasticities vary among the different categories and seem not to be constant. Electricity is not assumed to substitute for fossil fuels in the case of higher crude prices.

Fossil fuel demand does not react to oil price changes in some categories. In the case of the iron and steel industry, higher oil prices result under certain conditions in even higher fossil fuel energy demand due to resubstitution for oil and gas by coal.

Residential
The rather high deviation coefficients shown in Tables 7.20 and 7.21 indicate that in the residential sector the relationship of energy demand to the scenario descriptors cannot be properly described by a log-linear function. At any rate, the signs of the numbers give interesting information.

Economic growth does not lead simply to higher energy demand. In the case of fossil fuel demand for nonheating purposes, the demand decreases with economic growth, although the total effect is rather small. Electric energy demand is supposed to be rather sensitive to growth.

Higher oil prices clearly favor electric energy demand in the residential sector. The sign of the "cross-elasticity" is positive, in contrast to the industrial demand cross-elasticity.

The difference in policy choices between vigorous and restrained accounts for about 3.5 x 10^6 tons of coal equivalent, which is 5.6 percent of residential heating demand and about 4.5 percent of total residential demand.

Transportation
Economic growth has a heavy impact on nonindividual transportation, but a very weak impact on individual transportation (Table 7.22).

The energy demand for bunkers is estimated

Table 7.18
Impact of economic growth on major categories of industrial energy demand

	Fossil Fuels		Electricity	
	A–B	C–D	A–B	C–D
Iron and Steel	0.53	0.36	0.98	0.98
Nonferrous Metals	0.77	0.77	0.81	0.81
Paper and Allied	0.16	0.30	0.79	0.62
Chemicals	1.68	1.70	1.88	1.86
Mineral Products	0.20	0.21	0.29	0.29
Food and Related	0.06	0.06	0.50	0.50
Total	0.73	0.69	1.28	1.26

Table 7.19
Impact of oil price on major categories of industrial energy demand

	Fossil Fuels			Electricity		
	A–C	B–D	C–E	A–C	B–D	C–E
Iron and Steel	0.07	0	−0.26	0	0	−0.28
Nonferrous Metals	0	0	−0.08	0	0	−0.09
Paper and Allied	−0.23	−0.17	−0.18	0	−0.07	−0.06
Chemicals	−0.26	−0.25	−0.15	−0.05	−0.06	−0.02
Mineral Products	−0.12	−0.12	−0.14	0	0	0
Food and Related	−0.28	−0.28	−0.17	0	0	0
Total	−0.08	−0.09	−0.16	−0.01	−0.02	−0.03

Table 7.20
Impact of oil price and economic growth on residential energy demand

	Fossil Fuels			Electricity		
	α	γ	"Test"	α	γ	"Test"
Heating	−0.22	0.09	1.7%	0.18	0.89	10.3%
Other Uses	−0.39	−0.16	13.6%	0.14	0.83	1.0%
Total	−0.23	0.07	2.3%	0.15	0.85	2.3%

Table 7.21
Impact of oil price, economic growth, and energy policy on residential room heating

	α	γ	"Test"
Total Room Heating	−0.19	0.15	2.1%
Total Other Residential	0	0.57	3.2%

Table 7.22
Impact of oil price and economic growth on transportation energy demand

	α	γ	"Test"
Automobiles	−0.12	0.02	3.4%
Passenger Transportation	−0.11	0.59	2.6%
Freight Transportation	−0.11	0.54	0.01%

separately and independently from activity levels and energy coefficients because this demand is highly dependent on the results of the supply-demand integration. The reason for this is that about half of the demand is used for energy transport. The demand is also dependent on international price relations. An estimated demand of 200 x 10^{15} J has been inserted to account for this factor.

7.5 CONCLUSION

The final energy demand estimations based on the scenario assumptions reveal an uncertainty range of about 20 percent. For primary energy demand the range is assumed to be smaller (about 17 percent) because a change in fuel mix causes changes in the losses. The uncertainties in demand differ considerably among the sectors (Table 7.23).

The analysis applied to the estimated numbers in this study may give guidelines for further energy demand projections.

As indicated by the uncertainty range and by the relative weight of the sectors, a sector in which it is crucial to narrow the uncertainty range is the industrial sector. The determining factor for industrial energy demand is the economic growth (Tables 7.16 and 7.17). In the residential sector, as shown in Table 7.21, the energy demand for heating purposes depends heavily on oil price development, whereas other residential energy demand depends more on private income. Narrowing the uncertainty range of economic growth would, therefore, cause a considerable gain in certainty in this area.

Other sectors, particularly transportation, have a comparatively low range of uncertainty.

Table 7.23
Uncertainty range of the final demand projection

	E_E/E_B	Sectoral Weight
Industrial	1.26	33
Nonenergy Use	1.22	12
Residential	1.18	23
Commercial and Public	1.12	10
Transportation	1.10	17
Agriculture, Mining, Commercial	1.10	5
Total	1.20	100

7.6 DEMAND STUDY WORKSHEETS

This section contains summary tables and worksheets in the following order:

Energy Use Summary

Sectoral Percentages of Final Energy Demand

Energy Growth Rates Summary

Base Year (1972) data (5 sheets)

Worksheets for:

Case A (5 sheets)

Case B (5 sheets)

Case C (5 sheets)

Case D (5 sheets)

Case E (5 sheets).

WAES NATIONAL ENERGY DEMAND STUDIES, 1972-1985
ENERGY USE SUMMARY

Country Federal Republic of Germany

Scenario Assumptions

WAES Case	A	B	C	D	E	
GNP 1976-1985	High	Low	High	Low	High	
Oil Price*	\$11.50 → \$17.25	\$11.50 → \$17.25	\$11.50	\$11.50	\$11.50 → \$7.66	
Natl. Policy Response	Vigorous	Vigorous	Vigorous	Restrained	Restrained	
Natl. GNP Growth Rate 1976-1985 (%/yr.)	4,5%	2,5%	4,5%	2,5%	4,5%	1972 Base Year

Fuel Type	A Fossil Fuel	A Elec.	B Fossil Fuel	B Elec.	C Fossil Fuel	C Elec.	D Fossil Fuel	D Elec.	E Fossil Fuel	E Elec.	1972 Fossil Fuel	1972 Elec.
Units	10^6 tce	10^6 tce	10^6 tce	10^6 tce	10^6 tce	10^6 tce	10^6 tce	10^6 tce	10^6 tce	10^6 tce	10^6 tce	10^6 tce
Energy-Using Sectors												
Industrial (excluding feedstocks)	87,89	27,98	77,37	22,41	90,61	28,12	80,34	22,58	96,76	28,49	73,37	14,70
Transportation	58,72	1,53	56,51	1,53	61,27	1,56	59,23	1,56	62,53	1,53	48,55	1,05
Commercial and Public	26,96	9,97	25,57	9,36	27,57	9,97	26,62	9,36	28,97	9,97	21,73	4,15
Residential	61,13	15,33	60,35	13,26	67,12	14,42	67,35	12,65	73,13	13,50	61,74	6,87
Fishing, Agriculture, Mining, Construction	13,70	1,97	13,09	1,94	14,55	2,01	13,94	1,97	14,62	1,97	11,97	1,22
Feedstocks	39,17		36,69		41,89		38,39		44,95		26,01	
Final Energy Demand (excluding processing losses)	287,74	56,78	269,58	48,45	303,01	56,07	285,87	48,11	320,96	55,45	243,34	28,08
Processing Losses	17,41	127,77	16,22	109,75	17,78	126,14	17,78	108,97	19,24	124,64	31,86	57,63
Primary Energy Input	305,15	184,55	285,80	158,20	320,77	182,21	303,65	157,08	340,20	180,09	275,20	85,71

*Oil price 1975 US\$ per barrel Arabian light crude oil FOB Persian Gulf.

WAES NATIONAL ENERGY DEMAND STUDIES, 1972-1985
SECTORAL PERCENTAGES OF TOTAL

Country Federal Republic of Germany

Scenario Assumptions												
WAES Case	A		B		C		D		E			
GNP 1976-1985	High		Low		High		Low		High			
Oil Price*	$11.50 → $17.25		$11.50 → $17.25		$11.50		$11.50		$11.50 → $7.66			
Natl. Policy Response	Vigorous		Vigorous		Vigorous		Restrained		Restrained			
Natl. GNP Growth Rate 1976-1985 (%/yr.)	4,5%		2,5%		4,5%		2,5%		4,5%		1972 Base Year	
Fuel Type	Fossil Fuel	Elec.	Fossil Fuel	Elec.	Fossil Fuel	Elec.	Fossil Fuel	Elec.	Fossil Fuel	Elec.	Fossil Fuel	Elec.
Units												
Energy-Using Sectors												
Industrial (excluding feedstocks)	30,5	49,3	26,7	46,2	29,9	50,1	28,1	46,9	30,1	51,4	30,2	52,7
Transportation	20,4	2,7	21,0	3,2	20,2	2,8	20,7	3,3	19,5	2,8	19,9	3,7
Commercial and Public	9,4	17,5	9,5	19,3	9,1	17,8	9,3	19,4	9,0	18,0	8,9	14,8
Residential	21,3	27,0	22,4	27,3	22,2	25,7	23,6	26,3	22,8	24,3	25,4	24,4
Fishing, Agriculture, Mining, Construction	4,8	3,5	4,8	4,0	4,8	3,6	4,9	4,1	4,6	3,5	4,9	4,4
Feedstocks	13,6		13,6		13,8		13,4		14,0		10,7	
Final Energy Demand (excluding processing losses)	100	100	100	100	100	100	100	100	100	100	100	100
Processing Losses												
Primary Energy Input												

*Oil price 1975 US$ per barrel Arabian light crude oil FOB Persian Gulf.

WAES NATIONAL ENERGY DEMAND STUDIES, 1972-1985
ENERGY GROWTH RATES SUMMARY

Country Federal Republic of Germany

Scenario Assumptions

WAES Case	A		B		C		D		E	
GNP 1976-1985	High		Low		High		Low		High	
Oil Price*	$11.50 → $17.25		$11.50 → $17.25		$11.50		$11.50		$11.50 → $7.66	
Natl. Policy Response	Vigorous		Vigorous		Vigorous		Restrained		Restrained	
Natl. GNP Growth Rate 1976-1985 (%/yr.)	4,5%		2,5%		4,5%		2,5%		4,5%	
Fuel Type	Fossil Fuel	Elec.	Fossil Fuel	Elec.	Fossil Fuel	Elec.	Fossil Fuel	Elec.	Fossil Fuel	Elec.
Units (%/yr. 1972-1985)										
Energy-Using Sectors										
Industrial (excluding feedstocks)	1,40	5,03	0,42	3,25	1,64	5,07	0,71	3,31	2,15	5,17
Transportation	1,47	2,91	1,17	2,91	1,81	3,08	1,54	3,08	1,96	2,91
Commercial and Public	1,67	6,97	1,26	6,45	1,85	6,97	1,58	6,45	2,24	6,97
Residential	(0,08)	6,37	(0,18)	5,17	0,64	5,87	0,67	4,81	1,30	5,33
Fishing, Agriculture, Mining, Construction	1,10	4,68	0,69	3,60	1,52	3,87	1,18	4,68	1,55	4,68
Feedstocks	3,20		2,68		3,73		3,04		4,30	
Final Energy Demand (excluding processing losses)	1,30	5,56	0,79	4,28	1,70	5,46	1,25	4,23	2,14	5,37
Processing Losses										
Primary Energy Input										

*Oil price 1975 US$ per barrel Arabian light crude oil FOB Persian Gulf.

NATIONAL ENERGY DEMAND PROJECTIONS

Country Federal Republic of Germany

Year 1972

WAES Case Base Case

Transportation

	(A)	(B)	(C)	(D)	(E)	(F)	(G)	
	Population/Auto	(km) Distance/Auto/Yr.	Load Factor	Fraction Urban or Rural	(10^6 J/100 km) Energy/Distance	Fossil Fuel (10^{15} J)	Electricity	
Urban Auto	3,84	15890	1,4	37,4%	366,25	351,3		(1)
Rural Auto		15270	1,85	62,6%	322,3	498,4		(2)
	(10^6 pkm-tkm) Activity Level	% Electrified	Load Factor	Fossil Fuel/ Activity Level (10^6 J/pkm-tkm)	Electricity/Activity			
Urban Bus[1]	48305	22,5	22,8	0,377	0,596	14,1	6,48	(3)
Intercity Bus[1]	14053		30,8	0,3		4,2		(4)
Rail (Passenger)	39638	72,6		3,92	0,587	42,6	16,9	(5)
Air (Passenger, Domestic)	3474			3,52		12,2		(6)
Air (Passenger, International)	15983			3,52		56,1		(7)
Ship (Passenger)	15							(8)
Rail (Freight)	66700	66,4		1,27	0,17	28,4	7,6	(9)
Air (Freight, Domestic)	45			29,2		1,3		(10)
Air (Freight, International)	591			29,2		15,8		(11)
Truck (Local)	39700			2,33				(12)
Truck (Intercity)	49200			2,33		206,9		(13)
Ship (Freight, Domestic)	26971			1,149		31,0		(14)
Ship (Freight, International)	362839			0,192		165,4		(15)
Miscellaneous								(16)
					Total	1427,6	31,0	

1) Instead of "Urban" and "Intercity Bus," read "Scheduled" and "Unscheduled Public Road Transport."

NATIONAL ENERGY DEMAND PROJECTIONS

Country Federal Republic of Germany

Year 1972

WAES Case Base Case

Agriculture, Mining, and Construction

Agriculture, Forestry, and Fishing (ISIC Major Division 1)	(A) (10^6 \$) Value Added	(B) (10^6 J) Energy/Value Added	(C) (2,5x10^6 m^2) Area Harvested	(D) Energy/Area	(E) Physical Output	(F) (10^6 J) Energy/Output	(G) (10^{15} J) Fossil Fuel	(H) Electricity	
Farm Machinery	↓	↓	32890				54,8		(17)
Greenhouses									(18)
Other Agricultural Uses	22594	6,55	835				72,7	20,5	(19)
Forestry (ISIC 12)			17945		23,76x10^6 m^2		3,8	0,6	(20)
Fishing Boats (ISIC 13)									(21)
Other Fishing Use (ISIC 13)					405126 tons		8,9		(22)
Total Fertilizer (tons)		N ____	P ____		K ____	Total	140,2	21,1	

Mining	(A) Value Added	(B) Energy/Value Added	(C) Physical Output	(D) Energy/Physical Output	(E) Fossil Fuel	(F) Electricity	
Handling and Internal Transportation	258,6	98,2			21,3	4,1	(23)
Other Energy Use							(24)
				Total			

Construction	(A) Value Added	(B) Energy/Value Added	(C) Physical Output 10^3 m^3	(D) Energy/Physical Output	(E) Fossil Fuel	(F) Electricity	
Handling and Internal Transportation	25451	7,91	509,411		190,9	10,5	(25)
Other Energy Use							(26)
				Total			
				Agriculture, Mining, and Construction Total	452,4	35,7	

1) All monetary variables (Value Added and GDP-Contribution) are in constant 1972 U.S. dollars based on an exchange rate of US\$1 = 3,19 DM.

NATIONAL ENERGY DEMAND PROJECTIONS

Country Federal Republic of Germany

Year 1972

WAES Case Base Case

Commercial and Public

Commercial	(A) Floor Area	(B) Heat Gain or Loss/Area	(C) System Efficiency	(D) Fossil Fuel (10^{15}J)	(E) Electricity	
Space Heat (Fossil Fuel)						(27)
Space Heat (Electricity)						(28)
Air Conditioning (Fossil Fuel)						(29)
Air Conditioning (Electricity)						(30)
	Value Added (10^6\$)	Energy/Value Added (10^6J)	Electricity/Value Added (10^6J)			
Lighting						(31)
Miscellaneous	38287	10,75	1,91	414,7	73,2	(32)
Presently Unknown						(33)
			Total			

Public	($10^6 m^2$) Floor Area	(Gcal/m^2/year) Heat Gain or Loss/Area	System Efficiency	Fossil Fuel	Electricity	
Space Heat (Fossil Fuel)	77	1800	0,65	89,3		(34)
Space Heat (Electricity)						(35)
Air Conditioning (Fossil Fuel)						(36)
Air Conditioning (Electricity)						(37)
	(10^6\$) GDP-Contribution	Energy/Area	Electricity/GDP (10^6J/\$)			
Lighting & Power	26624		1,83		48,7	(38)
Miscellaneous				133,8		(39)
Presently Unknown						(40)
			Total	223,1	48,7	
			Commercial and Public Total	637,8	121,9	

1) All monetary variables (Value Added and GDP-Contribution) are in constant 1972 U.S. dollars based on an exchange rate of US\$1 = 3,19 DM.

NATIONAL ENERGY DEMAND PROJECTIONS

Country Federal Republic of Germany

Year 1972

WAES Case Base Case

Industrial

	(A) Value Added[1]	(B) Energy/Value Added	(C) Physical Output	(D) Energy/Output	(E) % Exported	(F) Fossil Fuel	(G) Electricity	(H) Feedstocks	
	(10^6 $)	(10^9 J/$)			(turnover)	(10^{15} J)	(10^{15} J)	(10^{15} J)	
Iron and Steel (ISIC 371)	3843	0,2163			27,3	766,0	65,2		(41)
Nonferrous Metals (ISIC 372)	1276	0,066			16,8	46,4	37,8	20,4	(42)
Paper and Allied (ISIC 34)	887	0,1091			16,4	74,1	22,7		(43)
Chemicals and Allied (ISIC 35 minus ISIC 353)	11436	0,0363			32,0	287,3	127,4	571,4	(44)
Mineral Products (ISIC 36)	3806	0,0874			3,5	305,9	26,6		(45)
Food and Related (ISIC 31)	9931	0,0174			3,1	155,0	18,4		(46)
Transport Eqpt. (ISIC 384)	7815	0,0107			40,9	60,5	23,4		(47)
Key Industry (N.E.C.[2])	4119	0,0205			13,7	68,2	16,2		(48)
Key Industry (N.E.C.)	1824 10671	0,052 0,007			19,9 37,3	86,6 60,4	8,3 15,1		(49)
Misc. Manufacturing	35025	0,0092				248,2	73,1		(50)
N.E.C. = Not Elsewhere Classified					**Total**	2158,6	434,6	591,8[3]	

1) Value Added in constant 1972 U.S. dollars based on an exchange rate of US$1 = 3,19 DM.
2) Key Industries: Textiles
Glass and Related
Machinery.
3) Additional non-energy use occurs in road construction, etc. (10^{15} J).

NATIONAL ENERGY DEMAND PROJECTIONS

Country Federal Republic of Germany

Year 1972

WAES Case Base Case

Residential

Population 61,7 x 10^6 Persons/Dwelling 2,86 Dwellings 21,6 x 10^6

Central Space Heating		(A) Dwellings (10^3)	(B) Heat Loss[7] (kcal/DW)	(C) System Efficiency	(D) Duration (Hrs/Y)	(E) Fossil Fuel (10^{15} J)	(F) Electricity	
Single Family	Fossil Fuels[2]	4764/78	160 x 87	0,6 /0,88	1650	766/8,5		(51)
	Electricity[3]	493	115 x 87	0,95	750		16,4	(52)
Town or Row Houses[1]	Fossil Fuels[2]	4336/ 859	130 x 62	0,6 /0,88	1650	404/54,5		(53)
	Electricity[3]	727	105 x 62	0,95	750		15,7	(54)
Apartments[1]	Fossil Fuels[2]	758	130 x 62	0,6 /0,88	1650	71/ 1,6		(55)
	Electricity[3]	78	105 x 62	0,95	750		1,5	(56)
		% Saturation (Fossil Fuels) (10^3)	% Saturation (Electricity) (10^3)	Energy/Unit (Fossil Fuels) (10^9)	Energy/Unit (Electricity) (kWh/y)			
Room Heating		9522	8200[4]	34,7	360	330,4	10,7	(57)
Air Conditioning (Window)								(58)
Air Conditioning (Central)								(59)
Water Heating[5]		14865	9738	7,4	1050	110	36,9	(60)
Refrigerators			19044		190		12,4	(61)
Lighting			21641		240		18,7	(62)
Cooking[5]		13850	13850	4,9	600	67,9	29,9	(63)
Dishwashers			1515		600		3,3	(64)
Freezers			6276		365		8,3	(65)
Television			19477		200		14,0	(66)
Other Key Appliances[6]			17096		400		24,6	(67)
Misc. Appliances						2,5	9,1	(68)
Potential Appliances								(69)
					Total	1816,8	201,5	

1) Instead of "Town or Row Houses" and "Apartments," read "Over 2-Family" and "Other" for the Base Case and "All Other than 1- and 2-Family Houses" for the 1985 cases.
2) Fossil Fuels/District Heat.
3) Electric Heat Storage.
4) Electric Direct Heat.
5) Saturation exceeds 100%.
6) Washer.
7) kcal/m^2 x average m^2/dwelling.

NATIONAL ENERGY DEMAND PROJECTIONS

Country Federal Republic of Germany

Year 1985

WAES Case A

Transportation

	(A)	(B)	(C)	(D)	(E)	(F)	(G)	
	Population/Auto	(km) Distance/Auto/Yr.	Load Factor	Fraction Urban or Rural	(10^6 J/100km) Energy/Distance	(10^{15} J) Fossil Fuel	Electricity	
Urban Auto	2,8	14200		35 %	350	377		(1)
Rural Auto		13650		65 %	310	596		(2)
	(10^6 pkm-tkm) Activity Level	% Electrified	Load Factor	Fossil Fuel/Activity Level (10^6 J/pkm-tkm)	Electricity/Activity			
Urban Bus [1]	50830	30		0,36	0,59	12,8	9,0	(3)
Intercity Bus [1]	16310			0,28		4,6		(4)
Rail (Passenger)	49375	83		3,6	0,58	30,2	23,8	(5)
Air (Passenger, Domestic)	4900			2,9		14,2		(6)
Air (Passenger, International)	32800			2,9		95,1		(7)
Ship (Passenger)								(8)
Rail (Freight)	86642	80		1,17	0,18	20,3	12,5	(9)
Air (Freight, Domestic)	50			29,0		1,5		(10)
Air (Freight, International)	2000			29,0		58,0		(11)
Truck (Local)	56084			2,2		123,3		(12)
Truck (Intercity)	69575			2,2		153,1		(13)
Ship (Freight, Domestic)	37156			1,1		40,8		(14)
Ship (Freight, International)						200		(15)
Miscellaneous								(16)
					Total	1726,9	45,3	

1) Instead of "Urban" and "Intercity Bus," read "Scheduled" and "Unscheduled Public Road Transport."

NATIONAL ENERGY DEMAND PROJECTIONS

Country Federal Republic of Germany

Year 1985

WAES Case A

Agriculture, Mining, and Construction

Agriculture, Forestry, and Fishing (ISIC Major Division 1)	(A) Value Added (10^6 \$)	(B) Energy/Value Added (10^6 J/\$)	(C) Area Harvested	(D) Energy/Area	(E) Physical Output	(F) Energy/Output	(G) Fossil Fuel (10^{15} J)	(H) Electricity (10^{15} J)	
Farm Machinery									(17)
Greenhouses									(18)
Other Agricultural Uses	26643	7,0					147	40	(19)
Forestry (ISIC 12)							4	1	(20)
Fishing Boats (ISIC 13)									(21)
Other Fishing Use (ISIC 13)							9		(22)
Total Fertilizer (tons)		N	P	K		Total	160	41	

Mining	(A) Value Added	(B) Energy/Value Added	(C) Physical Output	(D) Energy/Physical Output	(E) Fossil Fuel	(F) Electricity	
Handling and Internal Transportation	260	93			19,7	4,5	(23)
Other Energy Use							(24)
				Total			

Construction	(A) Value Added	(B) Energy/Value Added	(C) Physical Output	(D) Energy/Physical Output	(E) Fossil Fuel	(F) Electricity	
Handling and Internal Transportation	31388	7,5			222,9	12,6	(25)
Other Energy Use							(26)
				Total			
				Agriculture, Mining, and Construction Total	402,6	58,1	

1) All monetary variables (Value Added and GDP-Contribution) are in constant 1972 U.S. dollars based on an exchange rate of US\$1 = 3,19 DM.

NATIONAL ENERGY DEMAND PROJECTIONS

Country Federal Republic of Germany

Year 1985

WAES Case A

Commercial and Public

Commercial	(A) Floor Area	(B) Heat Gain or Loss/Area	(C) System Efficiency	(D) Fossil Fuel (10^{15}J)	(E) Electricity	
Space Heat (Fossil Fuel)						(27)
Space Heat (Electricity)						(28)
Air Conditioning (Fossil Fuel)						(29)
Air Conditioning (Electricity)						(30)
	Value Added (10^6\$)	Energy/ V A (10^6J/\$)	Electricity/V A (10^6J/\$)			
Lighting						(31)
Miscellaneous	56973	9,7	3,4	552	193,7	(32)
Presently Unknown						(33)
			Total			
Public	(10^6m^2) Floor Area	(Gcal/m^2/y) Heat Gain or Loss/Area	System Efficiency	Fossil Fuel	Electricity	
Space Heat (Fossil Fuel)	99	17oo	o,7	1o1		(34)
Space Heat (Electricity)						(35)
Air Conditioning (Fossil Fuel)						(36)
Air Conditioning (Electricity)						(37)
	(10^6\$) GDP-Contribution	Energy/	(10^6J/\$) Electricity/GDP			
Lighting & Power	33915		2,93		99,4	(38)
Miscellaneous				14o		(39)
Presently Unknown						(40)
			Total	241	99,4	
			Commercial and Public Total	796	293,1	

1) All monetary variables (Value Added and GDP-Contribution) are in constant 1972 U.S. dollars based on an exchange rate of US\$1 = 3,19 DM.

NATIONAL ENERGY DEMAND PROJECTIONS

Country Federal Republic of Germany

Year 1985

WAES Case A

Industrial

	(A) Value Added [1]	(B) Energy/Value Added	(C) Physical Output	(D) Energy/Output	(E) % Exported	(F) Fossil Fuel	(G) Electricity	(H) Feedstocks	
	(10^6 $)	(10^9 J/$)				(10^{15} J)	(10^{15} J)	(10^{15} J)	
Iron and Steel (ISIC 371)	5165	0,18				822	108		(41)
Nonferrous Metals (ISIC 372)	1954	0,06				64	53	24	(42)
Paper and Allied (ISIC 34)	1384	0,08				72	39		(43)
Chemicals and Allied (ISIC 35 minus ISIC 353)	26485	0,0275				437	291	950	(44)
Mineral Products (ISIC 36)	5052	0,072				323	40		(45)
Food and Related (ISIC 31)	15076	0,0145				183	36		(46)
Transport Eqpt. (ISIC 384)	8473	0,0105				57	32		(47)
Key Industry (N.E.C.) [2]	4604	0,0186				64	22		(48)
Key Industry (N.E.C.)	2342	0,045				92	13		(49)
	~~17482~~	~~0,0065~~				~~83~~	~~31~~		
Misc. Manufacturing	60615	0,09				388	158		(50)
					Total	2585	823	974 [3]	

N.E.C. = Not Elsewhere Classified

1) Value Added in constant 1972 U.S. dollars based on an exchange rate of US$1 = 3,19 DM.

2) Key Industries: Textiles
Glass and Related
Machinery.

3) Additional non-energy use occurs in road construction, etc. (10^{15} J).

NATIONAL ENERGY DEMAND PROJECTIONS

Country Federal Republic of Germany

Year 1985

WAES Case A

Residential

Population 60,7 x 10^6 Persons/Dwelling 2,43 Dwellings 24,97 x 10^6

Central Space Heating		(A) Dwellings (10^3)	(B) Heat Loss[7] (kcal/Dw)	(C) System Efficiency	(D) Duration (Hrs/y)	(E) Fossil Fuel (10^{15}J)	(F) Electricity	
Single Family	Fossil Fuels[2]	6324/478	130/95	0,65/0,9	1600	807/44		(51)
	Electricity[3]	1953	110/95	0,98	825		72	(52)
Town or Row Houses[1]	Fossil Fuels[2]	7256/3245	105/70	0,65/0,9	1600	551/178		(53)
	Electricity[3]	2163	102/70	0,98	825		55	(54)
Apartments	Fossil Fuels[2]							(55)
	Electricity[3]							(56)
		% Saturation (Fossil Fuels) (10^3)	% Saturation (Electricity) (10^3)	Energy/Unit (Fossil Fuels) (10^9J)	Energy/Unit (Electricity) (kWh/y)			
Room Heating		3550	8500[4]	32,2	400	114	12	(57)
Air Conditioning (Window)								(58)
Air Conditioning (Central)								(59)
Water Heating[5]		10000	16000	6,7	1250	67	72	(60)
Refrigerators			23800		190		19	(61)
Lighting			24970		380		34	(62)
Cooking[5]		7000	18000	4,5	550	32	36	(63)
Dishwashers			9000		650		21	(64)
Freezers			13000		400		19	(65)
Television			23500		200		17	(66)
Other Key Appliances[6]			20600		450		33	(67)
Misc. Appliances							20	(68)
Potential Appliances						3	30	(69)
					Total	1798	451	

1) Instead of "Town or Row Houses" and "Apartments," read "Over 2-Family" and "Other" for the Base Case and "All Other than 1- and 2-Family Houses" for the 1985 cases.
2) Fossil Fuels/District Heat.
3) Electric Heat Storage.
4) Electric Direct Heat.
5) Saturation exceeds 100%.
6) Washer.
7) kcal/m^2 x average m^2/dwelling.

NATIONAL ENERGY DEMAND PROJECTIONS

Country Federal Republic of Germany

Year 1985

WAES Case B

Transportation

	(A)	(B)	(C)	(D)	(E)	(F)	(G)	
	Population/Auto	(km) Distance/Auto/Yr.	Load Factor	Fraction Urban or Rural	(10^6 J/100km) Energy/Distance	(10^{15} J) Fossil Fuel	Electricity	
Urban Auto	2,81	14200		35 %	350	376		(1)
Rural Auto		13650		65 %	310	594		(2)
	(10^6 pkm-tkm) Activity Level	% Electrified	Load Factor	Fossil Fuel/ Activity Level	Electricity/Activity			
Urban Bus 1)	50600	30		(10^6 J/pkm-tkm) 0,36	0,59	12,8	9,0	(3)
Intercity Bus 1)	16300			0,28		4,6		(4)
Rail (Passenger)	49375	83		3,6	0,58	30,6	23,8	(5)
Air (Passenger, Domestic)	4700			2,9		13,6		(6)
Air (Passenger, International)	27000			2,9		78,3		(7)
Ship (Passenger)								(8)
Rail (Freight)	80910	80		1,17	0,19	18,9	12,3	(9)
Air (Freight, Domestic)	50			29		1,5		(10)
Air (Freight, International)	1800			29		52,2		(11)
Truck (Local)	51300			2,2		112,9		(12)
Truck (Intercity)	63600			2,2		139,9		(13)
Ship (Freight, Domestic)	31900			1,1		35,1		(14)
Ship (Freight, International)						190		(15)
Miscellaneous								(16)
					Total	1660	45,1	

1) Instead of "Urban" and "Intercity Bus," read "Scheduled" and "Unscheduled Public Road Transport."

NATIONAL ENERGY DEMAND PROJECTIONS

Country Federal Republic of Germany

Year 1985

WAES Case B

Agriculture, Mining, and Construction

Agriculture, Forestry, and Fishing (ISIC Major Division 1)	(A) (10^6 \$) Value Added	(B) (10^9 J/\$) Energy/Value Added	(C) Area Harvested	(D) Energy/Area	(E) Physical Output	(F) Energy/Output	(G) (10^{15} J) Fossil Fuel	(H) Electricity	
Farm Machinery									(17)
Greenhouses									(18)
Other Agricultural Uses	26643	7,0					147	40	(19)
Forestry (ISIC 12)							4	1	(20)
Fishing Boats (ISIC 13)									(21)
Other Fishing Use (ISIC 13)							9		(22)
Total Fertilizer (tons)		N	P	K		Total	160	41	

Mining	(A) Value Added	(B) Energy/Value Added	(C) Physical Output	(D) Energy/Physical Output	(E) Fossil Fuel	(F) Electricity	
Handling and Internal Transportation	260	93			19,7	4,5	(23)
Other Energy Use							(24)
				Total			

Construction	(A) Value Added	(B) Energy/Value Added	(C) Physical Output	(D) Energy/Physical Output	(E) Fossil Fuel	(F) Electricity	
Handling and Internal Transportation	28900	7,5			205,2	11,6	(25)
Other Energy Use							(26)
				Total			
				Agriculture, Mining, and Construction Total	384,9	57,1	

1) All monetary variables (Value Added and GDP-Contribution) are in constant 1972 U.S. dollars based on an exchange rate of US\$1 = 3,19 DM.

NATIONAL ENERGY DEMAND PROJECTIONS

Country Federal Republic of Germany

Year 1985

WAES Case B

Commercial and Public

	(A)	(B)	(C)	(D)	(E)	
Commercial	Floor Area	Heat Gain or Loss/Area	System Efficiency	Fossil Fuel (10^{15}J)	Electricity	
Space Heat (Fossil Fuel)						(27)
Space Heat (Electricity)						(28)
Air Conditioning (Fossil Fuel)						(29)
Air Conditioning (Electricity)						(30)
	Value Added (10^6\$)	Energy/ V A (10^6J/\$)	Electricity/ V A (10^6J/\$)			
Lighting						(31)
Miscellaneous	53140	9,7	3,4	515	180,7	(32)
Presently Unknown						(33)
			Total			

Public	(10^6m^2) Floor Area	(Gcal/m^2/y) Heat Gain or Loss/Area	System Efficiency	Fossil Fuel	Electricity	
Space Heat (Fossil Fuel)	97	1700	0,7	99		(34)
Space Heat (Electricity)						(35)
Air Conditioning (Fossil Fuel)						(36)
Air Conditioning (Electricity)						(37)
	(10^6\$) GDP-Contribution	Energy/Area	Electricity/ GDP(10^6J/\$)			
Lighting	32310		2,93		94,7	(38)
Miscellaneous				138		(39)
Presently Unknown						(40)
			Total	237	94,7	
			Commercial and Public Total	752	275,4	

1) All monetary variables (Value Added and GDP-Contribution) are in constant 1972 U.S. dollars based on an exchange rate of US\$1 = 3,19 DM.

NATIONAL ENERGY DEMAND PROJECTIONS

Country Federal Republic of Germany

Year 1985

WAES Case B

Industrial

	(A) Value Added [1]	(B) Energy/Value Added	(C) Physical Output	(D) Energy/Output	(E) % Exported	(F) Fossil Fuel	(G) Electricity	(H) Feedstocks	
	(10^6 $)	(10^9 J/$)				(10^{15} J)	(10^{15} J)	(10^{15} J)	
Iron and Steel (ISIC 371)	4545	0,185				750	91		(41)
Nonferrous Metals (ISIC 372)	1651	0,062				56	46	24	(42)
Paper and Allied (ISIC 34)	1223	0,085				70	34		(43)
Chemicals and Allied (ISIC 35 minus ISIC 353)	19133	0,028				326	210	880	(44)
Mineral Products (ISIC 36)	4799	0,073				312	38		(45)
Food and Related (ISIC 31)	14488	0,0148				181	33		(46)
Transport Eqpt. (ISIC 384)	8194	0,0105				55	31		(47)
Key Industry (N.E.C.) [2]	4119	0,0187				58	19		(48)
Key Industry (N.E.C.)	2242 14570	0,045 0,0065				89 69	12 26		(49)
Misc. Manufacturing	47710	0,009				310	119		(50)
N.E.C. = Not Elsewhere Classified					Total	2276	659	904 [3]	

1) Value Added in constant 1972 U.S. dollars based on an exchange rate of US$1 = 3,19 DM.

2) Key Industries: Textiles
Glass and Related
Machinery.

3) Additional non-energy use occurs in road construction, etc. (10^{15} J).

NATIONAL ENERGY DEMAND PROJECTIONS

Country Federal Republic of Germany

Year 1985

WAES Case B

Residential

Population $60,7 \times 10^6$ Persons/Dwelling 2,52 Dwellings $24,1 \times 10^6$

Central Space Heating		(A) Dwellings (10^3)	(B) Heat Loss[7] (kcal/Dw)	(C) System Efficiency	(D) Duration (hrs/y)	(E) Fossil Fuel (10^{15} J)	(F) Electricity	
Single Family	Fossil Fuels	[2] 6134/428	135 x 93	0,65/0,9	1600	796/40		(51)
	Electricity	[3] 1643	110 x 93	0,98	825		59	(52)
Town or Row Houses[1]	Fossil Fuels	[2] 6946/2995	110 x 68	0,65/0,9	1600	537/167		(53)
	Electricity	[3] 1973	102 x 68	0,98	825		48	(54)
Apartments	Fossil Fuels	[2]						(55)
	Electricity	[3]						(56)
		% Saturation (10^3) (Fossil Fuels)	% Saturation (10^3) (Electricity)	Energy/Unit (Fossil Fuels) (10^9 J)	Energy/Unit (Electricity) (kWh/y)			
Room Heating		3980	8000 [4]	32,2	400	128	12	(57)
Air Conditioning (Window)								(58)
Air Conditioning (Central)								(59)
Water Heating[5]		10500	14500	6,5	1300	68	68	(60)
Refrigerators			23300		220		18	(61)
Lighting			24100		350		31	(62)
Cooking[5]		7500	16700	4,5	600	34	36	(63)
Dishwashers			8500		650		20	(64)
Freezers			12500		400		18	(65)
Television			23000		200		17	(66)
Other Key Appliances[6]			20000		450		32	(67)
Misc. Appliances							15	(68)
Potential Appliances						5	15	(69)
					Total	1775	389	

1) Instead of "Town or Row Houses" and "Apartments," read "Over 2-Family" and "Other" for the Base Case and "All Other than 1- and 2-Family Houses" for the 1985 cases.
2) Fossil Fuels/District Heat.
3) Electric Heat Storage.
4) Electric Direct Heat.
5) Saturation exceeds 100%.
6) Washer.
7) kcal/m^2 x average m^2/dwelling.

NATIONAL ENERGY DEMAND PROJECTIONS

Country Federal Republic of Germany

Year 1985

WAES Case C

Transportation

	(A)	(B)	(C)	(D)	(E)	(F)	(G)	
	Population/Auto	(km) Distance/Auto/Yr.	Load Factor	Fraction Urban or Rural	(10^6 J/100km) Energy/Distance	(10^{15} J) Fossil Fuel	Electricity	
Urban Auto	2,8	14300		35 %	365	396,0		(1)
Rural Auto		13750		65 %	322	623,9		(2)
	(10^6 pkm-tkm) Activity Level	% Electrified	Load Factor	Fossil Fuel/ Activity Level	Electricity/Activity			
Urban Bus[1]	50830	30		(10^6 J/pkm-tkm) 0,365	0,59	13,0	9,0	(3)
Intercity Bus[1]	16310			0,29		4,7		(4)
Rail (Passenger)	49375	83		3,7	0,58	31,0	23,8	(5)
Air (Passenger, Domestic)	4900			3,1		15,2		(6)
Air (Passenger, International)	32800			3,1		101,7		(7)
Ship (Passenger)								(8)
Rail (Freight)	86642	80		1,2	0,18	20,8	12,8	(9)
Air (Freight, Domestic)	50			31		1,6		(10)
Air (Freight, International)	2000			31		62,0		(11)
Truck (Local)	56084			2,3		129,0		(12)
Truck (Intercity)	69575			2,3		160,0		(13)
Ship (Freight, Domestic)	37156			1,15		42,7		(14)
Ship (Freight, International)						200		(15)
Miscellaneous								(16)
					Total	1801,6	45,6	

1) Instead of "Urban" and "Intercity Bus," read "Scheduled" and "Unscheduled Public Road Transport."

NATIONAL ENERGY DEMAND PROJECTIONS

Country Federal Republic of Germany

Year 1985

WAES Case C

Agriculture, Mining, and Construction

	(A)	(B)	(C)	(D)	(E)	(F)	(G)	(H)	
Agriculture, Forestry, and Fishing (ISIC Major Division 1)	(10^6 \$) Value Added	(10^6 J/\$) Energy/Value Added	Area Harvested	Energy/Area	Physical Output	Energy/Output	(10^{15} J) Fossil Fuel	Electricity	
Farm Machinery									(17)
Greenhouses									(18)
Other Agricultural Uses	26643	7,5					160	40	(19)
Forestry (ISIC 12)							4	1	(20)
Fishing Boats (ISIC 13)									(21)
Other Fishing Use (ISIC 13)							9		(22)
Total Fertilizer (tons)		N ____	P ____	K ____		Total	173	41	

	(A)	(B)	(C)	(D)	(E)	(F)	
Mining	Value Added	Energy/Value Added	Physical Output	Energy/Physical Output	Fossil Fuel	Electricity	
Handling and Internal Transportation	260	96			20	5	(23)
Other Energy Use							(24)
				Total			

	(A)	(B)	(C)	(D)	(E)	(F)	
Construction	Value Added	Energy/Value Added	Physical Output	Energy/Physical Output	Fossil Fuel	Electricity	
Handling and Internal Transportation	31388	7,9			235,4	12,6	(25)
Other Energy Use							(26)
				Total			
				Agriculture, Mining, and Construction Total	428,4	58,6	

1) All monetary variables (Value Added and GDP-Contribution) are in constant 1972 U.S. dollars based on an exchange rate of US\$1 = 3,19 DM.

NATIONAL ENERGY DEMAND PROJECTIONS

Country Federal Republic of Germany

Year 1985

WAES Case C

Commercial and Public

Commercial	(A) Floor Area	(B) Heat Gain or Loss/Area	(C) System Efficiency	(D) Fossil Fuel (10^{15}J)	(E) Electricity	
Space Heat (Fossil Fuel)						(27)
Space Heat (Electricity)						(28)
Air Conditioning (Fossil Fuel)						(29)
Air Conditioning (Electricity)						(30)
	Value Added (10^6\$)	Energy/ V A (10^6 J/\$)	Electricity/ V A (10^6 J/\$)			
Lighting						(31)
Miscellaneous	56973	1o	3,4	57o	193,7	(32)
Presently Unknown						(33)
			Total			
Public	Floor Area (10^6 m^2)	Heat Gain or Loss/Area (Gcal/m^2/y)	System Efficiency	Fossil Fuel	Electricity	
Space Heat (Fossil Fuel)	99	17oo	o,7	1o1		(34)
Space Heat (Electricity)						(35)
Air Conditioning (Fossil Fuel)						(36)
Air Conditioning (Electricity)						(37)
	GDP-Contribution (10^6\$)	Energy/	Electricity/ GDP (10^6 J/\$)			
Lighting & Power	33915		2,93		99,4	(38)
Miscellaneous				14o		(39)
Presently Unknown						(40)
			Total	241	99,4	
			Commercial and Public Total	811	293,1	

1) All monetary variables (Value Added and GDP-Contribution) are in constant 1972 U.S. dollars based on an exchange rate of US\$1 = 3,19 DM.

NATIONAL ENERGY DEMAND PROJECTIONS

Country Federal Republic of Germany

Year 1985

WAES Case C

Industrial

	(A) Value Added [1]	(B) Energy/Value Added	(C) Physical Output	(D) Energy/Output	(E) % Exported	(F) Fossil Fuel	(G) Electricity	(H) Feedstocks	
	(10^6 \$)	(10^9 J/\$)				(10^{15} J)	(10^{15} J)	(10^{15} J)	
Iron and Steel (ISIC 371)	5165	0,175				798	108		(41)
Nonferrous Metals (ISIC 372)	1954	0,06				64	53	24	(42)
Paper and Allied (ISIC 34)	1384	0,085				79	39		(43)
Chemicals and Allied (ISIC 35 minus ISIC 353)	26978	0,029				485	297	1030	(44)
Mineral Products (ISIC 36)	5052	0,075				339	40		(45)
Food and Related (ISIC 31)	15076	0,016				205	36		(46)
Transport Eqpt. (ISIC 384)	8735	0,0107				62	33		(47)
Key Industry (N.E.C.) [2]	4604	0,019				66	21		(48)
Key Industry (N.E.C.)	2342 16999	0,047 0,067				97 83	13 31		(49)
Misc. Manufacturing	60343	0,009				387	156		(50)
N.E.C. = Not Elsewhere Classified					Total	2665	827	1054 [3]	

1) Value Added in constant 1972 U.S. dollars based on an exchange rate of US\$1 = 3,19 DM.
2) Key Industries: Textiles
Glass and Related
Machinery.
3) Additional non-energy use occurs in road construction, etc. (10^{15} J).

NATIONAL ENERGY DEMAND PROJECTIONS

Country Federal Republic of Germany

Year 1985

WAES Case C

Residential

Population $60{,}7 \times 10^6$ Persons/Dwelling 2,43 Dwellings $24{,}97 \times 10^6$

Central Space Heating		(A) Dwellings (10^3)	(B) Heat Loss (kcal/Dw)[7]	(C) System Efficiency	(D) Duration (hrs/y)	(E) Fossil Fuel (10^{15}J)	(F) Electricity	
Single Family	Fossil Fuels	[2] 6544/478	135 x 95	0,65/0,9	1650	895/47		(51)
	Electricity	[3] 1733	110 x 95	0,98	850		66	(52)
Town or Row Houses [1]	Fossil Fuels	[2] 7436/3145	110 x 70	0,65/0,9	1650	610/186		(53)
	Electricity	[3] 2083	102 x 70	0,98	850		54	(54)
Apartments	Fossil Fuels	[2]						(55)
	Electricity	[3]						(56)
		% Saturation (Fossil Fuels) (10^3)	% Saturation (Electricity) (10^3)	Energy/Unit (Fossil Fuels) (10^9J)	Energy/Unit (Electricity) (kWh/y)			
Room Heating		3551	7000[4]	32,2	350	114	9	(57)
Air Conditioning (Window)								(58)
Air Conditioning (Central)								(59)
Water Heating [5]		12000	15000	7,0	1200	84	65	(60)
Refrigerators			23800		220		19	(61)
Lighting			24970		400		36	(62)
Cooking [5]		7500	17600	4,5	550	33	35	(63)
Dishwashers			9000		650		21	(64)
Freezers			13000		400		19	(65)
Television			23500		200		17	(66)
Other Key Appliances [6]			20600		450		33	(67)
Misc. Appliances							20	(68)
Potential Appliances						5	30	(69)
					Total	1974	424	

1) Instead of "Town or Row Houses" and "Apartments," read "Over 2-Family" and "Other" for the Base Case and "All Other than 1- and 2-Family Houses" for the 1985 cases.
2) Fossil Fuels/District Heat.
3) Electric Heat Storage.
4) Electric Direct Heat.
5) Saturation exceeds 100%.
6) Washer.
7) kcal/m^2 x average m^2/dwelling.

NATIONAL ENERGY DEMAND PROJECTIONS

Country Federal Republic of Germany

Year 1985

WAES Case D

Transportation

	(A)	(B)	(C)	(D)	(E)	(F)	(G)	
	Population/Auto	(km) Distance/Auto/Yr.	Load Factor	Fraction Urban or Rural	(10^6 J/100km) Energy/Distance	(10^{15} J) Fossil Fuel	Electricity	
Urban Auto	2,81	14300		35 %	370	400		(1)
Rural Auto		13750		65 %	325	627		(2)
	(10^6 pkm-tkm) Activity Level	% Electrified	Load Factor	Fossil Fuel/ Activity Level (10^6 J/pkm-tkm)	Electricity/Activity			
Urban Bus [1]	50600	30		0,37	0,59	13,1	9,0	(3)
Intercity Bus [1]	16300			0,3		4,9		(4)
Rail (Passenger)	49375	83		3,8	0,59	31,9	24,2	(5)
Air (Passenger, Domestic)	4700			3,1		14,6		(6)
Air (Passenger, International)	27000			3,1		83,7		(7)
Ship (Passenger)								(8)
Rail (Freight)	80900	80		1,23	0,19	19,9	12,3	(9)
Air (Freight, Domestic)	50			31		1,6		(10)
Air (Freight, International)	1800			31		55,8		(11)
Truck (Local)	51300			2,3		118,0		(12)
Truck (Intercity)	63600			2,3		146,3		(13)
Ship (Freight, Domestic)	31900			1,15		36,7		(14)
Ship (Freight, International)						190		(15)
Miscellaneous								(16)
					Total	1743,5	45,5	

1) Instead of "Urban" and "Intercity Bus," read "Scheduled" and "Unscheduled Public Road Transport."

NATIONAL ENERGY DEMAND PROJECTIONS

Country Federal Republic of Germany

Year 1985

WAES Case D

Agriculture, Mining, and Construction

	(A)	(B)	(C)	(D)	(E)	(F)	(G)	(H)	
Agriculture, Forestry, and Fishing (ISIC Major Division 1)	(10^6\$) Value Added	(10^6J/\$) Energy/Value Added	Area Harvested	Energy/ Area	Physical Output	Energy/ Output	(10^{15}J) Fossil Fuel	Electricity	
Farm Machinery									(17)
Greenhouses									(18)
Other Agricultural Uses	26643	7,5					160	40	(19)
Forestry (ISIC 12)							4	1	(20)
Fishing Boats (ISIC 13)									(21)
Other Fishing Use (ISIC 13)							9		(22)
Total Fertilizer (tons)		N	P	K		Total	173	41	

	(A)	(B)	(C)	(D)	(E)	(F)	
Mining	Value Added	Energy/Value Added	Physical Output	Energy/Physical Output	Fossil Fuel	Electricity	
Handling and Internal Transportation	260	96			20	5	(23)
Other Energy Use							(24)
				Total			

	(A)	(B)	(C)	(D)	(E)	(F)	
Construction	Value Added	Energy/Value Added	Physical Output	Energy/Physical Output	Fossil Fuel	Electricity	
Handling and Internal Transportation	28900	7,9			216,8	11,6	(25)
Other Energy Use							(26)
				Total			
				Agriculture, Mining, and Construction Total	409,8	57,6	

1) All monetary variables (Value Added and GDP-Contribution) are in constant 1972 U.S. dollars based on an exchange rate of US\$1 = 3,19 DM.

NATIONAL ENERGY DEMAND PROJECTIONS

Country Federal Republic of Germany

Year 1985

WAES Case D

Commercial and Public

Commercial	(A) Floor Area	(B) Heat Gain or Loss/Area	(C) System Efficiency	(D) Fossil Fuel (10^{15} J)	(E) Electricity	
Space Heat (Fossil Fuel)						(27)
Space Heat (Electricity)						(28)
Air Conditioning (Fossil Fuel)						(29)
Air Conditioning (Electricity)						(30)
	Value Added (10^6 \$)	Energy/ VA (10^6 J/\$)	Electricity/ VA (10^6 J/\$)			
Lighting						(31)
Miscellaneous	53140	10	3,4	531	180,7	(32)
Presently Unknown						(33)
			Total			

Public	($10^6 m^2$) Floor Area	(Gcal/m^2/y) Heat Gain or Loss/Area	System Efficiency	Fossil Fuel	Electricity	
Space Heat (Fossil Fuel)	97	1750	0,7	102		(34)
Space Heat (Electricity)						(35)
Air Conditioning (Fossil Fuel)						(36)
Air Conditioning (Electricity)						(37)
	(10^6 \$) GDP-Contribution	Energy/Area	Electricity/ GDP (10^6 J/\$)			
Lighting	32310		2,93		94,7	(38)
Miscellaneous				150		(39)
Presently Unknown						(40)
			Total	252	94,7	
			Commercial and Public Total	783	275,4	

1) All monetary variables (Value Added and GDP-Contribution) are in constant 1972 U.S. dollars based on an exchange rate of US\$1 = 3,19 DM.

NATIONAL ENERGY DEMAND PROJECTIONS

Country Federal Republic of Germany

Year 1985

WAES Case D

Industrial

	(A)	(B)	(C)	(D)	(E)	(F)	(G)	(H)	
	Value Added[1)]	Energy/Value Added	Physical Output	Energy/Output	% Exported	Fossil Fuel	Electricity	Feedstocks	
	(10^6 $)	(10^9 J/$)				(10^{15} J)	(10^{15} J)	(10^{15} J)	
Iron and Steel (ISIC 371)	4545	0,185				750	91		(41)
Nonferrous Metals (ISIC 372)	1651	0,062				56	46	24	(42)
Paper and Allied (ISIC 34)	1223	0,09				75	35		(43)
Chemicals and Allied (ISIC 35 minus ISIC 353)	19523	0,0295				361	215	930	(44)
Mineral Products (ISIC 36)	4799	0,076				327	38		(45)
Food and Related (ISIC 31)	14488	0,0163				203	33		(46)
Transport Eqpt. (ISIC 384)	8447	0,0107				58	23		(47)
Key Industry (N.E.C.)[2)]	4119	0,0191				61	18		(48)
Key Industry (N.E.C.)	2242 14341	0,0475 0,0067				94 70	12 26		(49)
Misc. Manufacturing	47285	0,009				308	118		(50)
N.E.C. = Not Elsewhere Classified					Total	2363	664	954[3)]	

1) Value Added in constant 1972 U.S. dollars based on an exchange rate of US$1 = 3,19 DM.

2) Key Industries: Textiles
Glass and Related
Machinery.

3) Additional non-energy use occurs in road construction, etc. (10^{15} J).

NATIONAL ENERGY DEMAND PROJECTIONS

Country Federal Republic of Germany

Year 1985

WAES Case D

Residential

Population 60,7 x 10^6 Persons/Dwelling 2,52 Dwellings 24,1 x 10^6

Central Space Heating		(A) Dwellings (10^3)	(B) Heat Loss[7] (kcal/Dw)	(C) System Efficiency	(D) Duration (hrs/y)	(E) Fossil Fuel (10^{15} J)	(F) Electricity	
Single Family	Fossil Fuels	2) 6544/328	140 x 93	0,65/0,9	1650	908/33		(51)
	Electricity	3) 1333	110 x 93	0,98	850		50	(52)
Town or Row Houses 1)	Fossil Fuels	2) 7636/2495	115 x 68	0,65/0,9	1650	637/150		(53)
	Electricity	3) 1783	105 x 68	0,98	850		46	(54)
Apartments	Fossil Fuels	2)						(55)
	Electricity	3)						(56)

	% Saturation (Fossil Fuels) (10^3)	% Saturation (Electricity) (10^3)	Energy/Unit (Fossil Fuels) (10^9 J)	Energy/Unit (Electricity) (kWh/y)			
Room Heating	3980	7000 4)	32,2	400	128	13	(57)
Air Conditioning (Window)							(58)
Air Conditioning (Central)							(59)
Water Heating 5)	12000	13500	7,0	1250	84	61	(60)
Refrigerators		23300		220		18	(61)
Lighting		24100		370		32	(62)
Cooking 5)	7900	16300	4,5	600	36	35	(63)
Dishwashers		8500		650		20	(64)
Freezers		12500		400		18	(65)
Television		23000		200		17	(66)
Other Key Appliances 6)		20000		450		32	(67)
Misc. Appliances						15	(68)
Potential Appliances					5	15	(69)
				Total	1981	372	

1) Instead of "Town or Row Houses" and "Apartments," read "Over 2-Family" and "Other" for the Base Case and "All Other than 1- and 2-Family Houses" for the 1985 cases.
2) Fossil Fuels/District Heat.
3) Electric Heat Storage.
4) Electric Direct Heat.
5) Saturation exceeds 100%.
6) Washer.
7) kcal/m^2 x average m^2/dwelling.

NATIONAL ENERGY DEMAND PROJECTIONS

Country Federal Republic of Germany

Year 1985

WAES Case E

Transportation

	(A)	(B)	(C)	(D)	(E)	(F)	(G)	
	Population/Auto	(km) Distance/Auto/Yr.	Load Factor	Fraction Urban or Rural	(10^6 J/100km) Energy/Distance	(10^{15} J) Fossil Fuel	Electricity	
Urban Auto	2,8	14300		35 %	375	407		(1)
Rural Auto		13750		65 %	330	639		(2)
	(10^6 pkm÷tkm) Activity Level	% Electrified	Load Factor	Fossil Fuel/ Activity Level (10^6 J/pkm-tkm)	Electricity/Activity			
Urban Bus[1)]	50830	30		0,37	0,59	13,2	9,0	(3)
Intercity Bus[1)]	16310			0,3		4,9		(4)
Rail (Passenger)	49375	83		3,8	0,58	31,9	23,8	(5)
Air (Passenger, Domestic)	4900			3,2		15,7		(6)
Air (Passenger, International)	32800			3,2		105,0		(7)
Ship (Passenger)								(8)
Rail (Freight)	86642	80		1,23	0,18	21,3	12,5	(9)
Air (Freight, Domestic)	50			32		1,6		(10)
Air (Freight, International)	2000			32		64,0		(11)
Truck (Local)	56084			2,33		130,7		(12)
Truck (Intercity)	69575			2,33		162,1		(13)
Ship (Freight, Domestic)	37156			1,15		42,7		(14)
Ship (Freight, International)						200		(15)
Miscellaneous								(16)
					Total	1839,1	45,3	

1) Instead of "Urban" and "Intercity Bus," read "Scheduled" and "Unscheduled Public Road Transport."

NATIONAL ENERGY DEMAND PROJECTIONS

Country Federal Republic of Germany

Year 1985

WAES Case E

Agriculture, Mining, and Construction

Agriculture, Forestry, and Fishing (ISIC Major Division 1)	(A) Value Added (10^6 \$)	(B) Energy/Value Added (10^6 J/\$)	(C) Area Harvested	(D) Energy/Area	(E) Physical Output	(F) Energy/Output	(G) Fossil Fuel (10^{15} J)	(H) Electricity (10^{15} J)	
Farm Machinery									(17)
Greenhouses									(18)
Other Agricultural Uses	26643	7,5					160	40	(19)
Forestry (ISIC 12)							4	1	(20)
Fishing Boats (ISIC 13)									(21)
Other Fishing Use (ISIC 13)							9		(22)
Total Fertilizer (tons)		N	P	K		Total	173	41	

Mining	(A) Value Added	(B) Energy/Value Added	(C) Physical Output	(D) Energy/Physical Output	(E) Fossil Fuel	(F) Electricity	
Handling and Internal Transportation	260	98			21,3	4,2	(23)
Other Energy Use							(24)
				Total			

Construction	(A) Value Added	(B) Energy/Value Added	(C) Physical Output	(D) Energy/Physical Output	(E) Fossil Fuel	(F) Electricity	
Handling and Internal Transportation	31388	7,9			235,4	12,6	(25)
Other Energy Use							(26)
				Total			
				Agriculture, Mining, and Construction Total	429,7	57,8	

1) All monetary variables (Value Added and GDP-Contribution) are in constant 1972 U.S. dollars based on an exchange rate of US\$1 = 3,19 DM.

NATIONAL ENERGY DEMAND PROJECTIONS

Country Federal Republic of Germany

Year 1985

WAES Case E

Commercial and Public

Commercial	(A) Floor Area	(B) Heat Gain or Loss/Area	(C) System Efficiency	(D) Fossil Fuel (10^{15} J)	(E) Electricity	
Space Heat (Fossil Fuel)						(27)
Space Heat (Electricity)						(28)
Air Conditioning (Fossil Fuel)						(29)
Air Conditioning (Electricity)						(30)
	Value Added (10^6 $)	Energy/ VA (10^6 J/$)	Electricity/ VA (10^6 J/$)			
Lighting						(31)
Miscellaneous	56973	10,3	3,4	587	193,7	(32)
Presently Unknown						(33)
			Total			

Public	(10^6 m^2) Floor Area	(Gcal/m^2/y) Heat Gain or Loss/Area	System Efficiency	Fossil Fuel	Electricity	
Space Heat (Fossil Fuel)	99	1750	0,7	107		(34)
Space Heat (Electricity)						(35)
Air Conditioning (Fossil Fuel)						(36)
Air Conditioning (Electricity)						(37)
	(10^6 $) GDP-Contribution	Energy/Area	Electricity/ GDP(10^6 J/$)			
Lighting	33915		2,93		99,4	(38)
Miscellaneous				161		(39)
Presently Unknown						(40)
			Total	265	99,4	
			Commercial and Public Total	852	293,1	

1) All monetary variables (Value Added and GDP-Contribution) are in constant 1972 U.S. dollars based on an exchange rate of US$1 = 3,19 DM.

NATIONAL ENERGY DEMAND PROJECTIONS

Country Federal Republic of Germany

Year 1985

WAES Case E

Industrial

	(A) Value Added 1)	(B) Energy/Value Added	(C) Physical Output	(D) Energy/Output	(E) % Exported	(F) Fossil Fuel	(G) Electricity	(H) Feedstocks	
	(10^6 $)	(10^9 J/$)				(10^{15} J)	(10^{15} J)	(10^{15} J)	
Iron and Steel (ISIC 371)	5165	0,195				893	114		(41)
Nonferrous Metals (ISIC 372)	1954	0,062				66	55	24	(42)
Paper and Allied (ISIC 34)	1384	0,09				85	40		(43)
Chemicals and Allied (ISIC 35 minus ISIC 353)	27172	0,03				516	299	1120	(44)
Mineral Products (ISIC 36)	5052	0,079				359	40		(45)
Food and Related (ISIC 31)	15076	0,017				220	36		(46)
Transport Eqpt. (ISIC 384)	8910	0,0107				61	34		(47)
Key Industry (N.E.C.) 2)	4604	0,0195				69	21		(48)
Key Industry (N.E.C.)	2342 16835	0,05 0,007				104 88	13 30		(49)
Misc. Manufacturing	60138	0,009				385	156		(50)
					Total	2846	838	1144	

N.E.C. = Not Elsewhere Classified

1) Value Added in constant 1972 U.S. dollars based on an exchange rate of US$1 = 3,19 DM.
2) Key Industries: Textiles
Glass and Related
Machinery.
3) Additional non-energy use occurs in road construction, etc. (10^{15} J).

NATIONAL ENERGY DEMAND PROJECTIONS

Country Federal Republic of Germany

Year 1985

WAES Case E

Residential

Population $60,7 \times 10^6$ Persons/Dwelling 2,43 Dwellings $24,97 \times 10^6$

Central Space Heating		(A) Dwellings (10^3)	(B) Heat Loss[7] (kcal/Dw)	(C) System Efficiency	(D) Duration (hrs/y)	(E) Fossil Fuel (10^{15} J)	(F) Electricity	
Single Family	Fossil Fuels	[2] 7094/378	140/95	0,65/0,9	1650	1006/39		(51)
	Electricity	[3] 1283	110/95	0,98	850		49	(52)
Town or Row Houses[1]	Fossil Fuels	[2] 8186/2645	115/70	0,65/0,9	1650	703/164		(53)
	Electricity	[3] 1833	105/70	0,98	850		49	(54)
Apartments	Fossil Fuels	[2]						(55)
	Electricity	[3]						(56)
		% Saturation (Fossil Fuels) (10^3)	% Saturation (Electricity) (10^3)	Energy/Unit (Fossil Fuels) (10^9 J)	Energy/Unit (Electricity) (kWh/y)			
Room Heating		3550	7000[4]	32,2	350	114	9	(57)
Air Conditioning (Window)								(58)
Air Conditioning (Central)								(59)
Water Heating[5]		12000	15000	7,0	1200	84	65	(60)
Refrigerators			23800		220		19	(61)
Lighting			24970		400		36	(62)
Cooking[5]		8000	17200	4,5	550	36	34	(63)
Dishwashers			9000		650		21	(64)
Freezers			13000		400		19	(65)
Television			23500		200		17	(66)
Other Key Appliances[6]			20600		450		33	(67)
Misc. Appliances							20	(68)
Potential Appliances						5	35	(69)
					Total	2151	397	

1) Instead of "Town or Row Houses" and "Apartments," read "Over 2-Family" and "Other" for the Base Case and "All Other than 1- and 2-Family Houses" for the 1985 cases.
2) Fossil Fuels/District Heat.
3) Electric Heat Storage.
4) Electric Direct Heat.
5) Saturation exceeds 100%.
6) Washer.
7) kcal/m^2 x average m^2/dwelling.

8 Italy

8.1 INTRODUCTION

8.1.1 National Structure and Reference Group

The Italian WAES study is being carried out under the supervision of two Participants: Prof. Umberto Colombo, Director, Research and Development Division, Montedison, and Prof. Sergio Vaccà, Director, Istituto di Economia delle Fonti di Energia, Università Bocconi. A team of three Associates—Dr. Riccardo Galli, Dr. Oliviero Bernardini, and Dr. William Mebane—organizes and coordinates activities in Italy related to WAES.

Key Italian organizations active in the energy context form the backbone of the support reference group. These include the main energy industries, financial holdings, and industrial firms and several institutes active in the socioeconomic and energy fields.

Contributions on specific problems requiring technical expertise also come from Ministries, industrial and other category associations, municipal authorities, and a large number of industries and companies.

Activities are organized in five working groups that meet periodically and function both as expertise trusts and control committees focusing, respectively, on socioeconomic development, energy supply, energy demand in industry, energy demand for civilian uses, and energy demand in transport.

The first and most arduous problem tackled by the working groups on energy demand was establishing a data base in the highly disaggregated form necessary for the demand projections and moreover extending sufficiently far back in time as to permit incorporation of the fundamental dynamic components of change. The data collected by the working groups and partly presented in this work are to be considered on the whole completely original and should be useful both as a starting point for improving our knowledge of the Italian energy system and as a basis for defining rational energy policies.

Overall the data collected amounts to about 20,000 pieces of information on activity and energy demand in some 250 sectors over the period 1953 to 1973, an average of 80 pieces of information per sector. This is generally sufficient to permit fairly accurate projections over periods of the range involved in this study.

Through lack of space it has not been possible to include all the details involved in the projections of energy demand and the underlying data on which these are based. A more complete description of the Italian energy system and projections will be published shortly in Italian.

8.1.2 Background Energy Situation

Total energy delivered for final consumption in Italy during 1972 amounted to 953.7×10^{12} kcal, an average of 17.4×10^{6} kcal/resident. Electricity accounted for 10.4 percent of this energy or 114.7×10^{9} kWh.

During the period 1953 to 1972 total delivered energy increased at an average annual rate which was among the highest in the world: 8.7 percent. Petroleum products, which accounted for most of the growth, increased their share in total delivered energy from 33.3 percent in 1953 to 71.6 percent in 1972 (Table 8.1).

Italy is one of the few countries for which delivered electricity grew slower than fossil fuels: 7.9 as compared to 8.8 percent per annum during 1953 to 1972. This is basically a consequence of the high share of industry in total energy consumption. Electricity delivered to industry grew at an average annual rate of 7.3 percent, as compared to 8.9 percent for fossil fuels (Table 8.2), reflecting the predominance of chemicals and petrochemicals in energy consumption. Over forty percent of all energy delivered to industry went to this sector.

The rapid growth of energy consumption is not likely to continue into the future quite apart from effects of economic growth. The elasticity of ener-

Table 8.1
Fueling of the Italian energy system in 1953 and 1972 (%)

	1953	1972
Wood	15.8	1.3
Coal	29.4	5.1
Petroleum	33.3	71.6
Natural Gas	9.8	11.7
Electricity	11.7	10.3
Total	100	100

Table 8.2
Energy growth rates during the period 1953 to 1972 (%)

	Electricity	Fossil Fuels	Total	Sector Share (%)	
				1953	1972
Industry	7.3	8.9	8.7	49.4	49.3
Transportation	2.7	8.9	8.7	18.5	18.2
Civilian	11.0	8.7	8.9	29.4	30.2
Agriculture	6.1	7.1	7.1	2.7	2.3
Total	7.9	8.8	8.7	100	100

gy growth with respect to gross national product actually increased from 1.55 to 1.62 in the decades 1953 to 1963 and 1963 to 1973. This increase is, however, largely due to phenomena such as private automobile travel and conversion of household heating systems, which, though by no means over, are rapidly reaching equilibrium.

While it is tempting to extrapolate historic trends into the future, it must be recognized that since it is a nonequilibrium system the projection of Italian energy demand can be quite misleading unless tied at a disaggregated level to the time-dependent phenomena that determine it.

Italy is poor in natural resources. In 1972 it produced only 15.5% of its primary energy input. Most of this was natural gas and hydroelectric power (respectively 7.6 and 3.0 percent).

The 1973 oil crisis caused considerable damage to the Italian economy. The balance of payments deficit, which had reached 1075 billion Lire in 1972 and had increased to 3559 billion Lire in 1973, jumped to 7218 billion Lire in 1974 in the face of a general economic deflator of 1.66 between 1972 and 1974.

The subsequent drop to 2672 billion Lire in 1975 was due largely to the profound economic crisis that hit the country (gross national product decreased some 3.9 percent and industrial production 9.7 percent), reducing energy consumption and petroleum imports.

In the face of increasing demand for energy and in a world of increasing competition for remaining energy sources, there is a clear necessity to understand the Italian energy system and to enact adequate energy policies and long-term programs.

The essential objectives of a national energy policy are:

- to decrease the relative importance of oil as an energy source;
- to maximize the domestic supply of primary energy;
- to guarantee geographic diversification of assured energy sources;
- to promote energy savings through rationalization of use;
- to favor the most appropriate allocation of energy sources with an eye to ecological externalities;
- to increase research and development on primary and secondary energy sources and on the energy system as a whole.

8.2 METHODOLOGY

8.2.1 The Determinants of Energy Demand

Three basic variables—economic development G, the price of oil P, and policy response R—are assumed to exhaust all degrees of freedom of the energy demand system. In this approximation all other determinants of demand V must either have negligible effect on demand or must be forced into some functional relation with the basic demand variables:

$$V \approx h(G,P,R),$$

so that total energy demand E is

$$E = g(G,P,R, \{ V \}) \approx f(G,P,R).$$

Economic development as measured by gross national product, the price of oil, and policy response are assumed not to influence one another in any way and to be separable in the relation

$$E \approx a(G)\, b(P)\, c(R),$$

which to a good approximation can be written

$$E \approx K\, G^{\alpha}\, P^{\beta}\, R^{\gamma}.$$

This expression provides a convenient way of sounding the sensitivity of energy demand to its different determinants.

The coefficients α, β, γ are estimated through projections of energy demand from the base year 1972 to 1985 for five different sets of values of the basic variables (scenarios) allowing linear equations with α, β, γ as unknowns.

For each set of values (G,P,R) energy demand is projected at the subsystem and sector level as the product of the sector activity A_i and the technical coefficient T_i, both determined as functions of the three basic variables:

$$E = \sum_i E_i = \sum_i A_i(G,P,R)\, T_i(G,P,R).$$

Policy Response

A vigorous policy response is defined as the maximum possible within technical, financial, and life-style constraints. This definition does not imply an implementation judgment. Insofar as a vigorous policy response as defined here places secondary emphasis on institutional and political constraints, it does not necessarily have any great likelihood of materializing.

A restrained policy response is defined as the least expected, independently of the oil price and economic assumptions of the scenario.

The choice of definitions is suggested by the desire to determine the complete sensitivity of the energy system to the policy response variable. In any event, the likelihood of implementation of a vigorous energy policy is not necessarily small since institutional and political constraints are less binding than technical, financial, and life-style constraints and are subject to change even in the short term given adequate boundary conditions.

Policy measures are considered both at the general system level where they impact only on activities and in subsystem- and sector-specific applications where they impact only on technical coefficients.

The Price of Energy

The price of delivered fuel is determined as a function of the price of oil and policy response, the former acting on industrial costs, the latter on sector-specific fiscal exactions.

Economic Development

After recovery to 1973 levels by the end of 1976, the Italian economy is projected to develop to 1985 at an average annual rate of 4.5 and 2.5 percent in response to high and low world growth, respectively.

Considering that gross national product grew 6.5 percent during 1973, these figures translate into an average of 3.6 and 2.2 percent over the 13-year period extending from the base year of the analysis.

Whereas policy response is defined not to influence overall economic development in any way, the allocation of gross national product among consumption sectors of the economy is assumed to be fundamentally a function of this variable. Specifically, a vigorous energy policy is taken to favor public over private consumption and investment, whereas a restrained energy policy leaves the share of gross national product among these two components practically unchanged relative to historic values.

The rationale behind this choice is that a reduction in growth of private in favor of public consumption will slow down the diffusion of more energy-intensive consumer durables, at the same time allowing for increased spending in energy-saving public services such as urban mass transit, railways, and buses and government-subsidized incentives for saving, especially in household energy use.

At the general system level a vigorous response is also assumed to be consistent with a more general policy of increased productive investment in the South of Italy as opposed to the North. This has appreciable by-product savings effects both for transport and space heating.

In order to maintain identical growth in gross national product in scenarios with different oil prices (the separability hypothesis), a compensatory monetary policy is introduced favoring growth in the services sector at the expense of the industrial sector with increasing petroleum prices.

Besides the specific action of policy response at the general system level and the application of the separability hypothesis, the most important trends and assumptions underlying the scenarios are as follows. Relative to gross national product:

- private consumption will grow at less than historic rates. The economy is increasingly driven by exports rather than by private consumption. Higher inflation rates cut private consumption because of time lags between price and income rises. There is some substitution of public for private consumption.

• public consumption will tend to grow slower than in the past. There is a historic trend for slower growth in public expenditure. Large public debts, presently exceeding 10 percent of gross national product, will cut down growth. Projected reforms concentrate on making the already existing public services more efficient with some change in the mix of services.

• investment rates will remain slightly above the average of the period 1964 to 1974 but significantly below that of the period 1951 to 1963. The heavy industrialization phase of the fifties is not likely to be repeated, but new investments are needed for industrial conversion.

• exports will continue to grow rapidly but somewhat slower than in the previous periods. Exports will be the main driving force behind economic development. However, production costs are rising in terms of both labor and raw materials. Thus growth will be contained because of increasing competition in export markets, from both more developed economies with higher productivity and less developed economies with lower labor costs, in products with low and medium technological content.

• growth of imports will slow down slightly from the historical average. The economy needs to reach a more satisfactory balance of payments. The slowdown in private consumption is expected to help bring this about.

• stocks will increase at the same rate. The present ratio of stocks to sales is satisfactory and is assumed to remain stable in the long term.

• growth in agriculture will increase slightly. It is assumed that the long-awaited agricultural reforms will be passed, but that because of the structural nature of the problems involved (for example, extensive new irrigation) the increased growth will be moderate through 1985.

• industrial growth is projected to be lower than in the past but not substantially. There are higher costs of production and increased competition from countries with lower costs and more advanced technology.

• growth in construction will increase considerably over the period 1964 to 1974. This compensates for the very low average growth of that decade.

• the services sector will grow slightly below the historic rate. In general, disposable income is expected to grow more slowly than in the past, and existing services will be somewhat improved by rationalization and modernization.

The growth rates applied to quantify these hypotheses were tested for consistency with the aid of the econometric model "Link-Prometeia" developed at the University of Bologna. This model is of a short-term nature but nevertheless enables a test for consistency and allows adjustments to be made to reflect medium-term phenomena.

Historic growth rates of the basic components of gross national product are compared with the scenario projections in Table 8.3.

8.2.2 Activity Level

With the scenarios specified in terms of these basic components, the Italian energy system resolves into 25 essentially noninteracting subsystems. In the text these are grouped under the following headings:

urban passenger transport

long-distance passenger transport (two subsystems)

long-distance freight transport (two subsystems)

local freight transport

manufacturing and mining

agriculture, livestock, forestry, and fishing

construction

cooling (three subsystems)

street lighting

appliances (nine subsystems)

space heating (three subsystems).

The activity levels A_h of these subsystems are projected independently on the basis of functional relations with the scenario variables obtained through phenomenological analysis over the period

Table 8.3
Economic development hypotheses: Average annual growth rates of the basic components of gross national product, 1977-1985 (%)

Scenario Name	1954–1964	1964–1974	B	D	A	C	E
Gross World Product (%/yr.)			3.5	3.5	6.0	6.0	6.0
Energy Policy			Vigorous	Restrained	Vigorous	Vigorous	Restrained
Oil Price (1975 $/bbl)			11.50→17.25	11.50	11.50→17.25	11.50	11.50→7.66
Agriculture	3.1	1.2	1.5	1.5	2.5	2.5	2.5
Industry	7.5	5.3	2.4	2.7	4.9	5.1	5.3
Construction	6.6	1.3	1.85	1.85	2.4	2.4	2.4
Services	5.0	4.8	2.9	2.6	4.4	4.2	4.0
GNP	5.7	4.7	2.5	2.5	4.5	4.5	4.5
Private Consumption	5.5	5.0	1.65	2.0	3.4	3.45	3.9
Public Consumption	3.9	3.5	2.3	1.8	2.9	2.4	1.9
Investment	7.4	3.7	2.9	2.9	4.8	4.8	4.8
Exports	13.9	10.8	5.8	5.4	10.0	9.8	9.25
Imports	12.4	9.6	4.8	4.8	8.5	8.2	8.2
Change in Stocks	–	–	2.5	2.5	4.5	4.5	4.5

1953 to 1973:

$$A_h \approx \phi_h\,(G,P,R).$$

Further decomposition, if possible and when data are available, leads to subsystems that are not mutually independent. (Thus industry is held together by an input-output matrix; agriculture, forestry and fishing are specified together as one scenario variable; urban private and public transport are coupled systems; etc.)

The activity levels of these subsystems are projected on the basis of a simple substitution model which can be expressed in terms of the $\frac{1}{2}n_h(n_h-1)$ relations:

$$\frac{A_{hi}}{A_{hj}} = s_{hij}\,(G,P,R),$$

$$h = 1,2,\ldots 25,$$

$$i,j = 1,2,\ldots n_h,$$

$$\frac{A_{hi}}{A_n} = \sum_j \left(\frac{A_{hj}}{A_{hi}}\right)^{-1},$$

where n_h is the number of interactive subsystems and where the function s_{hij} is obtained by regression analysis on the determinants of substitution over a historic period of reasonable length.[1] Adjustments are introduced where necessary to correct for political decisions taken in recent years or other variations altering the historic trend.

Considerable simplification of treatment is possible by identifying structural hierarchies within the interactive systems. This identification permits the activity and substitution corresponding to different levels of the hierarchy to be projected independently of the other levels.[2]

System resolution is generally determined by the availability of data. Other cut-off criteria are the inadequacy of the scenario variables as activity specifiers and also their negligible contribution to total energy demand.

Under these constraints the Italian energy demand system is described and projected as 253 sectors distributed as follows:

urban passenger transport	18
long-distance passenger transport	15
long-distance freight transport	14

1. It can be shown that the projection methodology used is equivalent to cross-elasticity technique.
2. The reader is referred to articles to be published for further details, which cannot be illustrated in any detail in this report through limitations on space.

local freight transport	6
manufacturing and mining	69
agriculture, livestock, forestry, and fishing	4
construction	1
cooling	6
street lighting	1
appliances	27
space heating	92

8.2.3 Fuel Substitution

Fuel substitution is projected at the sector level as a function of policy response acting basically through relative fuel prices but also through legislation and other measures, within possible constraints such as peak load posed by the supply system.

Fuels considered are:

wood
coal and coke
fuel oil
gas oil
kerosene
jet fuel
gasoline
light distillates
liquid petroleum gas
natural gas
electricity.

8.2.4 Technical Coefficient

Technical coefficients are projected at the sector level fuel by fuel. Over the period to 1985 they are assumed to be determined essentially by policy response and the price of oil.

8.2.5 Energy Demand Projections in the WAES Format

Simple linear mappings permit reaggregation of activity and energy demand in the more highly aggregate standard WAES format.

8.2.6 A Remark on Methodology

The hierarchical systems methodology at the basis of the demand projections does not consider feedback effects between levels. The error introduced in this approximation is estimated to be quite negligible on the whole and anyway no greater than the uncertainty deriving from other approximations.

8.3 PRESENTATION AND ANALYSIS OF RESULTS

8.3.1 Transportation

Urban Passenger Transport

Urban passenger transport, defined to include commuter as well as metropolitan traffic, accounted for 39.4 percent of energy consumption in transport and 7.2 percent of total delivered energy in 1972.

The energy necessary to deliver the 146.7×10^9 passenger-km of urban travel registered during this year (7.4 km/day per inhabitant) was 68.5×10^{12} kcal, corresponding to 467 kcal/passenger-km.

Demand for urban passenger travel correlates with gross national product. Short-term cycles in the two growth rates observed over the period 1953 to 1972 match without delay through a ratio that has been rapidly decreasing and was already approaching zero in 1972.

Even considering population growth and continuing migration to the cities, urban travel demand should saturate by 1990. Under the assumption of a vigorous policy at the general system level—interpreted as decentralization of hospitals, schools, commercial centers, and industry over the coming years—it is estimated that urban transport demand could even begin decreasing around 1982.

Increases in oil price within the limits considered are not expected to have any depressing effect whatever on total travel demand, but only to determine shifts between travel modes.

The share of urban transport delivered by private means, which was only 31.4 percent in 1953, increased rapidly during the fifties and sixties and reached a maximum of 69.6 percent in 1970, after which it began showing signs of giving way again to public transport. This development coincided with the rapid diffusion of the automobile, determined by unprecedented growth in private income, and subsequent traffic congestion in cities, which forced many travelers to revert back to public transport.[3]

Historically, substitution between modes has been driven largely by private consumption, and more recently also by public spending and investment in mass transit, and is described through the differential cost of the passenger-km to the consumer in passing between private modes and to the municipal authorities financing public transport.

Projections to 1985 assume a continuation of the behavioral equations of the period 1953 to 1972 with oil-price-adjusted coefficients. Oil price changes impact directly on the cost of the passenger-km and are therefore important in determining differential modal shifts.

The results indicate considerable variation in substitution between scenarios. Increased public spending and high oil prices in the vigorous policy scenarios are responsible for an abrupt downturn in private urban travel. In the restrained policy scenarios, the increased price of oil has some effect in favoring public over private transport; this is, however, minor in the case of the high growth scenario E.

The share of urban transport going to private modes is estimated to change from its 1972 value of 66.9 percent as follows:

scenario A	47 percent
scenario B	55 percent
scenario C	43 percent
scenario D	59 percent
scenario E	70 percent.

The slowdown in private consumption growth and the high oil price is effective in delaying substitution in favor of larger automobiles in the vigorous policy scenarios and also in the low growth scenario D. In these cases average (activity-weighted) automobile capacity increases only very slightly (less than 5 percent) from its 1972 level of 1038 cc. In scenario E, however, the relatively lower oil price and the continued high growth in private consumption allow a move to larger automobiles, with average capacity increasing to 1166 cc.

Interfuel substitution is basically between gasoline and diesel and is a function of policy response. A vigorous policy response is defined to favor diesel over gasoline motors in buses and in automobiles above the 1000–1500 cc capacity class.

Improved traffic control using advanced computer techniques is introduced as a vigorous policy measure to decrease traffic congestion. Reorganization of public transport networks using systems techniques is estimated to increase the average load factor by 30 percent. Other vigorous policy measures include specific guidelines to industry for improvement in motors and suspension systems.

The average energy necessary to deliver a passenger-km of activity in 1985 decreases considerably relative to the 1972 value of 467 kcal in all scenarios including Case E:

scenario A	271 kcal/passenger-km
scenario B	269 kcal/passenger-km
scenario C	263 kcal/passenger-km
scenario D	397 kcal/passenger-km
scenario E	428 kcal/passenger-km.

Scenarios A, B, and C actually bring the energy efficiency of the system back to levels characteristic of the fifties.

Long-Distance Passenger Transport

In 1972, 49.5×10^{12} kcal were consumed to deliver 195.0×10^{9} passenger-km over long distances in Italy, an average of 254 kcal/passenger-km. This corresponds to 28.5 percent of energy consumption in transport and 5.1 percent of total energy delivered in that year.

Since 1953, long-distance passenger travel has grown at an average annual rate of 9.9 percent, almost exactly twice as fast as growth in private income, to which it is closely correlated in terms of short-term cycles.

Taking population growth into account, this amounts to a fivefold increase in average distance traveled by Italians over long distances: from 680 km in 1953 to 3525 km in 1972. This remarkable increase is due largely to the diffusion of pleasure and holiday travel, which was enjoyed by just over 13 percent of the population in 1953 and still had not reached one out of two Italians as late as 1972.

Even under the most optimistic projections of private income growth specified in the scenarios, not more than 60 percent of Italians are expected to benefit from this type of travel by 1985, leaving the system far from saturation and allowing for a hypothesis of continuation of the dynamics of

3. In the period 1953 to 1973 automobile diffusion per family increased from 5 to 82 percent.

growth experienced in the past.

In these projections it is assumed that increases in the price of oil within the limits considered have no effect on total travel demand, which is determined only by private income. On the other hand they are expected to result in shifts between travel modes.

Travel demand is projected as two essentially independent subsystems: travel requiring oversea crossings (travel between the islands or between the mainland and the islands) and travel in competition with terrestrial modes.

Travel requiring oversea crossings amounted to only 1.8×10^9 passenger-km in 1972 but is much faster growing and more energy-intensive than the bulk of long-distance travel delivered in competition with terrestrial modes.[4] Over the decade 1975 to 1985 it is projected to grow some three times as fast as gross national product, to which it has been correlated in the past, as a result of faster than average economic development on the islands and a pronounced increase in tourism.

In 1972 private means accounted for 79.9 percent of long-distance travel in competition with terrestrial modes. After a very rapid increase in share from 38.2 percent in 1953, however, private travel began showing signs of slowing down relative to public forms of travel, especially after increased investments in railways from 1967 onwards made railway travel more attractive to long-distance travelers. A move to air and private bus travel determined by increases in private income is also responsible for this slowdown, but these modes were still relatively unimportant in 1972.

Substitution between modes is projected to 1985 as a function of oil price, private income, and public expenditures through the differential cost of the passenger-km, much as for urban transport.

Growth in long-distance passenger travel continues to be so rapid up to 1985 that the most optimistic public expenditure scenarios are unable to effect a truly significant shift to the less energy-intensive terrestrial transport modes such as buses and trains. The share of terrestrial long-distance travel going to private means is projected as:

scenario A	76 percent
scenario B	81 percent
scenario C	73 percent
scenario D	82 percent
scenario E	87 percent.

This is to be compared with a share of 79.9 percent in 1972.

Air travel in competition with land travel is effectively slowed down in the vigorous policy scenarios that penalize private income in favor of public spending and investment. In trips requiring sea crossings, however, air travel is not so effectively inhibited, because of the practical disadvantages (especially travel time) of the other competing mode. Total air travel is therefore projected to develop rapidly over the decade in question, with an average annual rate as high as 13 percent.

The high oil price of the vigorous policy scenarios is expected to lead to a decrease in the energy intensiveness of this mode by between 35 and 50 percent, largely through technological development, thus partly counteracting the energy effect of the increase in activity. Scenario E, however, which benefits from neither a particularly high oil price nor a penalization of private income, results in a considerable increase—almost 500 percent over the 1972 levels—in energy delivered to air travel.

Policy measures envisaged in the projections are largely related to improvements in energy efficiency through fuel substitution (between gasoline and diesel) and guidelines to industry for improved technology. Some savings also come from the application of speed limits on normal highways.

The average energy necessary to power a passenger-km over long distances decreases somewhat relative to 1972 (254 kcal) in the vigorous policy scenarios, but not as spectacularly as for urban travel:

scenario A	199 kcal/passenger-km
scenario B	198 kcal/passenger-km
scenario C	202 kcal/passenger-km
scenario D	246 kcal/passenger-km
scenario E	287 kcal/passenger-km.

The substantial increase in scenario E is due basically to the shift to private modes, to more powerful automobiles, and to air travel. This scenario brings the energy efficiency of the system back to the level of the early fifties when steam trains still delivered 12.0 percent of all long-distance travel, consuming 41.8 percent of the total energy.

4. The average energy intensiveness of the two long-distance subsystems in 1972 was 1360 and 239 kcal/passenger-km, respectively.

Long-Distance Freight Transport

Freight travel over distances greater than 100 km amounted to 11.3×10^9 ton-km in 1972, consuming 38.7×10^{12} kcal with an average efficiency of 347 kcal/ton-km. In that year long-distance freight transport accounted for 22.3 percent of energy consumed for transport and 4.0 percent of total delivered energy.

Over the period 1953 to 1972 it increased at an average annual rate of 6.1 percent, only slightly faster than the growth of agriculture and industry to which it is closely related in terms of short-term cycles through a one-year delay.

In the long term the growth of long-distance freight travel relative to industry and agriculture is slowing down due to a decrease in average travel distance determined by an increase in industrial location density. Over the period 1975 to 1985 the ratio of the growth rates is projected as slightly lower for the vigorous policy scenarios (0.76 against 0.88) as a result of industrial siting policies at the general system level with emphasis on new investments in the South.

Increases in the price of oil are assumed not to affect either demand for freight travel or the modal mix since cost increases are passed on as service prices.

Freight transport demand is also projected as two essentially independent subsystems: travel requiring oversea crossings and travel in competition with terrestrial modes.

The former, which accounts for all travel between the islands and between the mainland and the islands, amounted to 3.0 percent of all long-distance freight travel in 1972: a mere 3.4×10^9 ton-km. Over the decade 1975 to 1985 it is, however, projected to grow almost two times as fast as the growth of industry and agriculture as a result of faster than average economic development on the islands, bringing its share of total freight transport to between 4.0 and 5.3 percent.

Historically, rail transport has consistently lost share (from 30.4 percent in 1953 to 14.7 percent in 1972), largely to sea travel and pipelines, which together more than tripled their share of total activity in competition with terrestrial modes during this period. Road travel gained share at the expense of railroads during the fifties, but after reaching a maximum of 64.7 percent in 1963, it began losing to sea and pipeline. In 1972 it represented 58.1 percent of all long-distance activity.

Substitution between modes is driven largely by gross national product, and projections to 1985 assume a continuation of the historic dynamics that tend to favor higher-capacity carriers.

The vigorous policy scenarios contemplate rapid growth in river and canal traffic at the expense of road carriers, largely as a consequence of investments related to the Po River network.

Results of the projection to 1985 indicate a continuation of rapid growth in sea travel under all scenarios, increasing the share of this mode to values as high as 23.2 percent. Road travel maintains its share but with considerable change in mix in favor of the larger and more efficient carriers. In particular, tractor-semitrailers benefiting from the particular role they play in relation to sea travel are boosted from 9.6 to over 20 percent of all road travel (20.8 and 26.5 percent, respectively, in the low growth and high growth scenarios). Railway traffic is projected to grow only very slowly and even to decrease in the low growth scenario B.

The energy efficiency of the long-distance freight system has tended to increase in the long term from values greater than 450 kcal/ton-km during the early fifties. This can be attributed partly to the penetration by sea transport, which more than doubled its share in activity from 7.9 percent to 16.2 percent between 1953 and 1973, but largely to the phasing out of very low efficiency steam trains. In 1953 these accounted for 38.8 percent of the energy consumed but delivered only 13.2 percent of freight activity over long distances.

The increase in efficiency would have been greater but for the simultaneous loss in efficiency of freight travel by road.

Average carrying capacity of trucks increased from 9.5 to 13.6 tons during the period 1953 to 1972, with a corresponding increase in energy intensity from 2690 to 3290 kcal/vehicle-km: a 22 percent theoretical improvement in energy efficiency. Average load factor, on the other hand, increased only from 8.2 to 9.2 tons, corresponding to an actual 9 percent deterioration.

Factors determining the 21 percent decrease in capacity utilization are largely related to the highly dispersed structure of the trucking sector—an average of 1.5 trucks/company in 1972!—and to legislation favoring lower-efficiency carriers since 1963.

Policy measures to increase the energy efficiency

of long-distance freight travel are principally envisaged as increasing capacity utilization. This can be achieved by abolishing the 1963 law prohibiting issue of new truck licenses to third-customer trucking companies and by favoring the creation of consortia providing computer services for optimal (time- and space-optimized) location of transport facilities.

Improvements in technology can increase efficiency by at most some 5–7 percent by 1985, but an increase in theoretical efficiency is expected anyway in the medium to long term due to an increase in average carrier capacity: 5 percent average over the whole stock but as much as 18 percent for the lower-capacity classes.

Prohibition of gasoline in favor of diesel fuel also allows for some energy savings.

Energy demand projections indicate savings in absolute terms, even with respect to 1972, for the vigorous policy scenarios, notwithstanding sustained increases in activity in Cases A and C.

The energy efficiency of the system improves slightly even in the restrained policy scenarios due largely to a change in modal mix in favor of sea travel and pipeline:

scenario A	190 kcal/ton-km
scenario B	191 kcal/ton-km
scenario C	191 kcal/ton-km
scenario D	334 kcal/ton-km
scenario E	335 kcal/ton-km.

This is just sufficient to counteract the deterioration of road transport efficiency (down from 498 kcal/ton-km in 1972 to 518 and 513 in scenarios D and E) due to the increase in average carrier capacity without a corresponding increase in capacity utilization.

Local Freight Transport

Local freight transport, defined as transport over distances less than 100 km, accounted for only 9.9 percent of energy consumption in transport and 1.8 percent of total delivered energy in 1972. The energy expended, 17.2×10^{12} kcal, resulted in delivery of only 13.5×10^{9} ton-km, an average of 1270 kcal/ton-km, almost four times as intensive as long-distance freight transport.

Historically, the dynamics of growth of local freight transport correlates with the aggregate of the construction, industrial, and agriculture sectors through an elasticity that is close to 0.5 and has been slowly declining. The projections of total demand assume a continuation of the historic relation with decreasing elasticity.

Cost effects due to changes in the price of oil are assumed to be passed on in services without ulterior effects on the level of either total transport demand or modal mix.

With the exception of three-wheeled vans, modes competing for local freight transport also deliver freight over long distances. Freight carriers in this system, however, tend to be concentrated in the low-capacity and therefore high-energy-intensity range.

Substitution between modes is driven by gross national product and systematically favors the higher-capacity road carriers.

The modal mix projected to 1985 shows increased emphasis on road as against railway transport, a continuing decrease in three-wheeled vans, and an increase in traffic delivered by the heavier trucks.

Policy measures contemplated specially for local freight transport relate to the structure and organization of trucking services, which constitute a highly dispersed sector with very low average capacity utilization and high propensity to waste. It is estimated that measures enacted either to create consortia between freight carriers or to establish special services such as provincial computer networks to assist in locating and scheduling transport facilities can increase average capacity utilization 1.5 times over the present low average value of 41 percent.

Legislation to discourage the use of gasoline in trucks in favor of diesel also increases the efficiency of the system.

The energy efficiency of the system improves over the 1972 level of 1270 kcal/ton-km by a factor of two in the vigorous policy scenarios and deteriorates in the restrained policy scenarios due to an increase in average capacity without a corresponding increase in capacity utilization:

scenario A	625 kcal/ton-km
scenario B	643 kcal/ton-km
scenario C	624 kcal/ton-km
scenario D	1560 kcal/ton-km
scenario E	1500 kcal/ton-km.

8.3.2 Manufacturing and Mining

Manufacturing and mining accounted for 49.3 percent of total delivered energy in 1972, corresponding to 470.4×10^{12} kcal, and contributed 29.3 percent or $17{,}763 \times 10^{9}$ Lire (value added expressed in 1972 Lire at factor cost) to gross national product.

The resulting average energy intensity of Italian industry in 1972 was therefore relatively low (26.5 kcal/Lira), but it is almost unique among industrialized countries in that it has shown a consistent increase over time.

During the period 1953 to 1973, growth in manufacturing and mining averaged 6.6 percent per year, considerably more rapid during the first decade (8.3 percent) than in the second, when it had slowed down to 5.5 percent. This represents a drop even in terms of growth relative to gross national product. The elasticities during the two decades were 1.51 and 1.17, respectively, attesting to a considerable variation in the structure of gross national product during the period.

Growth in industrial production is specified directly as one of the scenario variables. In all scenarios except Case E, growth in industry relative to gross national product continues the declining historic trend. In the most extreme scenario, Case B, industry grows slower than gross national product.

The composition of industrial production has changed appreciably over the period since 1953 (Table 8.4).

Mechanical industries, followed by textiles and clothing, have always ranked as top contributors to total industrial value added. Food and related has historically occupied third position but was being effectively displaced by chemicals and allied up to 1972.

Three sectors—chemicals and allied, iron and steel, and transport equipment—have penetrated with considerable vigor, each roughly doubling its share in total value added since 1953. In 1972, however, iron and steel still ranked behind paper and allied and mineral products, which has tended to hold on to its share of total industrial value added.

Most of this penetration has been at the expense of four sectors: textiles and clothing, paper and allied, mechanical industries, and mining.

The variation in mix is complex to explain, and no attempt is made to do so in this context. Suffice it to say that the substantial increase in average energy intensity of Italian industry between 1953 and 1972 (from 18.8 to 26.5 kcal/Lira) is largely, though not entirely, explained by variation in mix.

Prospects for the future indicate major components (lower investment per capita and greater employment potential of higher technologies) pushing for a shift away from the low and medium technology that presently characterizes Italian industry. This is, however, necessarily a slow pro-

Table 8.4
Composition of industrial value added under preembargo oil prices (%)

				1985	
Sector	1953	1963	1972	A,C,E	B,D
Mining	3.28	2.57	2.08	1.59	1.78
Iron and Steel	2.20	3.98	4.56	5.41	5.13
Nonferrous Metals	1.65	1.39	1.26	1.18	1.23
Paper and Allied	6.53	5.65	5.31	4.76	4.98
Chemicals and Allied	5.10	9.06	10.55	12.39	11.79
Mineral Products	4.55	5.63	5.29	5.03	5.16
Food and Related	11.88	11.75	11.12	10.34	10.60
Transport Equipment	4.29	6.78	7.54	8.36	7.99
Mechanical	20.34	20.41	19.37	19.36	19.66
Textiles and Clothing	17.72	14.58	13.09	11.53	12.02
Miscellaneous Manufacturing and Energy Industries	22.46	18.20	19.83	20.05	19.66
Total	100	100	100	100	100

Note: Value added expressed in 1972 Lire at factor cost.

cess and not likely to be far-reaching within the time scale of the projections. In the projections the assumption is made that whatever shifts may materialize before 1985 will take place within the aggregate sectors and not between them.

Historically, substitution between the aggregate sectors can be described as driven by gross national product:

$$\frac{A_i}{A_j} = K\,G^{a_{ij}}\,.$$

Because of relatively stable oil prices up till 1972, the oil price dependence of substitution does not appear explicity in this relation.

Projections to 1985 taking different oil prices P into account are made by factoring out the oil price dependence:

$$\frac{A_i}{A_j} = a(G)b(P).$$

The cross-substitution ratios are first projected with the historic relation, which assumes preembargo oil prices (Table 8.4). A fundamental assumption is that the pattern of exports and imports implicit in the substitution coefficients does not change appreciably from the historic period.

The most noticeable change in industrial structure under these assumptions is the acceleration in growth of chemicals and allied, iron and steel, and transport equipment at the expense of paper and allied, food and related, and textiles and clothing, while mechanical and mineral products roughly retain their 1972 shares.

The cross-substitution ratios are subsequently adjusted for the oil price effect by applying binary activity shifts $\pi_{ij}(P)$ determined from a 30-sector input-output model of Italian industry:

$$\frac{A_i}{A_j} = K\,G^{a_{ij}}\,(1 + \pi_{ij}).$$

The results of these projections (Table 8.5) indicate that oil price increases impact quite selectively on industrial sectors. In terms of share in total value added, paper and allied, textiles and clothing, and miscellaneous manufacturing are practically insensitive to price increases of the magnitude considered. Again in terms of share, iron and steel, nonferrous metals, and transport equipment are affected positively to a minor extent. The main interplay is between chemicals and allied and mineral products on the one hand and mechanical and food and related on the other.

The strongest depressing effect is observed for chemicals and allied, which loses as much as 1.8 percent of its share in the highest price scenarios. The greatest beneficiary from oil price increases is the mechanical industry, which acquires up to 1.1 percent in total share.

The impacts of oil price increases within sectors are available to a certain extent from the model and are treated under specific sector headings.

Table 8.5
Composition of industrial value added in 1985 (%)

		1985				
	1972	A	B	C	D	E
Mining	2.08	1.69	1.88	1.67	1.87	1.65
Iron and Steel	4.56	5.62	5.31	5.58	5.27	5.55
Nonferrous Metal	1.26	1.22	1.27	1.22	1.26	1.21
Paper and Allied	5.31	4.75	4.99	4.75	4.99	4.76
Chemicals and Allied	10.55	10.59	10.06	10.95	10.40	11.15
Mineral Products	5.29	4.63	4.76	4.72	4.84	4.78
Food and Related	11.12	10.84	11.11	10.74	11.00	10.67
Transport Equipment	7.54	8.60	8.21	8.55	8.16	8.53
Mechanical	19.37	20.46	20.77	20.23	20.55	20.08
Textiles and Clothing	13.09	11.44	11.91	11.46	11.94	11.49
Miscellaneous Manufacturing and Energy Industries	19.83	20.16	19.73	20.13	19.72	20.13
Total	100	100	100	100	100	100

Note: Value added expressed in 1972 Lire at factor cost.

Formally, it would be a simple matter to make intersectoral and intrasectoral substitution a function of policy response acting through sector-specific energy prices. This has not been attempted, however, because the complexity of the system and a possible collision of objectives make it very difficult if not impossible to isolate energy prices that will reduce energy demand on the one hand and maintain the industrial system viable on the other.

Increases in oil price can result in appreciable savings effects. The channels through which price acts on technical coefficients can be summarized as: process substitution, improvements in plant and plant operation, waste heat recovery, and general housekeeping measures.

Because of the greater weight of investment costs and the lack of adequate expertise in the smaller size firms, and moreover the technical applicability of some measures only above a certain threshold size, response to oil price increases is fundamentally a function of firm size.

Policy response is given an important role over and above that played by price by securing greater energy-saving investments even by smaller size firms, through advisory channels and adequately dosed government subsidies or credit incentives.

Since distribution by firm size is for many sectors peaked in the lower firm-size range, energy demand by the industrial system is highly sensitive to different combinations of oil price and policy response.

Mining

Mining contributed 2.1 percent to total value added of manufacturing and mining in 1972, consuming just under 0.5 percent of the total delivered energy (2.3×10^{12} kcal) with a specific energy consumption of 6.1 kcal/Lira.

The largest single contribution to the value added (47 percent) came from quarrying, mainly limestone and marble. Fossil fuels, mostly natural gas, contributed 25 percent. Only 6 percent of the value added came from metal-bearing ores.

Mining is projected to continue its declining historic trend in share of value added (Tables 8.4, 8.5). The average specific energy consumption is also estimated to remain stable as in the past.

Mechanical

In 1972 mechanical industries contributed 19.4 percent to total value added in manufacturing and mining, consuming just over 5.3 percent of the energy delivered to this system (25.0×10^{12} kcal) with an average specific energy consumption of 7.3 kcal/Lira.

Nonelectrical machinery constituted the bulk (47 percent) of this production, followed by electrical machinery (33 percent), metal products (11 percent), and precision machinery (9 percent).

Metal products, consisting largely of metal hardware (62 percent) but also including metal stampings, plumbing, pipes, etc. (38 percent), was naturally the most energy-intensive, consuming 45.0 percent of all the energy delivered to the mechanical sector with a specific energy consumption of 29.7 kcal/Lira. Most of this energy was coke used in smelting ovens.

Nonelectrical machinery with an average specific energy consumption of 6.3 kcal/Lira contributed 40.8 percent to the total energy consumed by the mechanical sector. Energy delivered to this sector went largely (42 percent) to the production of industrial machinery, with an average specific energy consumption of 7.2 kcal/Lira. At the other end of the spectrum, engines and turbines required only 2.1 kcal to produce 1 Lira of added value.

Electrical machinery accounted for 11.8 percent of energy consumed by the mechanical industries, averaging 2.6 kcal/Lira. Electronic components, communications equipment, and lighting materials contributed most (44 percent) of the value added in this sector. The remainder was made up of electrical motors, generators, and industrial equipment (34 percent) and home appliances (22 percent).

At the bottom of the scale, precision machinery consumed on average only 1.9 kcal/Lira.

Historically, the composition of value added in the mechanical industry has been relatively stable over time, at least since the early sixties. There has been no marked tendency to move away from the heavy industrial machinery, in which it has traditionally specialized, toward higher technology sectors. Although electronic components and communications equipment have been growing for some years at twice the average rate of the mechanical industry, in 1974 they still accounted for only 5.8 percent of the value added.

During the ten years preceding the energy crisis, the mechanical industry lost considerable share in total manufacturing and mining value added due to the inroads made by faster growers such as chemicals and allied, iron and steel, and transport equipment.

This displacement had, however, lost most of its vigor before 1972, and the projections assuming preembargo oil prices have the mechanical industry holding on to its share of total manufacturing and mining value added (Table 8.4).

The sudden increase in the price of oil is estimated in these projections to allow the mechanical industry to win back most of the share it has lost to chemicals and allied since the early sixties, and perhaps even to increase its share.

The elasticity trends derived from the projections indicate that lower economic development as well as higher oil prices favor the mechanical industry over manufacturing and mining in general.

Over the period to 1985, a significant shift is expected in favor of production with higher value added content. This will take place mostly in electrical machinery and particularly in electronic components and communications equipment, which are projected to capture as much as 8 percent of total value added in the mechanical industry. Precision machinery is also projected to increase its share slightly: from 9 percent in 1972 to 11 percent in 1985. Most of this penetration takes place at the expense of metal products, whose share falls to 9 percent in 1985.

There is very little space for energy savings in the mechanical industry except possibly for heat recovery in the metal products sector. It is expected that this will save, at most, 10 percent of nonelectrical energy used in this sector as a result of the price effect.

The fuel mix for the sector as a whole has been moving away from coal and coke, which in 1972 accounted for only 7.3 percent of the energy consumed. This is, however, close to a lower limit because of the present use of coke in some of the metallurgical processes of the mechanical sector.

Electricity is expected to increase slightly over its 1972 share of 21.2 percent. Fuel oil and other petroleum derivatives, which contributed 56.3 percent to total energy consumed, are projected as losing considerable share to natural gas. By 1985 it is expected that natural gas will account for close to 30 percent of the energy consumed by the mechanical sector.

Combining the above projections indicates that the energy intensity of the sector as a whole drops from 7.3 in 1972 to as low as 6.5 kcal/Lira, mostly as a result of change in product mix.

Textiles and Clothing

In 1972 textiles and clothing consumed $23.5x10^{12}$ kcal or 5.0 percent of the energy delivered to manufacturing and mining. Its share in the total value added of this sector was 13.1 percent, corresponding to an average specific energy consumption of 10.1 kcal/Lira.

Within the sector, energy intensiveness varies considerably from a low of 2.2 kcal/Lira in the production of wearing apparel to 30.7 kcal/Lira in the cultivation and treatment of silk.

During the sixties and early seventies substitution within textiles and clothing in favor of the less energy-intensive sectors was responsible for a substantial decrease in energy intensiveness: from 15.1 kcal/Lira in 1966 to 10.1 kcal/Lira in 1972.

During these six years wearing apparel more than doubled its share in value added to the sector, passing from 14.3 to 31.3 percent.

At the same time cotton ginning, wool, and silk, taken together, dropped from 62.4 to 37.4 percent. Other sectors to gain from this displacement were woolens and stockings, carpets and rugs, and other products included in the miscellaneous textiles and clothing sector.

As late as 1972 the change in composition was still a highly dynamic process promising to bring the energy intensiveness of the sector to even lower values.

Projections to 1985 assuming preembargo oil prices have wearing apparel contributing close to 40 percent to the value added in textiles and clothing. Miscellaneous textiles, including carpets and rugs and textiles for decoration, come a close second with 38.5 percent of the value added. All other sectors except weaving of synthetics (which holds on to a 4 percent share), continue the declining historic trend. Wool and cotton ginning drop to 7.5 and 7 percent, respectively, while silk and flax, linen, and jute together barely reach 3 percent of total value added of the sector. These sectors thus continue to experience decline even in absolute terms.

Most of this structural change is in the direction of higher value added products and responds largely to the difficulties experienced in recent years by the Italian textile industry in marketing traditional products under increasing international competition.

Oil price increases impact only slightly on the composition of the sector, favoring wearing apparel and other less energy-intensive sectors over textile industries. In the highest price scenario, wearing apparel gains only 0.5 percent over the preembargo trends.

The textile and clothing industry has tended to grow slowly relative to other sectors since the early fifties when it contributed in the neighborhood of 20 percent to value added in manufacturing and mining. The trend to displacement by faster-growing sectors is projected to continue to 1985 and is hardly affected by increases in the price of oil (Tables 8.4 and 8.5).

There is only limited space for energy savings in textiles and clothing, and response to price changes is expected to be very minor except possibly in the silk industry.

In 1972 energy delivered to the textile and clothing industry consisted chiefly (63 percent) of petroleum products (largely fuel oil), with electricity (23 percent) and natural gas (14 percent) making up the rest. The inroads made by natural gas during the late fifties came to an abrupt end as natural gas became increasingly scarce. Over the period to 1985, thanks to medium-term contracts, natural gas will again be plentiful and is estimated as at least tripling its share in total fuel. Electricity is also expected to increase its share somewhat over the 1972 levels, leaving petroleum products a maximum share of 32 percent.

Energy demand projections for the textile and clothing industry indicate a considerable drop in energy intensity from the 1972 level to between 5.6 and 6.3 kcal/Lira according to the scenario. Roughly 3 percent of the reduction in intensity is due to substitution by the higher-quality natural gas.

Transport Equipment

Transport equipment in 1972 was the second lowest energy consumer, accounting for barely over 1.9 percent of total energy delivered to manufacturing and mining or 9.0×10^{12} kcal. The sector is relatively low in energy intensiveness, contributing 7.5 percent to total value added during this year with an average specific energy consumption of 6.7 kcal/Lira.

The composition of value added in the sector has remained relatively stable over the last decade. Land vehicles accounted for most (85 percent) of the production, with automobiles and trucks contributing 64 and 15 percent, respectively. The remaining value added came almost exclusively from shipbuilding.

Over the period 1975 to 1985 a substantial shift in favor of shipbuilding is likely, largely on account of a rapid increase in freight transport in domestic waters projected under all scenarios. At the same time the automobile industry is expected to lose the impetus of the sixties. The automobile stock, which grew sixfold between 1960 and 1972, is estimated to increase by only another 60 percent by 1985 as a result of saturation in the home market and an increase in the average life of automobiles.

By 1985 it is estimated that shipbuilding will contribute over 20 percent to total value added in the sector.

Transport equipment is projected to continue growing faster than manufacturing and mining as a whole, gaining about one percentage point in total value added by 1985. Oil price increases favor the industry over others to a minor extent (Tables 8.4 and 8.5).

Oil price increases and policy response are expected to have only a limited role in increasing average energy efficiency of the industry.

The fuel mix, presently favoring petroleum products (47 percent in 1972), is expected to change in favor of natural gas as this fuel becomes increasingly available over the coming decade. By 1985 it is estimated that natural gas will contribute over 30 percent to total energy delivered to the sector, with petroleum products (almost exclusively fuel oil) dropping to 30 percent or lower. Coal and coke, which lost share during the fifties and sixties, can be considered as stabilizing to 7 percent of the total, mainly for metallurgical processes. Electricity has been relatively stable in the past and is not expected to rise much above its historic average of 28 percent.

Shipbuilding is four times less energy-intensive than land vehicle production: 1.8 against 8.0 kcal/Lira. As a consequence, the change in mix projected to 1985 will involve a slight drop in energy intensity of the sector as a whole: to between 5.4 and 6.3 kcal/Lira.

Food and Related

In 1972 food and related industries consumed $27.5x10^{12}$ kcal, corresponding to 5.8 percent of total energy delivered to manufacturing and mining. Their contribution to total value added was 11.1 percent, giving an average specific energy consumption of 13.9 kcal/Lira.

More than 60 percent of the value added in the sector came from bakery products (27.2 percent), canning and preserving (22.8 percent), and sugar refining (13.5 percent).

Sugar refining was also the most energy-intensive subsector, consuming over 39.6 percent of the total energy delivered at an average specific energy consumption of 40.8 kcal/Lira.

The composition of value added in food and related industries has not changed remarkably over the last decade. Basically, canning and preserving industries have gained some five percentage points at the expense mostly of tobacco. Within the bakery products subsector the more energy-intensive bread and spaghetti industries (11.7 kcal/Lira) have given way considerably to cake products (4.9 kcal/Lira). Sugar refining has maintained a relatively constant share in value added.

Projections to 1985 assume a continuation of the sectoral shifts of the past decade, which lead, however, to only insignificant effects on specific energy consumption of the sector as a whole.

The food and related sector is projected to continue the slow declining trend in total share of manufacturing and mining value added experienced over the period 1953 to 1973. Increases in the price of oil since 1972 postulated in the scenarios, however, favor the food and related industries over most other sectors, so that by 1985 the loss in share by this sector is minimal (Tables 8.4 and 8.5).

In 1972 the food and related industries were fueled largely by petroleum derivatives (65.6 percent, of which 98 percent was fuel oil), followed by natural gas (15.6 percent) and electricity (11.7 percent). Solid fuels, especially wood, which were important fuels in the early sixties, still contributed 7.1 percent to total energy delivered to the sector.

The availability of natural gas over the period to 1985 should displace fuel oil to a large extent, especially in some of the higher value added subsectors such as bakery products and canning and preserving. Electricity is expected to increase its share slightly from 12 percent and solid fuels to decline to less than 5 percent.

Average energy efficiency is estimated to increase some 2–3 percent as a result of penetration by natural gas. Increases in oil price over the 1972 values are likely to impact substantially in the more energy-intensive subsectors such as sugar refining and oil products. More extensive heat recovery in these industries should lead to 5–15 percent reduction in specific energy consumption as a consequence of the increased price of energy.

Space for specific policy actions aimed at reducing energy intensity of food and related is limited by the high degree of dispersion of the industry.

Average specific energy consumption in these estimates drops from the 1972 value of 13.9 kcal/Lira to 12.4, 13.2, and 13.6 kcal/Lira, respectively, with decreasing oil price. Most of this decrease is attributable to the price effect and to penetration by methane.

Iron and Steel

In 1972 iron and steel was the third largest energy consumer of Italian industry, with a final energy input of $66.5x10^{12}$ kcal, corresponding to 14.1 percent of total energy delivered to manufacturing and mining. Iron and steel, however, produced only 4.6 percent of the total value added, making it the third most energy-intensive sector.

During the fifties and sixties iron and steel was the fastest-growing industry, averaging 14.9 and 8.1 percent per annum, respectively, in the two decades beginning 1953. Projections to 1985 indicate it will continue to grow faster than manufacturing and mining as a whole and will be favored slightly by increases in oil price (Tables 8.4, 8.5).

The composition of value added in iron and steel has been essentially constant since the midfifties. The industry concentrates largely in blast furnaces and steel mills, with 55 percent of total production coming from basic metal and 45 percent from finished products.

Generally speaking, the value content of a ton of steel in 1972 did not vary substantially between subsectors. In all cases, however, there has been an appreciable increase over time, and this is most evident for special steels. This tendency, which is responsible for about 10 percent of the decline in average energy intensity of the industry since the midsixties, is expected to continue into the future with increasing specialization in more refined products.

Iron and steel is practically the only industrial sector to have shown a decrease in energy intensity over time. The decrease by almost a factor of two since 1953 (154 to 82.1 kcal/Lira) is almost entirely due to improvements in technology and has occurred notwithstanding the drop in recycled metal (mostly imported), which accounted for 79.3 percent of all steel produced in 1954 as compared to 61.2 percent in 1972.

Bessemer converters were displaced by the L/D. Coke in the blast furnace was partly replaced by more efficient reducing agents such as fuel oil and methane. Blast furnaces were lined with better refractory coatings. Starting from practically zero diffusion in 1965, continuous casting had spread to over 27 percent of all steel products by 1972. Beginning in the midsixties, automated process control has also contributed to reducing the energy intensity of the industry.

Because of the inherently high cost of energy to the Italian steel industry, improvements in technology are expected to continue decreasing the energy intensity of iron and steel even independently of increases in the price of oil. Under the scenario oil price hypotheses, the increase in efficiency is certain to be even more pronounced.

By 1985 it is estimated that continuous casting will have spread to between 40 and 50 percent of all finished products, determining considerable reduction in energy intensity. Improvements in refractory coatings and control of the blast furnace process are expected to decrease the energy intensity of pig iron production by around 10 percent. Heat recovery will also play a substantial role in reducing specific energy consumption throughout the iron and steel industry. A minor reduction will also come from an increase in recycling.

However, the major role in the reduction of energy intensity of the iron and steel industry is almost certain to come from a change in composition of the sector rather than from improvements in technology. By 1985, value added in the sector is expected to concentrate in finished products as basic steel is increasingly imported.

Considering that there are presently no plans for increasing blast furnace capacity, beyond one addition in 1977, the sectoral shift to 1985 is projected as quite pronounced.

Since the distribution by fuel varies considerably between subsectors, the change in product mix determines an appreciable change in the fueling of the industry away from solid fuels and toward petroleum derivatives and natural gas, with electricity gaining slightly in share.

On the basis of these considerations, average specific energy consumption drops from 82 kcal/Lira in 1972 to between 73 and 59 kcal/Lira in 1985.

Nonferrous Metals

In 1972 the nonferrous metals industries contributed 1.3 percent to total value added in manufacturing and mining and accounted for 2.1 percent of the total energy delivered (9.7×10^{12} kcal).

Average specific energy consumption was 43.4 kcal/Lira but ranged within the sector from a low of 3.6 kcal/Lira for tin to a high of 70.3 kcal/Lira for aluminum.

The relatively low average energy intensity of the aluminum industry is attributable partly to the fairly high percentage of value added coming from aluminum rolling and drawing (63.3 percent in 1972) but mostly from the very high contribution by secondary aluminum production. Recycling accounted for 51 percent of all the aluminum produced in Italy in 1972.

Notwithstanding a considerable increase in recycling in the aluminum industry in the decade beginning 1953, the energy intensity of the nonferrous metals sector as a whole almost doubled. This was due partly to a shift in composition in favor of the more energy-intensive aluminum production, but mainly to mechanization of the industry, which increased energy consumption and at the same time decreased value added per unit of production due to a substitution of capital for labor. This trend is expected to continue to 1985.

During the sixties the share of value added among the subsectors of nonferrous metals continued moving steadily toward aluminum and zinc and away from lead and copper. This trend is expected to continue into the future along with the trend within sectors for metalworking as opposed to metal production. Aluminum production, for example, is not expected in the medium term to increase substantially above the 1972 levels.

The change in composition of sectors and products of the nonferrous metals industry by 1985 is expected to decrease energy intensity by 3-4 percent from the 1972 value of 43.4 kcal/Lira.

A further 3-7 percent decrease in energy intensity is expected as a result of increases in the price of oil acting through general housekeeping measures, extensive application of process control, and improvements of a more technical nature.

New processes, such as the Alcoa process for aluminum, are unlikely to be introduced before 1990 because of the high investment costs involved and because most basic metal production plants are relatively new.

The nonferrous metals industries are driven mainly by electricity (66.3 percent), which is largely (63 percent) self-produced, followed by methane (14.0 percent) and petroleum derivatives (12.0 percent). This situation is not expected to change substantially over the period to 1985.

Paper and Allied

Paper and allied industries including printing and publishing contributed 5.3 percent to total value added in manufacturing and mining in 1972 and accounted for 3.7 percent of the total energy delivered (17.3×10^{12} kcal) resulting in an average specific energy consumption of 18.3 kcal/Lira. This, however, does not include the energy content of the raw materials input to the industries which amounted to 10.5×10^{12} kcal.

The relatively low average technical coefficient is mostly a consequence of the large share in total value added coming from printing and publishing (58.9 percent) with very low energy intensity: 1.0 kcal/Lira. The remaining value added came largely (36.3 percent) from the production of paper and paper products. Only 4.8 percent of the value was added in the more energy-intensive manufacture of pulp, thus explaining the low residual specific energy consumption of the pulp, paper, and paper products sector: 43.3 kcal/Lira.

The composition of the paper and allied industry has changed considerably over time. Paper and paper products grew most rapidly in the period up to 1966, capturing considerable share from printing and publishing, with pulp manufacture retaining roughly constant share.

The increase in energy intensity of the sector as a whole from 11.6 kcal/Lira in 1953 to 21.8 kcal/Lira in 1966 can be attributed almost entirely to the gain of the more energy-intensive paper industries over printing and publishing. Some of the increase (~4 kcal/Lira) is due, however, to mechanization of the industry and a shift to higher-quality papers.

After 1966 the energy intensity of the industry decreased again as printing and publishing gained over both pulp manufacture and paper products.

Over the period to 1985 pulp manufacture is expected to continue losing share as pulp input to the paper products industries is increasingly imported. In 1972 imports accounted for 58 percent of all pulp inputs, against 43 percent in 1966, and pulp production practically stopped growing in absolute terms after 1969.

Paper and paper products are expected to grow somewhat faster than printing and publishing. This fact together with the trend to higher-quality paper products makes for a higher energy intensity of the paper and allied sector as a whole.

In terms of share in total value added of manufacturing and mining, the paper and allied industries are practically insensitive to the variations in price specified in the scenarios and continue the declining preembargo trend.

Oil price increases are not expected to have any substantial effect on efficiency other than in pulp manufacture where some savings are available from heat recovery. Appreciable substitution of petroleum derivatives by natural gas is, however, expected to lead to greater energy efficiency.

Average specific energy consumption of the paper and allied sector in 1985 projected on the basis of these considerations comes to between 18 and 20 kcal/Lira.

Mineral Products

In 1972 mineral products consumed 79.3×10^{12} kcal or 16.9 percent of the energy delivered to manufacturing and mining, making it the second largest energy consumer of Italian industry. Its relatively low contribution to value added (5.3 percent) made it also the second most intensive sector, with an average of 84.5 kcal/Lira.

The average specific energy consumption of the

sector has been showing a tendency to decrease since the early fifties when it approached 100 kcal/Lira. Over the period beginning 1962 the decrease was attributable almost entirely to the change in composition of the sector in favor of the less energy-intensive stone and concrete products (respectively 1.7 and 6.2 kcal/Lira) and at the expense of the highly intensive cement manufacture (386 kcal/Lira in 1972).

A substantial part of the decrease was, however, a consequence of technological change in the cement industry, which has always been sensitive to the cost of energy.

The dry process for cement manufacture, with a specific energy consumption of 990 kcal/kg clinker, rapidly displaced the much more energy-intensive wet process (1485 kcal/kg), while the traditional vertical kilns (1150 kcal/kg) almost vanished during the sixties.

In 1974 the three processes represented, respectively, 86, 13, and 1 percent of all clinker produced. The further displacement of the wet process is expected to be sensitive to the price scenario, and projections are made assuming an extant 3–7 percent by 1985.

The increase in slag and pozzolanic cements over Portland cement during the sixties also contributed to the reduction in energy intensity, but this effect was largely counteracted by the trend to higher-quality products requiring longer grinding times.

Slag and pozzolanic cements containing up to 50 percent clinker require, on average, some 15 percent less energy than Portland cement per ton of product. Since they are practically equivalent to Portland cement (if not superior for some specific applications), a vigorous policy response is envisaged as penalizing Portland cement by means of a special energy sales tax. This is estimated to reduce the share of Portland in total cement production from 54 percent in 1974 to 25 percent in 1985, leading to a 15 kcal/Lira drop in energy intensity of cement manufacture in the vigorous policy scenarios.

General housekeeping measures and technological improvements brought about by increases in the price of oil coupled with a vigorous policy response are expected to induce up to 13 percent energy savings by 1985.

Specific energy consumption of cement manufacture projected on the basis of the above considerations lies between 334 and 376 kcal/Lira for the highest and lowest oil price scenarios, respectively.

Increases in the price of oil specified in the scenarios are also estimated to bring in considerable energy savings, over the period to 1985, in the structural clay products sector. These will come largely through technological changes such as substitution for the much more diffuse Hoffmann kilns by tunnel kilns in the Laking process. These are some 30 percent less energy-intensive and in 1974 accounted for only 37.5 percent of total production. Further reduction in energy intensity will also come through substitution by continuous-process driers, which are over 40 percent less intensive than the more diffuse static driers and flash driers.

Specific energy consumption in the structural clay products sector is projected to 1985 as 118, 125, and 130 kcal/Lira with decreasing energy price scenario.

During the late fifties and the sixties glass and glass products industries experienced a considerable increase in energy intensiveness (30 percent between 1961 and 1972), which can be attributed almost entirely to extensive mechanization, decreasing the labor component of value. Between 1966 and 1972 the value added per ton of product decreased on average almost 15 percent.

The mechanization phase of the glass and glass products industries is largely over, so that substantial increases in energy intensiveness on this account are not expected in the future. These increases will be compensated by energy savings coming from measures of various kinds responding to increases in the price of oil, so that the average energy intensiveness of the glass and glass products sector is expected to decrease appreciably between 1972 and 1985.

The increase in energy intensiveness of the pottery and ceramics industries (more than 30 percent) since the late fifties likewise contributed to slowing down the decrease in energy intensiveness of the mineral products sector as a whole. This increasing component due largely to mechanization of the industry is expected to slow down considerably in the future.

Counteracting the residual increase to 1985 are large energy savings coming from technological improvements responding at least partly to increases in the price of oil.

The introduction of unit-process pressing and enameling will be responsible for most of the reduction in energy intensiveness by 1985. In the high oil price scenario, savings are estimated as 20 percent, assuming about 50 percent replacement of the traditional process involving two bakings instead of one.

Substitution of natural gas for some petroleum derivatives (mostly fuel oil) in the glass and glass products industries, and especially in the pottery and ceramics industries, is expected to lead to a further slight reduction in energy intensiveness in these subsectors.

The composition of the mineral products sector is not expected to change substantially from 1972. Although construction is projected to recover from the low average growth of the sixties, this will impact equally on all components of mineral products that feed largely into housing.

On the basis of these considerations, the energy intensiveness of the mineral products sector in 1985 lies between 74 and 84 kcal/Lira according to scenario.

In terms of share in total value added of manufacturing and mining, the mineral products sector has been relatively stable since 1953. Projections to 1985 under increasing oil prices indicate it will lose quite heavily to other sectors (Tables 8.4 and 8.5).

Chemicals and Allied

In 1972 the chemicals industry represented 10.6 percent of the total valued added and consumed 41.4 percent of the energy delivered to manufacturing and mining, including energy used for chemical feedstocks. Energy per value added in 1972 was 103.6 kcal/Lira, considerably above the industrial average of 26.5.

Within the chemicals industry energy intensiveness varies considerably from a low of 8 kcal/Lira for matches and explosives to a high of 327 kcal/Lira for carbon electrodes. Heavy petrochemicals, the most intensive major subsector, consumed 55 percent of the delivered energy. Heavy petrochemicals and heavy inorganics together accounted for 84.4 percent of the energy, as opposed to 15.6 percent for fine chemicals and specialty products.

The product mix of chemicals and allied has tended to evolve toward heavy petrochemicals partly as a result of substitution for natural materials by plastics and synthetic fibers and rubbers. This contrasts with the more gradual expansion of heavy inorganic chemicals, a more mature product group. Fine chemicals grew slightly less rapidly than the industry as a whole, reflecting the concentration of Italian chemical production in the heavy chemicals sector. Within fine chemicals, those consumed primarily by industry have increased somewhat faster than those consumed primarily by households.

The chemicals industry as a whole is the hardest hit by the oil price increases specified in the scenarios. The decrease in growth rate relative to preembargo trends is between 68 percent for the high oil price scenario B and 18 percent for the lower oil price scenario E. Part of the strong impact in Case B is due to the low consumer spending specified in this scenario.

Value added for each major subsector is projected through average historic elasticities relative to appropriate scenario variables. For example, private consumption is used to project fine chemicals for household consumption, and industrial production to project fine chemicals for industrial use.

Slight proportional adjustments are applied to bring projections of total value added obtained by this technique in line with the results of the industry cross-substitution projections, maintaining the same product mix.

The results indicate a continued strong growth in heavy petrochemicals and fine chemicals for industry. Petrochemical derivatives increase faster than the industry as a whole, whereas fine chemicals for households and heavy inorganic chemicals are below the industry average.

Within heavy chemicals, savings are expected to obtain mostly through the integration of processes—for example, waste heat recovery using gas turbines[5]—and by improved combustion. In the heavy petrochemical sectors, savings possibly amount to 15 percent of the petroleum input. All energy-saving measures with payback periods of less than three years are assumed to be implemented with probabilities that vary with the subsector, the firm-size class, the oil price, and the policy response.

5. See *The Application of a Gas Turbine to a Steam Reforming Ammonia Plant,* in Task 1, Energy Saving Project, Working Papers, Workshop on Alternative Energy Strategies, October 1974.

Substantial energy savings are possible even in the fine chemical sectors but are expected to obtain only under vigorous policy response because of the low average firm size.

Fueling (energy and nonenergy) of chemicals and allied changes relative to that of the early seventies with coal and petroleum declining and natural gas increasing. The decline in petroleum, providing 74.8 percent of the delivered energy in 1972, is due largely to energy savings occurring mainly in this area. Coal continues its historic decline because of replacement by petroleum and in part by natural gas. The proportion of electricity, which in 1972 was almost entirely (89 percent) self-produced, remains roughly constant. Natural gas increases due to substitution for both petroleum and coal.

Notwithstanding substantial energy savings resulting from oil price increases and a vigorous policy response, the change in mix favoring the highest intensity subsector (heavy petrochemicals) is sufficiently powerful to increase the average energy intensity of the chemicals and allied sector in all but scenario B: up to 124 kcal/Lira.

Miscellaneous Manufacturing and Energy Industries

In 1972 energy industries and industries not elsewhere classified contributed 19.8 percent to total value added in manufacturing and mining but accounted for only 4.1 percent of the energy delivered (15.6×10^{12} kcal). This, however, does not include energy consumed by the energy industries, which is considered elsewhere under processing losses of the supply system.

Sectors contributing to the average specific energy consumption of 4.4 kcal/Lira ranged in energy intensity from the very low intensity footwear industries (1.0 kcal/Lira) to the more intense rubber and wood industries (12.2 and 12.9 kcal/Lira, respectively).

Projections of the contribution of value added coming from energy industries (48 percent in 1972) are made on the basis of total delivered energy.

Within miscellaneous manufacturing, increases in the price of oil are projected to favor wood industries over rubber manufacture.

Specific energy consumption in most of these sectors of miscellaneous manufacturing has increased in the past due to mechanization decreasing the labor content of value. A continuation of mechanization is expected to lead to a further substantial increase.

Electricity and natural gas are projected to increase their share in fuel delivered to miscellaneous manufacturing at the expense of petroleum derivatives.

Energy intensity of the sector as a whole under these considerations varies between 3.2 and 3.7 kcal/Lira according to scenario.

8.3.3 Agriculture, Livestock, Forestry, and Fishing

Agriculture, livestock, forestry, and fishing accounted together for 2.2 percent of total delivered energy, or 21.3×10^{12} kcal, in 1972.

Fishing was the sector with the highest energy intensity (16.5 kcal/Lira), followed by forestry (6.9 kcal/Lira) and by agriculture (5.4 kcal/Lira). Livestock had the lowest energy intensity, consuming only 0.3 kcal/Lira of value added.

Almost all the value added in this system in 1972 came from agriculture (53.7 percent) and livestock (41.2 percent).

Historically, the share in total value added has tended to move away from agriculture and forestry to livestock and fishing. The average annual growth rates over the period 1953 to 1973 were:

agriculture	1.8 percent
livestock	3.3 percent
forestry	−0.2 percent
fishing	3.1 percent.

Projections to 1985 assume a continuation of the historic substitution trends driven by gross national product as for manufacturing and mining. The sectoral shares in total system activity obtained in this way are given in Table 8.6.

Under the total system growth rates specified in the scenarios (Table 8.3), these translate into the following. Livestock continues to grow rapidly in the high growth scenarios but slows down considerably in the low growth scenarios. Agriculture does not recover from the slow growth of the period 1963 to 1973. After 1976 forestry becomes the fastest-growing sector, recovering much of the

Table 8.6
Share of value added among agriculture, livestock, forestry, and fishing in 1985 (%)

	1972	1985 A,C,E	1985 B,D
Agriculture	53.70	49.43	51.98
Livestock	41.23	46.38	43.62
Forestry	3.18	2.28	2.57
Fishing	1.89	1.91	1.83
Total	100	100	100

ground lost during the sixties. This is consistent with government projects for redevelopment. Fishing grows as fast as or faster than in the past.

During the period 1953 to 1972 energy used to create a Lira of value added in forestry, agriculture, and livestock increased by factors of 3.8, 3.3, and 2.9, respectively.

The trend to increasing energy intensiveness in all sectors except fishing due to mechanization and modernization is projected to continue well into the future, though at reduced pace. The 1985 energy intensities are estimated as 10.5, 0.8, 12.1, and 15.0 kcal/Lira for agriculture, livestock, forestry, and fishing, respectively.

8.3.4 Construction

Construction is a very low energy consumer, barely 0.1 kcal/Lira of value added in 1972. Its contribution to total energy consumption in this year was just over 0.05 percent: 0.5×10^{12} kcal. Most of this energy is electricity.

Under the scenario growth rate hypotheses (Table 8.3) and considering that specific energy consumption is increasing, energy delivered to construction in 1985 is projected as:

scenarios A, C, E 1.1×10^{12} kcal
scenarios B, D 1.0×10^{12} kcal.

8.3.5 Residential, Commercial, and Public

Cooling

Energy consumed for space cooling grew very rapidly during the sixties and early seventies. In 1972 cool air conditioning, both central and window, accounted for 0.20, 0.74, and 0.27×10^{12} kcal, respectively, in the residential, commercial, and public sectors: a total of 1.21×10^{12} kcal. Although this corresponds to a negligible fraction of total delivered energy (barely 0.12 percent), the very rapid diffusion of air conditioning is expected to bring its share in total electricity delivered to the residential, commercial, and public sectors to well above the 1972 level of 3.7 percent.

Policy measures aimed at adequate insulation of new buildings and retrofitting of old buildings, as described under space heating, are estimated to save some 20 percent in the residential, 18 percent in the commercial, and 15 percent in the public sector.

The effect of oil price increases is estimated to be minor.

Energy demand projections for cooling in 1985 are highly sensitive to the scenarios on account of the variation in growth of private consumption (Table 8.3), which is the basic determinant of cooling demand. The range in total energy demand for cooling is between 2.8 and 5.5×10^{12} kcal for scenarios B and E, respectively.

Street Lighting

In 1972 street lighting accounted for 1.7×10^{12} kcal final energy demand: barely 0.17 percent of total delivered energy, but 1.8 percent of total electricity, 47 percent of total energy delivered to the public sector, and 51 percent of electricity consumed by this sector.

Over the last twenty years energy consumed for street lighting has been growing on average 1.6 times as fast as the growth of gross national product, yet the level of street lighting in 1972 was still relatively low, corresponding to just over one lamp for every fifteen inhabitants and 9 W/inhabitant.

Over the forecast period street lighting is projected to continue growing at the same rate as in recent years, bringing electricity demand in this sector in 1985 to the following levels:

scenarios A, C, E 3.5×10^{12} kcal
scenarios B, D 2.6×10^{12} kcal.

Neither policy response nor the price of oil is expected to influence these figures in the long term since they still correspond to relatively low levels of lighting: 18 and 13 W/inhabitant, respectively.

Appliances

Appliances in the residential, commercial, and public sectors accounted for 5.6 percent of total delivered energy in 1972, corresponding to just under $53.0x10^{12}$ kcal, almost exactly $1.0x10^6$ kcal/person.[6]

The bulk of this energy went to the residential sector, with $39.5x10^{12}$ kcal total consumption of which 51.4 percent was in the form of electricity. The commercial sector consumed $11.6x10^{12}$ kcal and the public sector the remaining $1.9x10^{12}$ kcal. The share of electricity in these sectors was 31.0 and 69.3 percent, respectively. As an overall average, 47.6 percent of the energy delivered to the three sectors was in the form of electricity.

Over the period 1960 to 1972 final energy used for appliances in these sectors grew at an average annual rate of 5.9 percent (5.1 percent in the residential, 9.1 percent in the commercial, and 6.5 percent in the public sector).

Electricity was by far the fastest grower (11.5 percent), attesting to the rapid diffusion of new appliances such as televisions, refrigerators, and clotheswashers during the sixties. This growth breaks down into a 13.3 percent annual average for the residential, 9.8 percent for the commercial, and 9.4 percent for the public sector. Hardly any of the growth was due to fuel substitution.

Natural gas was also a fast grower (6.3 percent) due largely to expansion of the distribution network and migration to the cities (4.7 percent) but also due to substitution of town gas (1.6 percent).

Liquid petroleum gas, which enjoyed rapid growth (as high as 11.7 percent) before 1964 at the expense of more traditional fuels, began to lose share to natural gas as the distribution network expanded, and its consumption has been stationary at about $8.5x10^{12}$ kcal since the late sixties.

The distribution network, which reached 22.0 and 32.9 percent of all consumers in 1961 and 1971, respectively, is projected to capture as much as 46 percent of the market by 1985, so that liquid petroleum gas consumption by the residential, commercial, and public sectors should begin declining in the very near future. Notwithstanding that liquid petroleum gas reaches more households than natural gas, its consumption in appliances is lower than that of natural gas. This is partly a consequence of the easy wastefulness of natural gas, but largely because liquid petroleum gas is used almost solely for cooking, whereas natural gas also finds use in water heating. Moreover, liquid petroleum gas tends to be concentrated in lower-income regions.

Other fuels such as wood, coal, and charcoal, which were still important in the late fifties, had all but vanished in 1972 or were limited to certain very special categories of use: such as charcoal steaks (Table 8.7).

The appliances sector is projected as nine essentially independent function-specific subsystems:[7]

cooking
dishwashing
clotheswashing
personal hygiene
lighting
refrigeration
entertainment
general services
other.

The distribution by subsystem of total energy consumed has changed considerably since the late fifties and varies substantially between the residential, commercial, and public sectors.

The dynamics of change can be readily analyzed and projected by decoupling each subsystem into a three-level hierarchy. The first level describes competition for total activity between the residential, commercial, and public sectors. Within each sector substitution between appliance types serving the

Table 8.7
Energy consumed by appliances (10^{12} kcal)

Fuel	1960	1972
Electricity	6.7	26.2
Natural Gas and Town Gas	6.2	14.1
Liquid Petroleum Gas	5.1	8.7
Wood	4.7	1.5
Gas Oil	—	1.2
Fuel Oil	—	0.6
Coal and Coke	3.0	0.4
Charcoal	1.1	0.2
Kerosene	—	0.1
Total	26.8	53.0

6. The commercial sector is defined to include stores, hotels, restaurants and bars, business offices, cinemas and theatres, etc. The public sector includes schools, hospitals, public administration offices, military installations, railway stations, airports, etc.

7. Stoves and window air conditioning are not projected as appliances since they compete with other heating and cooling systems that evolve with the housing sector and thus require separate treatment.

function is described at the second level. Substitution between fuels within each appliance type is determined at the third level.

The three-level structure is suggested largely by the fact that some appliances serve more than one function. The main example is the hot water heater, which is used for dishwashing, clotheswashing, and personal hygiene. Projection of energy requirements for hot water heating cannot be made without reference to the complex phenomena that determine it: the demand for dishwashing, clotheswashing, and personal hygiene correlated with private consumption; population and household growth; competition between the various appliances serving these functions (for example, between automatic dishwashers, hot water heaters, and kitchen stoves), determined largely by private consumption; and finally the shift between the sectors in which these functions are served, usually characterized by different energy efficiencies.

The hierarchical structure also facilitates interpretation of the various changes that have taken place in the past.

The decrease in energy consumed for cooking by the residential sector between 1960 and 1972, for example, is due to a substantial shift from the residential to the commercial sector. In 1960 81 percent of all meals were cooked in the residential sector, and 16 percent in the commercial sector. In 1972 these figures had changed to 68 and 28 percent. Average 1972 energy efficiency for cooking in these two categories was 625 and 522 kcal/person-day, respectively.

Competition between the residential and commercial sector has been even more substantial in the entertainment subsystem due largely to the diffusion of television during the sixties. In 1960 the commercial sector entertained mainly through movies, for 60 percent of the total activity as against 40 percent in 1972. Average energy intensity for this subsystem in 1972 was some twenty times greater in the residential sector: 167 against 8 kcal/person-day.

Growth of total energy delivered varies considerably among appliances and among sectors.

Overall, cooking was the slowest grower, partly because of the substantial shift to the more efficient commercial sector but largely because cooking has reached saturation and grows only with population. Personal hygiene also grew slowly largely because habits in this subsystem (number of showers/person-day, hair washing and setting, etc.) change slowly.

The fastest growers are subsystems characterized by the diffusion phase of an appliance type without competition. Refrigerators and freezers, and television for entertainment, diffuse *within* the residential sector practically without competition.

On the other hand, automatic clotheswashers and dishwashers diffuse by displacing existing traditional appliances based on the use of lower-energy-intensity hot water heaters, stoves, and cookers or even just cold water warmed by sitting in the sun.

The fairly rapid growth of energy for lighting is basically a simple income phenomenon since electricity had diffused into over 99 percent of all households well before 1960.

OECD projections are used for population and numbers of households in 1985.

Substitution between categories at the first level and between appliance types at the second level is projected as driven by private income and gross national product, implying basically a continuation of historic trends. These were adjusted by expert opinion where necessary. In some cases, such as cooking, the projections opt for an improvement, and in other cases, such as entertainment, for a deterioration, in overall energy efficiency.

Technical coefficients are, in general, estimated as increasing substantially by 1985 as a result of economic development. In the residential sector this is a consequence of an increase in the number of units per households and in power output of appliance types rather than in the duration of use. An obvious example is the diffusion of color television, but the same general tendency is observed for all appliances in all three sectors. Indicators used to determine the variation in the residential, commercial, and public sectors are, respectively, private consumption, gross national product, and public consumption.

Distribution by fuel type is projected at the third and lowest level of the hierarchy by considering competition between natural gas, liquid petroleum gas, and electricity under constraints posed by the gas distribution network and peak load.

The use of solar energy for appliances is likely to remain negligible during the time scale of the projections. Solar water heating is not estimated as winning more than marginal market share by 1985 even though some prototype heating units are already commercial. Changes in oil price within the limits considered are estimated as having only very minor effects on intensity of use and then only in the residential sector.

As vigorous policy response, a special energy-use sales tax proportioned to energy intensity of appliances should be effective in slowing down diffusion of the larger, more powerful models.

Space Heating

Energy delivered for space heating to the residential, commercial, and public sectors in 1972 amounted to 232.0×10^{12} kcal or 24.3 percent of total delivered energy. The residential sector accounted for 77.4 percent of this energy, the commercial sector for 17.0 percent, and the public sector for the remaining 5.6 percent.

The uneven distribution of energy for space heating between northern (69.7 percent), central (17.3 percent), and southern Italy (13.0 percent) was due basically to the large variation in degree-days:

North 2275 deg-day
Center 1894 deg-day
South 872 deg-day;

to the regional distribution of the 15.6×10^6 households:

North 49.7 percent
Center 18.9 percent
South 31.4 percent;

and to different average household volumes:

North 312 m^3
Center 268 m^3
South 236 m^3.

Contributing to the variation in degree-days over and above the difference in average outside winter temperatures (North 8.8°C, Center 9.1°C, South 12.4°C) were substantial differences in heating systems characterized by very different average inside temperatures ranging from 23°C for central heating systems to 13°C for stoves and close to outside temperatures for no heating systems.

Practically, the only factor acting to even out the regional distribution of energy for space heating was the different average thermal volume conductivities of households:

North 0.69 kcal/h·m³°C
Center 0.64 kcal/h·m³°C
South 0.76 kcal/h·m³°C.

This was not, however, a consequence of differences in materials used in buildings but of the different distribution among household classes with very different average thermal conductivities:

1-family units 1.86 kcal/h·m³°C

2-family units 1.48 kcal/h·m³°C

3- to 15-family apartment buildings 1.07 kcal/h·m³°C

16- to 30-family apartment buildings 0.70 kcal/h·m³°C

>30-family apartment buildings 0.50 kcal/h·m³°C.

The distribution of energy for space heating by household type (Table 8.8) is largely explained by the regional variations in thermal conductivity and volume.

Energy consumed by single-family and two-family units is lower than would be expected from the larger average thermal conductivities and volumes because heating systems characterized by low degree-days tend to concentrate in these two household types.

The structure of space heating in Italy has undergone fundamental changes since the midfifties, and in 1972 was still in a high state of flux. These changes can be described as decisive shifts from traditional heating systems and stoves to single household radiators and central heating.

The basic reason for the change was increase in income determining substitution between heating

Table 8.8
Distribution of energy for space heating by household class in the residential sector in 1972 (10^{12}kcal)

	North	Center	South
1-Family Units	19.6	3.2	3.4
2-Family Units	27.6	4.6	5.5
3 to 15-Family Apartment Buildings	47.6	10.9	9.7
16 to 30-Family Apartment Buildings	20.2	8.2	3.2
>30-Family Apartment Buildings	10.2	4.2	1.5
Total	125.2	31.1	23.3

systems and in favor of larger houses with fewer shared surfaces. Migration to the cities and, to a smaller extent, from the South to the North were also partly responsible for determining shifts between household types and heating systems.

The impact of private income growth and migration during the fifties and sixties is summarized in Table 8.9, which compares growth in number of households with growth in energy consumption for space heating in the residential sector, indicating that structural change was fastest during the sixties and particularly after 1965.

Substitution between fuels is largely tied up with changes in heating systems. Wood (43 percent), followed by coal and coke (33 percent) and by fuel oil (17 percent), were the principal fuels in 1953 when heating systems were prevalently traditional.

Wood remained the predominating fuel until the late fifties when fuel oil took the lead as central heating began diffusing. Kerosene, used largely in stoves, also began growing rapidly in the early sixties as stoves displaced traditional heating systems.

Gas oil began substituting for fuel oil in 1968 after antipollution legislation confined the use of fuel oil for space heating either to very large buildings ($> 10^6 m^3$) or to less densely populated areas.

Finally, natural gas began to diffuse with extreme vigor both in single-household radiating systems and in central heating systems after 1973 as prices of gas oil increased much more than prices of natural gas.

For the future, wood, coal, and coke are likely to disappear as traditional heating systems and outdated coal boilers are replaced, and fuel oil will become roughly stabilized in importance. Kerosene, which had begun decreasing in absolute terms in the early seventies due to displacement of stoves, is expected to increase somewhat even though stoves are disappearing due to the growth of single-household radiators, especially in the North.

The main fuel competition over the period 1985 is between natural gas and gas oil and depends on the relative prices of these fuels. The diffusion of natural gas is naturally most favored but is subject to two constraints: the gas distribution network and the maximum capacity for natural gas at peak load.[8]

The residential space heating system is projected to 1985 as a four-level hierarchy. The first level determines the total number of households N through a simple elasticity relation with gross national product. The second level describes household flows between the three climatic zones. The third level describes on the one hand substitution between heating systems and on the other shifts between household classes. The fourth level describes substitution between fuels.

Energy delivered in region i to heating systems of type j, burning fuel k, in households of the class ℓ is calculated as:

$$E_{ijk\ell} = \omega_{ijk\ell} \cdot N \cdot \frac{N_i}{N} \cdot \frac{N_{ij}}{N_i} \cdot \frac{N_{ijk}}{N_{ij}} \cdot \frac{N_{i\ell}}{N_i} \cdot U_\ell \cdot V_{ijk\ell} \cdot \frac{G_{ij}}{\eta_{jk\ell}},$$

where U, V, G, and η are thermal conductivity, volume, degree-day, and system efficiency, respectively, where ω is a normalization constant, and where the household distributions N_i/N, N_{ij}/N_i, N_{ijk}/N_{ij}, and $N_{i\ell}/N_i$ are projected independently by application of a substitution model driven by gross national product with adjustments suggested by expert opinion.

8. The latter technical constraint requires natural gas delivered to residential and other nonindustrial users not to exceed 34 percent of total delivery.

Table 8.9
Development of space heating in the residential sector in the period 1953–1972

Year	Number of Households (millions)	Energy for Space Heating (10^{12} kcal)	Average Annual Energy Growth Rate (%)	Average Annual Household Growth Rate (%)
1953	11.2	30.3		
1960	12.8	48.9	7.1	1.9
1966	14.0	88.7	10.4	1.5
1969	14.8	127.3	12.8	1.9
1972	15.6	180.6	12.4	1.8

The oil price dependence appears basically in the projection of degree-days, for which price elasticity data are available from a comparison of pre- and post-crisis average inside temperatures.

Thermal conductivity projections are sensitive to increases in oil price only when combined with a vigorous policy response, interpreted as rigorous application of standards for new buildings and subsidies to help finance improvements in older buildings. Retrofitting is assumed to have no application in buildings built before 1951 because of the high insulating costs. Moreover, these buildings generally have considerably higher thermal capacity because of their thicker walls, making for higher overall heating efficiency. Double pricing of fuels geared to degree-day and thermal conductivity standards is expected to induce insulation by private homeowners in the vigorous policy scenarios. This does not discriminate between income classes if accompanied by specific government subsidies for retrofitting.

A vigorous policy response in the form of a capillary information campaign is expected to increase average boiler efficiencies by 13 and 9 percent in the high and medium oil price scenarios, respectively. This is not necessarily optimistic considering that the average efficiency of new boilers has increased some 18 percent since the early sixties and allowing for a nine-year lifetime for boilers.

Recent legislation (June 1976) goes only a very short way toward covering these vigorous policy assumptions; basically it focuses on insulation standards for new buildings and regulation and control of inside temperatures.

A vigorous policy response is also interpreted as breaking down political barriers to district heating. Cost estimates indicate that 600,000 households in northern and 200,000 in central Italy can be district-heated by 1985, with profits for the sponsoring municipal authorities.[9]

Solar energy for space heating is not expected to make significant headway by 1985 even though considerable activity has been triggered in this sector as a result of the increase in oil prices.

The results of the projections indicate the space heating system to be extremely sensitive to variations in the scenario in terms of both fuel mix and overall efficiency. The differences between scenarios are to be imputed to numerous effects.

Household growth is highest in the high growth scenarios ($19.2\text{x}10^6$ as compared to $18.2\text{x}10^6$ households in the low growth scenarios). Households actually heated are lowest in the vigorous policy scenarios due mainly to the assumption of district heating. This is not accounted for in the heat balance since district heating taken into consideration for 1985 is basically heat recovery from power plants.

The economic growth variable also introduces slight differences between scenarios in the distribution by household class and by heating system. In all cases, however, unheated houses and traditional heating systems essentially vanish, and even stoves drop from an average diffusion of 32.2 percent in 1972 to lower than 7.9 percent in 1985. The outcome is an increasing component of energy consumption for space heating in all scenarios because of an increase in degree-days.

The most interesting result of this displacement is, however, in the distribution of energy by household class, which moves decisively in favor of single-family households and two-family units as a result of the substitution for traditional heating systems and stoves by single-household radiators and central heating.

In scenario E, for example, single-family households and two-family units, which comprise only 21.3 percent of all households, account for 45.7 percent of the energy consumed. In contrast, households in buildings with over fifteen apartments—36.9 percent of all households—consume barely 18 percent of the energy.

Differences in the price of oil, especially if coupled to a vigorous policy response, have the greatest effect on total energy demand for space heating, as can be seen from a comparison of scenarios A, C, and E. Most of this effect is due to the price acting on degree-days, but much comes from improved insulation in the vigorous policy scenarios.

An increase in efficiency of boilers—due partly to replacement by more modern boilers but essen-

9. The city of Brescia is planning to be almost entirely district-heated by 1985. In 1975 some 10,000 households (10 percent of the 1985 total) were already connected to the hot water network.

tially to vigorous policy educating homeowners to adequate maintenance—is also responsible for considerable savings in energy use.

Fuel mix does not vary substantially from pre-embargo trends, except for liquid petroleum gas, which experiences fairly rapid growth, and, of course, competition between natural gas and gas oil, which plays a key role in this energy system.

Natural gas availability for space heating is determined by total natural gas supply, which is fairly well established as about 340×10^{12} kcal/yr. by 1985,[10] and by total consumption by industry, which is scenario-dependent. The projections assume that natural gas not delivered to industry goes to households, subject to the constraint that this be not more than 34 percent of total delivery (the peak load constraint).[11]

Under this assumption the supply of natural gas is always greater than, or at most (scenario E) equal to, total demand, implying that natural gas should be competitive with gas oil even in terms of price.

In no case does the substitution of natural gas for gas oil approach the gas distribution network constraint. Even in scenario A, where substitution is greatest, 15 percent of households reached by the distribution network are not heated by natural gas.

Substitution by natural gas is responsible for a considerable gain in overall efficiency in the vigorous policy scenarios over and above that due to improvements in boilers and boiler maintenance. The phasing out of residual coal and wood boilers is also partly responsible for the increase in overall efficiency in these cases: from 0.56 in 1972 to as much as 0.69 in scenario A.

In the restrained policy scenarios the increase in efficiency due to substitution by natural gas and a slight (~5 percent) improvement in average boiler efficiency is not sufficient to counteract the drop in efficiency due to the substitution of the lower-intensity traditional systems and stoves by single-system radiators and central heating, so that overall average efficiency remains essentially unchanged from 1972.

10. 230×10^{12} kcal/yr. under medium-term contracts with The Netherlands, USSR, Libya, and Algeria and 110×10^{12} kcal/yr. domestic production.

11. This is not to be understood as a statement of priority of use but as an allocation algorithm.

8.4 CONCLUSIONS

Total delivered energy, average growth rates over the period 1972 to 1985 and elasticities of energy growth are compared with historic values in Table 8.10.

It is necessary to draw a net distinction between the period up to 1972, when oil price was practically constant and policy response irrelevant, and the scenario hypotheses, which make price and policy response fundamental variables. A direct comparison between historical and projected elasticities with respect to economic development can be misleading.

In terms of the fundamental energy equation derived earlier in this study, not one but all three elasticities α, β, and γ are necessary to specify energy growth over the period to 1985:

$$E = K\, G^{\alpha}\, P^{\beta}\, R^{\gamma}.$$

In fact, no attempt is made to assign numerical values to policy response R, so that γ cannot be specified as an elasticity. In this study policy response elasticity is replaced by the more manageable concept of the energy-saving impact S of a vigorous over a restrained policy response and introduced as the multiplicative term 1–ZS, where Z=0 for a restrained policy, 1 for a vigorous policy. Moreover, three price elasticities are used: the first, β_1, for oil price variations in the range \$2.90 to \$7.66; the second, β_2, for the range \$7.66 to \$11.50; and the third, β_3, for the range \$11.50 to \$17.25.

These are all forward price elasticities, which implies an assumption of long-term reversibility of the effect of variations in the price of oil. The price terms corresponding to the different ranges are assumed multiplicative.

Application of the basic general energy equation thus obtained leads to the following results.

The average income elasticity of energy demand over the period 1972 to 1985 is 1.55, representing a downturn in the historic trend of 1.21, 1.55, and 1.83 for the periods 1953 to 1960, 1960 to 1966, and 1966 to 1972, respectively.

The price elasticity of energy demand decreases from -0.07 for oil price increases in the range \$2.90 to \$7.66 to -0.16 for the range \$7.66 to \$11.50 and -0.29 for the range \$11.50 to \$17.25.

Finally, the overall energy-saving impact of a

Table 8.10
Overall results of the energy demand projections for the period 1972–1985

	Total Delivered Energy (10^{12} kcal)	Average Growth Rates (%)	Apparent Income Elasticity
1953	197.2		
1963	479.9	9.3	1.55
1972	953.7	7.9	1.62
1985			
Scenario A	1279.9	2.3	0.64
Scenario B	1027.3	0.6	0.27
Scenario C	1437.7	3.2	0.89
Scenario D	1373.1	2.8	1.27
Scenario E	1821.8	5.1	1.42

vigorous policy response is 16.0 percent.

The sectoral distribution of total delivered energy changes considerably in all scenarios relative to 1972, in distinct contrast to the remarkable stability of the previous two decades (Table 8.11). The reason for the strong change in scenarios B and C must be imputed largely to the vigorous policy response reducing the energy intensity of transportation and space heating but having only negligible effect on industry. The share of industry in total delivered energy in fact declines in the restrained policy scenarios, while the residential sector acquires share due to increases in space heating.

Parameters describing energy demand growth in the various sectors are given in Table 8.12. These indicate the greatest energy-saving impact of a vigorous policy response is in the transportation sector, followed by the residential sector. Impact in the industrial sector is considerably smaller.

The price elasticities vary considerably between sectors as well as between price ranges. Transportation overacts in the lower price range and in the higher price range is not so sensitive to oil price increases. This is because most of the energy-saving effects of price (basically shifts from private to public transport) are induced by oil prices below \$11.50. The civilian sectors are the most sensitive to price on account of space heating.

Fueling of the Italian energy system varies considerably between scenarios (Table 8.13). In all cases there is a substantial drop in the petroleum share of total delivered energy that is most evident in the vigorous policy scenarios. The fuels to gain most are natural gas and, to a lesser extent, electricity. Most of the shift to natural gas and away from petroleum is the result of existing international contracts for natural gas supply over the period to 1985.

Total natural gas under long-term contract plus official estimates of domestic supply amount to about 340×10^{12} kcal/yr. in 1985. The scenario dependence of penetration is, however, not determined by total supply since in all cases except scenario E total natural gas supply is in excess of demand (Table 8.14).

Most of the difference in penetration is due to rather complex systems effects resulting from the

Table 8.11
Sector shares in total delivered energy (%)

			1985				
	1953	1972	A	B	C	D	E
Industrial	49.2	49.1	57.6	52.9	56.1	45.9	49.6
Transportation	18.5	18.2	12.8	13.2	11.6	15.8	16.3
Commercial and Public	6.7	7.1	6.1	6.9	6.6	7.8	7.0
Residential	22.7	23.1	19.3	22.0	21.9	26.7	24.1
Agriculture, Mining, and Construction	2.9	2.5	4.2	5.0	3.8	3.8	3.0
Total	100	100	100	100	100	100	100

Table 8.12
Sensitivity of the Italian energy system to economic development, oil price, and policy response

	α	β_1	β_2	β_3	S(%)
Industrial	1.56	-0.07	-0.13	-0.22	2.6
Transportation	1.72	-0.26	-0.31	-0.06	36.2
Commercial and Public	1.48	-0.05	-0.19	-0.48	19.5
Residential	1.66	-0.07	-0.23	-0.60	21.6
Agriculture, Mining, and Construction	1.79	~0	~0	-0.01	0.1
Total	1.55	-0.07	-0.16	-0.29	16.0

interplay between economic growth, oil price, and policy response under constraints set largely by the supply system.[12]

The lower electricity penetration in the restrained policy scenarios is due not to lower growth of electricity but largely to unabated growth of energy consumed for space heating, basically in the form of petroleum derivatives. Similarly, penetration in vigorous policy response scenarios is due mostly to a decrease in the consumption of oil derivatives rather than to enhanced growth in electricity consumption.

Electricity is projected to grow faster than in the past relative to economic growth, in contrast to overall energy growth which decreases considerably.

The elasticities given in Table 8.15 contain price and policy response components which, when separated out, leave a pure income elasticity of electricity growth of 1.69 compared to 1.24, 1.48, and 1.59 in the periods 1953 to 1960, 1960 to 1966, and 1966 to 1972, respectively.

12. These effects are most important in space heating under which heading they are illustrated.

Table 8.14
Supply and demand of natural gas in 1985 (10^{12} kcal)

	Supply	Demand[1]	Excess
Scenario A	340	303	37
Scenario B	340	224	116
Scenario C	340	328	12
Scenario D	340	252	88
Scenario E	340	356	-16

1. Includes natural gas used in secondary energy conversion processes and is calculated by means of the supply-demand integration model.

The oil price elasticities of electricity growth are moderately low (β_1 = -0.02, β_2 = -0.05, and β_3 = -0.10), and the energy-saving impact of a vigorous policy response is very small (5.6 percent).

The composition of petroleum derivatives for final consumption is also highly sensitive to scenario (Table 8.16). The main impact is on the balance between gas oil and fuel oil, which in no case retains its historic value. In the vigorous policy scenarios, fuel oil delivered for final use maintains supremacy over gas oil. In the restrained policy scenarios, however, the balance is completely inverted in favor of gas oil, possibly implying changes in refinery structure by 1985 in favor of lighter fractions. Differences are also marked for light distillates and gasoline.

The change in composition is determined by a complicated interplay between economic growth, oil price, and policy response. Most of the effect is, however, attributable to the vigorous policy decreasing gasoline consumption in favor of diesel in transportation and the restrained policy favoring gas oil in space heating, with fuel oil and light distillates remaining relatively indifferent to policy response insofar as they are determined mostly by industry.

Table 8.13
Fueling of the Italian energy system (% of total delivered energy)

		1985				
	1972	A	B	C	D	E
Wood	1.3	0.1	0.2	0.2	0.2	0.2
Coal	5.1	4.7	4.7	4.6	4.3	4.6
Oil	71.6	58.0	60.9	59.9	66.9	65.0
Gas	11.7	22.5	20.5	21.7	17.3	18.6
Electricity	10.3	14.7	13.7	13.6	11.3	11.6
Total	100	100	100	100	100	100

Table 8.15
Projections of electricity demand in the period 1972 to 1985

	Total Delivered Energy (10^9 kWh)	Average Growth Rates (%)	Apparent Income Elasticity
1953	26.9		
1963	60.9	8.5	1.42
1972	114.8	7.3	1.46
1985			
Scenario A	218.4	5.1	1.42
Scenario B	164.2	2.8	1.27
Scenario C	227.1	5.4	1.50
Scenario D	180.7	3.6	1.64
Scenario E	245.3	6.0	1.67

Table 8.16
Composition of petroleum derivatives for final domestic consumption (%)

		1985				
	1972	A	B	C	D	E
Fuel Oil	38.7	32.4	33.0	31.5	27.0	26.1
Gas Oil	25.8	26.2	31.2	29.9	36.3	31.8
Kerosene	3.4	2.7	3.0	3.1	3.5	3.1
Jet Fuel	0.4	1.0	0.5	1.1	0.6	1.2
Gasoline	16.6	11.7	12.8	9.7	15.4	16.7
Light Distillates	12.0	22.8	17.4	21.6	14.0	18.0
Liquid Petroleum Gas	3.1	3.2	2.1	3.1	3.2	3.1
Total	100	100	100	100	100	100

8.5 DEMAND STUDY WORKSHEETS

The aggregate data for 1972 presented in these tables do not correspond exactly to the national energy accounts, *Bilanci Energetici*, published by the Ministry of Industry.

The difference lies almost entirely in the transportation sector and is due to the definition given this sector in this study. Transportation is defined to include only domestic travel, whereas the national accounts for final delivered energy in 1972 include air bunkerage. Moreover, the official accounts do not include sea bunkerage under final delivered energy, whereas energy consumption for sea travel in domestic waters is accounted for in the tables presented here.

Energy is in 10^{12} kcal.

Processing Losses calculated using Italian supply-demand integration model.

Gross input by fuel does not include bunkerage for nondomestic use, change in stocks, or imports and exports of refinery products.[13]

13. See *Italian Supply-Demand Integration Methodology*, WAES Working Papers, September 1975. Assumptions regarding the supply of primary energy used in the integration will be published in Italian.

WAES NATIONAL ENERGY DEMAND STUDIES, 1972-1985
ENERGY USE SUMMARY

Country Italy

Scenario Assumptions

WAES Case	A		B		C		D		E			
GNP 1976-1985	High		Low		High		Low		High			
Oil Price*	\$11.50 → \$17.25		\$11.50 → \$17.25		\$11.50		\$11.50		\$11.50 → \$7.66			
Natl. Policy Response	Vigorous		Vigorous		Vigorous		Restrained		Restrained			
Natl. GNP Growth Rate 1976-1985 (%/yr.)	4.5		2.5		4.5		2.5		4.5		1972 Base Year	
Fuel Type	Fossil Fuel	Elec.	Fossil Fuel	Elec.	Fossil Fuel	Elec.	Fossil Fuel	Elec.	Fossil Fuel	Elec.	Fossil Fuel	Elec.
Units 10^{12} Kcal												
Energy-Using Sectors												
Industrial (excluding feedstocks)	434.5	107.1	338.1	80.4	478.4	112.7	394.5	87.7	535.7	119.6	313.7	60.5
Transportation	157.3	5.6	131.3	4.0	160.7	6.2	213.0	4.3	292.4	4.4	170.3	3.3
Commercial and Public	55.2	23.3	52.3	18.7	72.0	23.5	87.8	19.5	102.8	25.1	57.4	10.6
Residential	201.4	46.0	192.8	32.8	268.0	47.1	327.9	38.5	384.7	56.1	198.2	21.7
Fishing, Agriculture, Mining, Construction	48.5	5.8	46.2	5.3	48.8	5.8	46.4	5.4	48.9	5.8	21.4	2.6
Feedstocks	195.2		125.4		214.5		148.1		246.3		94.0	
Final Energy Demand (excluding processing losses)	1092.1	187.8	886.1	141.2	1242.4	195.3	1217.7	155.4	1610.8	211.0	855.0	98.7
Processing Losses	53.3	296.7	44.4	216.5	61.7	307.3	61.3	242.5	78.1	353.3	56.0	165.1
Primary Energy Input	1145.4	484.5	930.5	357.7	1304.1	502.6	1279.0	397.9	1688.9	564.3	911.0	263.8

*Oil price 1975 US\$ per barrel Arabian light crude oil FOB Persian Gulf.

WAES NATIONAL ENERGY DEMAND STUDIES, 1972-1985
SECTORAL PERCENTAGES OF TOTAL

Country Italy

Scenario Assumptions												
WAES Case	A		B		C		D		E			
GNP 1976-1985	High		Low		High		Low		High			
Oil Price*	$11.50 → $17.25		$11.50 → $17.25		$11.50		$11.50		$11.50 → $7.66			
Natl. Policy Response	Vigorous		Vigorous		Vigorous		Restrained		Restrained			
Natl. GNP Growth Rate 1976-1985 (%/yr.)	4.5		2.5		4.5		2.5		4.5		1972 Base Year	
Fuel Type	Fossil Fuel	Elec.	Fossil Fuel	Elec.	Fossil Fuel	Elec.	Fossil Fuel	Elec.	Fossil Fuel	Elec.	Fossil Fuel	Elec.
Units												
Energy-Using Sectors												
Industrial (excluding feedstocks)	37.9	22.1	36.3	22.5	36.7	22.4	30.8	22.0	31.7	21.2	34.5	22.9
Transportation	13.8	1.2	14.1	1.1	12.3	1.2	16.7	1.1	17.3	0.8	18.7	1.3
Commercial and Public	4.8	4.8	5.6	5.2	5.5	4.7	6.9	4.9	6.1	4.4	6.3	4.0
Residential	17.6	9.5	20.7	9.2	20.6	9.4	25.6	9.7	22.8	10.0	21.8	8.2
Fishing, Agriculture, Mining, Construction	4.2	1.2	5.0	1.5	3.7	1.2	3.6	1.4	2.9	1.0	2.3	1.0
Feedstocks	17.0		13.5		16.4		11.6		14.6		10.3	
Final Energy Demand (excluding processing losses)	95.3	38.8	95.2	39.5	95.2	38.9	95.2	39.1	95.4	37.4	93.9	37.4
Processing Losses	4.7	61.2	4.8	60.5	4.8	61.1	4.8	60.9	4.6	62.6	6.1	62.6
Primary Energy Input	100.0	100.0	100.0	100.0	100.0	100.0	100.0	100.0	100.0	100.0	100.0	100.0

*Oil price 1975 US$ per barrel Arabian light crude oil FOB Persian Gulf.

WAES NATIONAL ENERGY DEMAND STUDIES, 1972-1985
ENERGY GROWTH RATES SUMMARY

Country Italy

Scenario Assumptions

WAES Case	A		B		C		D		E	
GNP 1976-1985	High		Low		High		Low		High	
Oil Price*	$11.50 → $17.25		$11.50 → $17.25		$11.50		$11.50		$11.50 → $7.66	
Natl. Policy Response	Vigorous		Vigorous		Vigorous		Restrained		Restrained	
Natl. GNP Growth Rate 1976-1985 (%/yr.)	4.5		2.5		4.5		2.5		4.5	
Fuel Type	Fossil Fuel	Elec.	Fossil Fuel	Elec.	Fossil Fuel	Elec.	Fossil Fuel	Elec.	Fossil Fuel	Elec.
Units (%/yr. 1972-1985)										
Energy-Using Sectors										
Industrial (excluding feedstocks)	2.5	4.5	0.6	2.2	3.3	4.9	1.8	2.9	4.2	5.4
Transportation	-0.6	4.1	-2.0	1.6	-0.4	5.0	1.7	2.1	4.3	2.2
Commercial and Public	-0.4	6.2	-0.7	4.5	1.8	6.3	3.3	4.8	4.6	6.9
Residential	0.1	6.0	-0.2	3.2	2.3	6.1	3.9	4.5	5.2	7.6
Fishing, Agriculture, Mining, Construction	6.5	6.4	6.1	5.6	6.5	6.4	6.1	5.8	6.6	6.4
Feedstocks	5.8		2.2		6.6		3.6		7.7	
Final Energy Demand (excluding processing losses)	1.9	5.1	0.3	2.8	2.9	5.4	2.8	3.6	5.0	6.0
Processing Losses	-0.4	4.6	-1.8	2.1	0.7	4.9	0.7	3.0	2.6	6.0
Primary Energy Input	1.8	4.8	0.2	2.4	2.8	5.1	2.6	3.2	4.9	6.0

*Oil price 1975 US$ per barrel Arabian light crude oil FOB Persian Gulf.

NATIONAL ENERGY DEMAND

Country Italy

Year 1972

Summary

Category	Oil	Gas	Coal	Electricity	Other	Total by Category (Excluding Processing Losses)
Transportation	168.0	1.1	1.2	3.3		173.6
Agriculture, Mining, and Construction	21.1	0.3		2.6		24.0
Commercial	35.8	5.6	1.2	7.0	1.7	51.3
Public	11.3	1.0	0.4	3.6	0.4	16.7
Industrial	280.9	82.6	40.8	60.5	3.4	468.2
Residential	165.1	20.9	4.9	21.7	7.3	219.9
Total Delivered Energy by Fuel	682.2	111.5	48.5	98.7	12.8	953.7
Processing Losses by Fuel	38.1	1.4	16.5	165.1	∿ 0	221.1
Gross Input by Fuel	720.3	112.9	65.0	263.8	12.8	1174.8

Gross Energy Input (Add across bottom row)

NATIONAL ENERGY DEMAND

Country Italy

Year 1972

Transportation

	(A)	(B)	(C)	(D)	(E)	(F)	(G)	
	Population/Auto	Distance/Auto/Yr.	Load Factor	Fraction Urban or Rural	Energy/Distance	Fossil Fuel	Electricity	
Urban Auto	4.36	10370	1.24	0.52	882	59.6		(1)
Rural Auto			2.40	0.48	635	39.6		(2)
	Activity Level	% Electrified	Load Factor	Fossil Fuel/ Activity Level	Electricity/Activity			
Urban Bus	40.9	9.8	27	135	69.0	5.0	0.3	(3)
Intercity Bus	8.2		24	108		0.9		(4)
Rail (Passenger)	37.8	82.7	165	248	47.4	1.6	1.5	(5)
Air (Passenger, Domestic)	1.9		49	1530		2.9		(6)
Air (Passenger, International)								(7)
Ship (Passenger)	2.4			1320		3.2		(8)
Rail (Freight)	17.2	94.1	274	1050	84.4	1.1	1.4	(9)
Air (Freight, Domestic)	0.015			8480		0.1		(10)
Air (Freight, International)								(11)
Truck (Local)	12.6		1.6	1320		16.6		(12)
Truck (Intercity)	64.7		9.4	498		32.2		(13)
Ship (Freight, Domestic)	18.0			220		4.0		(14)
Ship (Freight, International)								(15)
Miscellaneous						3.5	0.1	(16)
					Total	170.2	3.3	

Distance/Auto/Yr. in km.

Load factor for air travel is % occupancy. For all other modes it is passenger/vehicle (tons/vehicle).

Energy/Distance in Kcal/km.

Fossil Fuel in 10^{12} Kcal.

Electricity in 10^{12} Kcal.

Activity Level in 10^{9} passenger·km (ton ·km)

% Electrified is the percentage of the total activity delivered by electrified mode.

Fossil Fuel/Activity in Kcal/passenger·km (Kcal/ton·km)

Electricity/Activity in Kcal/passenger·km (Kcal/ton·km)

Energy consumed for international travel is accounted for under bunkerage in the Italian Supply/Demand integration.

NATIONAL ENERGY DEMAND

Country Italy

Year 1972

Agriculture, Mining, and Construction

Agriculture, Forestry, and Fishing (ISIC Major Division 1)	(A) Value Added	(B) Energy/Value Added	(C) Area Harvested	(D) Energy/Area	(E) Physical Output	(F) Energy/Output	(G) Fossil Fuel	(H) Electricity	
Farm Machinery / Greenhouses	3170	5.3	131	129	90.6	187	16.2	0.7	(17) (18)
Livestock	2430	0.29			12.6	60	0.5	0.2	(19)
Forestry (ISIC 12)	170	7.6	62	20.2	6.5	192	1.2	0.1	(20)
Fishing Boats (ISIC 13) / Other Fishing Use (ISIC 13)	138	16.7			0.26	8850	2.3		(21) (22)
Total Fertilizer (10^6 tons)		N 0.62	P 0.57	K 0.25		Total	20.2	1.0	

Mining	(A) Value Added	(B) Energy/Value Added	(C) Physical Output	(D) Energy/Physical Output	(E) Fossil Fuel	(F) Electricity	
Handling and Internal Transportation / Other Energy Use	370	6.2	301	7.6	1.2	1.1	(23) (24)
				Total	1.2	1.1	

Construction	(A) Value Added	(B) Energy/Value Added	(C) Physical Output	(D) Energy/Physical Output	(E) Fossil Fuel	(F) Electricity	
Handling and Internal Transportation / Other Energy Use	4860	0.10	226	2.2	~0	0.5	(25) (26)
				Total	~0	0.5	
				Agriculture, Mining, and Construction Total	21.4	2.6	

Value Added in 10^9 Lire (1972).

Energy/Value Added in Kcal/Lira.

Area Harvested in 10^9 m^2

Energy/Area in Kcal/m^2

Physical Output in 10^6 tons for Agriculture and Mining, 10^6 m^3 for Consturction.

Energy/Output in 10^3 Kcal/ton (10^3 Kcal/m^3 for Construction).

Fossil fuel in 10^{12} Kcal.

Electricity in 10^{12} Kcal.

NATIONAL ENERGY DEMAND

Country Italy

Year 1972

Commercial and Public

	(A)	(B)	(C)	(D)	(E)	
Commercial	Floor Area	Heat Gain or Loss/Area	System Efficiency	Fossil Fuel	Electricity	
Space Heat (Fossil Fuel)	172	120	0.56	37.0		(27)
Space Heat (Electricity)	6	52.3	0.95		0.3	(28)
Air Conditioning (Fossil Fuel)	21	13.8	0.58	0.5		(29)
Air Conditioning (Electricity)	54	31.1	0.93		1.8	(30)
		Energy/Area	Electricity/Area			
Lighting	199	11.6	11.6		2.3	(31)
Miscellaneous	199	47.2	13.1	6.8	2.6	(32)
Presently Unknown						(33)
			Total	44.3	7.0	
Public	Floor Area	Heat Gain or Loss/Area	System Efficiency	Fossil Fuel	Electricity	
Space Heat (Fossil Fuel)	54	129	0.56	12.4		(34)
Space Heat (Electricity)	2	47.5	0.95		0.1	(35)
Air Conditioning (Fossil Fuel)	4	13.8	0.58	0.1		(36)
Air Conditioning (Electricity)	15	31.1	0.93		0.5	(37)
		Energy/Area	Electricity/Area			
Lighting					2.1	(38)
Miscellaneous	60	25.7	15.7	0.6	0.9	(39)
Presently Unknown						(40)
			Total	13.1	3.6	
			Commercial and Public Total	57.4	10.6	

Floor Area in 10^6 m^2.

Heat Gain or Loss/Area in 10^3 $Kcal/m^2$.

Fossil Fuel in 10^{12} Kcal.

Electricity in 10^{12} Kcal.

Energy/Area in 10^3 $Kcal/m^2$ includes fossil fuel and electricity.

Electricity/Area in 10^3 $Kcal/m^2$

Public Lighting includes street lighting.

NATIONAL ENERGY DEMAND

Country Italy

Year 1972

Industrial

	(A) Value Added	(B) Energy/Value Added	(C) Physical Output	(D) Energy/Output	(E) % Exported	(F) Fossil Fuel	(G) Electricity	(H) Feedstocks	
Iron and Steel (ISIC 371)	810	82.1	53.3	1250		55.3	11.2		(41)
Nonferrous Metals (ISIC 372)	223	43.4	1.5	6470		3.3	6.4		(42)
Paper and Allied (ISIC 34)	943	18.3	4.5	3930		13.3	4.0		(43)
Chemicals and Allied (ISIC 35 minus ISIC 353)	1880	103.6	28.3	6880		87.8	13.0	94.0	(44)
Mineral Products (ISIC 36)	939	84.5	53.2	1490		73.4	5.9		(45)
Food and Related (ISIC 31)	1980	13.9	18.5	1490		24.3	3.2		(46)
Transport Eqpt. (ISIC 384)	1340	6.7	2.3	3910		6.4	2.6		(47)
Mechanical	3440	7.3	2.8	8930		19.7	5.3		(48)
Textiles and Clothing	2330	10.1	1.3	18100		18.2	5.3		(49)
Misc. Manufacturing	3520	4.4				12.0	3.6		(50)
					Total	313.7	60.5	94.0	

N.E.C. = Not Elsewhere Classified

Value Added in 10^9 Lire (1972).

Energy/Value Added in Kcal/Lira.

Physical Output in 10^6 tons.

Energy/Output in 10^3 Kcal/ton.

Fossil Fuel in 10^{12} Kcal.

Electricity in 10^{12} Kcal.

Feedstocks in 10^{12} Kcal.

NATIONAL ENERGY DEMAND

Country ITALY

Year 1972

Residential

Population $54.6x10^6$ Persons/Dwelling 3.50 Dwellings $15.6x10^6$

		(A)	(B)	(C)	(D)	(E)	(F)	
Central Space Heating		Dwellings	Heat Loss	System Efficiency		Fossil Fuel	Electricity	
Single Family	Fossil Fuels	1.7	8.2	0.59		23.5		(51)
	Electricity							(52)
Town or Row Houses	Fossil Fuels	8.0	6.4	0.54		95.1		(53)
	Electricity							(54)
Apartments	Fossil Fuels	5.5	4.5	0.58		42.8		(55)
	Electricity							(56)
		% Saturation (Fossil Fuels)	% Saturation (Electricity)	Energy/Unit (Fossil Fuels)	Energy/Unit (Electricity)			
Room Heating		37.0	5.8	3000	1330	17.3	1.2	(57)
Air Conditioning (Window)			0.7		916		0.1	(58)
Air Conditioning (Central)		0.6		1070	3210	0.1	0.3	(59)
Water Heating		9.6	33.4	3270	1020	4.9	5.3	(60)
Refrigerators			89.0		215		3.0	(61)
Lighting			~100		199		3.1	(62)
Cooking		96.6	3.4	962	377	14.5	0.2	(63)
Dishwashers			7.6		965		1.1	(64)
Freezers			10.5		342		0.6	(65)
Television			84.3		144		1.9	(66)
Clotheswasher			63.4		411		4.1	(67)
Misc. Appliances							0.8	(68)
Potential Appliances								(69)
					Total	198.2	21.7	

Dwellings in 10^6 include only heated dwellings.

Heat Loss in 10^6 Kcal/dwelling.

Fossil Fuel in 10^{12} Kcal.

Electricity in 10^{12} Kcal.

Town or Row Houses defined as 2-family units plus 3/15 family apartment buildings. Apartments defined as > 15 family apartment buildings.

% Saturation is relative to total dwellings.

Energy/Unit (Fossil Fuels) in 10^3 Kcal.

Energy/Unit (Electricity) in 10^3 Kcal.

Water heating includes hot water heating on kitchen cookers and stoves .

NATIONAL ENERGY DEMAND PROJECTIONS

Country: Italy

Year: 1985

WAES Case: A

Summary

Category	Oil	Gas	Coal	Electricity	Other	Total by Category (Excluding Processing Losses)
Transportation	157.3			5.6		162.9
Agriculture, Mining, and Construction	48.2	0.3		5.8		54.3
Commercial	24.1	16.8	0.5	16.1	0.3	57.8
Public	8.3	5.0	0.1	7.2	0.1	20.7
Industrial	384.3	188.1	57.3	107.1		736.8
Residential	120.3	77.6	2.0	46.0	1.5	247.4
Total Delivered Energy by Fuel	742.5	287.8	59.9	187.8	1.9	1279.9
Processing Losses by Fuel	25.5	5.3	22.4	296.7	0.1	350.0
Gross Input by Fuel	768.0	293.1	82.3	484.5	2.0	1629.9

Gross Energy Input (Add across bottom row)

NATIONAL ENERGY DEMAND PROJECTIONS

Country Italy

Year 1985

WAES Case A

Transportation

	(A)	(B)	(C)	(D)	(E)	(F)	(G)	
	Population/Auto	Distance/Auto/Yr.	Load Factor	Fraction Urban or Rural	Energy/Distance	Fossil Fuel	Electricity	
Urban Auto	3.08	9190	1.15	0.26	680	30.9		(1)
Rural Auto			2.40	0.74	487	63.0		(2)
	Activity Level	% Electrified	Load Factor	Fossil Fuel/ Activity Level	Electricity/Activity			
Urban Bus	69.3	5.5	40	87	77	5.7	0.3	(3)
Intercity Bus	34.6		35	75		2.6		(4)
Rail (Passenger)	71.5	94.8	188	185	44.5	0.7	3.0	(5)
Air (Passenger, Domestic)	8.5		59	764		6.5		(6)
Air (Passenger, International)								(7)
Ship (Passenger)	5.2			1310		6.8		(8)
Rail (Freight)	17.8	98.4	312	350	80	0.1	1.4	(9)
Air (Freight, Domestic)	0.095			4180		0.4		(10)
Air (Freight, International)								(11)
Truck (Local)	15.3		3.6	652		10.0		(12)
Truck (Intercity)	76.7		12.5	270		20.7		(13)
Ship (Freight, Domestic)	37.9			180		6.8		(14)
Ship (Freight, International)								(15)
Miscellaneous						3.3	0.9	(16)
					Total	157.5	5.6	

Distance/Auto/Yr. in km.

Load factor for air travel is % occupancy. For all other modes it is passenger/vehicle (tons/vehicle).

Energy/Distance in Kcal/km.

Fossil Fuel in 10^{12} Kcal.

Electricity in 10^{12} Kcal.

Activity Level in 10^{9} passenger·km (ton ·km)

% Electrified is the percentage of the total activity delivered by electrified mode.

Fossil Fuel/Activity in Kcal/passenger·km (Kcal/ton·km)

Electricity/Activity in Kcal/passenger·km (Kcal/ton·km)

Energy consumed for international travel is accounted for under bunkerage in the Italian Supply/Demand integration.

NATIONAL ENERGY DEMAND PROJECTIONS

Country Italy

Year 1985

WAES Case A

Agriculture, Mining, and Construction

	(A)	(B)	(C)	(D)	(E)	(F)	(G)	(H)	
Agriculture, Forestry, and Fishing (ISIC Major Division 1)	Value Added	Energy/Value Added	Area Harvested	Energy/ Area	Physical Output	Energy/ Output	Fossil Fuel	Electricity	
Farm Machinery	4020	10.6	129	330	115.0	370	40.6	2.0	(17)
Greenhouses									(18)
Livestock	3770	0.80			19.5	154	2.0	1.0	(19)
Forestry (ISIC 12)	168	11.9	64	31.3	6.0	333	1.8	0.2	(20)
Fishing Boats (ISIC 13)	192	15.1			0.35	8290	2.9		(21)
Other Fishing Use (ISIC 13)									(22)
Total Fertilizer (10^6 tons)		N 0.80	P 0.73	K 0.32		Total	47.3	3.2	

	(A)	(B)	(C)	(D)	(E)	(F)	
Mining	Value Added	Energy/Value Added	Physical Output	Energy/Physical Output	Fossil Fuel	Electricity	
Handling and Internal Transportation	518	5.2	421	6.4	1.2	1.5	(23)
Other Energy Use							(24)
				Total	1.2	1.5	

	(A)	(B)	(C)	(D)	(E)	(F)	
Construction	Value Added	Energy/Value Added	Physical Output	Energy/Physical Output	Fossil Fuel	Electricity	
Handling and Internal Transportation	7670	0.14	356	3.1	~0	1.1	(25)
Other Energy Use							(26)
				Total	~0	1.1	
				Agriculture, Mining, and Construction Total	48.5	5.8	

Value Added in 10^9 Lire (1972).

Energy/Value Added in Kcal/Lira.

Area Harvested in 10^9 m^2

Energy/Area in Kcal/m^2

Physical Output in 10^6 tons for Agriculture and Mining, 10^6 m^3 for Consturction.

Energy/Output in 10^3 Kcal/ton (10^3 Kcal/m^3 for Construction).

Fossil fuel in 10^{12} Kcal.

Electricity in 10^{12} Kcal.

NATIONAL ENERGY DEMAND PROJECTIONS

Country Italy

Year 1985

WAES Case A

Commercial and Public

Commercial	(A) Floor Area	(B) Heat Gain or Loss/Area	(C) System Efficiency	(D) Fossil Fuel	(E) Electricity	
Space Heat (Fossil Fuel)	167	139	0.68	34.2		(27)
Space Heat (Electricity)	2	53.1	0.95		0.1	(28)
Air Conditioning (Fossil Fuel)	60	10.1	0.76	0.8		(29)
Air Conditioning (Electricity)	184	22.8	0.93		4.5	(30)
		Energy/Area	Electricity/Area			
Lighting	229	17.9	17.9		4.1	(31)
Miscellaneous	229	59.8	32.3	6.3	7.4	(32)
Presently Unknown						(33)
			Total	41.3	16.1	
Public	Floor Area	Heat Gain or Loss/Area	System Efficiency	Fossil Fuel	Electricity	
Space Heat (Fossil Fuel)	60	141	0.68	12.5		(34)
Space Heat (Electricity)						(35)
Air Conditioning (Fossil Fuel)	15	10.1	0.76	0.2		(36)
Air Conditioning (Electricity)	57	22.8	0.93		1.4	(37)
		Energy/Area	Electricity/Area			
Lighting					4.1	(38)
Miscellaneous	75	32.1	22.3	0.7	1.7	(39)
Presently Unknown						(40)
			Total	13.4	7.2	
			Commercial and Public Total	54.7	23.3	

Floor Area in 10^6 m^2.

Heat Gain or Loss/Area in 10^3 Kcal/m^2.

Fossil Fuel in 10^{12} Kcal.

Electricity in 10^{12} Kcal.

Energy/Area in 10^3 Kcal/m^2 includes fossil fuel and electricity.

Electricity/Area in 10^3 Kcal/m^2

Public Lighting includes street lighting.

NATIONAL ENERGY DEMAND PROJECTIONS

Country Italy

Year 1985

WAES Case A

Industrial

	(A) Value Added	(B) Energy/Value Added	(C) Physical Output	(D) Energy/Output	(E) % Exported	(F) Fossil Fuel	(G) Electricity	(H) Feedstocks	
Iron and Steel (ISIC 371)	1730	58.6	97.5	1040		80.2	21.2		(41)
Nonferrous Metals (ISIC 372)	372	38.5	2.3	6220		4.4	9.9		(42)
Paper and Allied (ISIC 34)	1460	18.1	11.6	2280		19.8	6.6		(43)
Chemicals and Allied (ISIC 35 minus ISIC 353)	3260	108.5	47.1	7510		133.5	25.0	195.2	(44)
Mineral Products (ISIC 36)	1420	73.8	80.2	1310		94.8	10.0		(45)
Food and Related (ISIC 31)	3338	12.4	30.5	1360		35.5	5.9		(46)
Transport Eqpt. (ISIC 384)	2640	5.4	3.9	3670		9.3	5.0		(47)
Mechanical	6300	6.5	4.5	9110		29.5	11.5		(48)
Textiles and Clothing	3510	5.6	1.4	14100		14.7	5.0		(49)
Misc. Manufacturing	6200	3.2				12.8	7.0		(50)
					Total	434.5	107.1	195.2	

Value Added in 10^9 Lire (1972).

Energy/Value Added in Kcal/Lira.

Physical Output in 10^6 tons.

Energy/Output in 10^3 Kcal/ton.

Fossil Fuel in 10^{12} Kcal.

Electricity in 10^{12} Kcal.

Feedstocks in 10^{12} Kcal.

NATIONAL ENERGY DEMAND PROJECTIONS

Country Italy

Year 1985

WAES Case A

Residential

Population $58.6x10^6$ Persons/Dwelling 3.05 Dwellings $19.2x10^6$

		(A)	(B)	(C)	(D)	(E)	(F)	
Central Space Heating		Dwellings	Heat Loss	System Efficiency		Fossil Fuel	Electricity	
Single Family	Fossil Fuels	2.1	15.5	0.66		49.2		(51)
	Electricity							(52)
Town or Row Houses	Fossil Fuels	9.5	7.0	0.67		99.1		(53)
	Electricity							(54)
Apartments	Fossil Fuels	6.8	3.3	0.76		29.5		(55)
	Electricity							(56)
		% Saturation (Fossil Fuels)	% Saturation (Electricity)	Energy/Unit (Fossil Fuels)	Energy/Unit (Electricity)			
Room Heating		8.6	1.1	3200	1420	5.3	0.3	(57)
Air Conditioning (Window)			2.8		558		0.3	(58)
Air Conditioning (Central)		2.1		744	1490	0.3	0.6	(59)
Water Heating		23.4	43.5	1380	1020	6.2	8.5	(60)
Refrigerators			97.9		318		6.0	(61)
Lighting			~100		219		4.2	(62)
Cooking		94.3	5.7	652	274	11.8	0.3	(63)
Dishwashers			33.8		1430		9.3	(64)
Freezers			39.5		435		3.3	(65)
Television			95.8		223		4.1	(66)
Clotheswasher			90.6		454		7.9	(67)
Misc. Appliances							1.2	(68)
Potential Appliances								(69)
					Total	201.4	46.0	

Dwellings in 10^6 include only heated dwellings.

Heat Loss in 10^6 Kcal/dwelling.

Fossil Fuel in 10^{12} Kcal.

Electricity in 10^{12} Kcal.

Town or Row Houses defined as 2-family units plus 3/15 family apartment buildings. Apartments defined as $>$ 15 family apartment buildings.

% Saturation is relative to total dwellings.

Energy/Unit (Fossil Fuels) in 10^3 Kcal.

Energy/Unit (Electricity) in 10^3 Kcal.

Water heating includes hot water heating on kitchen cookers and stoves.

NATIONAL ENERGY DEMAND PROJECTIONS

Country Italy

Year 1985

WAES Case B

Summary

Category	Oil	Gas	Coal	Electricity	Other	Total by Category (Excluding Processing Losses)
Transportation	131.3			4.0		135.3
Agriculture, Mining, and Construction	45.9	0.3		5.3		51.5
Commercial	25.7	12.6	0.5	12.5	0.3	51.6
Public	9.4	3.6	0.1	6.2	0.1	19.4
Industrial	280.9	137.1	45.5	80.4		543.9
Residential	132.2	56.9	2.1	32.8	1.6	225.6
Total Delivered Energy by Fuel	625.4	210.5	48.2	141.2	2.0	1027.3
Processing Losses by Fuel	20.1	3.4	20.9	216.5	~0	260.9
Gross Input by Fuel	645.5	213.9	69.1	357.7	2.0	1288.2

Gross Energy Input (Add across bottom row)

NATIONAL ENERGY DEMAND PROJECTIONS

Country Italy

Year 1985

WAES Case B

Transportation

	(A)	(B)	(C)	(D)	(E)	(F)	(G)	
	Population/Auto	Distance/Auto/Yr.	Load Factor	Fraction Urban or Rural	Energy/Distance	Fossil Fuel	Electricity	
Urban Auto	3.20	8040	1.20	0.32	670	31.6		(1)
Rural Auto			2.50	0.68	479	48.0		(2)
	Activity Level	% Electrified	Load Factor	Fossil Fuel/ Activity Level	Electricity/Activity			
Urban Bus	58.2	3.4	37	92.1	61.5	5.2	0.1	(3)
Intercity Bus	18.9		35	75.7		1.4		(4)
Rail (Passenger)	49.7	92.6	188	185	44.5	0.7	2.1	(5)
Air (Passenger, Domestic)	4.1		59	764		3.1		(6)
Air (Passenger, International)								(7)
Ship (Passenger)	3.8			1310		5.0		(8)
Rail (Freight)	15.5	98.3	312	350	80	0.1	1.2	(9)
Air (Freight, Domestic)	0.038			4180		0.2		(10)
Air (Freight, International)								(11)
Truck (Local)	14.0		3.5	689		9.6		(12)
Truck (Intercity)	60.5		12.8	291		17.6		(13)
Ship (Freight, Domestic)	25.1			180		4.5		(14)
Ship (Freight, International)								(15)
Miscellaneous						4.3	0.6	(16)
					Total	131.3	4.0	

Distance/Auto/Yr. in km.

Load factor for air travel is % occupancy. For all other modes it is passenger/vehicle (tons/vehicle).

Energy/Distance in Kcal/km.

Fossil Fuel in 10^{12} Kcal.

Electricity in 10^{12} Kcal.

Activity Level in 10^9 passenger·km (ton ·km)

% Electrified is the percentage of the total activity delivered by electrified mode.

Fossil Fuel/Activity in Kcal/passenger·km (Kcal/ton·km)

Electricity/Activity in Kcal/passenger·km (Kcal/ton·km)

Energy consumed for international travel is accounted for under bunkerage in the Italian Supply/Demand integration.

NATIONAL ENERGY DEMAND PROJECTIONS

Country Italy

Year 1985

WAES Case B

Agriculture, Mining, and Construction

	(A)	(B)	(C)	(D)	(E)	(F)	(G)	(H)	
Agriculture, Forestry, and Fishing (ISIC Major Division 1)	Value Added	Energy/Value Added	Area Harvested	Energy/Area	Physical Output	Energy/Output	Fossil Fuel	Electricity	
Farm Machinery	3710	11.1	129	318	111	369	39.2	1.8	(17)
Greenhouses									(18)
Livestock	3110	0.84			16.8	155	1.9	0.7	(19)
Forestry (ISIC 12)	167	12.6	64	32.8	6.2	339	1.5	0.6	(20)
Fishing Boats (ISIC 13)	163	15.3			0.32	7815	2.5		(21)
Other Fishing Use (ISIC 13)									(22)
Total Fertilizer (10^6 tons)		N 0.73	P 0.67	K 0.30		Total	45.1	3.1	

	(A)	(B)	(C)	(D)	(E)	(F)	
Mining	Value Added	Energy/Value Added	Physical Output	Energy/Physical Output	Fossil Fuel	Electricity	
Handling and Internal Transportation	444	5.2	360	6.4	1.1	1.2	(23)
Other Energy Use							(24)
				Total	1.1	1.2	

	(A)	(B)	(C)	(D)	(E)	(F)	
Construction	Value Added	Energy/Value Added	Physical Output	Energy/Physical Output	Fossil Fuel	Electricity	
Handling and Internal Transportation	7270	0.14	338	3.0	~0	1.0	(25)
Other Energy Use							(26)
				Total	~0	1.0	
				Agriculture, Mining, and Construction Total	46.2	5.3	

Value Added in 10^9 Lire (1972).

Energy/Value Added in Kcal/Lira.

Area Harvested in 10^9 m^2

Energy/Area in Kcal/m^2

Physical Output in 10^6 tons for Agriculture and Mining, 10^6 m^3 for Consturction.

Energy/Output in 10^3 Kcal/ton (10^3 Kcal/m^3 for Construction).

Fossil fuel in 10^{12} Kcal.

Electricity in 10^{12} Kcal.

NATIONAL ENERGY DEMAND PROJECTIONS

Country Italy

Year 1955

WAES Case B

Commercial and Public

	(A)	(B)	(C)	(D)	(E)	
Commercial	Floor Area	Heat Gain or Loss/Area	System Efficiency	Fossil Fuel	Electricity	
Space Heat (Fossil Fuel)	162	135	0.68	32.3		(27)
Space Heat (Electricity)	2	53.1	0.95		0.1	(28)
Air Conditioning (Fossil Fuel)	52	10.1	0.75	0.7		(29)
Air Conditioning (Electricity)	126	22.8	0.93		3.1	(30)
		Energy/Area	Electricity/Area			
Lighting	216	15.7	15.7		3.4	(31)
Miscellaneous	216	55.6	27.3	6.1	5.9	(32)
Presently Unknown						(33)
			Total	39.1	12.5	
Public	Floor Area	Heat Gain or Loss/Area	System Efficiency	Fossil Fuel	Electricity	
Space Heat (Fossil Fuel)	66	127	0.68	12.3		(34)
Space Heat (Electricity)						(35)
Air Conditioning (Fossil Fuel)	7	10.1	0.75	0.1		(36)
Air Conditioning (Electricity)	49	22.8	0.93		1.2	(37)
		Energy/Area	Electricity/Area			
Lighting					3.5	(38)
Miscellaneous	73	30.8	20.5	0.8	1.5	(39)
Presently Unknown						(40)
			Total	13.2	6.2	
			Commercial and Public Total	52.3	18.7	

Floor Area in 10^6 m^2.

Heat Gain or Loss/Area in 10^3 Kcal/m^2.

Fossil Fuel in 10^{12} Kcal.

Electricity in 10^{12} Kcal.

Energy/Area in 10^3 Kcal/m^2 includes fossil fuel and electricity.

Electricity/Area in 10^3 Kcal/m^2

Public Lighting includes street lighting.

NATIONAL ENERGY DEMAND PROJECTIONS

Country Italy

Year 1985

WAES Case B

Industrial

	(A) Value Added	(B) Energy/Value Added	(C) Physical Output	(D) Energy/Output	(E) % Exported	(F) Fossil Fuel	(G) Electricity	(H) Feedstocks	
Iron and Steel (ISIC 371)	1260	61.1	70.8	1090		61.8	15.2		(41)
Nonferrous Metals (ISIC 372)	300	38.9	1.9	6160		3.6	8.1		(42)
Paper and Allied (ISIC 34)	1180	18.2	9.4	2290		16.2	5.3		(43)
Chemicals and Allied (ISIC 35 minus ISIC 353)	2380	102.7	34.4	7100		101.4	17.6	125.4	(44)
Mineral Products (ISIC 36)	1120	74.7	63.5	1320		75.8	7.9		(45)
Food and Related (ISIC 31)	2630	12.4	24.0	1360		27.9	4.7		(46)
Transport Eqpt. (ISIC 384)	1930	5.4	2.8	3710		6.8	3.6		(47)
Mechanical	4880	6.5	3.5	9060		22.9	8.8		(48)
Textiles and Clothing	2810	5.7	1.1	14500		12.0	4.0		(49)
Misc. Manufacturing	4650	3.2				9.7	5.2		(50)
					Total	338.1	80.4	125.4	

Value Added in 10^9 Lire (1972).

Energy/Value Added in Kcal/Lira.

Physical Output in 10^6 tons.

Energy/Output in 10^3 Kcal/ton.

Fossil Fuel in 10^{12} Kcal.

Electricity in 10^{12} Kcal.

Feedstocks in 10^{12} Kcal.

NATIONAL ENERGY DEMAND PROJECTIONS

Country ITALY

Year 1985

WAES Case B

Residential

Population 58.6×10^6 Persons/Dwelling 3.22 Dwellings 18.2×10^6

		(A)	(B)	(C)	(D)	(E)	(F)	
Central Space Heating		Dwellings	Heat Loss	System Efficiency		Fossil Fuel	Electricity	
Single Family	Fossil Fuels	2.0	15.0	0.66		45.6		(51)
	Electricity							(52)
Town or Row Houses	Fossil Fuels	9.0	6.8	0.67		92.2		(53)
	Electricity							(54)
Apartments	Fossil Fuels	6.4	3.3	0.75		27.7		(55)
	Electricity							(56)
		% Saturation (Fossil Fuels)	% Saturation (Electricity)	Energy/Unit (Fossil Fuels)	Energy/Unit (Electricity)			
Room Heating		14.6	1.9	3200	1420	8.5	0.5	(57)
Air Conditioning (Window)			1.4		392		0.1	(58)
Air Conditioning (Central)		1.3		845	1690	0.2	0.4	(59)
Water Heating		19.2	38.6	1570	1070	5.5	7.5	(60)
Refrigerators			95.7		271		4.7	(61)
Lighting			~100		203		3.7	(62)
Cooking		94.5	5.5	762	300	13.1	0.3	(63)
Dishwashers			19.4		1220		4.3	(64)
Freezers			22.7		363		1.5	(65)
Television			92.7		178		3.0	(66)
Clotheswasher			75.7		421		5.8	(67)
Misc. Appliances							1.0	(68)
Potential Appliances								(69)
					Total	192.8	32.8	

Dwellings in 10^6 include only heated dwellings.

Heat Loss in 10^6 Kcal/dwelling.

Fossil Fuel in 10^{12} Kcal.

Electricity in 10^{12} Kcal.

Town or Row Houses defined as 2-family units plus 3/15 family apartment buildings. Apartments defined as > 15 family apartment buildings.

% Saturation is relative to total dwellings.

Energy/Unit (Fossil Fuels) in 10^3 Kcal.

Energy/Unit (Electricity) in 10^3 Kcal.

Water heating includes hot water heating on kitchen cookers and stoves .

NATIONAL ENERGY DEMAND PROJECTIONS

Country Italy

Year 1985

WAES Case C

Summary

Category	Oil	Gas	Coal	Electricity	Other	Total by Category (Excluding Processing Losses)
Transportation	160.7			6.2		166.9
Agriculture, Mining, and Construction	48.4	0.4		5.8		54.6
Commercial	35.6	17.7	0.6	16.5	0.4	70.8
Public	12.3	5.1	0.2	7.0	0.1	24.7
Industrial	422.7	207.0	63.2	112.7		805.6
Residential	181.8	81.6	2.7	47.1	1.9	315.1
Total Delivered Energy by Fuel	861.5	311.8	66.7	195.3	2.4	1437.7
Processing Losses by Fuel	29.9	6.0	25.8	307.3	~0	369.0
Gross Input by Fuel	891.4	317.8	92.5	502.6	2.4	1806.7

Gross Energy Input (Add across bottom row)

NATIONAL ENERGY DEMAND PROJECTIONS

Country Italy

Year 1985

WAES Case C

Transportation

	(A)	(B)	(C)	(D)	(E)	(F)	(G)	
	Population/Auto	Distance/Auto/Yr.	Load Factor	Fraction Urban or Rural	Energy/Distance	Fossil Fuel	Electricity	
Urban Auto	3.13	9220	1.15	0.25	690	29.8		(1)
Rural Auto			2.40	0.75	490	63.4		(2)
	Activity Level	% Electrified	Load Factor	Fossil Fuel/ Activity Level	Electricity/Activity			
Urban Bus	74.8	6.4	41	86	113	6.0	0.5	(3)
Intercity Bus	40.4		36	74		3.0		(4)
Rail (Passenger)	80.2	94.9	188	185	44.5	0.8	3.4	(5)
Air (Passenger, Domestic)	9.4		55	994		9.3		(6)
Air (Passenger, International)								(7)
Ship (Passenger)	5.3			1310		6.9		(8)
Rail (Freight)	18.1	98.4	312	350	80	0.1	1.4	(9)
Air (Freight, Domestic)	0.098			5450		0.5		(10)
Air (Freight, International)								(11)
Truck (Local)	15.4		3.6	649		10.0		(12)
Truck (Intercity)	77.9		12.5	270		21.0		(13)
Ship (Freight, Domestic)	38.6			180		7.0		(14)
Ship (Freight, International)								(15)
Miscellaneous						2.9	0.9	(16)
					Total	160.7	6.2	

Distance/Auto/Yr. in km.

Load factor for air travel is % occupancy. For all other modes it is passenger/vehicle (tons/vehicle).

Energy/Distance in Kcal/km.

Fossil Fuel in 10^{12} Kcal.

Electricity in 10^{12} Kcal.

Activity Level in 10^{9} passenger·km (ton ·km)

% Electrified is the percentage of the total activity delivered by electrified mode.

Fossil Fuel/Activity in Kcal/passenger·km (Kcal/ton·km)

Electricity/Activity in Kcal/passenger·km (Kcal/ton·km)

Energy consumed for international travel is accounted for under bunkerage in the Italian Supply/Demand integration.

NATIONAL ENERGY DEMAND PROJECTIONS

Country Italy

Year 1985

WAES Case C

Agriculture, Mining, and Construction

	(A)	(B)	(C)	(D)	(E)	(F)	(G)	(H)	
Agriculture, Forestry, and Fishing (ISIC Major Division 1)	Value Added	Energy/Value Added	Area Harvested	Energy/Area	Physical Output	Energy/Output	Fossil Fuel	Electricity	
Farm Machinery / Greenhouses	4020	10.6	129		115	370	40.6	2.0	(17) (18)
Livestock	3770	0.80			19.5	154	2.0	1.0	(19)
Forestry (ISIC 12)	168	11.9	64		6.0	333	1.8	0.2	(20)
Fishing Boats (ISIC 13) / Other Fishing Use (ISIC 13)	192	15.1			0.35	8290	2.9		(21) (22)
Total Fertilizer (10^6 tons)		N 0.80	P 0.73	K 0.32		Total	47.3	3.2	

	(A)	(B)	(C)	(D)	(E)	(F)	
Mining	Value Added	Energy/Value Added	Physical Output	Energy/Physical Output	Fossil Fuel	Electricity	
Handling and Internal Transportation / Other Energy Use	522	5.7	423	7.1	1.5	1.5	(23) (24)
				Total	1.5	1.5	

	(A)	(B)	(C)	(D)	(E)	(F)	
Construction	Value Added	Energy/Value Added	Physical Output	Energy/Physical Output	Fossil Fuel	Electricity	
Handling and Internal Transportation / Other Energy Use	7670	0.14	356	3.1	~0	1.1	(25) (26)
				Total	~0	1.1	
				Agriculture, Mining, and Construction Total	48.8	5.8	

Value Added in 10^9 Lire (1972).

Energy/Value Added in Kcal/Lira.

Area Harvested in 10^9 m^2

Energy/Area in Kcal/m^2

Physical Output in 10^6 tons for Agriculture and Mining, 10^6 m^3 for Consturction.

Energy/Output in 10^3 Kcal/ton (10^3 Kcal/m^3 for Construction).

Fossil fuel in 10^{12} Kcal.

Electricity in 10^{12} Kcal.

NATIONAL ENERGY DEMAND PROJECTIONS

Country Italy

Year 1985

WAES Case C

Commercial and Public

	(A)	(B)	(C)	(D)	(E)	
Commercial	Floor Area	Heat Gain or Loss/Area	System Efficiency	Fossil Fuel	Electricity	
Space Heat (Fossil Fuel)	168	173	0.62	46.9		(27)
Space Heat (Electricity)	2	55.2	0.95		0.1	(28)
Air Conditioning (Fossil Fuel)	59	13.2	0.71	1.1		(29)
Air Conditioning (Electricity)	160	29.7	0.93		5.1	(30)
		Energy/Area	Electricity/Area			
Lighting	229	17.9	17.9		4.1	(31)
Miscellaneous	229	59.0	31.4	6.3	7.2	(32)
Presently Unknown						(33)
			Total	54.3	16.5	
Public	Floor Area	Heat Gain or Loss/Area	System Efficiency	Fossil Fuel	Electricity	
Space Heat (Fossil Fuel)	62	167	0.62	16.7		(34)
Space Heat (Electricity)						(35)
Air Conditioning (Fossil Fuel)	11	13.2	0.71	0.2		(36)
Air Conditioning (Electricity)	44	29.7	0.93		1.4	(37)
		Energy/Area	Electricity/Area			
Lighting					4.1	(38)
Miscellaneous	73	31.5	21.0	0.8	1.5	(39)
Presently Unknown						(40)
			Total	17.7	7.0	
			Commercial and Public Total	72.0	23.5	

Floor Area in 10^6 m^2.

Heat Gain or Loss/Area in 10^3 $Kcal/m^2$.

Fossil Fuel in 10^{12} Kcal.

Electricity in 10^{12} Kcal.

Energy/Area in 10^3 $Kcal/m^2$ includes fossil fuel and electricity.

Electricity/Area in 10^3 $Kcal/m^2$

Public Lighting includes street lighting.

NATIONAL ENERGY DEMAND PROJECTIONS

Country Italy

Year 1985

WAES Case C

Industrial

	(A) Value Added	(B) Energy/Value Added	(C) Physical Output	(D) Energy/Output	(E) % Exported	(F) Fossil Fuel	(G) Electricity	(H) Feedstocks	
Iron and Steel (ISIC 371)	1730	64.4	98.2	1130		89.4	22.0		(41)
Nonferrous Metals (ISIC 372)	380	40.9	2.4	6460		4.9	10.6		(42)
Paper and Allied (ISIC 34)	1480	19.2	11.7	2430		21.6	6.8		(43)
Chemicals and Allied (ISIC 35 minus ISIC 353)	3410	114.1	49.4	7880		147.9	26.7	214.5	(44)
Mineral Products (ISIC 36)	1460	79.1	83.3	1380		104.4	10.7		(45)
Food and Related (ISIC 31)	3340	13.1	30.6	1430		37.6	6.2		(46)
Transport Eqpt. (ISIC 384)	2650	5.9	3.9	4000		10.5	5.1		(47)
Mechanical	6300	6.9	4.5	9670		31.6	11.9		(48)
Textiles and Clothing	3560	5.8	1.4	14700		15.2	5.4		(49)
Misc. Manufacturing	6270	3.6				15.3	7.3		(50)
					Total	478.4	112.7	214.5	

N.E.C. = Not Elsewhere Classified

Value Added in 10^9 Lire (1972).

Energy/Value Added in Kcal/Lira.

Physical Output in 10^6 tons.

Energy/Output in 10^3 Kcal/ton.

Fossil Fuel in 10^{12} Kcal.

Electricity in 10^{12} Kcal.

Feedstocks in 10^{12} Kcal.

NATIONAL ENERGY DEMAND PROJECTIONS

Country Italy

Year 1985

WAES Case C

Residential

Population $58.6x10^6$ Persons/Dwelling 3.05 Dwellings $19.2x10^6$

		(A)	(B)	(C)	(D)	(E)	(F)	
Central Space Heating		Dwellings	Heat Loss	System Efficiency		Fossil Fuel	Electricity	
Single Family	Fossil Fuels	2.1	18.7	0.59		66.6		(51)
	Electricity							(52)
Town or Row Houses	Fossil Fuels	9.5	8.8	0.61		136.8		(53)
	Electricity							(54)
Apartments	Fossil Fuels	6.8	4.3	0.71		40.9		(55)
	Electricity							(56)
		% Saturation (Fossil Fuels)	% Saturation (Electricity)	Energy/Unit (Fossil Fuels)	Energy/Unit (Electricity)			
Room Heating		7.7	1.0	3380	1500	5.0	0.3	(57)
Air Conditioning (Window)			2.8		558		0.3	(58)
Air Conditioning (Central)		2.2		1180	2370	0.5	1.0	(59)
Water Heating		23.6	43.7	1410	1030	6.4	8.6	(60)
Refrigerators			97.9		318		6.0	(61)
Lighting			~100		224		4.3	(62)
Cooking		94.3	5.7	652	274	11.8	0.3	(63)
Dishwashers			34.2		1430		9.4	(64)
Freezers			40.0		443		3.4	(65)
Television			95.9		223		4.1	(66)
Clotheswasher			90.8		465		8.1	(67)
Misc. Appliances							1.3	(68)
Potential Appliances								(69)
					Total	268.0	47.1	

Dwellings in 10^6 include only heated dwellings.

Heat Loss in 10^6 Kcal/dwelling.

Fossil Fuel in 10^{12} Kcal.

Electricity in 10^{12} Kcal.

Town or Row Houses defined as 2-family units plus 3/15 family apartment buildings. Apartments defined as > 15 family apartment buildings.

% Saturation is relative to total dwellings.

Energy/Unit (Fossil Fuels) in 10^3 Kcal.

Energy/Unit (Electricity) in 10^3 Kcal.

Water heating includes hot water heating on kitchen cookers and stoves .

NATIONAL ENERGY DEMAND PROJECTIONS

Country Italy

Year 1985

WAES Case D

Summary

Category	Oil	Gas	Coal	Electricity	Other	Total by Category (Excluding Processing Losses)
Transportation	211.9	0.8	0.3	4.3		217.3
Agriculture, Mining, and Construction	46.0	0.4		5.4		51.8
Commercial	50.1	13.8	0.7	13.6	0.5	78.7
Public	18.1	4.1	0.3	5.9	0.2	28.6
Industrial	331.5	156.5	54.6	87.7		630.3
Residential	260.0	61.9	3.5	38.5	2.5	366.4
Total Delivered Energy by Fuel	917.6	237.5	59.4	155.4	3.2	1373.1
Processing Losses by Fuel	30.9	4.0	26.4	242.5	~0	303.8
Gross Input by Fuel	948.5	241.5	85.8	397.9	3.2	1676.9

Gross Energy Input (Add across bottom row)

NATIONAL ENERGY DEMAND PROJECTIONS

Country Italy

Year 1985

WAES Case D

Transportation

	(A)	(B)	(C)	(D)	(E)	(F)	(G)	
	Population/Auto	Distance/Auto/Yr.	Load Factor	Fraction Urban or Rural	Energy/Distance	Fossil Fuel	Electricity	
Urban Auto	3.18	9030	1.20	0.34	887	50.2		(1)
Rural Auto			2.50	0.66	635	69.7		(2)
	Activity Level	% Electrified	Load Factor	Fossil Fuel/ Activity Level	Electricity/Activity			
Urban Bus	54.6	6.04	29	128	103	6.6	0.3	(3)
Intercity Bus	19.7		35	80		1.6		(4)
Rail (Passenger)	49.8	82.9	188	185	44.5	1.6	1.8	(5)
Air (Passenger, Domestic)	4.8		55	994		4.8		(6)
Air (Passenger, International)								(7)
Ship (Passenger)	3.9			1310		5.1		(8)
Rail (Freight)	18.1	94.7	312	585	80	0.6	1.4	(9)
Air (Freight, Domestic)	0.044			5450		0.2		(10)
Air (Freight, International)								(11)
Truck (Local)	14.0		1.8	1610		22.5		(12)
Truck (Intercity)	79.6		10.0	518		41.2		(13)
Ship (Freight, Domestic)	28.8			180		5.2		(14)
Ship (Freight, International)								(15)
Miscellaneous						3.7	0.8	(16)
					Total	213.0	4.3	

Distance/Auto/Yr. in km.

Load factor for air travel is % occupancy. For all other modes it is passenger/vehicle (tons/vehicle).

Energy/Distance in Kcal/km.

Fossil Fuel in 10^{12} Kcal.

Electricity in 10^{12} Kcal.

Activity Level in 10^{9} passenger·km (ton ·km)

% Electrified is the percentage of the total activity delivered by electrified mode.

Fossil Fuel/Activity in Kcal/passenger·km (Kcal/ton·km)

Electricity/Activity in Kcal/passenger·km (Kcal/ton·km)

Energy consumed for international travel is accounted for under bunkerage in the Italian Supply/Demand integration.

NATIONAL ENERGY DEMAND PROJECTIONS

Country Italy

Year 1985

WAES Case D

Agriculture, Mining, and Construction

Agriculture, Forestry, and Fishing (ISIC Major Division 1)	(A) Value Added	(B) Energy/Value Added	(C) Area Harvested	(D) Energy/Area	(E) Physical Output	(F) Energy/Output	(G) Fossil Fuel	(H) Electricity	
Farm Machinery / Greenhouses	3710	11.1	129		111	369	39.2	1.8	(17) (18)
Livestock	3110	0.84			16.8	155	1.9	0.7	(19)
Forestry (ISIC 12)	167	12.6	64		6.2	339	1.5	0.6	(20)
Fishing Boats (ISIC 13) / Other Fishing Use (ISIC 13)	163	15.3			0.32	7815	2.5		(21) (22)
Total Fertilizer (10^6 tons)		N 0.73	P 0.67	K 0.30		Total	45.1	3.1	

Mining	(A) Value Added	(B) Energy/Value Added	(C) Physical Output	(D) Energy/Physical Output	(E) Fossil Fuel	(F) Electricity	
Handling and Internal Transportation / Other Energy Use	459	5.7	374	7.0	1.3	1.3	(23) (24)
				Total	1.3	1.3	

Construction	(A) Value Added	(B) Energy/Value Added	(C) Physical Output	(D) Energy/Physical Output	(E) Fossil Fuel	(F) Electricity	
Handling and Internal Transportation / Other Energy Use	7270	0.14	338	3.0	~0	1.0	(25) (26)
				Total	~0	1.0	
				Agriculture, Mining, and Construction Total	46.4	5.4	

Value Added in 10^9 Lire (1972).

Energy/Value Added in Kcal/Lira.

Area Harvested in 10^9 m^2

Energy/Area in Kcal/m^2

Physical Output in 10^6 tons for Agriculture and Mining, 10^6 m^3 for Consturction.

Energy/Output in 10^3 Kcal/ton (10^3 Kcal/m^3 for Construction).

Fossil fuel in 10^{12} Kcal.

Electricity in 10^{12} Kcal.

NATIONAL ENERGY DEMAND PROJECTIONS

Country Italy

Year 1985

WAES Case D

Commercial and Public

	(A)	(B)	(C)	(D)	(E)	
Commercial	Floor Area	Heat Gain or Loss/Area	System Efficiency	Fossil Fuel	Electricity	
Space Heat (Fossil Fuel)	164	196	0.56	57.5		(27)
Space Heat (Electricity)	2	55.2	0.95		0.1	(28)
Air Conditioning (Fossil Fuel)	50	15.0	0.62	1.2		(29)
Air Conditioning (Electricity)	123	33.9	0.93		4.5	(30)
		Energy/Area	Electricity/Area			
Lighting	216	15.3	15.3		3.3	(31)
Miscellaneous	216	56.0	26.4	6.4	5.7	(32)
Presently Unknown						(33)
			Total	65.1	13.6	
Public	Floor Area	Heat Gain or Loss/Area	System Efficiency	Fossil Fuel	Electricity	
Space Heat (Fossil Fuel)	65	187	0.56	21.7		(34)
Space Heat (Electricity)						(35)
Air Conditioning (Fossil Fuel)	8	15.0	0.62	0.2		(36)
Air Conditioning (Electricity)	38	33.9	0.93		1.4	(37)
		Energy/Area	Electricity/Area			
Lighting					3.1	(38)
Miscellaneous	73	29.5	19.2	0.8	1.4	(39)
Presently Unknown						(40)
			Total	22.7	5.9	
			Commercial and Public Total	87.8	19.5	

Floor Area in 10^6 m^2.

Heat Gain or Loss/Area in 10^3 $Kcal/m^2$.

Fossil Fuel in 10^{12} Kcal.

Electricity in 10^{12} Kcal.

Energy/Area in 10^3 $Kcal/m^2$ includes fossil fuel and electricity.

Electricity/Area in 10^3 $Kcal/m^2$

Public Lighting includes street lighting.

NATIONAL ENERGY DEMAND PROJECTIONS

Country Italy

Year 1985

WAES Case D

Industrial

	(A) Value Added	(B) Energy/Value Added	(C) Physical Output	(D) Energy/Output	(E) % Exported	(F) Fossil Fuel	(G) Electricity	(H) Feedstocks	
Iron and Steel (ISIC 371)	1290	71.1	73.0	1260		75.3	16.4		(41)
Nonferrous Metals (ISIC 372)	308	42.1	1.9	6840		4.3	8.7		(42)
Paper and Allied (ISIC 34)	1230	19.5	9.7	2470		18.4	5.6		(43)
Chemicals and Allied (ISIC 35 minus ISIC 353)	2550	112.5	36.9	7780		119.3	19.5	148.1	(44)
Mineral Products (ISIC 36)	1180	81.2	66.9	1430		87.0	8.8		(45)
Food and Related (ISIC 31)	2690	13.3	24.7	1450		30.6	5.2		(46)
Transport Eqpt. (ISIC 384)	2000	6.1	2.9	4210		8.3	3.9		(47)
Mechanical	5020	7.1	3.6	9890		26.1	9.5		(48)
Textiles and Clothing	2930	6.1	1.2	14900		13.5	4.4		(49)
Misc. Manufacturing	4830	3.6				11.7	5.7		(50)
					Total	394.5	87.7	148.1	

Value Added in 10^9 Lire (1972).

Energy/Value Added in Kcal/Lira.

Physical Output in 10^6 tons.

Energy/Output in 10^3 Kcal/ton.

Fossil Fuel in 10^{12} Kcal.

Electricity in 10^{12} Kcal.

Feedstocks in 10^{12} Kcal.

NATIONAL ENERGY DEMAND PROJECTIONS

Country ITALY

Year 1985

WAES Case D

Residential

Population $58.6x10^6$ Persons/Dwelling 3.22 Dwellings $18.2x10^6$

		(A)	(B)	(C)	(D)	(E)	(F)	
Central Space Heating		Dwellings	Heat Loss	System Efficiency		Fossil Fuel	Electricity	
Single Family	Fossil Fuels	2.1	20.8	0.54		81.1		(51)
	Electricity							(52)
Town or Row Houses	Fossil Fuels	9.4	9.8	0.55		168.4		(53)
	Electricity							(54)
Apartments	Fossil Fuels	6.7	4.9	0.62		52.7		(55)
	Electricity							(56)
		% Saturation (Fossil Fuels)	% Saturation (Electricity)	Energy/Unit (Fossil Fuels)	Energy/Unit (Electricity)			
Room Heating		11.7	1.5	3380	1500	7.2	0.4	(57)
Air Conditioning (Window)			2.1		523		0.2	(58)
Air Conditioning (Central)		1.5		1100	2200	0.3	0.6	(59)
Water Heating		19.6	39.6	1570	1249	5.6	9.0	(60)
Refrigerators			96.3		285		5.0	(61)
Lighting			∼ 100		225		4.1	(62)
Cooking		94.3	5.7	734	289	12.6	0.3	(63)
Dishwashers			21.9		1250		5.0	(64)
Freezers			31.2		423		2.4	(65)
Television			93.4		206		3.5	(66)
Clotheswasher			84.6		442		6.8	(67)
Misc. Appliances							1.2	(68)
Potential Appliances								(69)
					Total	327.9	38.5	

Dwellings in 10^6 include only heated dwellings.

Heat Loss in 10^6 Kcal/dwelling.

Fossil Fuel in 10^{12} Kcal.

Electricity in 10^{12} Kcal.

Town or Row Houses defined as 2-family units plus 3/15 family apartment buildings. Apartments defined as > 15 family apartment buildings.

% Saturation is relative to total dwellings.

Energy/Unit (Fossil Fuels) in 10^3 Kcal.

Energy/Unit (Electricity) in 10^3 Kcal.

Water heating includes hot water heating on kitchen cookers and stoves.

NATIONAL ENERGY DEMAND PROJECTIONS

Country Italy

Year 1985

WAES Case E

Summary

Category	Oil	Gas	Coal	Electricity	Other	Total by Category (Excluding Processing Losses)
Transportation	291.2	0.9	0.3	4.4		296.8
Agriculture, Mining, and Construction	48.5	0.4		5.8		54.7
Commercial	57.4	18.9	0.8	18.1	0.5	95.7
Public	19.3	5.4	0.3	7.0	0.2	32.2
Industrial	478.6	228.4	78.1	119.6		904.7
Residential	291.8	86.3	3.9	56.1	2.7	440.8
Total Delivered Energy by Fuel	1186.8	340.3	83.4	211.0	3.4	1824.9
Processing Losses by Fuel	40.6	5.9	31.6	353.3	~0	431.4
Gross Input by Fuel	1227.4	346.2	115.0	564.3	3.4	2256.3

Gross Energy Input (Add across bottom row)

NATIONAL ENERGY DEMAND PROJECTIONS

Country Italy

Year 1985

WAES Case E

Transportation

	(A)	(B)	(C)	(D)	(E)	(F)	(G)	
	Population/Auto	Distance/Auto/Yr.	Load Factor	Fraction Urban or Rural	Energy/Distance	Fossil Fuel	Electricity	
Urban Auto	3.01	12170	1.15	0.32	965	73.2		(1)
Rural Auto			2.40	0.68	664	107.0		(2)
	Activity Level	% Electrified	Load Factor	Fossil Fuel/ Activity Level	Electricity/Activity			
Urban Bus	40.9	8.1	31	128	106	4.8	0.4	(3)
Intercity Bus	21.0		36	80		1.7		(4)
Rail (Passenger)	43.1	83.1	188	185	44.5	1.3	1.6	(5)
Air (Passenger, Domestic)	10.9		52	1220		13.3		(6)
Air (Passenger, International)								(7)
Ship (Passenger)	5.0			1310		6.6		(8)
Rail (Freight)	19.5	94.8	312	565	80	0.6	1.5	(9)
Air (Freight, Domestic)	0.10			6600		0.7		(10)
Air (Freight, International)								(11)
Truck (Local)	15.2		1.9	1560		23.7		(12)
Truck (Intercity)	95.7		9.8	513		49.1		(13)
Ship (Freight, Domestic)	41.2			180		7.4		(14)
Ship (Freight, International)								(15)
Miscellaneous						3.2	0.9	(16)
					Total	292.6	4.4	

Distance/Auto/Yr. in km.

Load factor for air travel is % occupancy. For all other modes it is passenger/vehicle (tons/vehicle).

Energy/Distance in Kcal/km.

Fossil Fuel in 10^{12} Kcal.

Electricity in 10^{12} Kcal.

Activity Level in 10^9 passenger·km (ton ·km)

% Electrified is the percentage of the total activity delivered by electrified mode.

Fossil Fuel/Activity in Kcal/passenger·km (Kcal/ton·km)

Electricity/Activity in Kcal/passenger·km (Kcal/ton·km)

Energy consumed for international travel is accounted for under bunkerage in the Italian Supply/Demand integration.

NATIONAL ENERGY DEMAND PROJECTIONS

Country Italy

Year 1985

WAES Case E

Agriculture, Mining, and Construction

Agriculture, Forestry, and Fishing (ISIC Major Division 1)	(A) Value Added	(B) Energy/Value Added	(C) Area Harvested	(D) Energy/Area	(E) Physical Output	(F) Energy/Output	(G) Fossil Fuel	(H) Electricity	
Farm Machinery									(17)
Greenhouses	4020	10.6	129		115	370	40.6	2.0	(18)
Livestock	3770	0.80			19.5	154	2.0	1.0	(19)
Forestry (ISIC 12)	168	11.9	64		6.0	333	1.8	0.2	(20)
Fishing Boats (ISIC 13)									(21)
Other Fishing Use (ISIC 13)	192	15.1			0.35	8290	2.9		(22)
Total Fertilizer (10^6 tons)		N 0.80	P 0.73	K 0.32		Total	47.3	3.2	

Mining	(A) Value Added	(B) Energy/Value Added	(C) Physical Output	(D) Energy/Physical Output	(E) Fossil Fuel	(F) Electricity	
Handling and Internal Transportation							(23)
Other Energy Use	525	5.9	429	7.2	1.6	1.5	(24)
				Total	1.6	1.5	

Construction	(A) Value Added	(B) Energy/Value Added	(C) Physical Output	(D) Energy/Physical Output	(E) Fossil Fuel	(F) Electricity	
Handling and Internal Transportation							(25)
Other Energy Use	7670	0.14	356	3.1	~0	1.1	(26)
				Total	~0	1.1	
				Agriculture, Mining, and Construction Total	48.9	5.8	

Value Added in 10^9 Lire (1972).

Energy/Value Added in Kcal/Lira.

Area Harvested in 10^9 m^2

Energy/Area in Kcal/m^2

Physical Output in 10^6 tons for Agriculture and Mining, 10^6 m^3 for Consturction.

Energy/Output in 10^3 Kcal/ton (10^3 Kcal/m^3 for Construction).

Fossil fuel in 10^{12} Kcal.

Electricity in 10^{12} Kcal.

NATIONAL ENERGY DEMAND PROJECTIONS

Country Italy

Year 1985

WAES Case E

Commercial and Public

	(A)	(B)	(C)	(D)	(E)	
Commercial	Floor Area	Heat Gain or Loss/Area	System Efficiency	Fossil Fuel	Electricity	
Space Heat (Fossil Fuel)	172	226	0.56	69.4		(27)
Space Heat (Electricity)						(28)
Air Conditioning (Fossil Fuel)	57	17.5	0.62	1.6		(29)
Air Conditioning (Electricity)	163	39.4	0.93		6.9	(30)
		Energy/Area	Electricity/Area			
Lighting	229	17.9	17.9		4.1	(31)
Miscellaneous	229	59.8	31.0	6.6	7.1	(32)
Presently Unknown						(33)
			Total	77.6	18.1	
Public	Floor Area	Heat Gain or Loss/Area	System Efficiency	Fossil Fuel	Electricity	
Space Heat (Fossil Fuel)	60	225	0.56	24.1		(34)
Space Heat (Electricity)						(35)
Air Conditioning (Fossil Fuel)	11	17.5	0.62	0.3		(36)
Air Conditioning (Electricity)	38	39.4	0.93		1.6	(37)
		Energy/Area	Electricity/Area			
Lighting					4.0	(38)
Miscellaneous	71	31.0	19.2	0.8	1.4	(39)
Presently Unknown						(40)
			Total	25.2	7.0	
			Commercial and Public Total	102.8	25.1	

Floor Area in 10^6 m^2.

Heat Gain or Loss/Area in 10^3 $Kcal/m^2$.

Fossil Fuel in 10^{12} Kcal.

Electricity in 10^{12} Kcal.

Energy/Area in 10^3 $Kcal/m^2$ includes fossil fuel and electricity.

Electricity/Area in 10^3 $Kcal/m^2$

Public Lighting includes street lighting.

NATIONAL ENERGY DEMAND PROJECTIONS

Country Italy

Year 1985

WAES Case E

Industrial

	(A) Value Added	(B) Energy/Value Added	(C) Physical Output	(D) Energy/Output	(E) % Exported	(F) Fossil Fuel	(G) Electricity	(H) Feedstocks	
Iron and Steel (ISIC 371)	1770	73.3	100.0	1300		106.6	23.1		(41)
Nonferrous Metals (ISIC 372)	384	43.1	2.4	6920		5.6	11.0		(42)
Paper and Allied (ISIC 34)	1520	20.0	12.1	2510		23.2	7.2		(43)
Chemicals and Allied (ISIC 35 minus ISIC 353)	3560	124.1	51.6	8560		166.9	28.5	246.3	(44)
Mineral Products (ISIC 36)	1520	83.5	86.3	1470		115.2	11.7		(45)
Food and Related (ISIC 31)	3400	13.6	31.2	1480		39.6	6.6		(46)
Transport Eqpt. (ISIC 384)	2720	6.3	4.0	4280		11.7	5.4		(47)
Mechanical	6400	7.2	4.6	10000		34.0	12.1		(48)
Textiles and Clothing	3680	6.3	1.5	15500		17.0	6.2		(49)
Misc. Manufacturing	6410	3.7				15.9	7.8		(50)
					Total	535.7	119.6	246.3	

Value Added in 10^9 Lire (1972).

Energy/Value Added in Kcal/Lira.

Physical Output in 10^6 tons.

Energy/Output in 10^3 Kcal/ton.

Fossil Fuel in 10^{12} Kcal.

Electricity in 10^{12} Kcal.

Feedstocks in 10^{12} Kcal.

NATIONAL ENERGY DEMAND PROJECTIONS

Country Italy

Year 1985

WAES Case E

Residential

Population 58.6×10^6 Persons/Dwelling 3.05 Dwellings 19.2×10^6

Central Space Heating		(A) Dwellings	(B) Heat Loss	(C) System Efficiency	(D)	(E) Fossil Fuel	(F) Electricity	
Single Family	Fossil Fuels	2.2	23.4	0.54		95.1		(51)
	Electricity							(52)
Town or Row Houses	Fossil Fuels	9.9	11.2	0.55		202.1		(53)
	Electricity							(54)
Apartments	Fossil Fuels	7.1	5.6	0.62		65.1		(55)
	Electricity							(56)
		% Saturation (Fossil Fuels)	% Saturation (Electricity)	Energy/Unit (Fossil Fuels)	Energy/Unit (Electricity)			
Room Heating		5.4	0.7	3540	1570	3.7	0.2	(57)
Air Conditioning (Window)			4.1		508		0.4	(58)
Air Conditioning (Central)		2.7		1350	2510	0.7	1.3	(59)
Water Heating		24.1	44.9	1450	1280	6.7	11.0	(60)
Refrigerators			98.3		371		7.0	(61)
Lighting			$\sim$100		250		4.8	(62)
Cooking		93.0	7.0	633	298	11.3	0.4	(63)
Dishwashers			38.7		1500		11.1	(64)
Freezers			45.2		481		4.2	(65)
Television			96.5		264		4.9	(66)
Clotheswasher			92.2		526		9.3	(67)
Misc. Appliances							1.5	(68)
Potential Appliances								(69)
					Total	384.7	56.1	

Dwellings in 10^6 include only heated dwellings.

Heat Loss in 10^6 Kcal/dwelling.

Fossil Fuel in 10^{12} Kcal.

Electricity in 10^{12} Kcal.

Town or Row Houses defined as 2-family units plus 3/15 family apartment buildings. Apartments defined as > 15 family apartment buildings.

% Saturation is relative to total dwellings.

Energy/Unit (Fossil Fuels) in 10^3 Kcal.

Energy/Unit (Electricity) in 10^3 Kcal.

Water heating includes hot water heating on kitchen cookers and stoves.

9 Japan

9.1 OVERVIEW

This report contains both high and low economic growth cases derived from the WAES boundary assumptions. As an annual average growth rate, 8.0 percent is assumed for the period 1977 to 1985 for the WAES high case of 6.0 percent, and 4.5 percent for the WAES low case of 3.5 percent. The price of oil will affect economic growth; however, this factor has been ignored so far in deriving the economic growth rate. Presented here are only some preliminary simulation results of the macroeconomic model concerning this issue. Given these growth rates, energy demand is estimated for the five WAES cases using the energy demand and supply forecasting model and taking into account the effect of energy conservation. The energy demand is estimated to be around 700 x 10^{13} kcal (2940 x 10^{16} joules) in cases of high economic growth (A,C,E), and 550 x 10^{13} kcal (2310 x 10^{16} joules) in cases of low growth (B,D). The detailed results are shown in the WAES demand study summary sheets and worksheets. In these sheets the figures for electricity include only purchased electricity; the real consumption of electricity is 10–15 percent larger due to the existence of industrial power plants. These latter figures are subsumed into the equivalent fossil fuel demand and are included under gross fossil fuel consumption in the industrial sector.

9.2 NATIONAL STRUCTURE

This study has been done by the WAES Japan team, which consists of the following research institutes:

The Industrial Research Institute

The Institute of Energy Economics

The Institute of Future Technology

The Japan Computer Usage Development Institute

The Japan Economic Research Center

The Japan Techno-Economics Society

The Mitsubishi Research Institute

The National Institute for Research Advancement

The Nomura Research Institute.

The National Institute for Research Advancement, several industrial organizations, and the Japan EXPO Fund have provided the funds necessary to carry through this study, including expenses incurred in attending the sponsoring meetings. For the administration of the project, the Japan Alternative Energy Strategies Organization was founded.

9.3 METHODOLOGY

9.3.1 Energy Demand and Supply Forecasting Econometric Model

We have developed an econometric model to estimate energy demand and supply. In this model energy demand is calculated basically by multiplying a projected activity level by a specific energy consumption per unit of activity.

In the industrial sector, the activity levels of the various subsectors are first estimated for a given GNP size and makeup, and these activity levels are then multiplied by separately estimated specific energy consumptions obtained through extrapolation.

The trend of improvement in specific energy consumption has saturated in the last three to five years, and the extrapolated specific energy consumptions are therefore very close to 1972 levels. For this reason we have taken the 1972 figures as our base figures, and also as our reference for estimating the effects of conservation. Possible changes in these specific energy consumptions due to expected energy conservation measures are analyzed.

In the transportation sector, the projections of energy demand are based on estimates of the following three items: (1) transportation volume by passenger and freight sector; (2) shares of volume by each means of transportation; (3) energy efficiency of each means of transportation. Here we estimate the transportation volume simply by relating it to private consumption in the case of passenger traffic and to GNP in the case of freight. With regard to the share of volume by each transportation means, two alternative assumptions are made: (1) the share will develop according to past historical trends; (2) the present share is stable. The energy efficiency of this sector is assumed not to change because (1) the possible room for improvement of efficiency in this sector seems very limited, and (2) this limited improvement will be offset by factors such as antipollution measures.

In the residential and commercial sectors the energy demand is forecasted using simple correlation equations in which GNP and private consumption are taken as governing parameters. For the residential sector, in order to comply with the WAES format, a submodel has been developed.

The basic concepts of the energy forecasting model are shown schematically in Figure 9.1.

9.3.2 World GNP Growth Rate and Japan

The gross national product of Japan has grown with an elasticity of 1.9 to the world GNP growth rate in the last 15 years. It is thought, however, that the Japanese economy will grow more slowly in the future, and a report of the Japan Economic Research Center suggests that the elasticity will be reduced to around 1.3 in the coming decade.[1] Accordingly, the Japanese gross national product is assumed here to grow at 8.0 percent per annum between 1977 and 1985, which corresponds to the WAES high growth case of 6.0 percent per annum. For the WAES low case of 3.5 percent, a growth rate of 4.5 percent is assumed.

1. The Japan Economic Research Center, "Future of World Economy and Japan," March 1975.

The recent economic projections done by the government and several economic institutions are concentrating around a 6 percent annual growth rate as most probable in the coming ten years.

9.3.3 Oil Price and GNP Growth Rate

The foregoing GNP growth rate assumption is made without taking into account the influence of high oil prices on economic activity. It is thought that the price of oil will have a significant effect on economic activity, and we therefore have doubts about the consistency of the oil price and GNP assumptions, especially in Case A, for Japan.

According to the macroeconomic model of the Institute of Energy Economics, assuming an annual economic growth rate of 5.7 percent in the coming ten years and a constant currency exchange rate, a 40 percent increase in the price of oil, in real terms, would produce a US$40 billion deficit

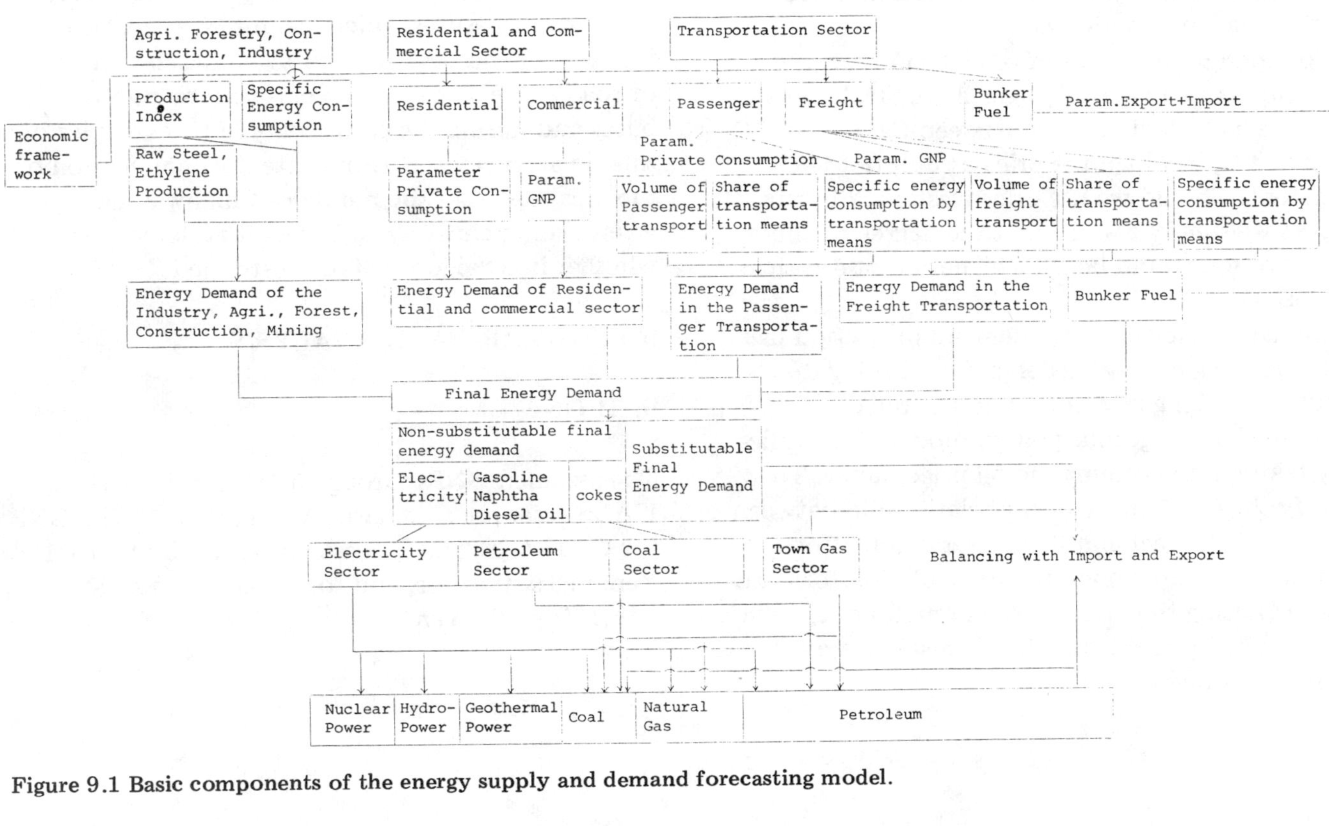

Figure 9.1 Basic components of the energy supply and demand forecasting model.

in the trade balance in 1985, which would not be acceptable in the Japanese economy. If this deficit is to be reduced to a reasonable range through changes in the exchange rate, the yen must be devalued to about US$1 = ¥380, and the average annual economic growth rate would then fall to 4.9 percent. We think, furthermore, that under these conditions, not only would the rate of exchange be altered, but other changes in national economic policy would be introduced and the growth rate would fall even further.

9.3.4 GNP, Oil Price, and National Policy Response

Our energy forecasting model is based principally on the economic activity level and the given or extrapolated specific energy consumption. It does not explicitly include the price factor as a governing parameter. Modifying the results of this model to cope with the price assumption and the national policy response introduces serious methodological problems.

One possible approach would be first to calculate GNP and economic activity, treating oil price as one of the chief governing parameters, then to calculate the energy demand using this economic activity, and finally to recalibrate these results by introducing the conservation effort.

To separate the effect of price from that of conservation is also methodologically difficult. How shall we make a difference between Cases A and C, or B and D, in which the price assumptions are different? Should we use price elasticity? In this report we simply assume that the degree of energy conservation would be a bit higher in Case A than in Case C.

We have done an energy demand projection using the foregoing energy model and modifying the results by taking into consideration the effects of restrained and vigorous policy, though not explicitly taking into account the oil price factor. To clarify the complexity of the relationship between the oil price effect and the conservation effort, however, we show one simple calculation for Cases B and D using the price elasticity method, though the assumed price elasticities themselves are bound to be hypothetical.

9.3.5 National Policy Response

The main factors in the energy demand projections of this report are economic activity and national policy response. We project the energy demand first, multiplying the estimated economic activity by 1972 energy efficiency, and then we modify the demand according to the assumed policy response.

Here we define restrained and vigorous policy as follows. On the supply side we cannot make any significant difference between restrained and vigorous policy. In either case the effect on domestic production would be negligibly small because our domestic energy reserves are so limited.

On the other hand, there is a wide range of possible energy conservation policies. We interpret the restrained policy in the demand sector as including those measures which would bring about reduced energy consumption per unit of economic activity but which are not obligatory by law and are carried out principally through guidance, target setting, and appeals for conservation. The essence of the vigorous policy, conversely, lies in the use of legal force and punishment in case of violation; the level of conservation would in some cases be set higher than that used in the restrained policy. There are, of course, measures which cannot be easily placed into one or the other category. Increasing the share of road space to be used exclusively by public buses and raising the car tax would be examples of this; in such cases, it is the extent of application of the measure (e.g., the number of highway lanes given over exclusively to buses) that determines whether it is "vigorous" or "restrained."

9.4 SUMMARY AND ANALYSIS

9.4.1 GNP

The assumed GNPs through 1985 for the various WAES cases are summarized in Table 9.1. The GNP for Cases A, C, and E is estimated at 192,870 x 10^9 Yen, in 1970 prices, and for Cases B and D, at 148,170 x 10^9 Yen.

Table 9.1
GNP assumptions (1970 prices, billions of Yen)

	1970	1975	1977	1985		
		(estimate)	(estimate)	A,C[1]	E[1]	B,D[2]
GNP	70,634.5	90,183.0	104,200	192,870	192,870	148,170
Private Consumption	36,258.9	48,515.8	53,660	96,030	97,390	82,240
Government Consumption	5,796.1	7,603.0	8,330	12,150	12,540	11,850
Housing Investment	4,760.9	6,392.0	8,330	15,820	16,390	10,370
Private Investment	14,195.3	14,928.3	17,710	36,650	36,650	20,740
Government Investment	5,811.2	8,841.0	10,420	19,290	20,250	12,590
Stock Investment	3,031.1	616.0	3,130	5,210	4,820	4,450
Export	8,271.7	14,032.3	16,670	36,650	36,650	26,670
Import	7,490.7	10,745.0	14,050	28,930	31,820	20,740

1. Assumed GNP growth rate, 1977-1985, of 8.0 percent.
2. Assumed GNP growth rate, 1977-1985, of 4.5 percent.

9.4.2 Energy Conservation—Restrained Policy

The energy conservation policy currently being implemented in Japan would fall into the category of restrained national response. The details of this policy are summarized in Table 9.2.

The conservation rate figures in the case of restrained policy are derived by summing the effects of the conservation measures outlined below. The concrete figures of this section are derived from our case study for the most plausible economic growth rate; they do not correspond exactly to any of the WAES cases but do correspond roughly to the median of the Cases C and D. To get rough figures for the conservation rate, the results of such a case study would be applicable.

Industrial Sector

In 1974, approximately 5 percent of electricity consumption and 7 percent of petroleum products consumption were conserved as a result of price response and some mandatory government measures.[2] This short-term effect will probably last into the future since it is unlikely that present conditions of instability of oil supply and high oil price will change quickly.

We therefore use this conservation rate for all industry groups. Then we add the following effects of energy conservation in the four energy-intensive industries (iron and steel, aluminum, chemicals, and cement).

2. Central Research Institute of Electricity, "Estimates of the Effect of the Energy Conservation Regulation," 13 September 1974.

Iron and Steel Industry In this subsector there are two realistic methods to conserve energy:

1. Enlargement of blast furnace. The large modern blast furnace attains a coke ratio of 370-390 kg/ton of pig iron, and it is expected that all blast furnaces to be built in the future will be of this large type. The production of raw steel in 1985 is estimated to reach 175 million tons. To produce this amount of raw steel we must add capacity for 40 million tons. If we assume that all of this new capacity is accounted for by large blast furnaces, we can expect that the coke ratio will decrease from the present level of 438 kg/ton of pig iron to 413 kg/ton. Pig iron production corresponding to the production of 175 million tons of raw steel would amount to some 133.1 million tons. We can thus expect that the reduction of coke consumption will be about 3.3 million tons, which is equivalent to 2.26×10^{13} kcal.

2. Introduction of continuous casting. In the existing slubbing method, hot metal from the converter at 1600°C is cooled into ingots of 900°C and heated again in the soaking furnace to 1270°C to become slubb for rolling. In this process, 18 kg of fuel oil per ton of hot metal is required at the soaking furnace. In continuous casting we can avoid this process and thereby conserve 14.5 liters of fuel oil per ton of raw steel produced.

At present the share of continuous casting is about 23 percent. If we assume that all new plants will use this method, this share will reach about 50 percent in 1985. Furthermore, we can expect that some 10 percent of the existing slubbing plant capacity will be replaced by contin-

Table 9.2
Components of national policy response

Sector	Restrained Policy	Vigorous Policy
General Principles	Guidance, recommendations, setting of targets, and appeals for conservation.	Obligation, tightening of standards, punishment in case of violation.
Industrial	Setting voluntary targets for energy conservation in each factory and each industrial sector. Making these targets public. Setting guidelines for thermal control in factories. Requiring large-lot energy consumers to report yearly on the status of their specific energy consumption. Subsidizing industrial investment for energy conservation.	Making the targets and guidelines legally binding. Tightening in many cases the targets and standards. Increasing subsidies and other types of incentives for energy conservation investments.
Transportation	Suggesting to service stations that they not open for service on Sundays and holidays. Standardizing the method of measuring mileage per unit of fuel. Appealing for less use of private cars and greater use of public transportation. Limiting private cars to certain lanes during rush hours in order to make bus transportation smoother.	
Residential, Commercial, Public	Offering guidelines for keeping room temperature under 20°C in cold weather and over 26°C in warm weather. Offering guidance for more efficient use of energy in buildings. Requiring an earlier end to the TV broadcasting day. Offering guidelines for reducing the amount of time neon signs are lighted. Offering guidelines for earlier closing times for places of amusement such as cinemas, bowling alleys, bars, night clubs, etc.	
Price, Tax System	Changing the structure of electricity rates to require a 20% higher charge for electricity consumed over 120 kWh/month and 30% higher for consumption over 200 kWh/month. The amount of electricity consumed over the level consumed in April 1974 is charged a 20% higher tariff. The gasoline tax was raised in April 1974, thereby increasing the price of gasoline by 5%.	

uous casting. Assuming a 60 percent share for continuous casting in 1985, about 0.85×10^{13} kcal (867×10^3 kl fuel oil equivalent) would be conserved as compared to the projected energy consumption for the existing distribution between the two methods. Through these two means we can expect energy conservation of 3.11×10^{13} kcal in 1985. Added to this amount we can expect about 3.23×10^{13} kcal conservation through other general conservation efforts, including the effects of the above-mentioned lasting short-term conservation effect realized in recent years.

Chemicals Industry In this subsector we can expect a fall in the specific naphtha consumption in the making of ethylene from the present 6.02 kl/ton. In 1985 ethylene production will reach 7.4 million tons. If we assume this fall of specific naphtha consumption, in making 7.4 million tons of ethylene some 2.02×10^6 kl of naphtha will be conserved. This is equal to about 2.51×10^{13} kcal. As in the iron and steel industry, we can add 4.52×10^{13} kcal of energy conservation through other conservation efforts including the lasting short-term effect.

Aluminum Industry Average specific electricity consumption in making primary aluminum is expected to be improved from the present level of 16,000 kWh/ton to 14,750 kWh/ton by 1985 through better control of electrical voltage in smelting furnaces and through scrapping old plants and replacing them with new ones using the existing aluminum smelting methods (the Hall-Héroult process). Processes other than Hall-Héroult would not be realized in Japan by 1985. In 1985 the production of primary aluminum is projected to reach 0.72 million tons.

Producing this amount of aluminum with improved efficiency will conserve 2.72×10^9 kWh of electricity consumption, which is equivalent to 0.23×10^{13} kcal of energy conservation, including, however, conservation not only in the aluminum smelting process but in other processes of the nonferrous metal industry as a whole.

Cement Industry This sector has realized substantial improvements in energy-use efficiency in the past and is expected to carry out further improvements through the introduction of the Presuspension Heating Kiln. Through this measure the present energy consumption of 108.7 liters of fuel oil equivalent per ton of cement would fall to 85 liters in 1985. As the production of cement is expected to reach 140 million tons in 1985, the conservation of energy through this improvement is estimated at about 3.32×10^6 kl in fuel oil equivalent (or 3.29×10^{13} kcal). In the cement clay industry we can add another 0.10×10^{13} kcal of energy conservation through other methods.

Other Industry In other industry groups we can expect energy conservation of about 5 percent in electricity and 7 percent in fuel oil as a short-term effect and an additional 5 percent in fuel oil as a long-term effect, this latter figure being derived from our analysis of many different kinds of reports. Taken together, these effects will account for a 6.5×10^{13} kcal conservation of fossil fuel and 0.7×10^{13} kcal of electricity.

Transportation Sector

Energy use in the transportation sector in Japan has been and is very efficient. Our cars are small, and railway service plays an important role even today. There is room for more efficient use of energy in this sector through technical improvement and other measures such as control of speed. We judge, however, that the possible conservation effects through such means are limited and will be offset or even surpassed by losses due to satisfaction of emerging environmental concerns, which will require more energy.

One thing that would contribute significantly to the reduction of energy use in this sector would be a shift in transportation means to forms that consume less energy per passenger-mile.

From this point of view we calculate the possibilities for energy conservation in the transportation sector by estimating the difference in energy consumption between the cases assuming changing shares and fixed shares for each transportation mode. For the fixed share case we use the 1972 data. This case assumes that the share of the car, an inefficient means of transportation, will not increase in the future, and that the currently declining share of the train will stop. Whether the distribution among the various modes of transportation corresponds to the assumed price of oil is difficult to determine. (This is to be understood as one hypothetical case among many possible responses to high energy prices and uncertainties in energy supply.) However, taking into account emerging factors such as increasing car prices and road taxes, an increasing insurance burden, parking difficulties, gasoline price hikes, and traffic jams, all of which would check the increasing reliance on the car, the assumed fixed share case could be realized by adding rather small financial and other policy incentives. A 1975 study suggests that the existing transportation shares of each means will not change drastically.[3]

The estimated difference of energy consumption between the historical trend case and the fixed share case amounts to 14.4 x 10^{13} kcal.

Residential and Commercial Sector

In the residential sector, energy conservation is expected mainly in the area of heating and cooling, which will account for about 50 percent of total energy consumption in this sector in 1985. Energy conservation will be brought about by providing insulation. Here we assume that all houses constructed between 1976 and 1985 will provide insulation capable of conserving up to 20 percent of the energy that would otherwise be consumed, and we further assume the share of apartment houses, which are more efficient in energy use than single-family houses, will grow to 30 percent from the 24.5 percent that is expected for 1985 in the non-conservation case.

The total effect of these assumptions is about 2.6 x 10^{13} kcal conservation of energy. This is included in the WAES format in the two items Air Conditioning (Window) and Room Heating (Kerosine). We can also add 0.4 x 10^{13} kcal energy conservation through the lasting short-term effect noted previously.

In the commercial sector, we cannot analyze the possibility for conservation because of our present lack of statistics and information. We therefore calculate only the short-term effect of 7 percent of electricity and 5 percent of petroleum as conservation. This is estimated to amount to 3.2 x 10^{13} kcal.

3. R. Kakumoto, "The End of the High Speed Era," Nihon Keizai Shimbunsha, 1975.

9.4.3 Energy Conservation—Vigorous Policy

Now turning to the effects of a vigorous policy, we have had to rely on very rough judgments brought about by comparisons of the reports of several experts. Overall, 3 percent more conservation is assumed in the case of a vigorous policy in the industrial sector. In Case A, however, we add 2 percent more conservation.

In the transportation sector, we assume that a vigorous policy would lead to a 5 percent decrease in the share of cars and trucks and a similar increase in the share of trains and buses. In Case A we assume a figure of 7 percent.

In the residential and commercial sectors, energy conservation is to be achieved mainly in the areas of heating and cooling. We add 5 percent more conservation for a vigorous policy, with 8 percent in Case A.

The effects of energy conservation under restrained and vigorous policies are summarized in Tables 9.3–9.5.

9.4.4 Energy Demand

Detailed energy demand estimates for all cases can be found in the form of WAES worksheets in Section 9.6. The primary energy demand in 1985 will fall in a range of 545 x 10^{13} kcal (Case B) to 740 x 10^{13} kcal (Case E). Energy demand will be around 550 x 10^{13} kcal in the case of low economic growth and around 700 x 10^{13} kcal in the high boundary case of economic growth.

If we take Case C as most plausible, we shall need to import some 490 million tons of oil, and dependency on oil in the primary energy supply will still remain about 70 percent. In Case B, the smallest energy demand case, the necessary volume of oil imports will be reduced to 350 million TOE, and in Case E, the largest energy demand case, this volume would reach 530 million TOE.

9.4.5 Energy Demand and Price Elasticity

The energy demand forecasts for all cases in this report take into account an estimated response to the relevant national policy, although the oil price effect is not explicitly identified.

Here we shall try to show the changes in energy demand due to energy price changes, applying

Table 9.3
Effects of conservation by type of policy response: Residential and commercial sectors[1]

	Restrained Policy (Cases D and E)	Vigorous Policy (Cases B and C)	Vigorous Policy (Case A)
Residential			
Fossil Fuels	6.1	11.1	14.1
Electricity	3.0	8.0	11.0
Total	5.5	10.5	13.5
Commercial			
Fossil Fuels	5.7	10.7	13.7
Electricity	7.0	12.0	15.0
Total	6.0	11.0	14.0

1. Figures are percentages of the total expected energy demand without conservation that would be saved by application of the given policy.

Table 9.4
Effects of conservation by type of policy response: Transportation sector

	Trend Share	Restrained Policy (Cases D and E)	Vigorous Policy (Cases B and C)	Vigorous Policy (Case A)
Freight[1]				
Rail	6.6	17.4	22.4 (+5.0)	24.4 (+7.0)
Truck	51.4	44.7	39.7 (–5.0)	37.7 (–7.0)
Ship	42.0	37.9	37.9	37.9
Total	100	100	100	100
Passenger[2]				
Rail	26.2	46.4	48.9 (+2.5)	49.4 (+3.0)
Bus	18.8	16.7	19.2 (+2.5)	20.7 (+4.0)
Ship	0.6	0.9	0.9	0.9
Air	7.5	2.0	2.0	2.0
Car	46.9	34.0	29.0 (–5.0)	27.0 (–7.0)
Total	100	100	100	100
Energy Conservation[2]		13.5	20.3	32.0

1. Figures are percentage shares of total transportation under the given policy assumptions.

2. Figures are percentages of total expected energy demand without conservation that would be saved by application of the given policy.

price elasticity factors and comparing Cases D and B as an example. These two cases differ in terms of price and national policy response. Our question is whether the difference in energy demand between Cases D and B can be explained by the price effect. Are the price effect and the conservation policy effect separable or identical?

First, we must calculate the effect of price. But how? One way would be to use an approach that is somewhat analogous to that of the OECD Long-Term Energy Assessment in 1974. In applying this approach, however, we face a serious problem in Japan, because our past econometric analysis has not produced sufficiently significant results to make this kind of analysis meaningful. For this reason we could not help relying on somewhat hypothetical elasticity figures that are very similar to those applied by the OECD study. So far this study has been quite limited.

Price Assumptions

We take as our base the mid-1975 energy prices for the $11.50 oil price case. For the $17.25 case, the price of each form of energy is derived by assuming that the existing margin between the crude oil price and each energy carrier's price in the $11.50 case will be maintained. This means that even if the price of crude oil is raised, the absolute margin per unit of energy delivered for energy suppliers will not change in real terms. The assumed prices of

Table 9.5
Effects of conservation by type of policy response: Industrial sector

	Restrained Policy (Cases D and E)	Vigorous Policy (Cases B and C)	Vigorous Policy (Case A)
Agriculture, Forestry	7.9	10.9	12.9
Fishery	7.7	10.7	12.7
Mining	3.3	6.3	8.3
Construction	4.1	7.1	9.1
Food	8.3	11.3	13.3
Textiles	9.5	12.5	14.5
Paper, Pulp	9.5	12.5	14.5
Chemicals	7.1	10.1	12.1
Cement, Clay	13.0	16.0	18.0
Iron and Steel	6.8	9.8	11.8
Nonferrous Metal	7.1	10.1	12.1
Metal, Machinery	8.5	11.5	13.5
Other Manufacturing	1.5	4.5	6.5
Total	8.7	11.7	13.7

Figures are percentages of total expected energy demand without conservation that would be saved by application of the given policy.

energy by energy type are summarized in Table 9.6. Here we take into account only the main energy carriers and also do not take account of coal.

Price Elasticity

The price elasticities applied here are shown in Table 9.7. These figures are slight modifications of those applied by the OECD study. We assume here two price elasticity cases for each energy type in order to compare the effects of the change in elasticity. The lower price elasticity case is Case B1 and the higher price elasticity case is Case B2.

Results

Table 9.8 summarizes the results. Primary energy demand differs by 3.8 percent from Case B to Case D. But the difference between Case D and Cases B1 and B2 amounts to about 7–8 percent. This shows that the energy demand in Case B, which takes into account the effect of the difference between restrained and vigorous policies, would underestimate the price effect.

We might also note in this case that only in the transportation sector is the effect of a vigorous policy estimated as being bigger than the effect of price. However, we again face the problem of what numbers are reasonable for price elasticity.

Relation between Demand and Price Elasticity

Ceteris paribus, demand D_1 under price P_1 and demand D_1 under price P_2 are related by

$$D_2 = D_1 \frac{P_2}{P_1} \eta, \qquad \text{where } \eta \text{ is price elasticity.}$$

Given a 50 percent price increase in real terms, the demand will change depending on the value of price elasticity as shown in Table 9.9.

9.5 CONCLUSION

If the world economy grows at 3.5 or 6.0 percent per annum between 1977 and 1985, the Japanese economy will grow at 4.5 or 8.0 percent per annum, respectively, in this period. Japan's primary energy requirement will amount to about 550 x 10^{13} kcal (550 million TOE) in the low growth case and about 680–740 x 10^{13} kcal (680–740 million TOE) in the high growth case.

The energy-GNP elasticity between 1972 and 1985 is calculated to be in the range 0.82-0.94.

The share of the energy sector, including conversion and distribution losses, is projected to be about 20 percent in all cases.

Table 9.6
1985 energy price assumptions

	Constant Price Case	Rising Price Case
Middle East Oil		
FOB Price (¥)	3,450	5,175
Freight (¥)	240	240
Import Tax (¥)	84	84
Net Japan Price per Barrel (¥/bbl)	3,774	5,499
Net Japan Price per Kiloliter (¥/kl)	23,740	34,590
Residential, Commercial		
Kerosine (¥/kl)	33,800	44,650
Gas (¥/10^4 kcal)	90.59	102.14
Electricity (¥/kWh)	14.89	21.48
Transportation		
Gasoline (¥/kl)	103,400	114,250
Diesel (¥/kl)	55,900	66,750
Electricity (¥/kWh)	9.37	14.00
Industry		
Heavy Fuel Oil (¥/kl)	19,000	29,850
Light Fuel Oil (¥/kl)	55,900	66,750

Note: For determining the FOB oil prices, a conversion rate of US$1 = ¥300 was used.

Table 9.7
Price elasticity assumptions

	Case B1	Case B2
Residential, Commercial		
Kerosine	−0.35	−0.40
Gas	−0.35	−0.40
Electricity	−0.30	−0.35
Transportation		
Gasoline and Diesel	−0.50	−0.55
Electricity	0	0
Industry		
Fuel Oil	−0.30	−0.35
Electricity	−0.30	−0.35

Table 9.8
The effect of price on energy demand

	Case D (10^{13} kcal)	Case B (10^{13} kcal)	Difference from Case D (%)	Case B1 (10^{13} kcal)	Difference from Case D (%)	Case B2 (10^{13} kcal)	Difference from Case D (%)
Primary Energy	566.5	545.1	3.8	525.9	7.2	521.2	8.0
Energy Sector Loss and Self-Consumption	107.2	104.7	2.3	99.0	7.6	97.8	8.8
Final Energy	459.3	440.4	4.1	426.9	7.1	423.4	7.8
Industry	224.7	211.5	3.7	204.7	8.9	201.6	10.3
Chemical Feedstocks	30.6	30.6		30.6		30.6	
Residential, Commercial	88.8	84.1	5.3	80.8	9.0	80.8	9.0
Transportation	76.1	70.1	7.9	71.7	6.8	71.3	6.3
Bunker Fuel	30.1	30.1		30.1		30.1	
Nonenergy	9.0	9.0		9.0		9.0	

Table 9.9
Demand change and price elasticity in the case of a 50 percent price increase

Price Elasticity	Demand (% Change)
0	0
−0.1	− 4.0
−0.2	− 7.8
−0.3	−11.5
−0.5	−15.0
−0.6	−21.6
−0.7	−24.7
−0.8	−27.7
−0.9	−30.6
−1.0	−33.4
−2.0	−55.6
−3.0	−70.4

The share of the industrial sector, including agriculture, mining, and construction, will go down by about 2–6 points from its 1972 value of 62.2 percent.

The share of the residential and commercial sectors, which was 16.8 percent in 1972, will increase by 2.0–2.5 points in the low growth cases (B, D) and remain at about the same level or decrease slightly in the high growth cases.

The share of the transportation sector will remain at about the same level, though it shows a slight gain in Case A and a slight decline in Case D.

The bunker requirement will grow about 2.5–3.0 points from its 1972 level of 4.3 percent.

9.6 ENERGY DEMAND WORKSHEETS

Included in this section are summary tables and worksheets in the following order:

Energy Use Summary

Energy Growth Rates Summary

Sectoral Percentages of Total Final Energy Input

1985 Economic Activity of Major Energy-Use Categories

Final Energy Demand of Major Energy-Use Categories

1985 Case Comparisons

1972 Base Year Data (6 sheets)

Case A worksheets (6 sheets)

Case B worksheets (6 sheets)

Case C worksheets (6 sheets)

Case D worksheets (6 sheets)

Case E worksheets (6 sheets).

WAES NATIONAL ENERGY DEMAND STUDIES, 1972-1985
ENERGY USE SUMMARY

Country JAPAN

Scenario Assumptions

WAES Case	A	B	C	D	E	
GNP 1976-1985	High	Low	High	Low	High	
Oil Price*	\$11.50 → \$17.25	\$11.50 → \$17.25	\$11.50	\$11.50	\$11.50 → \$7.66	
Natl. Policy Response	Vigorous	Vigorous	Vigorous	Restrained	Restrained	
Natl. GNP Growth Rate ~~1976-1985~~ (%/yr.) (1977-1985)	8.0	4.5	8.0	4.5	8.0	1972 Base Year

Fuel Type	A Fossil Fuel	A Elec.	B Fossil Fuel	B Elec.	C Fossil Fuel	C Elec.	D Fossil Fuel	D Elec.	E Fossil Fuel	E Elec.	1972 Fossil Fuel	1972 Elec.
Units												
Energy-Using Sectors												
Industrial (excluding feedstocks)	236.7	34.9	175.3	25.6	243.0	35.7	182.2	26.4	254.9	37.3	120.1	16.8
Transportation	81.1	2.3	68.2	1.9	84.1	2.2	74.4	1.7	92.3	2.0	42.0	1.0
Commercial and Public	42.2	7.3	40.0	5.7	43.6	7.6	42.3	6.0	46.0	8.0	18.6	5.0
Residential	35.1	9.2	30.4	8.0	36.3	9.5	32.1	8.4	39.1	10.2	19.5	3.9
Fishing, Agriculture, Mining, Construction	17.0	1.0	14.8	0.8	17.3	1.0	15.2	0.9	17.8	1.0	12.2	0.7
Feedstocks	38.2	-	30.6	-	38.2	-	30.6	-	38.2	-	23.8	-
Others incld. Bunkers	52.8	-	39.1	-	52.8	-	39.1	-	54.6	-	15.6	-
Final Energy Demand (excluding processing losses)	503.1	54.7	398.4	42.0	515.3	56.0	415.9	43.4	542.9	58.5	251.8	27.4
Processing Losses	133.4		104.7		135.8		107.2		140.8		67.2	
Primary Energy Input	616.0	75.2	470.6	74.5	631.9	75.2	492.0	74.5	667.0	75.2	322.8	23.6

*Oil price 1975 US\$ per barrel Arabian light crude oil FOB Persian Gulf.

WAES NATIONAL ENERGY DEMAND STUDIES, 1972-1985
ENERGY GROWTH RATES SUMMARY

Country JAPAN

Scenario Assumptions

WAES Case	A		B		C		D		E	
GNP 1976-1985	High		Low		High		Low		High	
Oil Price*	$11.50 → $17.25		$11.50 → $17.25		$11.50		$11.50		$11.50 → $7.66	
Natl. Policy Response	Vigorous		Vigorous		Vigorous		Restrained		Restrained	
Natl. GNP Growth Rate ~~1976-1985~~ (%/yr.) (1977-1985)	8.0		4.5		8.0		4.5		8.0	
Fuel Type	Fossil Fuel	Elec.	Fossil Fuel	Elec.	Fossil Fuel	Elec.	Fossil Fuel	Elec.	Fossil Fuel	Elec.
Units (%/yr. 1972-1985)	%	%	%	%	%	%	%	%	%	%
Energy-Using Sectors										
Industrial (excluding feedstocks)	5.4	5.8	3.0	3.3	5.6	6.0	3.3	3.5	6.0	6.3
Transportation	5.2	6.6	3.8	5.1	5.5	6.3	4.5	4.2	6.2	5.5
Commercial and Public	6.5	3.0	6.1	1.0	6.8	3.3	6.5	1.4	7.2	3.7
Residential	4.6	6.8	3.5	5.7	4.9	7.1	3.9	6.1	5.5	7.7
Fishing, Agriculture, Mining, Construction	2.6	2.8	1.5	1.0	2.7	2.8	1.7	2.0	2.9	2.8
Feedstocks	3.7	–	2.0	–	3.7	–	2.0	–	3.7	–
Others incld. Bunker	9.8	–	7.3	–	9.8	–	7.3	–	9.8	–
Final Energy Demand (excluding processing losses)	5.5	5.5	3.6	3.3	5.7	5.7	3.9	3.6	6.1	6.0
Processing Losses	5.4		3.5		5.6		3.7		5.9	
Primary Energy Input	5.1	9.3	2.9	9.2	5.3	9.3	3.3	9.2	5.7	9.3

*Oil price 1975 US$ per barrel Arabian light crude oil FOB Persian Gulf.

WAES NATIONAL ENERGY DEMAND STUDIES, 1972-1985
SECTORAL PERCENTAGES OF TOTAL

Country JAPAN

Scenario Assumptions

WAES Case	A		B		C		D		E			
GNP 1976-1985	High		Low		High		Low		High			
Oil Price*	$11.50 → $17.25		$11.50 → $17.25		$11.50		$11.50		$11.50 → $7.66			
Natl. Policy Response	Vigorous		Vigorous		Vigorous		Restrained		Restrained			
Natl. GNP Growth Rate ~~1976-1985~~ (%/yr.) (1977-1985)	8.0		4.5		8.0		4.5		8.0		1972 Base Year	
Fuel Type	Fossil Fuel	Elec.	Fossil Fuel	Elec.	Fossil Fuel	Elec.	Fossil Fuel	Elec.	Fossil Fuel	Elec.	Fossil Fuel	Elec.
Units												
Energy-Using Sectors												
Industrial (excluding feedstocks)	47.0	63.8	44.0	61.0	47.2	63.8	43.8	60.8	47.0	63.8	47.7	61.3
Transportation	16.1	4.2	17.1	4.5	16.3	3.9	17.9	3.9	17.0	3.4	16.7	3.6
Commercial and Public	8.4	13.3	10.0	13.6	8.5	13.6	10.2	13.8	8.5	13.7	7.4	18.2
Residential	7.0	16.8	7.6	19.0	7.0	17.0	7.7	19.4	7.2	17.4	7.7	14.2
Fishing, Agriculture, Mining, Construction	3.4	1.8	3.7	1.9	3.4	1.8	3.7	2.1	3.3	1.7	4.8	2.6
Feedstocks	7.6	-	7.7	-	7.4	-	7.4	-	7.0	-	9.5	-
Others incld. Bunkers	10.5	-	9.8	-	10.2	-	9.4	-	10.1	-	6.2	-
Final Energy Demand (excluding processing losses)	100 (80.7)	100	100 (80.8)	100	100 (80.8)	100	100 (81.1)	100	100 (81.0)	100	100 (80.6)	100
Processing Losses	(19.3)		(19.2)		(19.2)		(18.9)		(19.0)		(19.4)	
Primary Energy Input	(100)		(100)		(100)		(100)		(100)		(100)	

*Oil price 1975 US$ per barrel Arabian light crude oil FOB Persian Gulf.

WAES NATIONAL ENERGY DEMAND STUDIES, 1972-1985
1985 ECONOMIC ACTIVITY OF MAJOR ENERGY-USE CATEGORIES

Country JAPAN

Scenario Assumptions

WAES Case	A	B	C	D	E	
GNP 1976-1985	High	Low	High	Low	High	
Oil Price*	\$11.50 → \$17.25	\$11.50 → \$17.25	\$11.50	\$11.50	\$11.50 → \$7.66	
Natl. Policy Response	Vigorous	Vigorous	Vigorous	Restrained	Restrained	
Natl. GNP Growth Rate ~~1976-1985~~ (%/yr.) (1977-1985)	8.0	4.5	8.0	4.5	8.0	1972 Base Year
Fuel Type	Fossil Fuel / Elec.	Fossil Fuel / Elec.	Fossil Fuel / Elec.	Fossil Fuel / Elec.	Fossil Fuel / Elec.	Fossil Fuel / Elec.
Units						
Major Energy-Use Categories						
~~Transportation (1-8)~~ Passenger (10^9 Pass. Km.)	1,402.8	1,203.1	1,402.8	1,203.1	1,422.7	647.4
Transportation (9-16) Freight (10^9 ton Km.)	753.6	587.0	753.6	587.0	753.6	343.6
Agri., Mining, Construction (17-26) (V.A. 10^9 ¥ 1965 Price)	15,436.9	12,023.2	15,436.9	12,023.2	15,437.1	8,258.3
Iron & Steel (41)	2,951.7	2,295.6	2,951.7	2,295.6	2,951.7	1,821.6
Paper & Allied (43)	1,680.9	1,381.7	1,680.9	1,381.7	1,710.9	578.3
Chemical & Allied (44)	6,497.1	4,468.5	6,497.1	4,418.5	6,586.6	2,233.8
Mineral Products (45)	1,745.2	1,193.3	1,745.2	1,193.3	1,766.0	899.2
Metal Machinery (47)	36,788.6	23,670.8	36,788.6	23,670.8	37,282.0	12,954.6
Textile (48)	2,242.0	2,031.0	2,242.0	2,031.0	2,270.2	1,227.3
Commercial (27-40)						
Residential (51-69)						

*Oil price 1975 US\$ per barrel Arabian light crude oil FOB Persian Gulf.

WAES NATIONAL ENERGY DEMAND STUDIES, 1972-1985
FINAL ENERGY DEMAND OF MAJOR ENERGY-USE CATEGORIES

Country JAPAN

Scenario Assumptions

WAES Case	A		B		C		D		E			
GNP 1976-1985	High		Low		High		Low		High			
Oil Price*	$11.50 → $17.25		$11.50 → $17.25		$11.50		$11.50		$11.50 → $7.66			
Natl. Policy Response	Vigorous		Vigorous		Vigorous		Restrained		Restrained			
Natl. GNP Growth Rate ~~1976-1985~~ (%/yr.) (1977-1985)	8.0		4.5		8.0		4.5		8.0		1972 Base Year	
Fuel Type	Fossil Fuel	Elec.	Fossil Fuel	Elec.	Fossil Fuel	Elec.	Fossil Fuel	Elec.	Fossil Fuel	Elec.	Fossil Fuel	Elec.
Units												
Major Energy-Use Categories												
Transportation Passenger (1-8)	32.5	1.5	29.0	1.3	32.5	1.5	32.0	1.2	37.9	1.4	17.2	0.8
Transportation Freight (9-16)	48.6	0.8	39.2	0.6	51.6	0.8	42.4	0.5	54.4	0.6	24.8	0.2
Agri., Mining, Construction (17-26)	17.0	1.0	14.8	0.8	17.3	1.0	15.2	0.9	17.8	1.0	12.2	0.7
Iron & Steel (41)	78.0	7.4	62.1	5.9	79.8	7.6	64.1	6.1	82.5	7.8	55.0	4.8
Paper & Allied (43)	19.0	2.0	16.0	1.7	19.4	2.1	16.5	1.7	20.5	2.2	7.7	0.8
Chemical & Allied (44)	107.6	7.4	75.7	5.2	110.0	7.5	78.2	5.3	115.2	7.9	42.6	2.3
Mineral Products (45)	21.3	2.1	14.9	1.5	21.8	2.1	15.5	1.5	22.9	2.2	13.5	1.2
Metal, Machinery (47)	12.5	5.1	8.1	3.4	12.8	5.2	8.2	3.5	13.0	5.5	5.1	2.1
Textile (48)	9.1	1.0	8.4	0.9	9.3	1.0	8.7	1.0	9.8	1.0	5.9	0.6
Commercial (27-40)	42.2	7.3	40.1	5.7	43.6	7.6	42.3	6.0	46.1	8.0	18.6	5.0
Residential (51-69)	35.1	9.2	30.4	8.0	36.3	9.5	32.1	8.4	39.0	10.2	19.5	3.9

*Oil price 1975 US$ per barrel Arabian light crude oil FOB Persian Gulf.

WAES NATIONAL ENERGY DEMAND STUDIES
1985 CASE COMPARISONS

Country JAPAN

Scenario Cases Compared	GNP	Oil Price	National Policy Response	Energy-Use Category	Comparison (total change) units = 10^{13} Kcal	
	Scenario Variables					
C	8.0	11.50	Vig.	Prim. Energy	$E_C - E_A$	15.9
A	8.0	17.25	Vig.			
C	8.0	11.50	Vig.	"	$E_E - E_C$	35.1
E	8.0	7.66	Restr.			
B	4.5	17.25	Vig.	"	$E_D - E_B$	21.5
D	4.5	11.50	Restr.			

NATIONAL ENERGY DEMAND PROJECTIONS

Country JAPAN

Year 1972

WAES Case

Summary

Unit: 10^{13}Kcal

Category	Oil	Gas	Coal	Electricity	Other	Total by Category (Excluding Processing Losses)
Transportation	42.0			1.0		43.0
Bunker	12.1					12.1
Agriculture, Mining, and Construction	11.5	0.1	0.6	0.7		12.9
Commercial } Public	16.7	1.0	0.9	5.0		23.6
Industrial	74.6	2.1	43.4	16.8		136.9
Residential	15.0	3.9	0.6	3.9		23.4
Others*	27.3					27.3
Total Delivered Energy by Fuel	199.2	7.1	45.5	27.4		279.2
Processing Losses by Fuel**						67.2
Gross Input by Fuel	261.6	4.0	57.2	23.6		346.4

*Chemical feedstock, non energy use

**Plus self consumption of energy sector

Gross Energy Input (Add across bottom row)

NATIONAL ENERGY DEMAND PROJECTIONS

Country JAPAN

Year 1972

WAES Case

Transportation

	(A)	(B)	(C)	(D)	(E)	(F)	(G)	
	Population/Auto	Distance/Auto/Yr.	Load Factor	Fraction Urban or Rural	Energy/Distance	Fossil Fuel	Electricity	
Urban Auto						(10^9 Kcal)	(10^9 Kcal)	(1)
Rural Auto	220.3[1)]			612[3)]		134,858		(2)
	Activity Level	% Electrified	Load Factor	Fossil Fuel/ Activity Level	Electricity/Activity			
Urban Bus	108.2[1)]			164[3)]		17,795		(3)
Intercity Bus								(4)
Rail (Passenger)	300.5[1)]			52.0[3)]		7,803	7,923	(5)
Air (Passenger, Domestic)	12.7[1)]			842[3)]		10,665		(6)
Air (Passenger, International)								(7)
Ship (Passenger)	5.7[1)]					*		(8)
Rail (Freight)	59.5[2)]			180[4)]		9,141	1,548	(9)
Air (Freight, Domestic)	0.1[2)]					Nil		(10)
Air (Freight, International)								(11)
Truck (Local)	153.6[2)]			1,218[4)]		187,097		(12)
Truck (Intercity)								(13)
Ship (Freight, Domestic)	130.4[2)]			406[4)]		52,923		(14)
Ship (Freight, International)								(15)
Miscellaneous								(16)
					Total	420,272	9,521	

1) 10^9 Pass. Km. 2) 10^9 ton Km. 3) Kcal/Pass. Km. 4) Kcal/ton Km.

*Included in the item Ship Freight D.

NATIONAL ENERGY DEMAND PROJECTIONS

Country JAPAN

Year 1972

WAES Case ______

Agriculture, Mining, and Construction

	(A)	(B)	(C)	(D)	(E)	(F)	(G)	(H)	
Agriculture, Forestry, and Fishing (ISIC Major Division 1)	Value Added	Energy/Value Added	Area Harvested	Energy/ Area	Physical Output	Energy/ Output	Fossil Fuel	Electricity	
Farm Machinery									(17)
Greenhouses									(18)
Other Agricultural Uses	2769.7	9.20					24,470	1,001	(19)
Forestry (ISIC 12)									(20)
Fishing Boats (ISIC 13)									(21)
Other Fishing Use (ISIC 13)	597.3	85.67					51,170		(22)
Total Fertilizer (tons)		N ______	P ______	K ______		Total	75,640	1,001	

	(A)	(B)	(C)	(D)	(E)	(F)	
Mining	Value Added	Energy/Value Added	Physical Output	Energy/Physical Output	Fossil Fuel	Electricity	
Handling and Internal Transportation	307.4	50.13			12,450	2,960	(23)
Other Energy Use							(24)
				Total	12,450	2,960	

	(A)	(B)	(C)	(D)	(E)	(F)	
Construction	Value Added	Energy/Value Added	Physical Output	Energy/Physical Output	Fossil Fuel	Electricity	
Handling and Internal Transportation	4583.9	8.05			34,160	2,723	(25)
Other Energy Use							(26)
(ISIC 500)				Total	34,160	2,723	
				Agriculture, Mining, and Construction Total	122,250	6,684	

Value Added given in the Price of 1965 10^9 ¥

NATIONAL ENERGY DEMAND PROJECTIONS

Country JAPAN

Year 1972

WAES Case

Commercial and Public

Unit: 10^9 Kcal

	Fossil Fuel	Electricity
Total	185,761	49,960

NATIONAL ENERGY DEMAND PROJECTIONS

Country JAPAN

Year 1972

WAES Case

Industrial

Unit: 10^9 Kcal

	(A) Value Added	(B) Energy/Value Added	(C) Physical Output	(D) Energy/Output	(E) % Exported	(F) Fossil Fuel	(G) Electricity	(H) Feedstocks	
Iron and Steel (ISIC 371)	1,821.6	328.19				550,090	47,743		(41)
Nonferrous Metals (ISIC 372)	375.8	130.22				27,690	21,249		(42)
Paper and Allied (ISIC 34)	578.3	146.19				76,600	7,941		(43)
Chemicals and Allied (ISIC 35 minus ISIC 353)	2,233.8	201.22				425,990	23,496		(44)
Mineral Products (ISIC 36)	899.2	163.56				135,400	11,668		(45)
Food and Related (ISIC 31)	2,265.1	21.61				43,370	5,589		(46)
Fabricated Metal Prod. ~~Transport Eqpt~~. ~~(ISIC 384)~~ (ISIC 38)	12,954.6	5.53				51,000	20,587		(47)
Textiles ~~Key Industry (N.E.C.)~~ (ISIC 321-324)	1,227.3	52.60				58,640	5,910		(48)
Key Industry (N.E.C.)									(49)
Misc. Manufacturing	1,694.8	55.95				70,710	24,108		(50)
Total						1,439,490	168,291		

N.E.C. = Not Elsewhere Classified

Value Added is given in 1965 prices 10^9 ¥

1) Including feedstocks

NATIONAL ENERGY DEMAND PROJECTIONS

Country JAPAN

Year 1972

WAES Case

Residential

Population ______ Persons/Dwelling ______ Dwellings ______

Central Space Heating		(A) Dwellings	(B) Heat Loss	(C) System Efficiency	(D)	(E) Fossil Fuel	(F) Electricity	
Single Family	Fossil Fuels							(51)
	Electricity							(52)
Town or Row Houses	Fossil Fuels							(53)
	Electricity							(54)
Apartments	Fossil Fuels							(55)
	Electricity							(56)

		% Saturation (Fossil Fuels)	% Saturation (Electricity)	Energy/Unit (Fossil Fuels)	Energy/Unit (Electricity)	(10^9 Kcal)	(10^9 Kcal)	
Room Heating	Electric		15.8		412.8		1,861	(57)
	Gas	17.5		1,035.0		5,168		
	Kerosine	139.2		2,509.8		99,674		
Air Conditioning (Window)			15.8		619.2		2,791	(58)
Air Conditioning (Central)			0.1	-	-	-	-	(59)
Water Heating	Water	60.9		1,490.0		25,888		(60)
	Bath	51.5		2,305.0		33,867		
Refrigerators			99.9		389.6		11,104	(61)
Lighting			100				10,715	(62)
Cooking		100	100			28,530		(63)
Dishwashers			0.6		825.6		141	(64)
Freezers								(65)
Television	White		82.5		23.2		5,323	(66)
	Color				204.7			
Other Key Appliances*			100		89.4		3,295	(67)
Misc. Appliances						2,122	3,770	(68)
Potential Appliances								(69)
					Total	195,249	39,000	

*Kotatsu (Heating)

NATIONAL ENERGY DEMAND PROJECTIONS

Country JAPAN

Year 1985

WAES Case A

Summary

Unit: 10^{13} Kcal

Category	Oil	Gas	Coal	Electricity	Other	Total by Category (Excluding Processing Losses)
Transportation	71.9			2.3		74.2
Bunker	41.7					41.7
Agriculture, Mining, and Construction	17.0			1.0		18.0
Commercial }	39.9	2.3		7.3		49.5
Public }						
Industrial	157.7	3.5	75.5	34.9		271.6
Residential	26.8	8.3		9.2		44.3
Others*	49.3					49.3
Total Delivered Energy by Fuel	404.3	14.1	75.5	54.7		548.6
Processing Losses ** by Fuel						133.4
Gross Input by Fuel	467.2	48.4	91.2	75.2		682.0

*Chemical feedstock, non energy use

**Plus self consumption of energy sector

Gross Energy Input (Add across bottom row)

NATIONAL ENERGY DEMAND PROJECTIONS

Country JAPAN

Year 1985

WAES Case A

Transportation

	(A)	(B)	(C)	(D)	(E)	(F)	(G)	
	Population/Auto	Distance/Auto/Yr.	Load Factor	Fraction Urban or Rural	Energy/Distance	Fossil Fuel	Electricity	
Urban Auto								(1)
Rural Auto	378.7[1]			612[3]		231,764		(2)
	Activity Level	% Electrified	Load Factor	Fossil Fuel/ Activity Level	Electricity/Activity			
Urban Bus	290.4[1]			164[3]		47,626		(3)
Intercity Bus								(4)
Rail (Passenger)	693.0[1]			52.3[3]		21,194	15,049	(5)
Air (Passenger, Domestic)	28.1[1]			842[3]		23,660		(6)
Air (Passenger, International)								(7)
Ship (Passenger)	12.6[1]					*		(8)
Rail (Freight)	183.9[2]			180[4]		24,843	8,259	(9)
Air (Freight, Domestic)								(10)
Air (Freight, International)								(11)
Truck (Local)	284.1[2]			1,218[4]		254,197		(12)
Truck (Intercity)								(13)
Ship (Freight, Domestic)	285.6[2]			406[4]				(14)
Ship (Freight, International)						115,954		(15)
Miscellaneous								(16)
					Total	811,075	23,308	

1) 10^9 Pass. Km. 2) 10^9 ton Km. 3) Kcal/Pass. Km. 4) Kcal/ton Km.

*Included in the item Ship Freight D.

NATIONAL ENERGY DEMAND PROJECTIONS

Country JAPAN

Year 1985

WAES Case A

Agriculture, Mining, and Construction

Agriculture, Forestry, and Fishing (ISIC Major Division 1)	(A) Value Added	(B) Energy/Value Added	(C) Area Harvested	(D) Energy/Area	(E) Physical Output	(F) Energy/Output	(G) Fossil Fuel	(H) Electricity	
Farm Machinery									(17)
Greenhouses									(18)
Other Agricultural Uses	3,593.5	8.01					27,642	1,157	(19)
Forestry (ISIC 12)									(20)
Fishing Boats (ISIC 13)									(21)
Other Fishing Use (ISIC 13)	772.9	74.79					57,807		(22)
Total Fertilizer (tons)		N ______	P ______	K ______		Total	85,449	1,157	

Mining	(A) Value Added	(B) Energy/Value Added	(C) Physical Output	(D) Energy/Physical Output	(E) Fossil Fuel	(F) Electricity	
Handling and Internal Transportation	304.5	45.97			11,050	2,949	(23)
Other Energy Use							(24)
				Total	11,050	2,949	

Construction	(A) Value Added	(B) Energy/Value Added	(C) Physical Output	(D) Energy/Physical Output	(E) Fossil Fuel	(F) Electricity	
Handling and Internal Transportation	10,766.0	7.31			73,014	5,728	(25)
Other Energy Use							(26)
(ISIC 500)				Total	73,014	5,728	
				Agriculture, Mining, and Construction Total	169,513	9,834	

Value Added given in the Price of 1965 10^9 ¥

NATIONAL ENERGY DEMAND PROJECTIONS

Country JAPAN

Year 1985

WAES Case A

Commercial and Public

Unit: 10^9 Kcal

		Fossil Fuel	Electricity
	Total	421,754	73,042

NATIONAL ENERGY DEMAND PROJECTIONS

Country JAPAN

Year 1985

WAES Case A

Industrial

Unit: 10^9 Kcal

	(A) Value Added	(B) Energy/Value Added	(C) Physical Output	(D) Energy/Output	(E) % Exported	(F) Fossil Fuel	(G) Electricity	(H) Feedstocks	
Iron and Steel (ISIC 371)	2,951.7	289.46				780,277	74,116		(41)
Nonferrous Metals (ISIC 372)	879.2	114.46				55,591	45,039		(42)
Paper and Allied (ISIC 34)	1,680.9	124.99				190,004	20,096		(43)
Chemicals and Allied (ISIC 35 minus ISIC 353)	6,497.1	176.88				1,075,520[1)]	73,687		(44)
Mineral Products (ISIC 36)	1,745.2	134.12				213,248	20,819		(45)
Food and Related (ISIC 31)	4,031.5	18.74				65,933	9,619		(46)
Fabricated Metal Prod. ~~Transport Eqpt. (ISIC 384)~~ (ISIC 38)	36,788.6	4.78				125,087	50,846		(47)
Textiles ~~Key Industry (N.E.C.)~~ (ISIC 321-324)	2,242.0	44.97				91,060	9,759		(48)
Key Industry (N.E.C.)									(49)
Misc. Manufacturing	3,780.4	52.31				152,272	45,499		(50)
					Total	2,748,992	349,480		

N.E.C. = Not Elsewhere Classified

Value Added is given in 1965 prices 10^9 ¥

1) Including feedstocks

NATIONAL ENERGY DEMAND PROJECTIONS

Country JAPAN

Year 1985

WAES Case A

Residential

Population ______ Persons/Dwelling ______ Dwellings ______

		(A)	(B)	(C)	(D)	(E)	(F)	
Central Space Heating		Dwellings	Heat Loss	System Efficiency		Fossil Fuel	Electricity	
Single Family	Fossil Fuels							(51)
	Electricity							(52)
Town or Row Houses	Fossil Fuels							(53)
	Electricity							(54)
Apartments	Fossil Fuels							(55)
	Electricity							(56)
		% Saturation (Fossil Fuels)	% Saturation (Electricity)	Energy/Unit (Fossil Fuels)	Energy/Unit (Electricity)	(10^9 Kcal)	(10^9 Kcal)	
Room Heating	Electric		15.0		344		1,981	(57)
	Gas	30.0		3,030		34,897		
	Kerosine	155.0		2,339		139,155		
Air Conditioning (Window)			35.0		619		8,317	(58)
Air Conditioning (Central)								(59)
Water Heating	Water	75.0		1,215		34,984		(60)
	Bath	70.0		2,355		63,287		
Refrigerators			100		344		13,206	(61)
Lighting			100		316		12,132	(62)
Cooking		100	100			34,503	10,762	(63)
Dishwashers			10.0		172		661	(64)
Freezers			20.0		516		3,962	(65)
Television	White		64.0		23		565	(66)
	Color		150.0		205		11,805	
Other Key Appliances*			130.0		138		6,887	(67)
Misc. Appliances								(68)
Potential Appliances						44,231 1)	21,617 1)	(69)
					Total	351,057	91,895	

*Kotatsu (Heating)

1) Including energy consumption for heating other than shown under Water Heating.

NATIONAL ENERGY DEMAND PROJECTIONS

Country JAPAN

Year 1985

WAES Case B

Summary

Unit: 10^{13} Kcal

Category	Oil	Gas	Coal	Electricity	Other	Total by Category (Excluding Processing Losses)
Transportation	68.2			1.9		70.1
Bunker	30.1					30.1
Agriculture, Mining, and Construction	14.8			0.8		15.6
Commercial / Public	38.2	1.8		5.7		45.7
Industrial	111.1	3.2	61.0	25.6		200.9
Residential	23.4	7.0		8.0		38.4
Others*	39.6					39.6
Total Delivered Energy by Fuel	325.4	12.0		42.0		440.4
Processing Losses** by Fuel						104.7
Gross Input by Fuel	351.4	45.6	73.6	74.5		545.1
						Gross Energy Input (Add across bottom row)

*Chemical feedstock, non energy use
**Plus self consumption of energy sector

NATIONAL ENERGY DEMAND PROJECTIONS

Country JAPAN

Year 1985

WAES Case B

Transportation

	(A)	(B)	(C)	(D)	(E)	(F)	(G)	
	Population/Auto	Distance/Auto/Yr.	Load Factor	Fraction Urban or Rural	Energy/Distance	Fossil Fuel	Electricity	
Urban Auto }								(1)
Rural Auto }	348.9[1]			612[3]		213,527		(2)
	Activity Level	% Electrified	Load Factor	Fossil Fuel/ Activity Level	Electricity/Activity			
Urban Bus	231.0[1]			164[3]		37,884		(3)
Intercity Bus }								(4)
Rail (Passenger) }	588.3[1]			52.3[3]		17,995	12,773	(5)
Air (Passenger, Domestic)	24.1[1]			842[3]		20,292		(6)
Air (Passenger, International)								(7)
Ship (Passenger)	10.8[1]					*		(8)
Rail (Freight)	131.5[2]			18.0[4]		17,764	5,906	(9)
Air (Freight, Domestic)								(10)
Air (Freight, International)								(11)
Truck (Local) }	233.0[2]			1,218[4]		283,794		(12)
Truck (Intercity) }								(13)
Ship (Freight, Domestic)	222.5[2]			406[4]		90,335		(14)
Ship (Freight, International)								(15)
Miscellaneous								(16)
					Total	681,591	18,679	

1) 10^9 Pass. Km. 2) 10^9 ton Km. 3) Kcal/Pass. Km. 4) Kcal/ton Km.

*Included in the item Ship Freight D.

NATIONAL ENERGY DEMAND PROJECTIONS

Country JAPAN

Year 1985

WAES Case B

Agriculture, Mining, and Construction

	(A)	(B)	(C)	(D)	(E)	(F)	(G)	(H)	
Agriculture, Forestry, and Fishing (ISIC Major Division 1)	Value Added	Energy/Value Added	Area Harvested	Energy/ Area	Physical Output	Energy/ Output	Fossil Fuel	Electricity	
Farm Machinery									(17)
Greenhouses									(18)
Other Agricultural Uses	3,481.5	8.18					27,279	1,184	(19)
Forestry (ISIC 12)									(20)
Fishing Boats (ISIC 13)									(21)
Other Fishing Use (ISIC 13)	748.8	76.50					57,283		(22)
Total Fertilizer (tons)		N ______	P ______	K ______		Total	84,562	1,184	

	(A)	(B)	(C)	(D)	(E)	(F)	
Mining	Value Added	Energy/Value Added	Physical Output	Energy/Physical Output	Fossil Fuel	Electricity	
Handling and Internal Transportation	304.5	46.98			11,291	3,013	(23)
Other Energy Use							(24)
				Total	11,291	3,013	

	(A)	(B)	(C)	(D)	(E)	(F)	
Construction	Value Added	Energy/Value Added	Physical Output	Energy/Physical Output	Fossil Fuel	Electricity	
Handling and Internal Transportation	7,488.4	7.49			51,991	4,078	(25)
Other Energy Use							(26)
(ISIC 500)				Total	51,991	4,078	
				Agriculture, Mining, and Construction Total	147,844	8,275	

Value Added is given in the price of 1965 10^9 ¥

NATIONAL ENERGY DEMAND PROJECTIONS

Country JAPAN

Year 1985

WAES Case B

Commercial and Public

Unit: 10^9 Kcal

	Fossil Fuel	Electricity
Total	400,610	56,519

NATIONAL ENERGY DEMAND PROJECTIONS

Country JAPAN

Year 1985

WAES Case B

Industrial

Unit: 10^9 Kcal

	(A) Value Added	(B) Energy/Value Added	(C) Physical Output	(D) Energy/Output	(E) % Exported	(F) Fossil Fuel	(G) Electricity	(H) Feedstocks	
Iron and Steel (ISIC 371)	2,295.6	296.02				620,603	58,949		(41)
Nonferrous Metals (ISIC 372)	544.3	117.06				35,204	28,514		(42)
Paper and Allied (ISIC 34)	1,381.7	127.92				159,838	16,905		(43)
Chemicals and Allied (ISIC 35 minus ISIC 353)	4,468.5	180.90				756,539[1)]	51,798		(44)
Mineral Products (ISIC 36)	1,193.3	137,39				149,366	14,583		(45)
Food and Related (ISIC 31)	3,810.5	19.17				63,755	9,301		(46)
Fabricated Metal Prod. ~~Transport Eqpt.~~ ~~(ISIC 384)~~ (ISIC 38)	23,670.8	4.89				81,499	34,251		(47)
Textiles ~~Key Industry (N.E.C.)~~ (ISIC 321-324)	2,031.0	46.02				84,420	9,047		(48)
Key Industry (N.E.C.)									(49)
Misc. Manufacturing	2,642.4	53.43				108,705	32,481		(50)
Total						2,059,929	255,829		

N.E.C. = Not Elsewhere Classified

Value Added is given in 1965 prices 10^9 ¥

1) Including feedstocks

NATIONAL ENERGY DEMAND PROJECTIONS

Country JAPAN

Year 1985

WAES Case B

Residential

Population ______ Persons/Dwelling ______ Dwellings ______

		(A)	(B)	(C)	(D)	(E)	(F)	
Central Space Heating		Dwellings	Heat Loss	System Efficiency		Fossil Fuel	Electricity	
Single Family	Fossil Fuels							(51)
	Electricity							(52)
Town or Row Houses	Fossil Fuels							(53)
	Electricity							(54)
Apartments	Fossil Fuels							(55)
	Electricity							(56)
		% Saturation (Fossil Fuels)	% Saturation (Electricity)	Energy/Unit (Fossil Fuels)	Energy/Unit (Electricity)	(10^9 Kcal)	(10^9 Kcal)	
Room Heating	Electric		15.0		344		1,981	(57)
	Gas	30.0		3,030		34,897		
	Kerosine	155.0		1,550		92,182		
Air Conditioning (Window)			35.0		619		8,317	(58)
Air Conditioning (Central)								(59)
Water Heating	Water	75.0		1,215		34,984		(60)
	Bath	70.0		2,355		63,287		
Refrigerators			100		344		13,206	(61)
Lighting			100		316		12,132	(62)
Cooking		100.0	100			34,503	10,762	(63)
Dishwashers			10.0		172		661	(64)
Freezers			20.0		516		3,962	(65)
Television	White		64.0		23		565	(66)
	Color		150.0		205		11,805	
Other Key Appliances *			130.0		138		6,887	(67)
Misc. Appliances						44,231[1)]	9,847[1)]	(68)
Potential Appliances								(69)
					Total	304,084	80,125	

*Kotatsu (Heating)

1) Including energy consumption for heating other than shown under Water Heating.

NATIONAL ENERGY DEMAND PROJECTIONS

Country JAPAN

Year 1985

WAES Case C

Summary

Unit: 10^{13} Kcal

Category	Oil	Gas	Coal	Electricity	Other	Total by Category (Excluding Processing Losses)
Transportation	84.1			2.2		86.3
Bunker	41.7					41.7
Agriculture, Mining, and Construction	17.3			1.0		18.3
Commercial / Public	41.3	2.3		7.6		51.2
Industrial	164.0	3.5	75.5	35.7		278.7
Residential	28.0	8.3		9.5		45.8
Others*	49.3					49.3
Total Delivered Energy by Fuel	425.7	14.1	75.5	56.0		571.3
Processing Losses ** by Fuel						135.8
Gross Input by Fuel	492.3	48.4	91.2	75.2		707.1

*Chemical feedstock, non energy use

**Plus self consumption of energy sector

Gross Energy Input (Add across bottom row)

NATIONAL ENERGY DEMAND PROJECTIONS

Country JAPAN

Year 1985

WAES Case C

Transportation

	(A)	(B)	(C)	(D)	(E)	(F)	(G)	
	Population/Auto	Distance/Auto/Yr.	Load Factor	Fraction Urban or Rural	Energy/Distance	Fossil Fuel	Electricity	
Urban Auto }								(1)
Rural Auto }	406.8 1)			612 3)		248,962		(2)
	Activity Level	% Electrified	Load Factor	Fossil Fuel/ Activity Level	Electricity/Activity			
Urban Bus }	269.4 1)			164 3)		44,182		(3)
Intercity Bus }								(4)
Rail (Passenger)	686.0 1)			52.3 3)		20,980	14,898	(5)
Air (Passenger, Domestic)	28.1 1)			842 3)		23,660		(6)
Air (Passenger, International)								(7)
Ship (Passenger)	12.6 1)					*		(8)
Rail (Freight)	168.8 2)			180 4)		22,803	7,581	(9)
Air (Freight, Domestic)								(10)
Air (Freight, International)								(11)
Truck (Local) }	299.2 2)			1,218 4)		364,426		(12)
Truck (Intercity) }								(13)
Ship (Freight, Domestic)	285.6 2)			406 4)		115,954		(14)
Ship (Freight, International)								(15)
Miscellaneous								(16)
					Total	840,967	22,479	

1) 10^9 Pass. Km. 2) 10^9 ton Km. 3) Kcal/Pass. Km. 4) Kcal/ton Km.

*Included in the item Ship Freight D.

NATIONAL ENERGY DEMAND PROJECTIONS

Country JAPAN

Year 1985

WAES Case C

Agriculture, Mining, and Construction

	(A)	(B)	(C)	(D)	(E)	(F)	(G)	(H)	
Agriculture, Forestry, and Fishing (ISIC Major Division 1)	Value Added	Energy/Value Added	Area Harvested	Energy/Area	Physical Output	Energy/Output	Fossil Fuel	Electricity	
Farm Machinery									(17)
Greenhouses									(18)
Other Agricultural Uses	3,593.5	8.20					28,277	1,184	(19)
Forestry (ISIC 12)									(20)
Fishing Boats (ISIC 13)	772.9	76.51					59,134		(21)
Other Fishing Use (ISIC 13)									(22)
Total Fertilizer (tons)		N ______	P ______	K ______		Total	87,411	1,184	

	(A)	(B)	(C)	(D)	(E)	(F)	
Mining	Value Added	Energy/Value Added	Physical Output	Energy/Physical Output	Fossil Fuel	Electricity	
Handling and Internal Transportation	304.5	46.98			11,291	3,013	(23)
Other Energy Use							(24)
				Total	11,291	3,013	

	(A)	(B)	(C)	(D)	(E)	(F)	
Construction	Value Added	Energy/Value Added	Physical Output	Energy/Physical Output	Fossil Fuel	Electricity	
Handling and Internal Transportation	10,766.0	7.47			74,620	5,854	(25)
Other Energy Use							(26)
(ISIC 500)				Total	74,620	5,854	
				Agriculture, Mining, and Construction Total	173,322	10,051	

Value Added is given in the Price of 1965 10^9 ¥

NATIONAL ENERGY DEMAND PROJECTIONS

Country JAPAN

Year 1985

WAES Case C

Commercial and Public

Unit: 10^9 Kcal

	Fossil Fuel	Electricity
Total	436,415	75,619

NATIONAL ENERGY DEMAND PROJECTIONS

Country JAPAN

Year 1985

WAES Case C

Industrial

Unit: 10^9 Kcal

	(A) Value Added	(B) Energy/Value Added	(C) Physical Output	(D) Energy/Output	(E) % Exported	(F) Fossil Fuel	(G) Electricity	(H) Feedstocks	
Iron and Steel (ISIC 371)	2,951.7	296.02				797,970	75,797		(41)
Nonferrous Metals (ISIC 372)	879.2	117.05				56,856	46,052		(42)
Paper and Allied (ISIC 34)	1,680.9	127.92				194,449	20,566		(43)
Chemicals and Allied (ISIC 35 minus ISIC 353)	6,497.1	180.90				1,100,051[1)]	75,274		(44)
Mineral Products (ISIC 36)	1,745.2	137.39				218,449	21,327		(45)
Food and Related (ISIC 31)	4,031.5	19.17				67,454	9,841		(46)
Fabricated Metal Prod. ~~Transport Eqpt.~~ ~~(ISIC 384)~~ (ISIC 38)	36,788.6	4.89				127,979	52,022		(47)
Textiles ~~Key Industry (N.E.C.)~~ (ISIC 321-324)	2,242.0	46.02				93,190	9,987		(48)
Key Industry (N.E.C.)									(49)
Misc. Manufacturing	3,780.4	53.43				155,519	46,469		(50)
Total						2,811,917	357,335		

N.E.C. = Not Elsewhere Classified

Value Added is given in 1965 prices 10^9 ¥

1) Including feedstocks

NATIONAL ENERGY DEMAND PROJECTIONS

Country JAPAN

Year 1985

WAES Case C

Residential

Population ______ Persons/Dwelling ______ Dwellings ______

		(A)	(B)	(C)	(D)	(E)	(F)	
Central Space Heating		Dwellings	Heat Loss	System Efficiency		Fossil Fuel	Electricity	
Single Family	Fossil Fuels							(51)
	Electricity							(52)
Town or Row Houses	Fossil Fuels							(53)
	Electricity							(54)
Apartments	Fossil Fuels							(55)
	Electricity							(56)
		% Saturation (Fossil Fuels)	% Saturation (Electricity)	Energy/Unit (Fossil Fuels)	Energy/Unit (Electricity)	(10^9 Kcal)	(10^9 Kcal)	
Room Heating	Electric		15.0		344		1,981	(57)
	Gas	30.0		3,030		34,897		
	Kerosine	155.0		2,545		151,415		
Air Conditioning (Window)			35.0		619		8,317	(58)
Air Conditioning (Central)								(59)
Water Heating	Water	75.0		1,215		34,984		(60)
	Bath	70.0		2,355		63,287		
Refrigerators			100		344		13,206	(61)
Lighting			100		316		12,132	(62)
Cooking		100.0	100			34,503	10,762	(63)
Dishwashers			10.0		172		661	(64)
Freezers			20.0		516		3,962	(65)
Television	White		64.0		23		565	(66)
	Color		150.0		205		11,805	
Other Key Appliances*			130.0		138		6,887	(67)
Misc. Appliances						44,231 1)	24,715 1)	(68)
Potential Appliances								(69)
					Total	363,317	94,993	

*Kotatsu (Heating)

1) Including energy consumption for heating other than shown under Water Heating.

NATIONAL ENERGY DEMAND PROJECTIONS

Country JAPAN

Year 1985

WAES Case D

Summary

Unit: 10^{13} Kcal

Category	Oil	Gas	Coal	Electricity	Other	Total by Category (Excluding Processing Losses)
Transportation	74.4			1.7		76.1
Bunker	30.1					30.1
Agriculture, Mining, and Construction	15.2			0.9		16.1
Commercial / Public	40.5	1.8		6.0		48.3
Industrial	118.0	3.2	61.0	26.4		208.6
Residential	25.1	7.0		8.4		40.5
Others*	39.6					39.6
Total Delivered Energy by Fuel	342.9	12.0	61.0	43.4		459.3
Processing Losses** by Fuel						107.2
Gross Input by Fuel	372.8	45.6	73.6	74.5		566.5

Gross Energy Input (Add across bottom row)

*Chemical feedstock, non energy use

**Plus self consumption of energy sector

NATIONAL ENERGY DEMAND PROJECTIONS

Country JAPAN

Year 1985

WAES Case D

Transportation

	(A)	(B)	(C)	(D)	(E)	(F)	(G)	
	Population/Auto	Distance/Auto/Yr.	Load Factor	Fraction Urban or Rural	Energy/Distance	Fossil Fuel	Electricity	
Urban Auto }								(1)
Rural Auto }	409.0[1)]			612[3)]		250,308		(2)
	Activity Level	% Electrified	Load Factor	Fossil Fuel/ Activity Level	Electricity/Activity			
Urban Bus }	200.9[1)]			164[3)]		32,948		(3)
Intercity Bus }								(4)
Rail (Passenger)	558.2[1)]			52.3[3)]		17,074	12,119	(5)
Air (Passenger, Domestic)	24.1[1)]			842[3)]		20,292		(6)
Air (Passenger, International)								(7)
Ship (Passenger)	10.8[1)]					*		(8)
Rail (Freight)	102.0[2)]			180[4)]		13,779	4,581	(9)
Air (Freight, Domestic)								(10)
Air (Freight, International)								(11)
Truck (Local) }	262.4[2)]			1,218[4)]		319,603		(12)
Truck (Intercity) }								(13)
Ship (Freight, Domestic)	222.5[2)]			406[4)]		90,335		(14)
Ship (Freight, International)								(15)
Miscellaneous								(16)
					Total	744,339	16,700	

1) 10^9 Pass. Km. 2) 10^9 ton Km. 3) Kcal/Pass. Km. 4) Kcal/ton Km.

*Included in the item Ship Freight D.

NATIONAL ENERGY DEMAND PROJECTIONS

Country JAPAN

Year 1985

WAES Case D

Agriculture, Mining, and Construction

Agriculture, Forestry, and Fishing (ISIC Major Division 1)	(A) Value Added	(B) Energy/Value Added	(C) Area Harvested	(D) Energy/Area	(E) Physical Output	(F) Energy/Output	(G) Fossil Fuel	(H) Electricity	
Farm Machinery									(17)
Greenhouses									(18)
Other Agricultural Uses	3,481.5	8.47					28,262	1,226	(19)
Forestry (ISIC 12)									(20)
Fishing Boats (ISIC 13)									(21)
Other Fishing Use (ISIC 13)	748.8	77.87					58,311		(22)
Total Fertilizer (tons)		N ______	P ______	K ______		Total	85,573	1,226	

Mining	(A) Value Added	(B) Energy/Value Added	(C) Physical Output	(D) Energy/Physical Output	(E) Fossil Fuel	(F) Electricity	
Handling and Internal Transportation	304.5	48.48			11,652	3,109	(23)
Other Energy Use							(24)
				Total	11,652	3,109	

Construction	(A) Value Added	(B) Energy/Value Added	(C) Physical Output	(D) Energy/Physical Output	(E) Fossil Fuel	(F) Electricity	
Handling and Internal Transportation	7,488.4	7.73			53,670	4,210	(25)
Other Energy Use							(26)
(ISIC 500)				Total	53,670	4,210	
				Agriculture, Mining, and Construction Total	151,895	8,545	

Value Added is given in the Price of 1965 10^9 ¥

NATIONAL ENERGY DEMAND PROJECTIONS

Country JAPAN

Year 1985

WAES Case D

Commercial and Public

Unit: 10^9 Kcal

	Fossil Fuel	Electricity
Total	423,040	59,730

NATIONAL ENERGY DEMAND PROJECTIONS

Country JAPAN

Year 1985

WAES Case D

Industrial

Unit: 10^9 Kcal

	(A) Value Added	(B) Energy/Value Added	(C) Physical Output	(D) Energy/Output	(E) % Exported	(F) Fossil Fuel	(G) Electricity	(H) Feedstocks	
Iron and Steel (ISIC 371)	2,295.6	305.87				641,244	60,910		(41)
Nonferrous Metals (ISIC 372)	544.3	120.97				36,379	29,466		(42)
Paper and Allied (ISIC 34)	1,381.7	132.30				165,318	17,485		(43)
Chemicals and Allied (ISIC 35 minus ISIC 353)	4,468.5	186.93				781,813[1)]	53,499		(44)
Mineral Products (ISIC 36)	1,193.3	142.30				154,701	15,104		(45)
Food and Related (ISIC 31)	3,810.5	19.82				65,912	9,616		(46)
Fabricated Metal Prod. ~~Transport Eqpt.~~ ~~(ISIC 384)~~ (ISIC 38)	23,670.8	4.94				82,345	34,607		(47)
Textiles ~~Key Industry (N.E.C.)~~ (ISIC 321-324)	2,031.0	47.60				87,314	9,357		(48)
Key Industry (N.E.C.)									(49)
Misc. Manufacturing	2,642.4	55.10				112,109	33,498		(50)
Total						2,127,135	263,542		

N.E.C. = Not Elsewhere Classified

Value Added is given in 1965 prices 10^9 ¥

1) Including feedstocks

NATIONAL ENERGY DEMAND PROJECTIONS

Country JAPAN

Year 1985

WAES Case D

Residential

Population ______ Persons/Dwelling ______ Dwellings ______

		(A)	(B)	(C)	(D)	(E)	(F)	
Central Space Heating		Dwellings	Heat Loss	System Efficiency		Fossil Fuel	Electricity	
Single Family	Fossil Fuels							(51)
	Electricity							(52)
Town or Row Houses	Fossil Fuels							(53)
	Electricity							(54)
Apartments	Fossil Fuels							(55)
	Electricity							(56)
		% Saturation (Fossil Fuels)	% Saturation (Electricity)	Energy/Unit (Fossil Fuels)	Energy/Unit (Electricity)	(10^9 Kcal)	(10^9 Kcal)	
Room Heating	Electric		15.0		344		1,981	(57)
	Gas	30.0		3,030		34,897		
	Kerosine	155.0		1,837		109,284		
Air Conditioning (Window)			35.0		619		8,317	(58)
Air Conditioning (Central)								(59)
Water Heating	Water	75.0		1,215		34,984		(60)
	Bath	70.0		2,355		63,287		
Refrigerators			100		344		13,206	(61)
Lighting			100		316		12,132	(62)
Cooking		100.0	100			34,503	10,762	(63)
Dishwashers			10.0				661	(64)
Freezers			20.0				3,962	(65)
Television	White		64.0				565	(66)
	Color		150.0				11,805	
Other Key Appliances *			130.0				6,887	(67)
Misc. Appliances						44,231[1)]	14,201[1)]	(68)
Potential Appliances								(69)
					Total	321,186	84,479	

*Kotatsu (Heating)

1) Including energy consumption for heating other than shown under Water Heating.

NATIONAL ENERGY DEMAND PROJECTIONS

Country JAPAN

Year 1985

WAES Case E

Summary

Unit: 10^{13} Kcal

Category	Oil	Gas	Coal	Electricity	Other	Total by Category (Excluding Processing Losses)
Transportation	92.3			2.0		94.3
Bunker	43.5					43.5
Agriculture, Mining, and Construction	17.8			1.0		18.8
Commercial } Public	43.8	2.2		8.0		54.0
Industrial	175.9	3.5	75.5	37.3		292.2
Residential	30.6	8.5		10.2		49.3
Others*	49.3					49.3
Total Delivered Energy by Fuel	453.2	14.2	75.5	58.5		601.4
Processing Losses ** by Fuel						140.8
Gross Input by Fuel	527.4	48.4	91.2	75.2		742.2

*Chemical feedstock, non energy use

**Plus self consumption of energy sector

Gross Energy Input (Add across bottom row)

NATIONAL ENERGY DEMAND PROJECTIONS

Country JAPAN

Year 1985

WAES Case E

Transportation

	(A)	(B)	(C)	(D)	(E)	(F)	(G)	
	Population/Auto	Distance/Auto/Yr.	Load Factor	Fraction Urban or Rural	Energy/Distance	Fossil Fuel	Electricity	
Urban Auto								(1)
Rural Auto	483.7[1]			612[3]		296,024		(2)
	Activity Level	% Electrified	Load Factor	Fossil Fuel/ Activity Level	Electricity/Activity			
Urban Bus	237.6[1]			164[3]		38,966		(3)
Intercity Bus								(4)
Rail (Passenger)	660.1[1]			52.3[3]		20,191	14,332	(5)
Air (Passenger, Domestic)	28.5[1]			842[3]		23,997		(6)
Air (Passenger, International)								(7)
Ship (Passenger)	12.8[1]					*		(8)
Rail (Freight)	131.1[2]			180[4]		17,710	5,888	(9)
Air (Freight, Domestic)								(10)
Air (Freight, International)								(11)
Truck (Local)	336.9[2]			1,218[4]		410,344		(12)
Truck (Intercity)								(13)
Ship (Freight, Domestic)	285.6[2]			406[4]		115,954		(14)
Ship (Freight, International)								(15)
Miscellaneous								(16)
					Total	923,186	20,220	

1) 10^9 Pass. Km. 2) 10^9 ton Km. 3) Kcal/Pass. Km. 4) Kcal/ton Km.

*Included in the item Ship Freight D.

NATIONAL ENERGY DEMAND PROJECTIONS

Country JAPAN

Year 1985

WAES Case E

Agriculture, Mining, and Construction

Agriculture, Forestry, and Fishing (ISIC Major Division 1)	(A) Value Added	(B) Energy/Value Added	(C) Area Harvested	(D) Energy/Area	(E) Physical Output	(F) Energy/Output	(G) Fossil Fuel	(H) Electricity	
Farm Machinery									(17)
Greenhouses									(18)
Other Agricultural Uses	3,608.9	8.34					28,905	1,210	(19)
Forestry (ISIC 12)									(20)
Fishing Boats (ISIC 13)									(21)
Other Fishing Use (ISIC 13)	776.1	77.88					60,444		(22)
Total Fertilizer (tons)		N ______	P ______	K ______		Total	89,349	1,210	

Mining	(A) Value Added	(B) Energy/Value Added	(C) Physical Output	(D) Energy/Physical Output	(E) Fossil Fuel	(F) Electricity	
Handling and Internal Transportation	304.5	48.48			11,652	3,109	(23)
Other Energy Use							(24)
				Total	11,652	3,109	

Construction	(A) Value Added	(B) Energy/Value Added	(C) Physical Output	(D) Energy/Physical Output	(E) Fossil Fuel	(F) Electricity	
Handling and Internal Transportation	10,747.6	7.73			77,030	6,043	(25)
Other Energy Use							(26)
(ISIC 500)				Total	77,030	6,043	
				Agriculture, Mining, and Construction Total	178,031	10,362	

Value Added is given in the Price of 1965 10^9 ¥

NATIONAL ENERGY DEMAND PROJECTIONS

Country JAPAN

Year 1985

WAES Case E

Commercial and Public

Unit: 10^9 Kcal

	Fossil Fuel	Electricity
Total	460,850	79,916

NATIONAL ENERGY DEMAND PROJECTIONS

Country JAPAN

Year 1985

WAES Case E

Industrial

Unit: 10^9 Kcal

	(A) Value Added	(B) Energy/Value Added	(C) Physical Output	(D) Energy/Output	(E) % Exported	(F) Fossil Fuel	(G) Electricity	(H) Feedstocks	
Iron and Steel (ISIC 371)	2,951.7	305.87				824,510	78,318		(41)
Nonferrous Metals (ISIC 372)	887.4	120.96				59,305	48,035		(42)
Paper and Allied (ISIC 34)	1,710.9	132.30				204,710	21,651		(43)
Chemicals and Allied (ISIC 35 minus ISIC 353)	6,586.6	186.93				1,152,321[1)]	78,912		(44)
Mineral Products (ISIC 36)	1,766.0	142.30				228,946	22,352		(45)
Food and Related (ISIC 31)	4,053.6	19.82				70,118	10,230		(46)
Fabricated Metal Prod. ~~Transport Eqpt.~~ ~~(ISIC 384)~~ (ISIC 38)	37,282.0	4.94				129,695	54,506		(47)
Textiles ~~Key Industry (N.E.C.)~~ (ISIC 321-324)	2,270.2	47.60				97.595	10,459		(48)
Key Industry (N.E.C.)									(49)
Misc. Manufacturing	3,868.1	55.10				164,097	49,035		(50)
Total						2,931,297	373,498		

N.E.C. = Not Elsewhere Classified

Value Added is given in 1965 prices 10^9 ¥

1) Including feedstocks

NATIONAL ENERGY DEMAND PROJECTIONS

Country JAPAN

Year 1985

WAES Case E

Residential

Population ______ Persons/Dwelling ______ Dwellings ______

		(A)	(B)	(C)	(D)	(E)	(F)	
Central Space Heating		Dwellings	Heat Loss	System Efficiency		Fossil Fuel	Electricity	
Single Family	Fossil Fuels							(51)
	Electricity							(52)
Town or Row Houses	Fossil Fuels							(53)
	Electricity							(54)
Apartments	Fossil Fuels							(55)
	Electricity							(56)
		% Saturation (Fossil Fuels)	% Saturation (Electricity)	Energy/Unit (Fossil Fuels)	Energy/Unit (Electricity)	(10^9 Kcal)	(10^9 Kcal)	
Room Heating	Electric		15.0		344		1,981	(57)
	Gas	30.0		3,030		34,897		
	Kerosine	155.0		2,992		178,013		
Air Conditioning (Window)			50.0		619		11,510	(58)
Air Conditioning (Central)								(59)
Water Heating	Water	75.0		1,215		34,984		(60)
	Bath	70.0		2,355		63,287		
Refrigerators			100.0		344		13,206	(61)
Lighting			100.0		316		12,132	(62)
Cooking		100.0	100.0			34,503	10,762	(63)
Dishwashers			10.0		172		661	(64)
Freezers			20.0		516		3,962	(65)
Television	White		64.0		23		565	(66)
	Color		150.0		205		11,805	
Other Key Appliances*			130.0		138		6,887	(67)
Misc. Appliances						44,231[1)]	28,230[1)]	(68)
Potential Appliances								(69)
					Total	389,915	101,701	

* Kotatsu (Heating)

1) Including energy consumption for heating other than shown under Water Heating.

10 Mexico

10.1 OVERVIEW

This report presents the results of a study on energy development, based on several scenarios. Traditionally, the analysis of future energy requirements has been carried out by means of an extrapolation of historical trends. No claims are made by the present study about the consistency or adequacy of the assumptions used for the different scenarios. These have been established essentially through subjective consideration of the behavior of several related parameters that are relevant to the international energy supply situation under conditions of increasing demand.

Case A, for example, which assumes high GWP and increasing oil prices, results in a higher energy growth rate than has been historically observed and represents a superior boundary based on an acceleration of growth.

The fact that international oil prices have had a minimal influence on the domestic energy situation does not necessarily mean that this situation will prevail in the future. Unfortunately, there is not enough information available to allow accurate measurements of the elasticities of demand to prices; therefore, it is not possible at present to handle these relationships by more sophisticated methods.

Energy conservation measures resulting from the vigorous national policies assumed in Cases A and B should be feasible within the time scale of the present study. Additional possibilities for conservation are foreseen in the transportation sector but, aside from vigorous policies, these will require a change of attitude on the part of users, coupled with technological developments such as improved mass transportation systems or small electric vehicles. These latter considerations imply time spans for saturation that go beyond the year 1985.

A major effort will obviously be required to improve and disaggregate the information. Data must be obtained on the energy efficiency of industrial processes in order to better assess the margins of savings and to develop the practical methods to achieve them.

The results in this report were obtained before the finish of the WAES supply-demand integration exercises, now in progress. Once the energy situation is analyzed from the supply point of view, it may be that some of the demand assumptions will have to be modified to ensure a more consistent set of projections.

10.2 INTRODUCTION

10.2.1 National Structure

This study of Mexico's energy demand, including the data for 1972 and five cases for 1985 under different sets of assumptions, has been prepared by the secretariat of the Energy Commission with the assistance of representatives of the Mexican energy agencies who have supplied basic data and support for the continuous participation in the Workshop.

One of the main tasks of the Energy Commission of Mexico is the analysis of the future of the energy sector. A very important aspect of this task is that ways and means must be found to modify the energy consumption patterns of the country in order to achieve a more balanced situation in which the economy becomes less dependent on oil and gas. These fuels presently satisfy about 90 percent of the total energy needs. As a result of these studies, it is expected that a set of policy guidelines will be established that will:

- promote the diversification of primary energy resources,
- promote the utilization of coordinated planning systems within the whole energy sector,
- coordinate the exploration of energy resources,
- develop the national industries to produce the capital goods required by the energy supply sector, and
- promote coordinated research and development work in the energy field.

10.2.2 Present Energy Situation

The energy supply sector represents approximately 6 percent of the GNP. The average energy growth rate during the last decade has been approximately 8 percent per year.

Energy consumption is distributed roughly in the following way: 28.0 percent by the transportation sector; 30.1 percent by the industrial sector; 11.2 percent by the residential and commercial sectors; 3.0 percent by nonenergy applications of energy resources; 26 percent by conversion and transmission losses, plus self-consumption of the energy supply agencies; 1.7 percent by others.

The overall energy demand for 1975 was 572 x 10^{12} kcal, 88 percent of which was supplied by oil and gas, 6 percent by hydroelectricity, and 6 percent by coal. This distribution has been roughly the same over the last twenty years, with a slight increase in favor of oil and gas due to a relative decrease in the participation of hydroelectric energy.

The high dependency on hydrocarbons is a logical consequence of the availability of oil and gas. Between 1901 and December 1975, the accumulated production amounts to 8.1 billion barrels of crude oil equivalent, with reported recoverable reserves in December 1975 of 6.3 billion barrels of crude oil equivalent.

The characteristics of the new production area in the southeastern part of the country, which started at the peak of the so-called oil crisis, give good grounds for optimism about our oil and gas resources for the future, and it is highly probable that Mexico will not only be able to meet its energy requirements with domestic oil and gas, but will, perhaps, have a small surplus for export.

Coal reserves are, in the opinion of some experts, at least of the same order of magnitude as those of oil and gas. The use of coal as an energy source has been negligible, and the limited coal production has been absorbed by the steel industry. Coal consumption has grown at a rate of more than 9 percent per year, and a very substantial increase in production is foreseen in order to substitute partially for oil and gas. Coal will not, however, play an important role prior to 1985 due to the lead time required for its development.

10.2.3 Data Basis

The data for this report were derived from available information within the energy industry as well as from published statistics by several government agencies and private institutions.

Since the level of disaggregation of consumption data is very limited, and its degree of accuracy very doubtful, it was decided to deal with more aggregated but reliable information. The Comisión del Transporte and the Instituto Mexicano del Petróleo are attempting a more detailed analysis of the consuming sectors, especially the transportation and industrial sectors, each of which accounts for nearly 30 percent of total demand. The results of these studies are not yet available; but some information from them has been incorporated in the 1972 data.

Main Data Sources for This Report

1. *Bases para la planeación del desarrollo de Pemex*, Petróleos Mexicanos, August 1974.

2. *Estudio del desarrollo a largo plazo de los sistemas eléctricos de Mexico*, Comisión Federal de Electricidad, 1975.

3. *Inventario de aprovechamientos hidroeléctricos*, Comisión de Energéticos, Plan Nacional Hidráulico, Comisión Federal de Electricidad. Reporte final, August 1975.

4. *Energéticos: Demanda Sectorial. Análisis y perspectivas*, Instituto Mexicano del Petróleo, 1975.

5. *Memoria de Labores*, Petróleos Mexicanos. Reportes anuales, 1965–1974.

6. *México en cifras*, Nacional Financiera, 1974.

7. Reports from the Comisión del Transporte.

8. *Anuario estadístico*, Secretaría de Industria y Comercio, 1973.

9. *Resultados de explotación, 1962–1972*, Comisión Federal de Electricidad.

10.3 METHODOLOGY

10.3.1 Scenario Variables

For the WAES high and low GWP growth cases (6 and 3.5 percent per year), 7.5 and 6 percent were selected as Mexico's target of GNP growth rates for the period between 1975 and 1985 (Table 10.1). These boundaries are consistent with the estimates of various groups dealing with projections of the future economy of the country.

10.3.2 Oil Prices

Mexico's internal energy prices have not been

Table 10.1
Adaptations to Mexico of the five WAES Scenarios to 1985

	WAES World Scenarios			Mexico Scenarios			
Case	Economic Growth	Oil Price	Government Response	Economic Growth	Energy Growth	Electricity Growth	Government Response
A	6%	rising	vigorous	7.5%	8.3%	10.4%	Increase transportation & industry efficiency by 10%
B	3.5%	rising	vigorous	6%	6.3%	8.8%	Increase transportation & industry efficiency by 10%
C	6%	constant	vigorous	7.5%	7.6%	9.8%	Increase transportation efficiency by 15%, industry by 10%
D	3.5%	constant	restrained	6%	6.9%	8.8%	
E[1]	6%	falling	restrained	6%	6.8%	8.8%	

1. The low economic growth assumed for Case E is a consequence of low GNP growth due to elimination of potential oil exports.

affected in the past by the international oil price. The country is self-sufficient in energy production, and reserves have been maintained at reasonable levels to ensure continuous supply and modest exports.

Considering the fact that Mexico is an importer of goods and equipment, especially for the energy supply industries, the oil price rises of the last years have imposed a serious burden upon our balance of payments. Internal energy prices have been adjusted to improve the financial situation of these agencies, and oil production has been raised so that revenues from exports match foreign currency expenditures for imports of goods and services. No reduction or stimulation of energy demand as a direct consequence of oil price has been considered in our projections to 1985. It has been assumed that increases in international oil prices will encourage conservation measures in order to allow increased exports. This will make available more capital for the development of the country. As a consequence, the energy consumption growth rate (before conservation) is maintained at a relatively high level in Case A.

10.3.3 National Policy

The energy supply sector in Mexico is fully state-owned. Mexican legislation explicitly precludes the participation of private corporations or individuals in the exploitation of oil, gas, and uranium resources.

The restrained national policy is interpreted as lack of action, by government or private industry, to modify energy demand or the composition of energy sources utilized. However, the restrained policy allows for rationalization in energy utilization as well as for optimization of supply. The vigorous policy is interpreted as including mandatory actions by consumers and producers to achieve certain preestablished goals. In Cases A and B such actions are estimated to cause a demand reduction of 10 percent in the industrial sector and 15 percent in the transportation sector by 1985. In Case B, due to the lower economic activity, an additional reduction of 5 percent occurs in the industrial sector.

Some specific actions involved in national policy are described in Table 10.2.

10.4 1985 ENERGY PROJECTIONS

10.4.1 Overview

Energy demand projections in units of 10^{18} joules are summarized in the first worksheet in Section 10.6 for the five WAES cases. Case B—low economic growth and high oil price—has the lowest final energy demand, while Case A—high economic growth and high oil price—has the highest final energy demand. Case B is lower than Case D—low economic growth and constant oil price—as a result of the vigorous national response (as mentioned previously, energy demand is not driven by the international price of oil). Case C is approximately midway in the range of energy use, and it is similar to estimates made by various organizations. Figure 10.1 illustrates these results.

The electricity growth rate has averaged, for the last twenty years, 11.4 percent per year. The residential electricity consumption has grown at a rate above 12.5 percent per year. In Case A it has been

Table 10.2
Vigorous and restrained national policy responses in the transportation and industrial sectors

	Restrained	Vigorous
Transportation	Encourage car pools Appeal for lower highway driving speed Appeal for greater utilization of public transportation system	Improve and enlarge public transportation system Impose mandatory highway speed limits Close certain urban areas to private cars
Industrial	Appeal for voluntary efficiency improvements Offer guidance on technology selection for industrial processes Make loans available for energy conservation investments	Set minimum efficiency standards Restrict the choice of energy sources (primary and secondary) to those that are in the national interest Subsidize energy conservation investments

considered that the standard of living of the rural population (40 to 45 percent of the total population in the country) will be improved at a faster rate than in the past, the reasoning being that Mexico would do better in a world that is doing well (Case A) than in a troubled world (Cases B and D).

10.4.2 Energy Demand

Figures 10.2 and 10.3 show demand by sectors for the five WAES cases. One can observe that the industrial and transport sectors consume almost 60 percent of the total, while the residential and commercial sectors, together, use only about 10 percent. The remainder is consumed by conversion losses and self-consumption in the energy sector.

Figure 10.4 shows the breakdown of demand according to energy sources. Oil and gas supply 88 percent, coal 6 percent, hydroelectricity 4 percent, nuclear and others 2 percent.

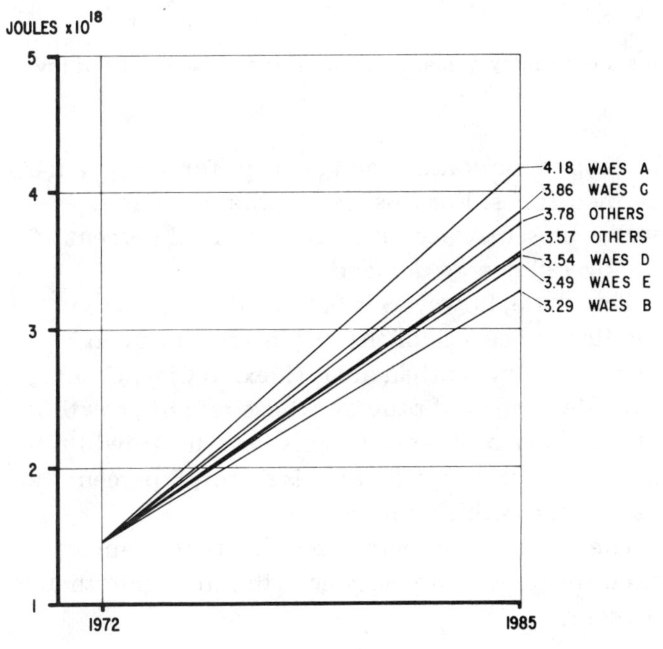

Figure 10.1 1985 final energy demand projections.

Industrial Sector

During the 1960 to 1972 period, the aggregate consumption of this sector has grown at an annual rate higher than 8 percent, whereas, during the same period, overall energy growth rate has been of the order of 7.5 percent.

Supply to the industrial sector has had the following distribution: 9 percent electricity; 20 percent coal; 71 percent oil and gas. The main industries, accounting for more than 50 percent of the total for this sector, are steel, chemicals and petrochemicals, cement, and sugar.

Figure 10.5 shows the WAES projections for final industrial sector energy demand in 1985.

Transportation Sector

This sector accounts for roughly 30 percent of the energy demand, and its rate of growth averages about 8.5 percent per year. Mexico's transportation system is based on automobiles, buses, and trucks. The railroad underwent an impressive development at the beginning of this century but has been in a state of stagnation since about twenty years ago. At present the railroad system has 28,000 km of track. The highway system has about 170,000 km and is the main transportation system. Mexico has two domestic airlines, covering fifty cities.

Because of the high proportion of automobiles

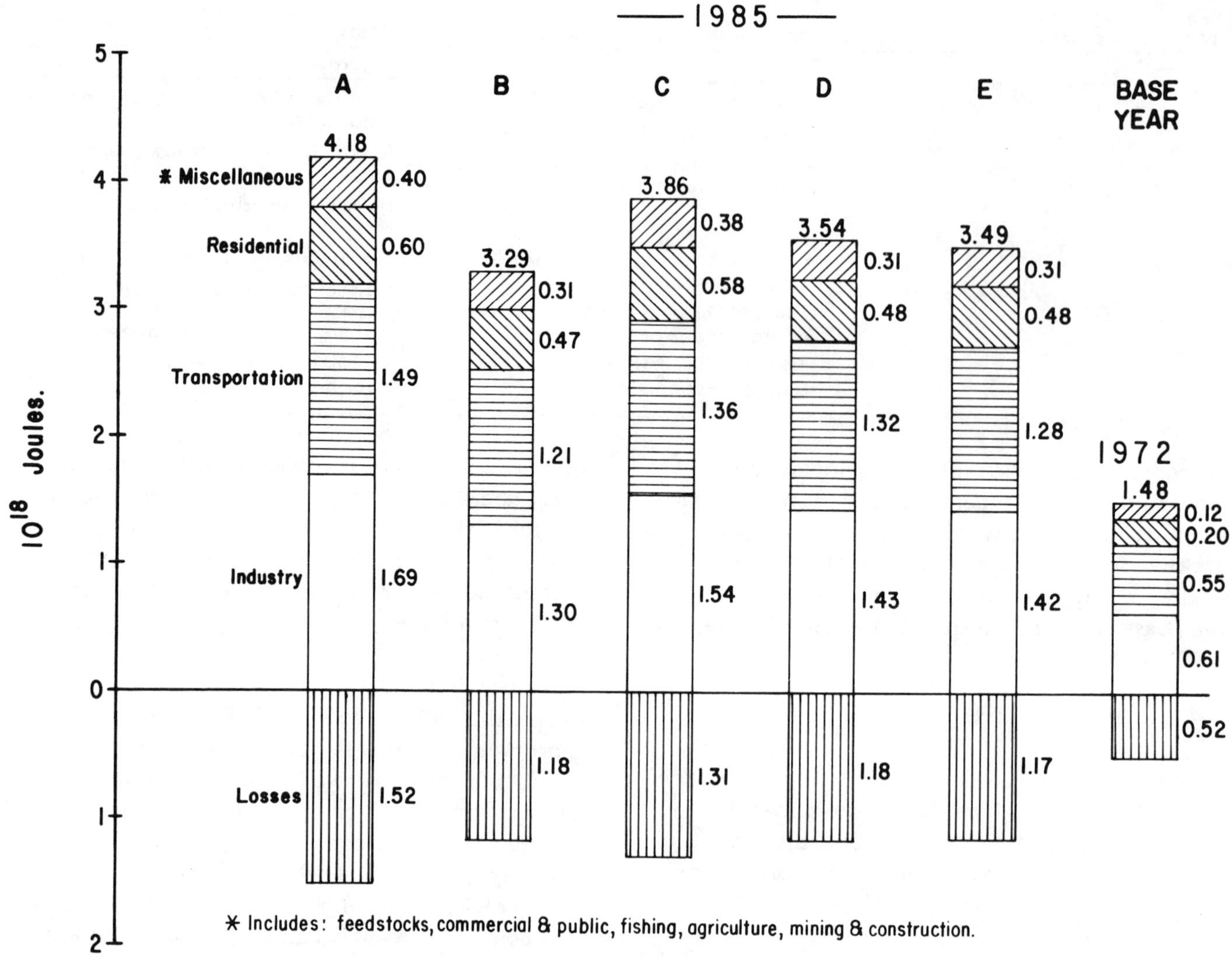

Figure 10.2 Summary of energy demand for the five WAES scenarios. Low energy demand in Case E is a consequence of low GNP growth due to elimination of potential oil exports.

and the possibility of using more fully the potential of the railroads, there is an ample margin for energy conservation in this sector. It has been estimated that a reduction of about 15 percent in consumption could be achieved by 1985.

Final energy demand within the transportation sector is shown in Figure 10.6.

Residential and Commercial Sectors

Mexico has one of the highest population growth rates in the world—3.5 percent per year. About 45 percent of this population lives in the rural areas, and their economic resources are rather limited. On the other hand, the relatively mild climate of the country makes it practically unnecessary to use heating or air conditioning, except for a few well-defined zones. For this reason, the residential sector presently accounts for only 11.2 percent of the overall energy demand.

This share, however, is projected to increase in the future because of the displacement of rural population into urban areas (Mexico City, Guadalajara, Mexicali, and others), whose rate of growth is of the order of 6 percent per year. The growth rate in energy consumption of this sector, 9 percent per year, reflects this phenomenon.

The commercial sector uses electricity almost exclusively, with a yearly growth rate higher than 9 percent.

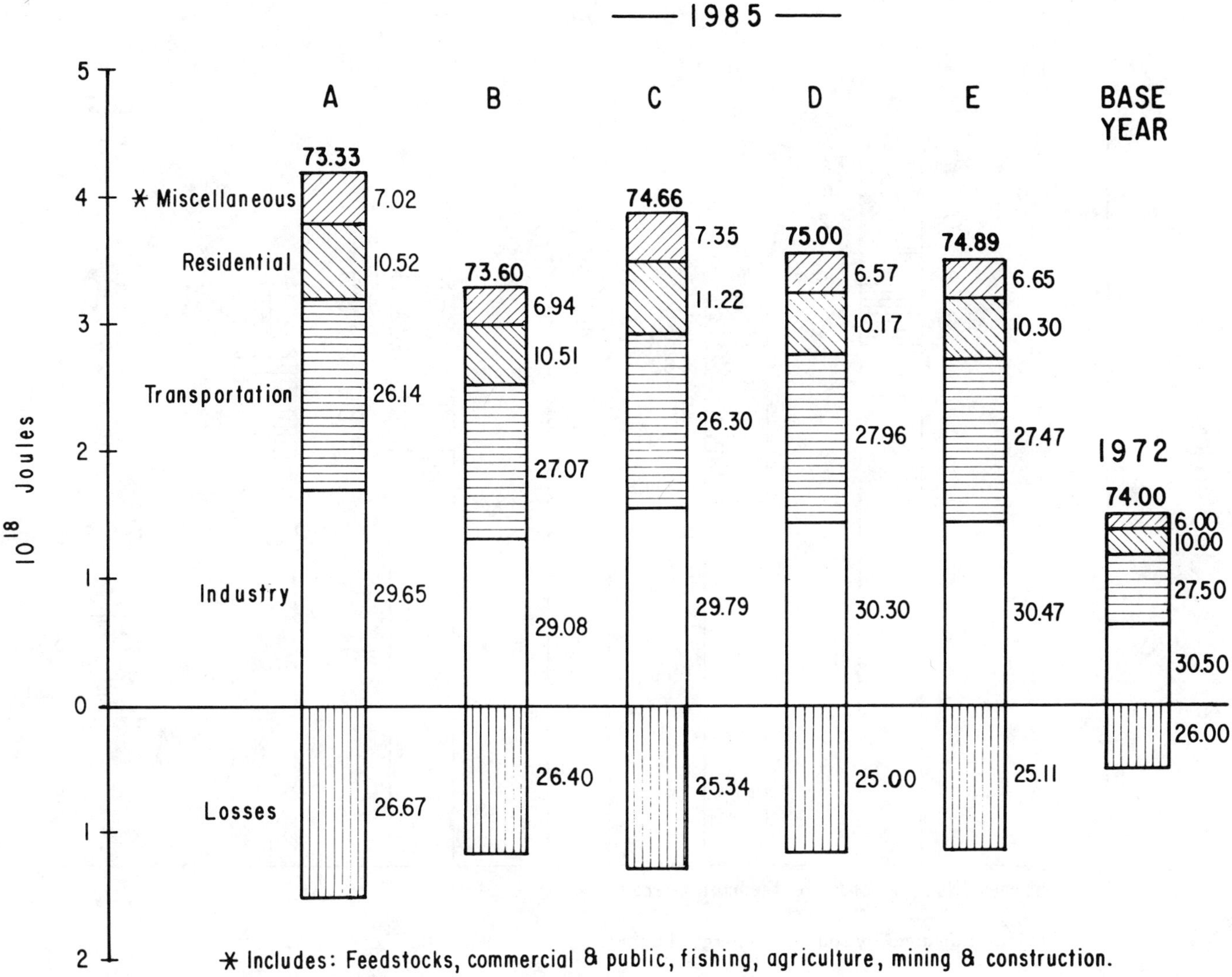

Figure 10.3 Summary of energy demand for the five WAES scenarios (in percent).

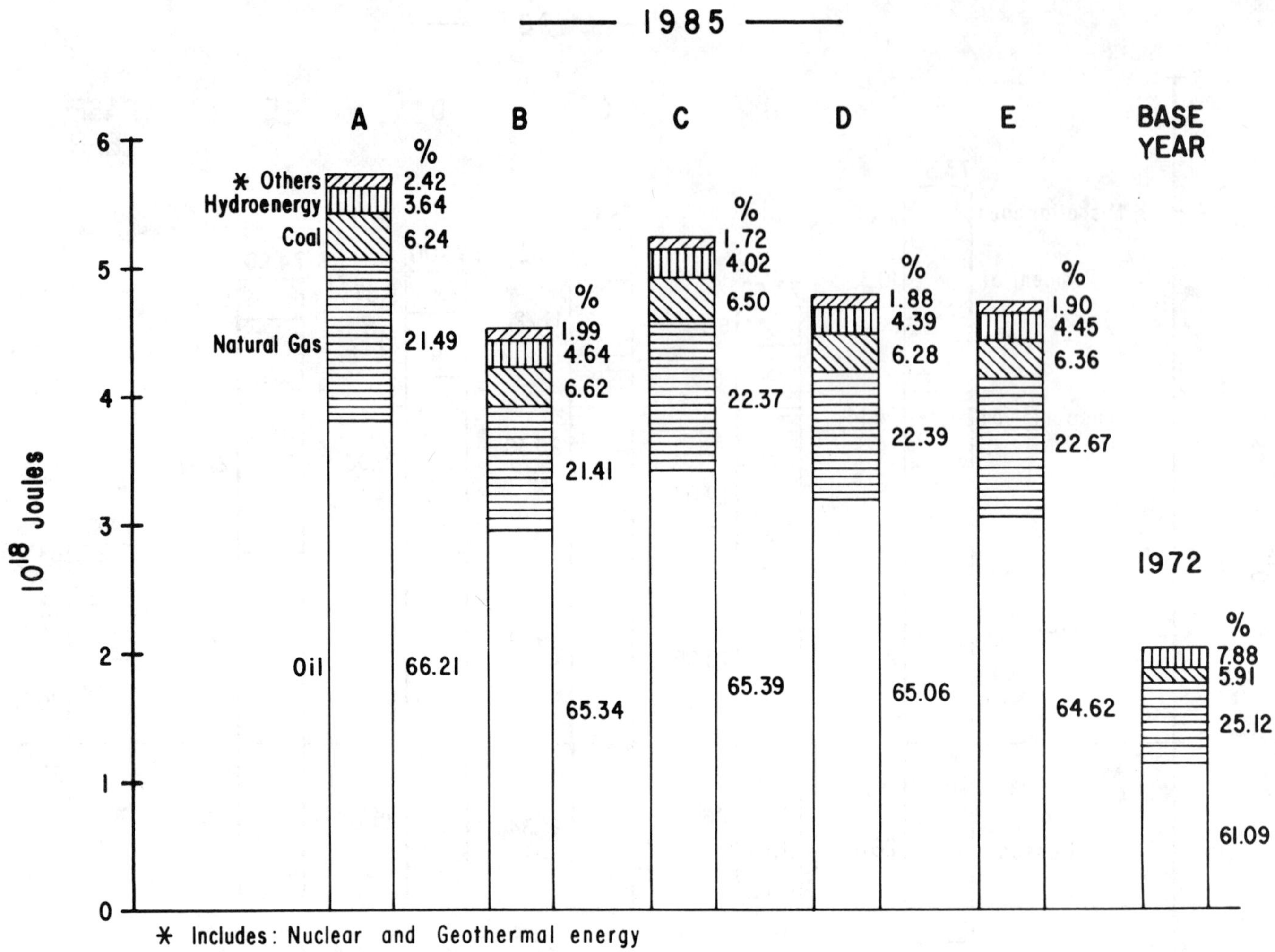

Figure 10.4 Distribution of demand by source of energy (in percent).

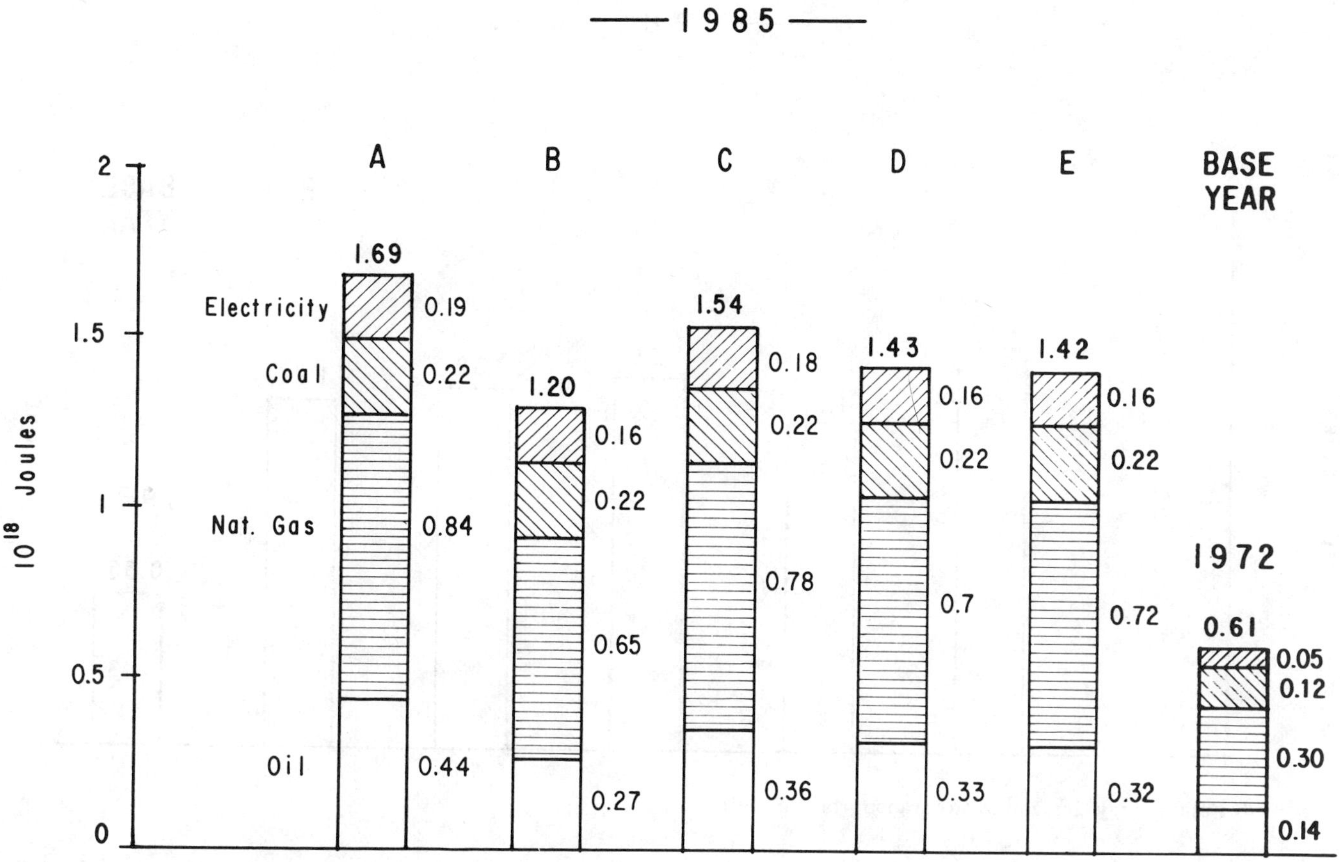

Figure 10.5 Final energy demand in the industrial sector.

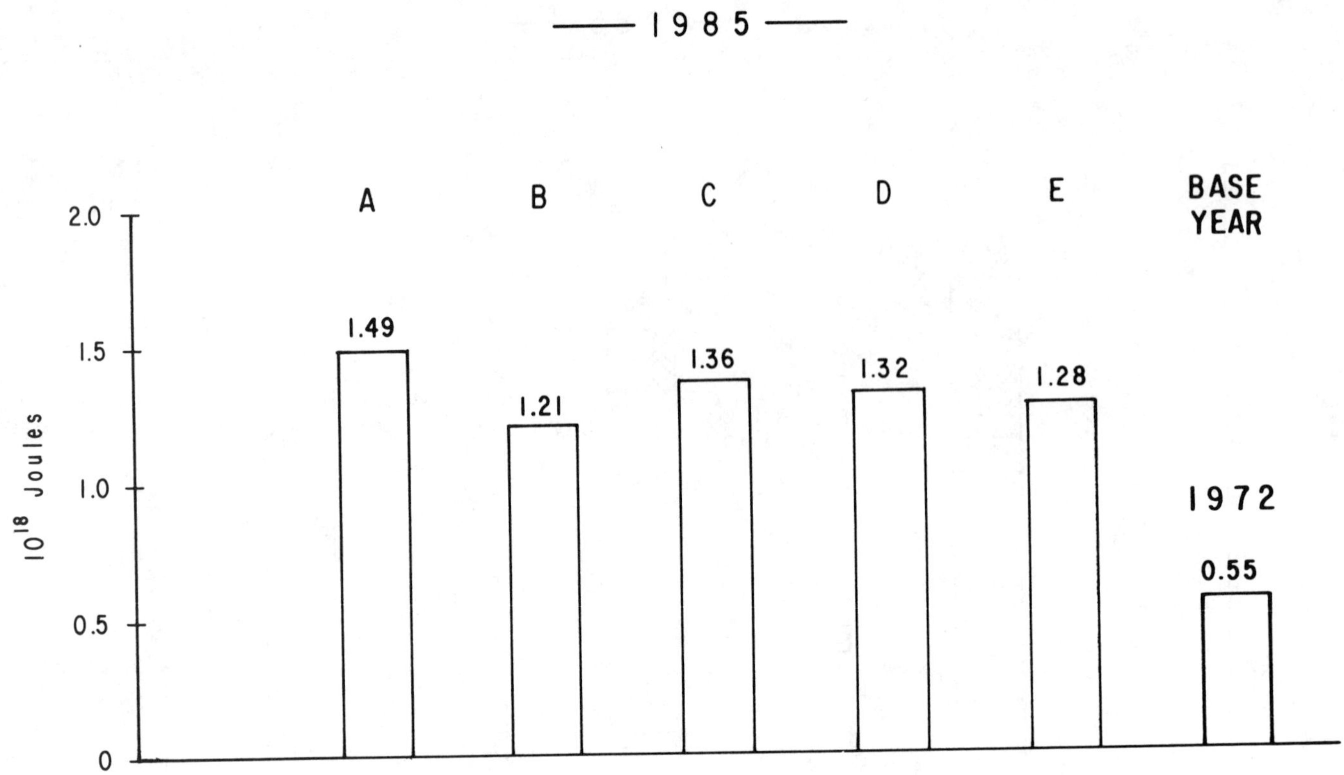

Figure 10.6 Final energy demand in the transportation sector.

10.5 DEMAND STUDY WORKSHEETS

This section includes WAES summary tables and worksheets in the following order:

Energy Use Summary

Energy Growth Rates Summary

Sectoral Percentages of Total Final Energy Demand

Sectoral Percentages of Total Final Energy Input

1985 Economic Activity of Major Energy-Use Categories

Final Energy Demand of Major Energy-Use Categories

1985 Case Comparisons (five sheets)

Base Year (1972) data (six sheets)

Summary tables for Cases A–E (five sheets).

WAES NATIONAL ENERGY DEMAND STUDIES, 1972-1985
ENERGY USE SUMMARY

Country MEXICO

Scenario Assumptions

WAES Case	A		B		C		D		E**			
GNP 1976-1985	High		Low		High		Low		High			
Oil Price*	\$11.50 → \$17.25		\$11.50 → \$17.25		\$11.50		\$11.50		\$11.50 → \$7.66			
Natl. Policy Response	Vigorous		Vigorous		Vigorous		Restrained		Restrained			
Natl. GNP Growth Rate 1976-1985 (%/yr.)	7.5%		6%		7.5%		6%		6%		1972 Base Year	
Fuel Type	Fossil Fuel	Elec.	Fossil Fuel	Elec.	Fossil Fuel	Elec.	Fossil Fuel	Elec.	Fossil Fuel	Elec.	Fossil Fuel	Elec.
Units	10^{18} Joules		10^{18} Joules		10^{18} Joules		10^{18} Joules		10^{18} Joules		10^{18} Joules	
Energy-Using Sectors												
Industrial (excluding feedstocks)	1.50	0.19	1.14	0.16	1.36	0.18	1.27	0.16	1.26	0.16	0.56	0.05
Transportation	1.49		1.21		1.36		1.32		1.28		0.55	
Commercial and Public		0.14		0.11		0.13		0.11		0.11		0.03
Residential	0.54	0.06	0.42	0.05	0.53	0.05	0.43	0.05	0.43	0.05	0.18	0.02
Fishing, Agriculture, Mining, Construction	0.01	0.03	0.01	0.02	0.01	0.03	0.01	0.02	0.01	0.02	0.01	0.01
Feedstocks	0.22		0.17		0.21		0.17		0.17		0.07	
Final Energy Demand (excluding processing losses)	3.76	0.42	2.95	0.34	3.47	0.39	3.20	0.34	3.15	0.34	1.37	0.11
Processing Losses	0.75	0.77	0.54	0.64	0.67	0.64	0.60	0.58	0.59	0.58	0.29	0.23
Primary Energy Input	4.51	1.19	3.49	0.98	4.14	1.03	3.80	0.92	3.74	0.92	1.66	0.34

*Oil price 1975 US\$ per barrel Arabian light crude oil FOB Persian Gulf.

**Low energy demand in Case E is a consequence of low GNP growth due to elimination of potential oil exports.

WAES NATIONAL ENERGY DEMAND STUDIES, 1972-1985
ENERGY GROWTH RATES SUMMARY

Country MEXICO

Scenario Assumptions										
WAES Case	A		B		C		D		E	
GNP 1976-1985	High		Low		High		Low		High	
Oil Price*	\$11.50 → \$17.25		\$11.50 → \$17.25		\$11.50		\$11.50		\$11.50 → \$7.66	
Natl. Policy Response	Vigorous		Vigorous		Vigorous		Restrained		Restrained	
Natl. GNP Growth Rate 1976-1985 (%/yr.)	7.5%		6%		7.5%		6%		6%	
Fuel Type	Fossil Fuel	Elec.	Fossil Fuel	Elec.	Fossil Fuel	Elec.	Fossil Fuel	Elec.	Fossil Fuel	Elec.
Units (%/yr. 1972-1985)	%		%		%		%		%	
Energy-Using Sectors										
Industrial (excluding feedstocks)	7.8	10.8	5.6	9.3	7.06	10.3	6.5	9.3	6.4	9.3
Transportation	7.9		6.2		7.2		6.9		6.7	
Commercial and Public		12.5		10.5		11.9		10.5		10.5
Residential	8.8	8.8	6.7	7.3	8.6	7.3	6.9	7.3	6.9	7.3
Fishing, Agriculture, Mining, Construction	0.0	8.8	0.0	5.4	0.0	8.8	0.0	5.4	0.0	5.4
Feedstocks	9.2		7.06		8.8		7.06		7.06	
Final Energy Demand (excluding processing losses)	8.08	10.86	6.08	9.07	7.41	10.23	6.74	9.07	6.61	9.07
Processing Losses	7.58	9.74	4.90	8.19	6.65	8.19	5.75	7.37	5.62	7.37
Primary Energy Input	7.99	10.12	5.88	8.48	7.28	8.90	6.58	7.96	6.45	7.96

*Oil price 1975 US\$ per barrel Arabian light crude oil FOB Persian Gulf.

WAES NATIONAL ENERGY DEMAND STUDIES, 1972-1985
SECTORAL PERCENTAGES OF TOTAL

Country MEXICO

Scenario Assumptions

WAES Case	A		B		C		D		E		1972 Base Year	
GNP 1976-1985	High		Low		High		Low		High			
Oil Price*	$11.50 → $17.25		$11.50 → $17.25		$11.50		$11.50		$11.50 → $7.66			
Natl. Policy Response	Vigorous		Vigorous		Vigorous		Restrained		Restrained			
Natl. GNP Growth Rate 1976-1985 (%/yr.)	7.5%		6%		7.5%		6%		6%			
Fuel Type	Fossil Fuel	Elec.	Fossil Fuel	Elec.	Fossil Fuel	Elec.	Fossil Fuel	Elec.	Fossil Fuel	Elec.	Fossil Fuel	Elec.
Units	%		%		%		%		%		%	
Energy-Using Sectors												
Industrial (excluding feedstocks)	39.8	45.2	38.6	47.06	39.1	46.1	39.6	47.06	40.0	47.06	40.8	45.4
Transportation	39.6		41.02		39.1		41.2		40.6		40.1	
Commercial and Public		33.3		32.3		33.3		32.3		32.3		27.2
Residential	14.3	14.2	14.2	14.7	15.2	12.8	13.4	14.7	13.6	14.7	13.1	18.1
Fishing, Agriculture, Mining, Construction	0.2	7.1	0.3	5.8	0.3	7.6	0.3	5.8	0.3	5.8	0.7	9.09
Feedstocks	5.8		5.7		6.05		5.3		5.4		5.1	
Final Energy Demand (excluding processing losses)	100.00	100.00	100.00	100.00	100.00	100.00	100.00	100.00	100.00	100.00	100.00	100.00
Processing Losses												
Primary Energy Input												

*Oil price 1975 US$ per barrel Arabian light crude oil FOB Persian Gulf.

WAES NATIONAL ENERGY DEMAND STUDIES, 1972-1985
SECTORAL PERCENTAGES OF TOTAL

Country MEXICO

Scenario Assumptions												
WAES Case	A		B		C		D		E			
GNP 1976-1985	High		Low		High		Low		High			
Oil Price*	$11.50 → $17.25		$11.50 → $17.25		$11.50		$11.50		$11.50 → $7.66			
Natl. Policy Response	Vigorous		Vigorous		Vigorous		Restrained		Restrained		1972 Base Year	
Natl. GNP Growth Rate 1976-1985 (%/yr.)	7.5%		6%		7.5%		6%		6%			
Fuel Type	Fossil Fuel	Elec.	Fossil Fuel	Elec.	Fossil Fuel	Elec.	Fossil Fuel	Elec.	Fossil Fuel	Elec.	Fossil Fuel	Elec.
Units	%		%		%		%		%		%	
Energy-Using Sectors												
Industrial (excluding feedstocks)	33.2	15.9	32.6	16.3	32.8	17.4	33.4	17.3	33.6	17.3	33.7	14.7
Transportation	33.04		34.6		32.8		34.7		34.2		33.1	
Commercial and Public		11.7		11.2		12.6		11.9		11.9		8.8
Residential	11.9	5.04	12.03	5.1	12.8	4.8	11.3	5.4	11.5	5.4	10.8	5.8
Fishing, Agriculture, Mining, Construction	0.2	2.5	0.2	2.04	0.2	2.9	0.2	2.1	0.2	2.1	0.6	2.9
Feedstocks	4.8		4.8		5.07		4.4		4.5		4.2	
Final Energy Demand (excluding processing losses)	83.37	35.29	84.53	34.69	83.82	37.86	84.21	36.96	84.22	36.96	82.53	32.35
Processing Losses	16.63	64.71	15.47	65.31	16.18	62.14	15.79	63.04	15.78	63.04	17.47	67.65
Primary Energy Input	100	100	100	100	100	100	100	100	100	100	100	100

*Oil price 1975 US$ per barrel Arabian light crude oil FOB Persian Gulf.

WAES NATIONAL ENERGY DEMAND STUDIES, 1972-1985
1985 ECONOMIC ACTIVITY OF MAJOR ENERGY-USE CATEGORIES

Country MEXICO

Scenario Assumptions

WAES Case	A		B		C		D		E			
GNP 1976-1985	High		Low		High		Low		High			
Oil Price*	\$11.50 → \$17.25		\$11.50 → \$17.25		\$11.50		\$11.50		\$11.50 → \$7.66			
Natl. Policy Response	Vigorous		Vigorous		Vigorous		Restrained		Restrained			
Natl. GNP Growth Rate 1976-1985 (%/yr.)	7.5%		6%		7.5%		6%		6%		1972 Base Year	
Fuel Type	Fossil Fuel	Elec.	Fossil Fuel	Elec.	Fossil Fuel	Elec.	Fossil Fuel	Elec.	Fossil Fuel	Elec.	Fossil Fuel	Elec.
Units	10^{18} Joules		10^{18} Joules		10^{18} Joules		10^{18} Joules		10^{18} Joules		10^{18} Joules	
Major Energy-Use Categories												
Industrial excluding feedstock	1.5	0.1	1.1	0.16	1.3	0.1	1.2	0.1	1.2	0.1	0.5	0.05
Transportation	1.4		1.2		1.3		1.3		1.2		0.5	
Commercial and Public		0.1		0.1		0.1		0.1		0.1		0.03
Feedstocks	0.2		0.1		0.2		0.1		0.1		0.07	
Residential	0.5	0.06	0.4	0.05	0.5	0.05	0.4	0.05	0.4	0.05	0.1	0.02

*Oil price 1975 US\$ per barrel Arabian light crude oil FOB Persian Gulf.

WAES NATIONAL ENERGY DEMAND STUDIES, 1972-1985
FINAL ENERGY DEMAND OF MAJOR ENERGY-USE CATEGORIES

Country MEXICO

Scenario Assumptions												
WAES Case	A		B		C		D		E			
GNP 1976-1985	High		Low		High		Low		High			
Oil Price*	$11.50 → $17.25		$11.50 → $17.25		$11.50		$11.50		$11.50 → $7.66			
Natl. Policy Response	Vigorous		Vigorous		Vigorous		Restrained		Restrained			
Natl. GNP Growth Rate 1976-1985 (%/yr.)	7.5%		6%		7.5%		6%		6%		1972 Base Year	
Fuel Type	Fossil Fuel	Elec.	Fossil Fuel	Elec.	Fossil Fuel	Elec.	Fossil Fuel	Elec.	Fossil Fuel	Elec.	Fossil Fuel	Elec.
Units	10^{18} Joules		10^{18} Joules		10^{18} Joules		10^{18} Joules		10^{18} Joules		10^{18} Joules	
Major Energy-Use Categories												
Industrial excluding feedstocks	1.50	0.19	1.14	0.16	1.36	0.18	1.27	0.16	1.26	0.16	0.56	0.05
Transportation	1.49		1.21		1.36		1.32		1.28		0.55	
Residential	0.54	0.06	0.42	0.05	0.53	0.05	0.43	0.05	0.43	0.05	0.18	0.02
Feedstocks	0.22		0.17		0.21		0.17		0.17		0.07	
Commercial and Public		0.14		0.11		0.13		0.11		0.11		0.03

*Oil price 1975 US$ per barrel Arabian light crude oil FOB Persian Gulf.

WAES NATIONAL ENERGY DEMAND STUDIES
1985 CASE COMPARISONS

Country MEXICO

Scenario Cases Compared	GNP (Scenario Variables)	Oil Price (Scenario Variables)	National Policy Response (Scenario Variables)	Energy-Use Category	Comparison elasticity units = %
C	7.5%	$11.50	Vigorous	INDUSTRIAL	0.195
A	7.5%	17.25	Vigorous		
C	7.5%	11.50	Vigorous	TRANSPORTATION	0.191
A	7.5%	17.25	Vigorous		
C	7.5%	11.50	Vigorous	COMMERCIAL AND PUBLIC	0.545
A	7.5%	17.25	Vigorous		
C	7.5%	11.50	Vigorous	RESIDENTIAL	0.553
A	7.5%	17.25	Vigorous		
C	7.5%	11.50	Vigorous	FINAL ENERGY DEMAND	0.541
A	7.5%	17.25	Vigorous		

$$\text{elasticity} = \frac{\dfrac{E_C - E_A}{E_A}}{\dfrac{P_C - P_A}{P_C}}$$

$$\%\ \text{change} = \frac{E_C - E_A}{E_C}$$

$$\text{total change} = E_C - E_A$$

WAES NATIONAL ENERGY DEMAND STUDIES
1985 CASE COMPARISONS

Country MEXICO

Scenario Cases Compared	Scenario Variables: GNP	Oil Price	National Policy Response	Energy-Use Category	Comparison (Total change) units = 10^{18} J
B	6%	$17.25	Vigorous	INDUSTRIAL	-0.39
A	7.5%	17.25	Vigorous		
B	6%	17.25	Vigorous	TRANSPORTATION	-0.28
A	7.5%	17.25	Vigorous		
B	6%	17.25	Vigorous	COMMERCIAL AND PUBLIC	-0.03
A	7.5%	17.25	Vigorous		
B	6%	17.25	Vigorous	RESIDENTIAL	-0.13
A	7.5%	17.25	Vigorous		
B	6%	17.25	Vigorous	FINAL ENERGY DEMAND	-0.89
A	7.5%	17.25	Vigorous		

$$\text{elasticity} = \frac{\dfrac{E_C - E_A}{E_A}}{\dfrac{P_C - P_A}{P_C}}$$

$$\%\ \text{change} = \frac{E_C - E_A}{E_C}$$

$$\text{total change} = E_C - E_A$$

WAES NATIONAL ENERGY DEMAND STUDIES
1985 CASE COMPARISONS

Country MEXICO

Scenario Cases Compared	Scenario Variables: GNP	Oil Price	National Policy Response	Energy-Use Category	Comparison Total change units = 10^{18} J
C	7.5%	$11.50	Vigorous	INDUSTRIAL	-0.15
A	7.5%	17.25	Vigorous		
C	7.5%	11.50	Vigorous	TRANSPORTATION	-0.13
A	7.5%	17.25	Vigorous		
C	7.5%	11.50	Vigorous	COMMERCIAL AND PUBLIC	-0.03
A	7.5%	17.25	Vigorous		
C	7.5%	11.50	Vigorous	RESIDENTIAL	-0.13
A	7.5%	17.25	Vigorous		
C	7.5%	11.50	Vigorous	FINAL ENERGY DEMAND	-0.89
A	7.5%	17.25	Vigorous		

$$\text{elasticity} = \frac{\frac{E_C - E_A}{E_A}}{\frac{P_C - P_A}{P_C}}$$

$$\%\ \text{change} = \frac{E_C - E_A}{E_C}$$

$$\text{total change} = E_C - E_A$$

WAES NATIONAL ENERGY DEMAND STUDIES
1985 CASE COMPARISONS

Country MEXICO

Scenario Cases Compared	GNP (Scenario Variables)	Oil Price (Scenario Variables)	National Policy Response (Scenario Variables)	Energy-Use Category	Comparison (% change) units = %
B	6%	$17.25	Vigorous	INDUSTRIAL	-0.3000
A	7.5%	17.25	Vigorous		
B	6%	17.25	Vigorous	TRANSPORTATION	-0.2314
A	7.5%	17.25	Vigorous		
B	6%	17.25	Vigorous	COMMERCIAL AND PUBLIC	-0.2727
A	7.5%	17.25	Vigorous		
B	6%	17.25	Vigorous	RESIDENTIAL	-0.2766
A	7.5%	17.25	Vigorous		
B	6%	17.25	Vigorous	FINAL ENERGY DEMAND	-0.2705
A	7.5%	17.25	Vigorous		

$$\text{elasticity} = \frac{\dfrac{E_C - E_A}{E_A}}{\dfrac{P_C - P_A}{P_C}}$$

$$\%\ \text{change} = \frac{E_C - E_A}{E_C}$$

$$\text{total change} = E_C - E_A$$

WAES NATIONAL ENERGY DEMAND STUDIES
1985 CASE COMPARISONS

Country MEXICO

Scenario Cases Compared	GNP	Oil Price	National Policy Response	Energy-Use Category	Comparison (% change) units = %
	Scenario Variables				
C	7.5%	$11.50	Vigorous	INDUSTRIAL	-0.0974
A	7.5%	17.25	Vigorous		
C	7.5%	11.50	Vigorous	TRANSPORTATION	-0.0956
A	7.5%	17.25	Vigorous		
C	7.5%	11.50	Vigorous	COMMERCIAL AND PUBLIC	-0.0769
A	7.5%	17.25	Vigorous		
C	7.5%	11.50	Vigorous	RESIDENTIAL	-0.0345
A	7.5%	17.25	Vigorous		
C	7.5%	11.50	Vigorous	FINAL ENERGY DEMAND	-0.0829
A	7.5%	17.25	Vigorous		

$$\text{elasticity} = \frac{\dfrac{E_C - E_A}{E_A}}{\dfrac{P_C - P_A}{P_C}}$$

$$\%\ \text{change} = \frac{E_C - E_A}{E_C}$$

$$\text{total change} = E_C - E_A$$

NATIONAL ENERGY DEMAND PROJECTIONS

Country MEXICO

Year 1972

WAES Case Base Year

Summary 10^{18} Joules

Category	Oil	Gas	Coal	Electricity	Other	Total by Category (Excluding Processing Losses)
Transportation	0.55					0.55
Agriculture, Mining, and Construction	0.01			0.01		0.02
Commercial				0.03		0.03
Public						
Industrial	0.14	0.30	0.12	0.05		0.61
Residential	0.18			0.02		0.20
Total Delivered Energy by Fuel	0.88	0.30	0.12	0.11		1.41
Processing Losses by Fuel	1.16	0.13		0.23		0.52
Gross Input by Fuel	1.04	0.43	0.12	0.34		1.93

Gross Energy Input (Add across bottom row)

NATIONAL ENERGY DEMAND PROJECTIONS

Country MEXICO

Year 1972

WAES Case Base Year

Transportation

	(A)	(B)	(C)	(D)	(E)	(F)	(G)	
	Population/Auto	Distance/Auto/Yr.	Load Factor	Fraction Urban or Rural	Energy/Distance	(10^{18}J) Fossil Fuel	Electricity	
Urban Auto	35.15	13 500		0.65	8.67 10^3J/Km.	0.18		(1)
Rural Auto				0.35				(2)
	Activity Level(1)	% Electrified	Load Factor	Fossil Fuel/ Activity Level(2)	Electricity/Activity			
Urban Bus						0.012		(3)
Intercity Bus	40.60			6.64		0.026		(4)
Rail (Passenger)	4.06			7.39		0.003		(5)
Air (Passenger, Domestic)	2.68							(6)
Air (Passenger, International)	9.28					0.021		(7)
Ship (Passenger)								(8)
Rail (Freight)								(9)
Air (Freight, Domestic)								(10)
Air (Freight, International)								(11)
Truck (Local)						0.106		(12)
Truck (Intercity)	29.80					0.197		(13)
Ship (Freight, Domestic)						0.009		(14)
Ship (Freight, International)								(15)
Miscellaneous								(16)
					Total	0.554		

(1) 10^9 Passenger kilometers or 10^9 ton kilometers

(2) 10^5 Joules/passenger kilometer

NATIONAL ENERGY DEMAND PROJECTIONS

Country MEXICO

Year 1972

WAES Case BASE YEAR

Agriculture, Mining, and Construction

	(A)	(B)	(C)	(D)	(E)	(F)	(G)	(H)	
Agriculture, Forestry, and Fishing (ISIC Major Division 1)	Value Added	Energy/Value Added	Area Harvested	Energy/ Area	Physical Output	Energy/ Output	Fossil Fuel	Electricity	
Farm Machinery							10^{18} Joules		(17)
Greenhouses			14.44 (1)		70.94 (2)				(18)
Other Agricultural Uses									(19)
Forestry (ISIC 12)					4.09 (3)				(20)
Fishing Boats (ISIC 13)									(21)
Other Fishing Use (ISIC 13)					0.31 (2)				(22)
Total Fertilizer (tons)		N 1.09 (2)	P 0.59 (2)	K		Total	0.0059	0.0059	

	(A)	(B)	(C)	(D)	(E)	(F)	
Mining	Value Added	Energy/Value Added	Physical Output	Energy/Physical Output	Fossil Fuel	Electricity	
Handling and Internal Transportation			12.59 (2)				(23)
Other Energy Use							(24)
				Total		0.0023	

	(A)	(B)	(C)	(D)	(E)	(F)	
Construction	Value Added	Energy/Value Added	Physical Output	Energy/Physical Output	Fossil Fuel	Electricity	
Handling and Internal Transportation							(25)
Other Energy Use							(26)
				Total	N.A	N.A	
				Agriculture, Mining, and Construction Total	0.0059	0.0082	

(1) 10^6 Ha.

(2) 10^6 Ton.

N.A. = Not available

NATIONAL ENERGY DEMAND PROJECTIONS

Country MEXICO

Year 1972

WAES Case BASE YEAR

Commercial and Public

	(A)	(B)	(C)	(D)	(E)	
Commercial	Floor Area	Heat Gain or Loss/Area	System Efficiency	Fossil Fuel	Electricity	
Space Heat (Fossil Fuel)	N				10^{18} Joules	(27)
Space Heat (Electricity)	N					(28)
Air Conditioning (Fossil Fuel)	N					(29)
Air Conditioning (Electricity)	N					(30)
		Energy/Area	Electricity/Area			
Lighting					N.A.	(31)
Miscellaneous					N.A.	(32)
Presently Unknown						(33)
			Total		0.034	
Public	Floor Area	Heat Gain or Loss/Area	System Efficiency	Fossil Fuel	Electricity	
Space Heat (Fossil Fuel)	N					(34)
Space Heat (Electricity)	N					(35)
Air Conditioning (Fossil Fuel)	N					(36)
Air Conditioning (Electricity)	N					(37)
		Energy/Area	Electricity/Area			
Lighting					0.003	(38)
Miscellaneous						(39)
Presently Unknown						(40)
			Total		0.003	
			Commercial and Public Total		0.037	

N = Negligible

N.A. = Not available

NATIONAL ENERGY DEMAND PROJECTIONS

Country MEXICO

Year 1972

WAES Case BASE YEAR

Industrial

	(A) Value Added	(B) Energy/Value Added	(C) Physical Output	(D) Energy/Output	(E) % Exported	(F) Fossil Fuel	(G) Electricity	(H) Feedstocks	
	$(10^6\$)$	$10^6 J/\$$	10^6Ton.			$10^{18}J$	$10^{18}J$	$10^{18}J$	
Iron and Steel (ISIC 371)	368	280	4.39			0.103			(41)
Nonferrous Metals (ISIC 372)									(42)
Paper and Allied (ISIC 34)	179		1.24			0.029			(43)
Chemicals and Allied (ISIC 35 minus ISIC 353) (1)			2.60			0.090		0.058	(44)
Mineral Products (ISIC 36) (2)			11.51			0.034			(45)
Food and Related (ISIC 31) (3)			6.26			0.063			(46)
Transport Eqpt. (ISIC 384)			8.75			0.007			(47)
Cement						0.063			(48)
Glass						0.050			(49)
Misc. Manufacturing						0.121			(50)
					Total	0.560	0.053	0.058	

N.E.C. = Not Elsewhere Classified

(1) Includes Petrochemistry

(2) Includes Nonferrous Metals

(3) Includes a high percentage of Sugar Industry

NATIONAL ENERGY DEMAND PROJECTIONS

Country MEXICO

Year 1972

WAES Case BASE YEAR

Residential

Population 53. 441 (1) Persons/Dwelling 9.06 Dwellings 5.9 (1)

Central Space Heating		(A) Dwellings	(B) Heat Loss	(C) System Efficiency	(D)	(E) Fossil Fuel	(F) Electricity	
Single Family	Fossil Fuels	N					10^{18} Joules	(51)
	Electricity	N						(52)
Town or Row Houses	Fossil Fuels	N						(53)
	Electricity	N						(54)
Apartments	Fossil Fuels	N						(55)
	Electricity	N						(56)
		% Saturation (Fossil Fuels)	% Saturation (Electricity)	Energy/Unit (Fossil Fuels)	Energy/Unit (Electricity)			
Room Heating		N	N					(57)
Air Conditioning (Window)		N	0.32					(58)
Air Conditioning (Central)		N	N					(59)
Water Heating		1.81	N					(60)
Refrigerators		0.0	4.74					(61)
Lighting								(62)
Cooking		9.08						(63)
Dishwashers			N					(64)
Freezers								(65)
Television			7.4					(66)
Other Key Appliances								(67)
Misc. Appliances			N					(68)
Potential Appliances			N					(69)
					Total	0.177	0.0160	

(1) 10^6

N = Negligible

NATIONAL ENERGY DEMAND PROJECTIONS

Country MEXICO

Year 1985

WAES Case A

Summary 10^{18} Joules

Category	Oil	Gas	Coal	Electricity	Other	Total by Category (Excluding Processing Losses)
Transportation	1.49					1.49
Agriculture, Mining, and Construction	0.01			0.03		0.04
Commercial				0.13		0.13
Public				0.01		0.01
Industrial	0.44	0.84	0.22	0.19		1.69
Residential	0.54			0.06		0.60
Total Delivered Energy by Fuel	2.48	0.84	0.22	0.42		3.96
Processing Losses by Fuel	0.49	0.26		0.77		1.52
Gross Input by Fuel	2.97	1.10	0.22	1.19		5.48

Gross Energy Input (Add across bottom row)

NATIONAL ENERGY DEMAND PROJECTIONS

Country MEXICO

Year 1985

WAES Case B

Summary 10^{18} Joules

Category	Oil	Gas	Coal	Electricity	Other	Total by Category (Excluding Processing Losses)
Transportation	1.21					1.21
Agriculture, Mining, and Construction	0.01			0.02		0.03
Commercial				0.10		0.10
Public				0.01		0.01
Industrial	0.27	0.65	0.22	0.16		1.30
Residential	0.42			0.05		0.47
Total Delivered Energy by Fuel	1.91	0.65	0.22	0.34		3.12
Processing Losses by Fuel	0.35	0.19		0.64		1.18
Gross Input by Fuel	2.26	0.84	0.22	0.98		4.30

Gross Energy Input (Add across bottom row)

NATIONAL ENERGY DEMAND PROJECTIONS

Country MEXICO

Year 1985

WAES Case C

Summary 10^{18} Joules

Category	Oil	Gas	Coal	Electricity	Other	Total by Category (Excluding Processing Losses)
Transportation	1.36					1.36
Agriculture, Mining, and Construction	0.01			0.03		0.04
Commercial				0.12		0.12
Public				0.01		0.01
Industrial	0.36	0.78	0.22	0.18		1.54
Residential	0.53			0.05		0.58
Total Delivered Energy by Fuel	2.26	0.78	0.22	0.39		3.65
Processing Losses by Fuel	0.42	0.25		0.64		1.31
Gross Input by Fuel	2.68	1.03	0.22	1.03		4.96

Gross Energy Input (Add across bottom row)

NATIONAL ENERGY DEMAND PROJECTIONS

Country MEXICO

Year 1985

WAES Case D

Summary 10^{18} Joules

Category	Oil	Gas	Coal	Electricity	Other	Total by Category (Excluding Processing Losses)
Transportation	1.32					1.32
Agriculture, Mining, and Construction	0.01			0.02		0.03
Commercial				0.10		0.10
Public				0.01		0.01
Industrial	0.33	0.72	0.22	0.16		1.43
Residential	0.42			0.05		0.48
Total Delivered Energy by Fuel	2.09	0.72	0.22	0.34		3.37
Processing Losses by Fuel	0.38	0.22		0.58		1.18
Gross Input by Fuel	2.47	0.94	0.22	0.92		4.55

Gross Energy Input (Add across bottom row)

NATIONAL ENERGY DEMAND PROJECTIONS

Country MEXICO

Year 1985

WAES Case E

Summary 10^{18} Joules

Category	Oil	Gas	Coal	Electricity	Other	Total by Category (Excluding Processing Losses)
Transportation	1.28					1.28
Agriculture, Mining, and Construction	0.01			0.02		0.03
Commercial				0.10		0.10
Public				0.01		0.01
Industrial	0.32	0.72	0.22	0.16		1.42
Residential	0.43			0.05		0.48
Total Delivered Energy by Fuel	2.04	0.72	0.22	0.34		3.32
Processing Losses by Fuel	0.37	0.22		0.58		1.17
Gross Input by Fuel	2.41	0.94	0.22	0.92		4.49

Gross Energy Input (Add across bottom row)

11 The Netherlands

11.1 INTRODUCTION

11.1.1 Organization

Participation in WAES was made possible by cooperation between the Ministry of Economic Affairs and the Future Shape of Technology Foundation, and by the support of Centraal Planbureau, ESTEL Hoesch-Hoogovens, AKZO-Zoutchemie, Paktank, OGEM Holding, and Verenigde Machinefabrieken. Supporting manpower and finance became available rather late in the program, and for this reason only two out of five scenarios could be worked out completely.

This report has been drawn up by a working party, consisting of:

A.G. Melman, ESTEL Hoesch-Hoogovens

R. Van Der Sande, ESTEL Hoesch-Hoogovens

A.C. Sjoerdsma, Future Shape of Technology Foundation

M.J. Stoffers, Centraal Planbureau.

11.1.2 DATA

The data and supporting assumptions represent the professional opinion of the working party and do not necessarily reflect official views.

The data have been partly derived from "Energy Conservation: Ways and Means" (Future Shape of Technology Publication no. 19, The Hague, 1974) and have been supplemented by data from diverse sources and by personal information from a number of authorities and specialists.

The energy demand data of Cases C and D can be considered "hard." Data on the fuel mix of total demand and processing losses are based on a preliminary manual supply-demand integration and may be subject to revision.

11.2 METHODOLOGY

11.2.1 General Outline

The WAES scenario approach assumes a dependence of energy demand on only three factors, viz. economic growth, oil price, and national policy response.

The high and low growth rates of gross world product adopted by WAES have been converted into high and low growth variants of The Netherlands' gross national product. These conversions have been made by means of the national econometric model of the Centraal Planbureau.

Only two demand projections (WAES Cases C and D) were developed to 1985 from base year 1972, using projections of activity levels and technical coefficients.

The activity levels (value added) were estimated by means of the sector activity model of the Centraal Planbureau. The other estimates were either derived from existing, independent projections or based on judgment.

Technical coefficients are the same as in base year 1972, except for those which are directly influenced by oil price and national policy response.

Total energy demand projections for WAES Cases A, B, and E have been derived from Cases C and D through the use of average European elasticity coefficients[1] and must be considered soft.

11.2.2 GNP Growth Rates

WAES scenarios assume for the gross world product between 1972 and 1985 a high growth variant (4.3 percent per year), and a low growth variant (3.2 percent per year), both in real terms.

In order to find the relationship between the growth of GWP and the growth of The Netherlands' GNP, it is necessary to allot the world growth to a large number of countries which are important importers of The Netherlands' products. Next, for each of those countries, the import growth must be calculated by means of estimated import elasticities for the GNP. Finally, these import growths are weighted with regard to the country's share of The Netherlands' export.

The weighted world trade thus obtained is an

1. O. Bernardini, "On the Consistency of WAES Demand Projections," Summary Report of WAES Associates' Meeting, Annex B, January 10, 1976, Cambridge, England.

important factor for the calculation of The Netherlands' export and GNP growth rates. With this method, the weighted world trade in the high and low growth cases showed a calculated growth of 6.8 and 4.7 percent per year, respectively, during the period 1972 to 1985.

Both WAES growth variants of GWP (4.3 and 3.2 percent) were compared with a central variant of the GNP growth to 1980 and a preliminary estimate to 1985 which were recently published by the Centraal Planbureau. The central variant assumes a GNP growth of 3.8 percent per year during the period 1975 to 1980. The growth between 1980 and 1985 is preliminarily assumed to be 3 percent per year. Using these assumptions, the average annual growth of GNP between 1972 and 1985 is calculated to be 3 percent. This average GNP growth rate goes with an average rate of industrial production growth of 3.4 percent per year.

The central variant to 1980 is based on an assumed growth of weighted world trade of 7.5–8 percent per year during the 1975 to 1980 period, whereas for the 1980 to 1985 period a world trade growth of approximately 6 percent per year is assumed. These assumptions can be combined into a growth figure for weighted world trade of 6.1 percent for the period 1972 to 1985.

The relationship between weighted world trade and industrial production can be described as follows. An average 1 percent increase (decrease) of world trade during a certain number of years implies an increase (decrease) of industrial production of 0.6 percent (0.7 percent) over this same period. This relationship can be used to derive the level of national industrial production from the WAES high and low growth assumptions as shown in Table 11.1. Or, in summary,

GWP growth of 4.3% → Netherlands GNP growth of 3.4%

GWP growth of 3.2% → Netherlands GNP growth of 2.2%.

11.2.3 Oil Price

In calculating the growth of GNP, the influence of oil price on economic growth has not been taken into account. This influence is difficult to assess, the more so where there is an interdependence with production and export of The Netherlands' natural gas.

The oil price has an influence on technical coefficients. A higher oil price will promote a more efficient utilization of energy, especially in those sectors where the share of accumulated energy costs in the product cost is relatively important. Such sectors are the primary metals sector, the chemicals sector, the transportation sector, electricity production, and the public sector.

The influence of oil price must be assessed by judgment, not by application of well-known price elasticities, since these are almost invariably based on marginal price fluctuations. In this study technical coefficients for all sectors have been developed for the medium oil price cases (C and D). In addition, technical coefficients for some of the sectors have been estimated for the high and low price cases (A, B, and E).

11.2.4 National Policy Response

The efficiency of energy use depends largely on the energy price level. As already pointed out, an increase in the price of energy would lead to a more efficient use of energy. If the price of energy is low, energy may not be used efficiently.

Policy response will anticipate these price effects. The potential for energy saving is often well-known—in The Netherlands, through the work done by the Future Shape of Technology Foundation—although the application of suitable policies may be rather difficult. In general, the intensity of policy measures varies with the energy price level and the economic situation. Vigorous policy, therefore, could be more vigorous for high price cases coupled with high growth rates than for high price cases at low growth rates.

The vigorous and restrained policies incorporated in this study are summarized in Table 11.2.

11.3 ANALYSIS OF DEMAND PROJECTIONS

11.3.1 General Outline

Two of the five WAES 1985 scenario cases have been analyzed in detail: Case C (high economic growth, medium oil price, and vigorous policy response) and Case D (low economic growth, medi-

Table 11.1
Derivation of relationship between GWP growth and Netherlands GNP growth for the WAES high and low world GWP growth assumptions, 1972–1985

	High Growth Case[1]	Low Growth Case[1]
WAES GWP Growth Rate	4.3	3.2
Netherlands WAES Weighted World Trade Growth	6.8	4.7
Netherlands Central Variant World Trade Growth	6.1	6.1
Netherlands Central Variant GNP Growth	3.0	3.0
Netherlands Central Variant Industrial Growth	3.4	3.4
Netherlands WAES Industrial Growth	3.4 + (6.8 - 6.1) × 0.6 = 3.8	3.4 - (6.1 - 4.7) × 0.7 = 2.5
Netherlands WAES GNP Growth	3.0 + (3.8 - 3.4) = 3.4	3.0 - (3.4 - 2.5) = 2.2

1. All figures are percent per year.

Table 11.2
Components of national policy response for the five WAES scenarios to 1985

	Vigorous Policy			Restrained Policy
	Case A	Case B	Case C	(Cases D and E)
General				
Information and advice	x	x	x	x
Regulations	x			
Transportation Sector				
Tax incentives in favor of diesel engines	xx	x	xx	
Heavier taxation of large cars	xx	x	xx	
Increase of gasoline (petrol) tax	x		x	
Speed limit of 100 km/h	xx	xx	xx	x
Encouragement of car pooling	xx	x	xx	
Information on driving styles and habits	xx	xx	xx	x
Residential, Public, and Commercial Sectors				
Review of building codes	x	x	x	
Subsidies for better insulation of existing buildings	x	x	x	
Sudsidies for improved systems in commercial buildings	x	x	x	
Changes in tariff structure	x	x	x	x
Energy labeling; regulations for electric and gas-heating appliances	x	x	x	x
Minimum qualifications for contractors	x	x	x	x
Information and advice to influence heating practice	x	x	x	x
Industrial				
Changes in tariff structure	x	x	x	x
Information and advice	x	x	x	x

um oil price, and restrained policy response). For the remaining cases—high growth A and E and low growth B—final energy demand has been calculated from Cases C and D using European price and growth elasticities α (= –0.15) and γ (=0.55) in the following equations:

$$E_A = E_C\left(\frac{P_A}{P_C}\right)^{\alpha}\left(\frac{G_A}{G_C}\right)^{\gamma},$$

$$E_B = E_C\left(\frac{P_B}{P_C}\right)^{\alpha}\left(\frac{G_B}{G_C}\right)^{\gamma},$$

$$E_E = E_D\left(\frac{P_E}{P_D}\right)^{\alpha}\left(\frac{G_E}{G_D}\right)^{\gamma},$$

in which E is final energy use, P is oil price, and G is economic growth.

Conclusions drawn from these calculations must be considered soft, the more so where the assumed European elasticities are not consistent with the elasticities implicit in Cases C and D.

The energy demand of the industrial sector has been calculated for all WAES 1985 scenario cases.

A summary of the energy demand projections for all five 1985 scenarios can be found in Section 11.5. It may be seen that economic growth is the most important parameter. The high growth cases show higher energy consumptions than the low growth cases.

Cases C and D can be considered midway between Cases B and E. In cases with the same economic growth, a lower energy price results in an increase of energy consumption (E vs. C and D vs. B).

A rising oil price combined with a vigorous policy response results in high energy consumption at high economic growth and low energy consumption at low economic growth due to the combined effects of activity level and thrift on energy use (A vs. B).

The effects of changing the various scenario variables on the final energy demand are summarized in Table 11.3. It should be noted that the processing losses were determined in a tentative way. These values are subject to revision after the final supply-demand integrations.

From the energy growth rates summary for 1972 to 1985 (in Section 11.5), it appears that the growth of energy demand will be similar to the economic growth in Case C. Only in the case of high economic growth and low oil price (Case E) will energy demand grow faster than the economy. Moreover, economic growth has its main influence on energy demand in the industrial sector, and electricity demand will grow faster than total fossil fuel demand, especially in the residential sector.

The energy use summary shows that, compared with the base year, 1972, there is a certain shift from the transportation and space heating sectors toward the industrial sector in high growth cases (for instance, Case C). Later on, it will be explained that this is caused by the decreasing importance of international bunkers as well as by the overproportional growth rate of industrial energy use.

11.3.2 Economic Activities

As already noted, energy demand under the WAES assumptions is a function of activity level and energy efficiency. For the purpose of this study, the activity levels in all scenarios are assumed not to vary with oil price.

A number of activity levels representing the major portion of energy end-users have been derived from the outputs of the sector activity model of the Centraal Planbureau. The results for both high and low economic growth are presented in Table 11.4.

Existing experience with export elasticity in each industrial branch with respect to the weighted world trade pattern has been used to calculate the resulting export volumes; and the existing input-output analysis has been used to calculate the effects of changing export volumes on production.

The table shows clearly the difference in sensitivity among industrial branches to the development of the world economy. It shows that the chemicals, metallurgical, and electrotechnical industries are very sensitive to world trade growth, while agriculture, construction, and various services are much less influenced by this factor. Some minor activities, all included in the transportation and space heating sectors, are directly or indirectly influenced by economic growth as well as by national policies. These activities could be derived from the major activities or from other existing

Table 11.3
Effects of scenario variables on final energy demand

Cases	Scenario Variable	Change	Impact: Trend	Impact: Resultant
C → D	GNP (%)	3.4 → 2.2	–	-17.9%
	price ($)	11.50		
	policy	vig. → restr.	+	
C → E	GNP (%)	3.4		+11.0%
	price ($)	11.50 → 7.66	+	
	policy	vig. → restr.	+	
D → B	GNP (%)	2.2		– 9.7%
	price ($)	11.50 → 17.25	–	
	policy	restr. → vig.	+	
C → A	GNP (%)	3.4		– 5.9%
	price ($)	11.50 → 17.25	–	
	policy	vig.		

Table 11.4
Gross value added (market prices) per sector in 1972 prices (DFl)[1]

	1972	1975	1985 High Growth	1985 Low Growth	1972–1985 High Growth (%)	1972–1985 Low Growth (%)
Agriculture, Forestry, and Fishing	8,038	9,258	13,027	12,840	3.8	3.7
Industry	43,113	42,944	76,054	5,993	4.5	2.0
Mining	2,744	3,898	3,872	3,585	3.3	2.7
Public Utilities	3,611	4,263	6,566	5,872	4.7	3.8
Building Industry	13,012	11,711	12,186	11,884	−0.5	−0.7
Distribution and Trade	22,901	24,394	37,534	33,202	3.8	2.9
Sea and Air Transport	1,873	2,152	4,134	2,787	6.3	3.1
Other Transport	8,691	9,049	13,852	11,722	3.7	2.3
Various Services	22,124	24,299	36,577	35,169	3.9	3.6

1. Because of the mechanical way of calculating and the preliminary nature of the results, these numbers may not be considered official estimates.

projections, modified by judgment or changed basic assumptions such as the rate of population growth.

Further details will be given in the next section, within the framework of a closer examination of energy demand by sector.

11.3.3 Energy Demand by Sector

Transportation

The most important transportation line items for The Netherlands are "auto" and "ship (freight, international)."

The number of automobiles will continue to increase until 1985. It will be higher in Case C (high economic growth) than in Case D (low economic growth): 5.6 million vs. 4.9 million. The number of passenger-kilometers, however, shows hardly any difference: 80 x 10^9 versus 75.5 x 10^9. This can be explained as public response to economic growth. The trend toward more efficient use of automobile fuel is caused by both higher energy price and government policy. It will be attained primarily by more thoughtful use, car pooling, and a further penetration of the diesel engine.

An energy-saving shift toward public transportation is not expected. Of the several modes of public transport, only rail transport will maintain its share, due to promotion measures such as higher frequency, new railways, etc. Technical coefficients are expected not to change very much from the 1972 figures.

The projections of energy use in air transport are based on calculations made by Royal Dutch

Airlines and are in accordance with other recent studies in this field. The growth rate of air travel will be considerably lower than earlier estimates. Nevertheless, it is expected that the number of passengers carried will increase by a factor of 2.85 to 16–18 million. The number of takeoffs—to which the energy demand is related—will increase from 88,500 in 1972 to 115,500 in 1985.

In freight traffic, trucks and inland shipping dominate the energy demand. Growth of these activities is based on production growth in the industrial and building sectors.

International ship freight consists of all bunkers for ocean-going trade taken on board within The Netherlands. They are included in the total end-use energy because of the historical position of The Netherlands in international traffic and trade. It is expected that bunker trade will grow at a considerably slower rate than in the past due to the anticipated slowdown in the growth of refinery capacity.

Table 11.5 shows the growth of energy demand in the transportation sector between 1972 and 1985. The numbers on bunkers are very rough estimates.

Commercial and Public

Energy use in this sector consists mainly of power and heating for a variety of buildings: offices, department stores, shops, hospitals, schools, etc. The number of employees has been chosen as indicative of the activity level because there are standards which prescribe a certain floor area per employee. Estimates have been made by the Centraal Planbureau for both the high and the low economic growth cases. The estimated numbers are 3.21 million employees in Case C and 2.96 million employees in Case D.

The present oil price and certain government policy incentives have already stimulated thermal insulation of buildings. Therefore, the energy coefficients will be improved in comparison with base year 1972. The improvement will amount to 10 percent for fossil fuel and 2 percent for electricity as compared with 1972 (64.6×10^3 J/employee vs. 16.3×10^3 J/employee). The total potential energy saving in this sector (20 percent on fossil fuel and 4 percent on electricity) has been allotted to Cases A and B (high oil price).

The growth of energy demand in the commercial and public sector from 1972 to 1985 has been summarized in Table 11.6.

Industry

The energy demand of the industrial sector has been completed for all five scenario cases. Activity levels have been presented earlier in Table 11.4.

In a separate study, the improvement of energy consumption in the different branches has been examined in the cases of a high and a medium oil price. A distinction has been made between energy for space heating and process heat and power. The results are summarized in Table 11.7.

Table 11.8 presents the development of total energy demand of the industrial sector to 1985. It appears that industrial energy consumption varies considerably with changes in economic growth.

The impact of prices on energy use is rather limited in general. The main contribution to the growth of energy demand comes from the chemicals industry.

Table 11.5
Energy demand growth in the transportation sector, 1972–1985

	1972	1985[1]	
	10^{15} J	Case C	Case D
Automobile	150.4	140	150
Rail (Passengers & Freight)	5.2	110	110
Air (Passengers & Freight)	36.9	140	140
Truck	55.9	150	120
Inland Shipping	41.3	130	120
International Bunkers	504.0	110	110
Total	1,937.0	120	120

1. Figures for 1985 are expressed as percentages of the 1972 values.

Table 11.6
Growth of commercial and public energy demand, 1972–1985

	1972 (10^{15} J)	1985[1] Case A	Case B	Case C	Case D	Case E
Fossil Fuel	155.1	110	100	120	110	130
Electricity	39.1	130	120	130	120	130

1. Figures for 1985 are expressed as percentages of the 1972 values.

Table 11.7
Industrial energy saving potential at high and medium oil prices

	% Share of Space Heat	% Savings at $17.25	$11.50
Iron and Steel	1.6	10	5
Nonferrous Metals	8	11.8	5.9
Paper and Allied	33	10.0	5.0
Chemicals	3	4.2	1.1
Mineral Products	8	0.8	−2.4
Food	21	4.5	1.1
Engineering	50	6.5	2.5
Other Industry		5.7	1.9
Total Industry		5.9	2.1

Residential

The principal variables that determine residential energy use in 1985 are shown in Table 11.9. It is anticipated that the rise in the number of dwellings will be determined not only by the growth of the population but also by the continuing decrease in the number of inhabitants per dwelling. In the longer term, the housing program will also be affected by the still dropping birth rate.

Almost all new dwellings will be fitted with central heating installations.

There is still a considerable potential for energy conservation in residential heating at the medium oil price level. This fact, coupled with policy measures such as a review of building standards and subsidies for the insulation of existing buildings, will lead to an improvement of heating efficiency which can be assessed as follows: In 50 percent of the centrally heated dwellings built before 1976, additional insulation will result in an average fuel saving of 20 percent for single-family homes, and 15 percent for apartments. All dwellings built after 1976 will have far better insulation, resulting in a saving of 30 percent for single-family homes and 20 percent for apartments.

Electricity consumption in the residential sector will continue to increase at a high rate due to the continuing penetration of electrical appliances. In Cases C and D the average energy consumption per appliance has been assumed to be the same as in 1972, except where a trend toward larger appliances with higher power rating is foreseen.

Potential energy savings at high oil prices could be gained as a result of technical improvements of gas heaters and electrical appliances coupled with energy labeling and official regulations.

Table 11.10 presents the growth of energy consumption in the residential sector from 1972 to 1985.

11.4 CONCLUSIONS

The period to 1985—which will probably not be one of real energy shortage—should be used to effect transitions from a former state where ener-

Table 11.8
Growth of industrial energy use, 1972–1985

	1972 (10^{15} J)	1985[1] Case A	Case B	Case C	Case D	Case E
Metals	132.1	190	120	200	150	210
Paper	30	150	120	160	130	170
Chemicals	438.5	290	170	300	170	300
Food	69	130	120	140	130	150
All Other	132.5	120	100	120	110	120

1. Figures for 1985 are expressed as percentages of the 1972 values.

gy was plentiful and cheap to a state where energy sources other than oil and natural gas will predominate. The more lead time we have, the smoother will the path be to a new energy structure. This time can be bought, at least partially, by the implementation of energy-saving policies. Energy saving in industry will probably amount to not much more than 10 percent. The specific structure of The Netherlands' industry (which leans heavily on energy-intensive process industries) will not allow much more.

Most of the possible energy savings are to be found in the residential, public, and commercial sectors: on the order of 15 percent of the demand projected without conservation measures can be saved. This can be done through better thermal insulation of buildings and through more thoughtful use of heat, light, and power. In the transportation sector, energy conservation can be effected by reducing the need for mobility, shifting transportation use to more efficient modes, and improving the construction and design of vehicles and engines. The Netherlands' traditional role as an international carrier of goods will be maintained, and for this reason the transportation sector will be very sensitive to world trade.

Finally, it must be emphasized that a technical and economic approach to our problem is only part of the story. Behavior research as well as insight into the real motives for the daily actions of people are just as important. Ultimately, it is the general public that will determine the total process of transition.

Table 11.9
Number of dwellings and penetration of central heating in 1985

	Upper Bound (Case C)	Lower Bound (Case D)
Number of Dwellings (thousands)	4950	4950
Central Heating Penetration (%)	85	75

Table 11.10
Growth of residential energy use, 1972–1985

	1972	1985	
	(10^{15} J)	Case C[1]	Case D[1]
Space Heating	427.7	140	130
Water Heating	44.1	210	210
Cooking	59.3	120	120
Other Electrical Appliances	29.6	220	220

1. Figures for 1985 are expressed as percentages of the 1972 values.

11.5 DEMAND STUDY WORKSHEETS

This section contains the detailed demand study worksheets in the following order:

Energy Use Summary

Sectoral Percentages of Total Final Energy Demand

Energy Growth Rates Summary

Base Year (1972) worksheets

Case A (industrial worksheet only)

Case B (industrial worksheet only)

Case C

Case D

Case E (industrial worksheet only).

WAES NATIONAL ENERGY DEMAND STUDIES, 1972-1985
ENERGY USE SUMMARY

Country THE NETHERLANDS

Scenario Assumptions

WAES Case	A		B		C		D		E			
GNP 1976-1985	High		Low		High		Low		High			
Oil Price*	\$11.50 → \$17.25		\$11.50 → \$17.25		\$11.50		\$11.50		\$11.50 → \$7.66			
Natl. Policy Response	Vigorous		Vigorous		Vigorous		Restrained		Restrained			
Natl. GNP Growth Rate 1976-1985 (%/yr.) **	3.4		2.2		3.4		2.2		3.4		1972 Base Year	
Fuel Type	Fossil Fuel	Elec.	Fossil Fuel	Elec.	Fossil Fuel	Elec.	Fossil Fuel	Elec.	Fossil Fuel	Elec.	Fossil Fuel	Elec.
Units (10^{15}J)												
Energy-Using Sectors	soft		soft						soft			
Industrial (excluding feedstocks)	979.8	127.4	647.3	83.2	1,028.4	133.6	683.6	88.0	1,038.3	135.4	498.3	63.9
Transportation					967.9	4.9	968.1	4.9			795.5	3.3
Commercial and Public	165.8	50.2	152.9	46.3	186.5	51.2	172.0	47.3	207.2	52.3	155.1	39.1
Residential					744.7	73.7	722.7	73.7			524.7	36.0
Fishing, Agriculture, Mining, Construction					231.9	11.9	225.4	11.4			152.9	9.8
Feedstocks	759.3	-	444.6	-	759.3	-	444.6	-	759.3	-	276.3	-
Final Energy Demand (excluding processing losses)	3,946.5		3,106.2		3,918.7	275.3	3,216.4	225.3	4,656.4		2,402.8	152.1
Processing Losses	486.3		404.9		173.8	342.8	168.2	281.3	597.5		170.9	304.0
Primary Energy Input	4,432.8		3.511.1		4,710.6		3,891.2		5,253.9		3,029.8	

*Oil price 1975 US\$ per barrel Arabian light crude oil FOB Persian Gulf.

** 1972-1985

WAES NATIONAL ENERGY DEMAND STUDIES, 1972-1985
SECTORAL PERCENTAGES OF TOTAL

Country THE NETHERLANDS

Scenario Assumptions

WAES Case	A		B		C		D		E			
GNP 1976-1985	High		Low		High		Low		High			
Oil Price*	\$11.50 → \$17.25		\$11.50 → \$17.25		\$11.50		\$11.50		\$11.50 → \$7.66			
Natl. Policy Response	Vigorous		Vigorous		Vigorous		Restrained		Restrained			
Natl. GNP Growth Rate 1976-1985 (%/yr.)**	3.4		2.2		3.4		2.2		3.4		1972 Base Year	
Fuel Type	Fossil Fuel	Elec.	Fossil Fuel	Elec.	Fossil Fuel	Elec.	Fossil Fuel	Elec.	Fossil Fuel	Elec.	Fossil Fuel	Elec.
Units	%		%		%		%		%		%	
Energy-Using Sectors												
Industrial (excluding feedstocks)					26.2	48.5	21.2	39.1			20.7	42.0
Transportation					24.7	1.8	30.1	2.2			33.1	2.2
Commercial and Public					4.8	18.6	5.3	21.0			6.5	25.7
Residential					19.0	26.8	22.6	32.7			21.8	23.7
Fishing, Agriculture, Mining, Construction					5.9	4.3	7.0	5.1			6.4	6.4
Feedstocks					19.4		13.8				11.5	
Final Energy Demand (excluding processing losses)					100	100	100	100			100	100
Processing Losses												
Primary Energy Input												

*Oil price 1975 US\$ per barrel Arabian light crude oil FOB Persian Gulf.

** 1972-1985

WAES NATIONAL ENERGY DEMAND STUDIES, 1972-1985
ENERGY GROWTH RATES SUMMARY

Country THE NETHERLANDS

Scenario Assumptions

WAES Case	A		B		C		D		E	
GNP 1976-1985	High		Low		High		Low		High	
Oil Price*	\$11.50 → \$17.25		\$11.50 → \$17.25		\$11.50		\$11.50		\$11.50 → \$7.66	
Natl. Policy Response	Vigorous		Vigorous		Vigorous		Restrained		Restrained	
Natl. GNP Growth Rate 1976-1985 (%/yr.)**	3.4		2.2		3.4		2.2		3.4	
Fuel Type	Fossil Fuel	Elec.	Fossil Fuel	Elec.	Fossil Fuel	Elec.	Fossil Fuel	Elec.	Fossil Fuel	Elec.
Units (%/yr. 1972-1985)										
Energy-Using Sectors	soft		soft						soft	
Industrial (excluding feedstocks)	5.4	5.5	2.0	2.1	5.7	5.6	2.4	2.5	5.8	5.9
Transportation					1.5	3.1	1.5	3.1		
Commercial and Public					1.4	2.1	1.1	1.5		
Residential					2.7	5.9	2.5	5.9		
Fishing, Agriculture, Mining, Construction					3.3	1.5	3.0	1.2		
Feedstocks	8.2	-	3.7	-	8.2	-	3.7	-	8.2	-
Final Energy Demand (excluding processing losses)	3.4		1.5		3.8	4.6	2.3	3.1	4.7	
Processing Losses					0.1	1.1	%0.1	%0.6		
Primary Energy Input	2.9		1.1		3.4		2.0		4.3	

*Oil price 1975 US\$ per barrel Arabian light crude oil FOB Persian Gulf.
** 1972-1985

NATIONAL ENERGY DEMAND PROJECTIONS

Country THE NETHERLANDS

Year 1972

WAES Case BASE YEAR

Summary

UNITS = 10^{15}J

Category	Oil	Gas	Coal	Electricity	Other	Total by Category (Excluding Processing Losses)
Transportation	802.0	-	0.2	3.3		805.5
Agriculture, Mining, and Construction	112.2	65.7	4.6	9.8		192.3
Commercial	95.8	54.6	4.7	39.1		194.2
Public						
Industrial	271.4	389.4	77.5	63.9		802.2
Residential	134.6	381.3	8.8	36.0		560.7
Total Delivered Energy by Fuel	1,416.0	891.0	95.8	152.1		
Processing Losses by Fuel	161.8	% 9.8	18.9	304.0		
Gross Input by Fuel	1,577.8	881.2	114.7	456.1 (1)		3,029.8

Gross Energy Input (Add across bottom row)

(1) primary fuel consumption in 10^{15}J for power generation

oil	97.5
gas	340.8
coal	14.2
nuclear	3.6

NATIONAL ENERGY DEMAND PROJECTIONS

Country THE NETHERLANDS

Year 1972

WAES Case BASE YEAR

Transportation

	(A)	(B)	(C)	(D)	(E)	(F)	(G)	
	Population/Auto	(km) Distance/Auto/Yr.	Load Factor	Fraction Urban or Rural	(10^6J/km) Energy/Distance	(10^{15}J) Fossil Fuel	(10^{15}J) Electricity	
Urban Auto	4.5	16,000	1.76		3.241	150.4		(1)
Rural Auto								(2)
	Activity Level (1)	% Electrified	Load Factor	Fossil Fuel/ (2) Activity Level	Electricity/Activity (2)			
Urban Bus[3]	1,769		25	0.9	0.3	0.4	0.4	(3)
Intercity Bus	7,090		20	0.6		4.2		(4)
Rail (Passenger)	8,643	90	38	1.1	0.35	1.0	2.6	(5)
Air (Passenger, Domestic) (4)	88.500 (5)			417 (6)		37.0		(6)
Air (Passenger, International)								(7)
Ship (Passenger)								(8)
Rail (Freight)	3,070	50	85	0.7	0.2	1.3	0.3	(9)
Air (Freight, Domestic)								(10)
Air (Freight, International)								(11)
Truck (Local)	13,973		69	4.0		55.9		(12)
Truck (Intercity)								(13)
Ship (Freight, Domestic)	29,334			1.4		41.3		(14)
Ship (Freight, International)						504.0		(15)
Miscellaneous (7)						6.7		(16)
					Total	802.2	3.3	

(1) 10^9 passenger km or 10^9 tonne km, if not stated otherwise

(2) 10^6J/passenger km or 10^6J/tonne km, if not stated otherwise

(3) includes bus, tram and metro

(4) includes air travel and freight

(5) number of take-offs

(6) 10^{12}J/take-off (fuel intake)

(7) lubricants (= feedstock/non-energy use)

NATIONAL ENERGY DEMAND PROJECTIONS

Country THE NETHERLANDS

Year 1972

WAES Case BASE YEAR

Agriculture, Mining, and Construction

Agriculture, Forestry, and Fishing (ISIC Major Division 1)	(A) (1972 Dfl x 10^6) Value Added	(B) (10^6J/Dfl) Energy/Value Added	(C) (10^6ha) Area Harvested	(D) Energy/ Area	(E) (tonnes) Physical Output	(F) Energy/ Output	(G) (10^{15}J) Fossil Fuel	(H) (10^{15}J) Electricity	
Farm Machinery									(17)
Greenhouses									(18)
Other Agricultural Uses	7,851		2,116.3				102.1	2.6	(19)
Forestry (ISIC 12)									(20)
Fishing Boats (ISIC 13)	187				333,800		3.3		(21)
Other Fishing Use (ISIC 13)									(22)
Total Fertilizer (tons)		N 375 x 10^3	P 44 x 10^3	K 104 x 10^3		Total	105.4	2.6	

Mining	(A) Value Added	(B) Energy/Value Added	(C) Physical Output (10^3 tonnes)	(D) Energy/Physical Output (10^9J/tonne)	(E) Fossil Fuel	(F) Electricity	
Handling and Internal Transportation	2,744	13.0/0.8	132,023	0.3	35.6	2.1	(23)
Other Energy Use							(24)
				Total	35.6	2.1	

Construction	(A) Value Added	(B) Energy/Value Added	(C) Physical Output	(D) Energy/Physical Output	(E) Fossil Fuel	(F) Electricity	
Handling and Internal Transportation	13,012	0.9/0.4			11.9	5.1	(25)
Other Energy Use (1)					29.6		(26)
				Total	41.5	5.1	
				Agriculture, Mining, and Construction Total	182.5	9.8	

(1) asphalts, bitumen (= feedstock/non-energy use)

NATIONAL ENERGY DEMAND PROJECTIONS

Country THE NETHERLANDS

Year 1972

WAES Case BASE YEAR

Commercial and Public

UNITS = 10^{15} J

		Fossil Fuel	Electricity
Commercial and Public Total	(1)	155.1	39.1

(1) in relation to 2.4 million employees in this sector:

64.6 J/empl. fuel

16.3 J/empl. power

NATIONAL ENERGY DEMAND PROJECTIONS

Country THE NETHERLANDS

Year 1972

WAES Case BASE YEAR

Industrial

	(A) Value Added	(B) Energy/Value Added	(C) Physical Output	(D) Energy/Output	(E) % Exported	(F) Fossil Fuel	(G) Electricity	(H) Feedstocks	
	(10^6 Dfl)	(10^6 J/Dfl)	(10^6 tonnes)	(10^9 J/tonne)		(10^{15} J)	(10^{15} J)	(10^{15} J)	
Iron and Steel (ISIC 371)	1,821	72.5	5.6	20.8		109.4	7.1		(41)
Nonferrous Metals (ISIC 372)			0.25	62.9		5.4	10.2		(42)
Paper and Allied (ISIC 34)	993	30.3	1.6			27.6	2.5		(43)
Chemicals and Allied (ISIC 35 minus ISIC 353)	5,280	37.6				177.2	21.3	240.0	(44)
Mineral Products (1) (ISIC 36)	107	467.2				46.6	3.4		(45)
Food and Related (ISIC 31)	8,165	8.5				62.8	6.2		(46)
Transport Eqpt. (ISIC 384)									(47)
Key Industry (N.E.C.) (2)	14,798	2.2				24.9	7.5		(48)
Key Industry (N.E.C.)									(49)
Misc. Manufacturing	9,246	5.4				44.4	5.7		(50)
					Total	498.3	63.9	240.0	

N.E.C. = Not Elsewhere Classified

(1) index-figure

(2) includes Engineering and other Metalworking Industry

NATIONAL ENERGY DEMAND PROJECTIONS

Country THE NETHERLANDS

Year 1972

WAES Case BASE YEAR

Residential

Population 13.3×10^6 Persons/Dwelling 3.3 Dwellings $4{,}042 \times 10^6$

Central Space Heating		(A) Dwellings (10^6)	(B) Heat Loss (10^9J/dw.yr)	(C) System Efficiency	(D)	(E) Fossil Fuel (10^{15}J)	(F) Electricity (10^{15}J)	
Single Family	Fossil Fuels							(51)
	Electricity							(52)
Town or Row Houses	Fossil Fuels	1.03	66.1	0.55		182.3		(53)
	Electricity							(54)
Apartments	Fossil Fuels	0.26	52.6	0.55				(55)
	Electricity							(56)
		% Saturation (Fossil Fuels)	% Saturation (Electricity)	Energy/Unit (Fossil Fuels) (10^9J/dw.yr)	Energy/Unit (Electricity) (kWh/dw.yr)			
Room Heating (1)				70.4	350	243.6	1.8	(57)
Air Conditioning (Window)			0		1,300			(58)
Air Conditioning (Central)			0					(59)
Water Heating (2)		70/5/4	16	12.3/15.8/28.9	1,750	40.5	3.6	(60)
Refrigerators			88		450		5.1	(61)
Lighting							9.0	(62)
Cooking		90	6	16.5	1,300	58.3	1.0	(63)
Dishwashers			4		900		0.5	(64)
Freezers			17		800		1.8	(65)
Television (3)			96/.		175/.		2.6	(66)
Other Key Appliances (4)			85/5		450/700		4.9	(67)
Misc. Appliances (5)			108/30		(660)/660		5.7	(68)
Potential Appliances								(69)
					Total	524.7	36.0	

(1) includes additional heating appliances with petroleum and electricity

(2) includes kitchen-/bathgeyser and hot water storage heater

(3) includes soundequipment

(4) includes washing machine and dryer

(5) includes vacuum cleaner and central heating pump

NATIONAL ENERGY DEMAND PROJECTIONS

Country THE NETHERLANDS

Year 1985

WAES Case A

Industrial

	(A)	(B)	(C)	(D)	(E)	(F)	(G)	(H)	
	Value Added	Energy/Value Added	Physical Output	Energy/Output	% Exported	Fossil Fuel	Electricity	Feedstocks	
	(10^6 Dfl)	(10^6 J/Dfl)	(10^6 tonnes)	(10^9 J/tonne)		(10^{15} J)	(10^{15} J)	(10^{15} J)	
Iron and Steel (ISIC 371)	3,798	65.1	12.0	18.7		215.1	32.1		(41)
Nonferrous Metals (ISIC 372)			0.54	55.5					(42)
Paper and Allied (ISIC 34)	1,695	27.3				42.4	3.8		(43)
Chemicals and Allied (ISIC 35 minus ISIC 353)	15,897	35.3				501.1	60.2	722.6	(44)
Mineral Products (1) (ISIC 36)	100	463.6				43.2	3.2		(45)
Food and Related (ISIC 31)	11,868	7.6				82.6	8.2		(46)
Transport Eqpt. (ISIC 384)									(47)
Key Industry (N.E.C.) (2)	27,937	2.1				44.3	13.3		(48)
Key Industry (N.E.C.)									(49)
Misc. Manufacturing	11,342	5.1				51.1	6.6		(50)
					Total	979.8	127.4	722.6	

N.E.C. = Not Elsewhere Classified

(1) index-figure

(2) includes Engineering and other Metalworking Industry

NATIONAL ENERGY DEMAND PROJECTIONS

Country THE NETHERLANDS

Year 1985

WAES Case B

Industrial

	(A) Value Added (10^6 Dfl)	(B) Energy/Value Added (10^6 J/Dfl)	(C) Physical Output (10^6 tonnes)	(D) Energy/Output (10^9 J/tonne)	(E) % Exported	(F) Fossil Fuel (10^{15} J)	(G) Electricity (10^{15} J)	(H) Feedstocks (10^{15} J)	
Iron and Steel (ISIC 371)	2,431	65.1	7.5	18.7		137.8	20.5		(41)
Nonferrous Metals (ISIC 372)			0.34	55.5					(42)
Paper and Allied (ISIC 34)	1,317	27.3				33.0	2.9		(43)
Chemicals and Allied (ISIC 35 minus ISIC 353)	8,990	35.3				283.4	34.0	408.6	(44)
Mineral Products (1) (ISIC 36)	98	463.6				39.5	2.9		(45)
Food and Related (ISIC 31)	11,141	7.6				77.5	7.7		(46)
Transport Eqpt. (ISIC 384)									(47)
Key Industry (N.E.C.) (2)	19,698	2.1				31.2	9.4		(48)
Key Industry (N.E.C.)									(49)
Misc. Manufacturing	9,970	5.1				44.9	5.8		(50)
					Total	647.3	83.2	408.6	

N.E.C. = Not Elsewhere Classified

(1) index-figure

(2) includes Engineering and other Metalworking Industry

NATIONAL ENERGY DEMAND PROJECTIONS

Country THE NETHERLANDS

Year 1985

WAES Case C

Summary

UNITS = 10^{15}J

Category	Oil	Gas	Coal	Electricity	Other	Total by Category (Excluding Processing Losses)
Transportation	976.9	-	-	4.9		981.8
Agriculture, Mining, and Construction	116.8	142.8	-	11.9		271.5
Commercial	40.1	146.4	-	51.2		237.7
Public						
Industrial	1,289.1	345.8	116.1	133.6		1,884.6
Residential	47.0	697.7	-	73.7		818.4
Total Delivered Energy by Fuel	2,469.9	1,332.7	116.1	275.3		
Processing Losses by Fuel	160.0	⁒ 12.8	26.6	342.8		
Gross Input by Fuel	2,629.9	1,319.9	142.7	618.1 (1)		4,710.6

Gross Energy Input (Add across bottom row)

(1) primary fuel consumption in 10^{15}J for power generation:

oil	362.5
gas	123.2
coal	73.2 (corresponding with 2.5 million tonnes of coal)
nuclear	59.2 (corresponding with 2,500 MW)

NATIONAL ENERGY DEMAND PROJECTIONS

Country THE NETHERLANDS

Year 1985

WAES Case C

Transportation

	(A)	(B)	(C)	(D)	(E)	(F)	(G)	
	Population/Auto	(km) Distance/Auto/Yr.	Load Factor	Fraction Urban or Rural	(10^6J/km) Energy/Distance	(10^{15}J) Fossil Fuel	(10^{15}J) Electricity	
Urban Auto	2.5	14,300	2.0		2,606	208.7		(1)
Rural Auto								(2)
	Activity Level (1)	% Electrified	Load Factor	Fossil Fuel/ Activity Level (2)	Electricity/Activity (2)			
Urban Bus (3)	1,800		25	0.9	0.3	0.5	0.5	(3)
Intercity Bus	7,000		20	0.6		4.2		(4)
Rail (Passenger)	13,100	90	38	1.1	0.35	1.4	4.1	(5)
Air (Passenger, Domestic) (4)	115,500 (5)			445 (6)		51.4		(6)
Air (Passenger, International)								(7)
Ship (Passenger)								(8)
Rail (Freight)	3,000	50	85	0.7	0.2	1.3	0.3	(9)
Air (Freight, Domestic)								(10)
Air (Freight, International)								(11)
Truck (Local)				4.0		81.4		(12)
Truck (Intercity)								(13)
Ship (Freight, Domestic)	38,400			1.4		54.0		(14)
Ship (Freight, International)						565.0		(15)
Miscellaneous (7)						9.0		(16)
					Total	976.9	4.9	

(1) 10^9 passenger km or 10^9 tonne km, if not stated otherwise

(2) 10^6J/passenger km or 10^6J/tonne km, if not stated otherwise

(3) includes bus, tram and metro

(4) includes air travel and freight

(5) number of take-offs

(6) 10^{12}J/take-off (fuel intake)

(7) lubricants (= feedstock/non-energy use)

NATIONAL ENERGY DEMAND PROJECTIONS

Country THE NETHERLANDS

Year 1985

WAES Case C

Agriculture, Mining, and Construction

Agriculture, Forestry, and Fishing (ISIC Major Division 1)	(A) (1972 Dfl x 10^6) Value Added	(B) (10^6J/Dfl) Energy/Value Added	(C) Area Harvested	(D) Energy/Area	(E) Physical Output	(F) Energy/Output	(G) (10^{15}J) Fossil Fuel	(H) (10^{15}J) Electricity	
Farm Machinery									(17)
Greenhouses									(18)
Other Agricultural Uses	13,027	13.1/0.3					170.6	4.2	(19)
Forestry (ISIC 12)									(20)
Fishing Boats (ISIC 13)									(21)
Other Fishing Use (ISIC 13)									(22)
Total Fertilizer (tons)		N	P	K		Total	170.6	4.2	

Mining	(A) Value Added	(B) Energy/Value Added	(C) Physical Output	(D) Energy/Physical Output	(E) Fossil Fuel	(F) Electricity	
Handling and Internal Transportation	3,872	13.0/0.8			50.2	3.0	(23)
Other Energy Use							(24)
				Total	50.2	3.0	

Construction	(A) Value Added	(B) Energy/Value Added	(C) Physical Output	(D) Energy/Physical Output	(E) Fossil Fuel	(F) Electricity	
Handling and Internal Transportation	12,186	0.9/0.4			11.1	4.7	(25)
Other Energy Use (1)					27.7		(26)
				Total	38.8	4.7	
				Agriculture, Mining, and Construction Total	259.6	11.9	

(1) asphalts, bitumen (= feedstock/non-energy use)

NATIONAL ENERGY DEMAND PROJECTIONS

Country THE NETHERLANDS

Year 1985

WAES Case C

Commercial and Public

UNITS = 10^{15} J

	Fossil Fuel	Electricity
Commercial and Public Total (1)	186.5	51.2

(1) calculated from the projected number of employees as the activity level (3.208 million) and technical coefficients of 58.1 x 10^3 J/empl. (fuel) and 15.9 x 10^3 J/empl. (electricity)

NATIONAL ENERGY DEMAND PROJECTIONS

Country THE NETHERLANDS

Year 1985

WAES Case C

Industrial

	(A) Value Added	(B) Energy/Value Added	(C) Physical Output	(D) Energy/Output	(E) % Exported	(F) Fossil Fuel	(G) Electricity	(H) Feedstocks	
	(10^6Dfl)	(10^6J/Dfl)	(10^6 tonnes)	(10^9J/tonne)		(10^{15}J)	(10^{15}J)	(10^{15}J)	
Iron and Steel (ISIC 371)	3,798	68.8	12.0	19.8		227.4	33.9		(41)
Nonferrous Metals (ISIC 372)			0.54	59.2					(42)
Paper and Allied (ISIC 34)	1,695	28.8				44.8	4.0		(43)
Chemicals and Allied (ISIC 35 minus ISIC 353)	15,897	37.2				527.8	63.4	722.6	(44)
Mineral Products(1) (ISIC 36)	100	478.2				41.6	3.0		(45)
Food and Related (ISIC 31)	11,868	8.1				87.3	8.6		(46)
Transport Eqpt. (ISIC 384)						46.2	13.9		(47)
Key Industry (N.E.C.)	27,937	2.15				46.2	13.9		(48)
Key Industry (N.E.C.)									(49)
Misc. Manufacturing(2)	11,342	5.3				53.3	6.8		(50)
N.E.C. = Not Elsewhere Classified					**Total**	1,028.4	133.6	722.6	

(1) index-figure

(2) includes Engineering and other Metalworking Industry

NATIONAL ENERGY DEMAND PROJECTIONS

Country THE NETHERLANDS

Year 1985

WAES Case C

Residential

Population 13.9×10^6 Persons/Dwelling 2.8 Dwellings 4.95×10^6

		(A)	(B)	(C)	(D)	(E)	(F)	
Central Space Heating		Dwellings (10^6)	Heat Loss (10^9J/dw.yr)	System Efficiency		Fossil Fuel (10^{15}J)	Electricity (10^{15}J)	
Single Family	Fossil Fuels							(51)
	Electricity							(52)
Town or Row Houses	Fossil Fuels	3.36	58.5	0.55		527.9		(53)
	Electricity							(54)
Apartments	Fossil Fuels	0.84	48.8	0.55				(55)
	Electricity							(56)
		% Saturation (Fossil Fuels)	% Saturation (Electricity)	Energy/Unit (Fossil Fuels) (10^9J/dw.yr)	Energy/Unit (Electricity) (kWh/dw.yr)			
Room Heating (1)		15.2	30	84.5	200	63.4	0.9	(57)
Air Conditioning (Window)			0		1,300			(58)
Air Conditioning (Central)								(59)
Water Heating (2)		50/40/5/10	16	12.3/15.8/28.9/44	1,750	87.1	4.1	(60)
Refrigerators			110		800		13.0	(61)
Lighting							15.5	(62)
Cooking		80	15	16.5	1,300	66.3	2.9	(63)
Dishwashers			50		900		6.6	(64)
Freezers			50		800		5.9	(65)
Television (3)			160/.		175/.		4.6	(66)
Other Key Appliances (4)			90/30		600/600		10.6	(67)
Misc. Appliances (5)			110/90		(660)/660		7.6	(68)
Potential Appliances (6)							2.0	(69)
					Total	744.7	73.7	

(1) includes additional heating appliances with petroleum and electricity

(2) includes kitchen-/bathgeyser, storage heater and combi-boiler

(3) includes soundequipment

(4) includes washing machine and dryer

(5) includes vacuum cleaner and central heating pump

(6) balance figure

NATIONAL ENERGY DEMAND PROJECTIONS

Country THE NETHERLANDS

Year 1985

WAES Case D

Summary

UNITS = 10^{15}J

Category	Oil	Gas	Coal	Electricity	Other	Total by Category (Excluding Processing Losses)
Transportation	977.1	-	-	4.9		982.0
Agriculture, Mining, and Construction	114.1	138.3	-	11.4		263.8
Commercial	24.0	148.0	-	47.3		219.3
Public						
Industrial	654.0	365.2	73.0	88.0		1,180.2
Residential	45.5	677.2	-	73.7		796.4
Total Delivered Energy by Fuel	1,814.7	1,328.7	73.0	225.3		
Processing Losses by Fuel	160.0	–8.7	16.9	281.3		
Gross Input by Fuel	1,974.7	1,320.0	89.9	506.6 (1)		3,891.2

Gross Energy Input (Add across bottom row)

(1) primary fuel consumption in 10^{15}J for power generation:

oil	265.5
gas	123.2
coal	58.5 (corresponding with 2 million tonnes of coal)
nuclear	59.2 (corresponding with 2,500 MW)

NATIONAL ENERGY DEMAND PROJECTIONS

Country THE NETHERLANDS

Year 1985

WAES Case D

Transportation

	(A)	(B)	(C)	(D)	(E)	(F)	(G)	
	Population/Auto	(km) Distance/Auto/Yr.	Load Factor	Fraction Urban or Rural	$(10^6 J/km)$ Energy/Distance	$(10^{15} J)$ Fossil Fuel	$(10^{15} J)$ Electricity	
Urban Auto	2.8	15,400	2.0		2,998	226.2		(1)
Rural Auto (→ included with Urban Auto)								(2)
	Activity Level [1]	% Electrified	Load Factor	Fossil Fuel/ Activity Level [2]	Electricity/Activity [2]			
Urban Bus [3]	1,800		25	0.9	0.3	0.5	0.5	(3)
Intercity Bus	7,000		20	0.6		4.2		(4)
Rail (Passenger)	13,100	90	38	1.1	0.35	1.4	4.1	(5)
Air (Passenger, Domestic) [4]	115,500 [5]			445 [6]		51.4		(6)
Air (Passenger, International) (→ included with Air (Passenger, Domestic))								(7)
Ship (Passenger)								(8)
Rail (Freight)	3,000	50	85	0.7	0.2	1.3	0.3	(9)
Air (Freight, Domestic)								(10)
Air (Freight, International)								(11)
Truck (Local)						69.1		(12)
Truck (Intercity) (→ included with Truck (Local))								(13)
Ship (Freight, Domestic)	34,800					49.0		(14)
Ship (Freight, International)						565.0		(15)
Miscellaneous [7]						9.0		(16)
					Total	977.1	4.9	

(1) 10^9 passenger km or 10^9 tonne km, if not stated otherwise

(2) 10^6J/passenger km or 10^6J/tonne km, if not stated otherwise

(3) includes bus, tram and metro

(4) includes air travel and freight

(5) number of take-offs

(6) 10^{12}J/take-off (fuel intake)

(7) lubricants (= feedstock/non-energy use)

NATIONAL ENERGY DEMAND PROJECTIONS

Country THE NETHERLANDS

Year 1985

WAES Case D

Agriculture, Mining, and Construction

Agriculture, Forestry, and Fishing (ISIC Major Division 1)	(A) (1972 Dfl x 10^6) Value Added	(B) (10^6 J/Dfl) Energy/Value Added	(C) Area Harvested	(D) Energy/Area	(E) Physical Output	(F) Energy/Output	(G) (10^{15} J) Fossil Fuel	(H) (10^{15} J) Electricity	
Farm Machinery									(17)
Greenhouses									(18)
Other Agricultural Uses	12,840	13.1/0.3					168.2	4.1	(19)
Forestry (ISIC 12)									(20)
Fishing Boats (ISIC 13)									(21)
Other Fishing Use (ISIC 13)									(22)
Total Fertilizer (tons)		N	P	K		Total	168.2	4.1	

(Values in column A for rows 17–22 are bracketed together as one entry.)

Mining	(A) Value Added	(B) Energy/Value Added	(C) Physical Output	(D) Energy/Physical Output	(E) Fossil Fuel	(F) Electricity	
Handling and Internal Transportation	3,585	13.0/0.8			46.5	2.7	(23)
Other Energy Use							(24)
				Total	46.5	2.7	

Construction	(A) Value Added	(B) Energy/Value Added	(C) Physical Output	(D) Energy/Physical Output	(E) Fossil Fuel	(F) Electricity	
Handling and Internal Transportation	11,884	0.9/0.4			10.7	4.6	(25)
Other Energy Use (1)					27.0		(26)
				Total	37.7	4.6	
				Agriculture, Mining, and Construction Total	252.4	11.4	

(1) asphalts, bitumen (= feedstock/non-energy use)

NATIONAL ENERGY DEMAND PROJECTIONS

Country THE NETHERLANDS

Year 1985

WAES Case D

Commercial and Public

UNITS = 10^{15} J

	Fossil Fuel	Electricity
Commercial and Public Total (1)	172.0	47.3

(1) calculated from the projected number of employees as the activity level (2.96 million) and technical coefficients of 58.1 x 10^3 J/empl. (fuel) and 16.0 x 10^3 J/empl. (electricity)

NATIONAL ENERGY DEMAND PROJECTIONS

Country THE NETHERLANDS

Year 1985

WAES Case D

Industrial

	(A) Value Added (10^6 Dfl)	(B) Energy/Value Added (10^6 J/Dfl)	(C) Physical Output (10^6 tonnes)	(D) Energy/Output (10^9 J/tonne)	(E) % Exported	(F) Fossil Fuel (10^{15} J)	(G) Electricity (10^{15} J)	(H) Feedstocks (10^{15} J)	
Iron and Steel (ISIC 371)	2,431	68.8	7.5	19.8		145.4	21.9		(41)
Nonferrous Metals (ISIC 372)			0.34	59.2					(42)
Paper and Allied (ISIC 34)	1,317	28.8				34.8	3.1		(43)
Chemicals and Allied (ISIC 35 minus ISIC 353)	8,990	37.2				298.4	35.9	408.6	(44)
Mineral Products (1) (ISIC 36)	98	478.2				43.7	3.2		(45)
Food and Related (ISIC 31)	11,141	8.1				81.9	8.1		(46)
Transport Eqpt. (ISIC 384)									(47)
Key Industry (N.E.C.) (2)	19,698	2.15				32.6	9.8		(48)
Key Industry (N.E.C.)									(49)
Misc. Manufacturing	9,970	5.3				46.8	6.0		(50)
					Total	683.6	88.0	408.6	

N.E.C. = Not Elsewhere Classified

(1) index-figures

(2) includes Engineering and other Metalworking Industry

NATIONAL ENERGY DEMAND PROJECTIONS

Country THE NETHERLANDS

Year 1985

WAES Case D

Residential

Population 13.9×10^6 Persons/Dwelling 2.8 Dwellings 4.95×10^6

Central Space Heating		(A) Dwellings (10^6)	(B) Heat Loss (10^9J/dw.yr)	(C) System Efficiency	(D)	(E) Fossil Fuel (10^{15}J)	(F) Electricity (10^{15}J)	
Single Family	Fossil Fuels							(51)
	Electricity							(52)
Town or Row Houses	Fossil Fuels	2.96	58.5	0.55		463.7		(53)
	Electricity							(54)
Apartments	Fossil Fuels	0.74	48.8	0.55				(55)
	Electricity							(56)
		% Saturation (Fossil Fuels)	% Saturation (Electricity)	Energy/Unit (Fossil Fuels) (10^9J/dw.yr)	Energy/Unit (Electricity)			
Room Heating (1)		25.3	30	84.5	200	105.6	0.9	(57)
Air Conditioning (Window)			0		1,300			(58)
Air Conditioning (Central)								(59)
Water Heating (2)		50/40/5/10	16	12.3/15.8/28.9/44	1,750	87.1	4.1	(60)
Refrigerators			110		800		13.0	(61)
Lighting							15.5	(62)
Cooking		80	15	16.5	1,300	66.3	2.9	(63)
Dishwashers			50		900		6.6	(64)
Freezers			50		800		5.9	(65)
Television (3)			160/.		175/.		4.6	(66)
Other Key Appliances (4)			90/30		600/600		10.6	(67)
Misc. Appliances (5)			110/90		(660)/660		7.6	(68)
Potential Appliances (6)							2.0	(69)
					Total	722.7	73.7	

(1) includes additional heating appliances with petroleum and electricity

(2) includes kitchen-/bathgeyser, storage heater and combi-boiler

(3) includes soundequipment

(4) includes washing machine and dryer

(5) includes vacuum cleaner and central heating pump

(6) balance figure

NATIONAL ENERGY DEMAND PROJECTIONS

Country THE NETHERLANDS

Year 1985

WAES Case E

Industrial

	(A) Value Added (10^6 Dfl)	(B) Energy/Value Added (10^6 J/Dfl)	(C) Physical Output (10^6 tonnes)	(D) Energy/Output (10^9 J/tonne)	(E) % Exported	(F) Fossil Fuel (10^{15} J)	(G) Electricity (10^{15} J)	(H) Feedstocks (10^{15} J)	
Iron and Steel (ISIC 371)	3,798	72.5	12.0	20.8		239.7	35.7		(41)
Nonferrous Metals (ISIC 372)			0.54	62.9					(42)
Paper and Allied (ISIC 34)	1,695	30.3				47.2	4.2		(43)
Chemicals and Allied (ISIC 35 minus ISIC 353)	15,897	37.6				517.5	62.2	722.6	(44)
Mineral Products (1) (ISIC 36)	100	467.2				40.6	3.0		(45)
Food and Related (ISIC 31)	11,868	8.5				91.8	9.1		(46)
Transport Eqpt. (ISIC 384)									(47)
Key Industry (N.E.C.) (2)	27,937	2.2				47.3	14.2		(48)
Key Industry (N.E.C.)									(49)
Misc. Manufacturing	11,342	5.4				54.2	7.0		(50)
N.E.C. = Not Elsewhere Classified					**Total**	1,038.3	135.4	722.6	

(Brackets drawn on the page join Nonferrous Metals (42) to the Iron and Steel figures in columns A and F.)

(1) index-figure

(2) includes Engineering and other Metalworking Industry

12.1 OVERVIEW

All of the scenarios analyzed for Norway assume quite high economic growth. The analysis shows that under such conditions the growth of energy demand will be of the same character as the economic growth; that is, if the rate of growth for the economy (percent per annum) is constant, then the rate of growth of the energy demand will be fairly constant also. The analysis also shows, however, that the rate of growth of the energy demand will be substantially lower than that of the economy in the period to 1985. During the period 1950 to 1972, the ratio between the rate of energy growth and the rate of GNP growth averaged approximately 1.2, and in the period 1960 to 1972 it was even higher. However, the scenarios analyzed for the period to 1985 all have ratios or "elasticities" in the range 0.7–0.8.

Figure 12.1 shows two projections of energy demand to 1985. In the projection marked "scenario" (which corresponds to WAES scenario C), net energy demand is estimated to increase from about 569 PJ (13.5 x 10^6 TOE) to about 900 PJ (20 x 10^6 TOE) in 1985, or by a factor of about 1.51.[1] It is important to note that the estimated energy demand includes the consumption of oil and gas used as industrial feedstocks.

The other projection, marked "traditional development," is based upon the same GNP growth rate as the "scenario" but assumes an elasticity of about 1.2. This elasticity is based upon energy delivered at the consumer end and GNP excluding extraction, refining, and primary transportation of oil and gas in order to make this development consistent with that of the period before 1972. The difference in demand between the two projections, which is quite significant in the long run, is due to:

1. Consumers' reactions to the higher energy prices that are assumed in the period to 1985.

2. Differing growth patterns, particularly in the metal-producing industries, where reduced growth has been assumed.

3. Government policies leading to improvements in energy efficiency beyond what is automatically obtained by consumer reaction.

1. 1 PJ = 10^3 TJ = 10^{15} J.

The results of the projections for the five WAES scenarios indicate that the growth rate for energy demand will be in the range of 2.5 percent per annum (for scenario B) to 3.2 percent per annum (for scenario E). The difference in demand between these extreme cases is relatively small—only about 8 percent—for the following reasons:

1. Economic activity related to the exploration for and production of oil and gas represents a significant part of the total economic activity in Norway. It has been assumed that this situation will make it possible to use the activity in the oil sector as a lever to control overall economic activity. For this reason, the economic growth rates for the high and low growth scenarios are only slightly different.

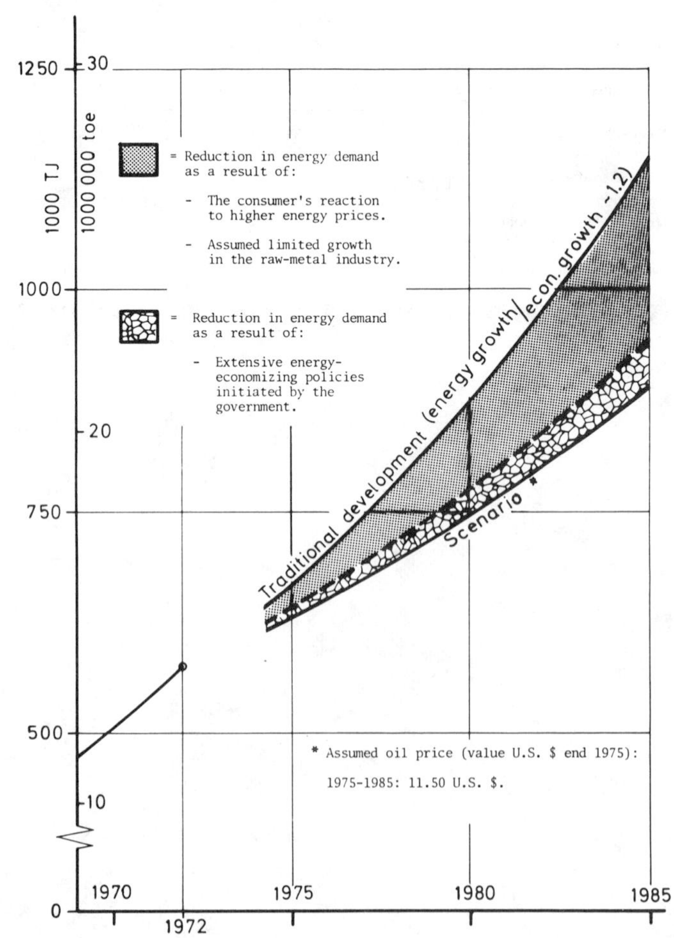

Figure 12.1 Development of energy demand at the consumer end, 1972–1985.

2. In Norway, electric energy has a share of the total demand that is much higher than in most other countries. Virtually all of this electric energy will be produced by hydro power even in 1985, and the price of electricity is therefore not directly influenced by the price of oil. A large and increasing share of the total amount of fossil fuels consumed will be used as feedstocks. For these reasons, the effect of oil prices on the total energy demand will be relatively small.

12.2 INTRODUCTION

12.2.1 Organization of WAES—Norway

The Norwegian Participant in WAES is Mr. Christian Sommerfelt, Chairman of Elkem-Spigerverket A/S, which is one of Norway's largest corporations. Elkem-Spigerverket has a special interest in energy questions because of its heavy involvement in the electrometallurgical sector.

The Norwegian secretariat for WAES is the engineering firm Scandpower A/S, which is involved in international consulting work in the energy sector. Scandpower A/S's former Managing Director, Mr. Henrik Ager-Hanssen, and Manager of Power Plant Engineering, Mr. Kai Killerud, are WAES Associates.

The Norwegian participation in WAES is financed mainly by Norges Teknisk-Naturvitenskapelige Forskningsråd (Royal Norwegian Council for Scientific and Industrial Research), NTNF.

In order to establish an effective channel of communication with policy makers and other people heavily involved in the energy field, a national reference group has been set up with high-level participation from the following government and industrial organizations:

Ministry of Finance

Ministry of Foreign Affairs

Ministry of Industry

NTNF

Den norske Creditbank (the largest private bank)

Federation of Norwegian Industries

Institutt for Atomenergi (Institute of Atomic Energy)

Norges vassdrags- og elektrisitetsvesen (the State Power Board), NVE

Elkem-Spigerverket A/S

Scandpower A/S.

In addition to the secretariat function, Scandpower has been managing and performing a detailed energy demand analysis for Norway as a separate WAES-related project. The total effort of this project took about 30 man-months. In addition to the main objective of analyzing the energy demand—serving the need for this information in WAES and in Norway itself—the project has been concerned with presenting a factual basis that can be used for determining energy research and development priorities in Norway. NTNF nominated a Steering Committee for this project, with participation from NVE, the Ministry of Industry, and the Federation of Norwegian Industries.

12.2.2 The Energy Situation Today

Compared to most other industrialized countries, the energy situation in Norway is favorable. Nevertheless, the country is facing a new situation since the abundant reserves of hydro power that the nation has enjoyed for many decades are nearing their end. Some two-thirds of the hydro power resources that can be economically utilized are already developed. Moreover, the last one-third will be more expensive to develop and will come into more serious conflict with environmental concerns. On the other hand, oil and gas discoveries on the Norwegian continental shelf represent a new and practically unlimited energy source for domestic consumption. Nuclear power, now becoming widespread among industrialized countries, may also become a serious alternative.

The objective of Norway's energy policies has been, up to quite recently, to ensure an abundant supply of energy, especially electricity. In the last few years the direction of policy has been changing: in general, the future supply of electricity to the electrometallurgical industries will be limited, and strict allocation by government will be practiced. The government stresses, however, the importance of safeguarding the many industrial communities that are built around the electrometallurgical industries by assuring the latter a long-term supply of energy. It is not yet clear whether oil and gas will be used to any large extent for production of electricity in Norway.

12.3 METHODOLOGY

Economic growth is an important factor in determining the growth of the energy demand. To be able to carry out fairly realistic analyses of future energy demand in Norway, it has been necessary to form a conception of the political and economic assumptions that will be most pertinent in the period up to 1985. Such assumptions will necessarily be filled with uncertainties; however, it has been assumed that the best signals can be obtained from official policy papers of different kinds.

12.3.1 Economic Assumptions

The latest available official document that provides basic figures for alternative economic developments up to 1985 and beyond is "Stortingsmelding No. 50, 1974–1975" (a proposition to the parliament), entitled "Natural Resources and Economic Development." This paper deals with subjects such as natural resources, international perspectives, population growth, social welfare programs, and environmental policy as they all relate to economic development.

In the working out of this proposition an attempt was made to take into consideration all of the important factors and constraints that will have a long-term influence on the development of Norwegian society. The economic trends outlined in this document should therefore provide a useful guide for our assumptions of a social, political, or resource-oriented nature.

It has been possible, with the aid of the judgment of economists, to match the alternatives presented in the proposition with the WAES global scenarios. Table 12.1 presents a summary of the economic assumptions that have served as a basis for the Norwegian demand analyses. It shows the sectoral data used, which have been taken from "Stortingsmelding No. 50." Small adjustments have been made only for the high alternative in the period 1974 to 1980.

The distribution of industry's total economic growth among different industrial groups is of great importance for calculations of the energy demand. Therefore, since the proposition did not present this kind of detailed data, it was necessary to estimate a distribution to be used in the analysis using professional judgment. Table 12.2 presents the rates of growth used for various industrial groups.

For the transportation sector the rates of growth for the different modes of transportation are based mainly upon work done by "Transportøkonomisk Institutt" (Institute of Transport Economics). Their figures give the basis for the determination of growth rates consistent with the data presented in Table 12.1. The resulting data for the transportation sector are presented in Table 12.3. In principle, the table is valid only for scenarios assuming no conservation policies other than those presently in effect.

12.3.2 National Energy Policy

As in the case of economic development, official policy documents have been consulted for signals about future energy policies. "Stortingsmelding No. 100, 1973–1974," entitled "Future Supply of Energy in Norway," is the latest official document dealing with energy questions. It deals primarily with energy supply, while this report focuses on demand. Certain points in the proposition have, however, influenced our demand analysis.

"Stortingsmelding No. 100" states that long-term marginal costs will, in principle, be the determining factor of electricity prices, unless special conditions indicate a different policy. In the analysis reported here, it has been assumed that prices corresponding to long-term marginal cost will be in effect by 1985 for the scenarios that require vigorous policy response. For scenarios with restrained policy response, it has been assumed that this goal will be reached somewhat later. Our analysis assumes—partly on the basis of indications in "Stortingsmelding No. 50"—that the growth of electricity supply for the metal-producing industries will be limited to 2 percent per year. It is supposed that there is a good correspondence between this rate of growth and the economic growth rates given in Table 12.2, which take into consideration improvements in energy efficiency and increases in energy prices.

Table 12.1
Sector level data for economic growth (%/yr.) for the low and high growth cases

Energy-Using Sectors	1974-1980 Low[1]	1974-1980 High[2]	1980-1985 Low	1980-1985 High
Agriculture, Forestry, Fishing	1.3	1.3	0.8	0.7
Mining, Industry, Power Production	3.2	3.4	4.4	4.7
Construction	4.8	5.0	4.0	4.4
Commercial	5.4	5.4	4.0	4.2
International Shipping	9.3	9.3	1.2	1.8
Other Transport	4.5	4.7	4.1	4.9
Misc. Public and Private Services	4.3	4.4	4.2	4.4
Subtotal	4.6	4.7	3.8	4.1
Drilling, Extraction, and Transport of Oil and Gas	87.5	87.5	-1.3	1.6
Total	7.8	7.9	3.4	4.2

1. This alternative is based directly on "St. meld. No. 50, 1974-1975."
2. Based on "St. meld. No. 50," but with small adjustments for certain activities.

Table 12.2
Growth rates (%/yr.) for different industrial groups under the low and high growth assumptions

Industrial Group	1974-1980 Low	1974-1980 High	1980-1985 Low	1980-1985 High
Iron and Steel	3.0	3.0	3.0	3.0
Nonferrous Metals	3.0	3.0	3.0	3.0
Paper and Allied	3.0	3.0	2.0	2.0
Chemicals and Allied	6.0	6.0	6.0	7.0
Mineral Products	3.0	3.0	4.0	4.0
Food and Related	1.5	1.5	1.5	1.5
Transport Equipment	2.0	3.0	3.0	4.0
Manufacture of Other Fabricated Metal Products	4.0	5.0	6.0	6.5
Other Industry	1.5	1.5	2.0	2.0
Total, Excluding Refining	3.0	3.2	4.3	4.6
Total Industry	3.2	3.4	4.4	4.7

12.3.3 The Effect of Oil Prices

Due to the production in the North Sea, Norway has become a net exporter of oil, and in less than ten years the production is expected to grow to about 90 million tons of oil equivalent per year, or about ten times Norway's own demand today. The economic activity related to exploration for and production of oil and gas represents a significant part of the total economic activity in Norway. It is assumed that it will be possible to use the activity in the oil sector as a lever to control overall economic activity and thus keep economic growth within acceptable limits. This will most probably be the case with oil prices at or above today's prices, and it has therefore been assumed that oil price will have no influence on economic activity. Although the situation may be different for the lowest oil price considered, the same economic figures have been used in this case. The price of oil will, to a certain extent, influence energy efficiency in some sectors: In oil-consuming processes, for example, higher oil prices will lead to improved specific consumptions. In the transportation sector, on the other hand, the price of oil will have only minor influence on energy efficiency because—due to refinery costs and heavy taxation on the fuel itself—the price of crude oil will have

Table 12.3
Activity growth (%/yr.) for various modes of transportation for the low and high growth cases (assuming no conservation policies other than those presently in effect)

Mode of Transportation	1972–1985 Low	High
Passenger Travel		
Automobile (number at end of period)	$1.6x10^6$	$1.7x10^6$
Bus	1.1	1.1
Rail	2.4	2.6
Airplane (Domestic)	6.8	8.3
Airplane (International)	8.2	9.2
Ship	1.9	2.0
Freight		
Rail	2.5	2.7
Airplane (Domestic)	8.0	10.1
Truck	6.4	7.6
Ship (Domestic)	5.7	7.0
Ship (International)	7.5	7.8

only negligible impact on the price of gasoline, diesel fuel, etc.

12.3.4 National Policy Response Studies

As a part of the demand analysis, an evaluation has been made of policies that can help to improve energy efficiency. The evaluation has been concentrated on the three sectors that have the largest energy demands: transportation, industrial, and residential.

In the transportation sector, passenger travel has been pinpointed as the subsector where the greatest energy savings can be achieved. In the industrial sector, the evaluation has been concentrated on the groups that are most energy-intensive: iron and steel, nonferrous metals, paper and allied, and chemicals and allied. Policies within the residential sector have been directed mainly toward the achievement of more efficient heating systems and better building insulation.

The policies evaluated may be divided into categories according to the type of government action required to obtain improved energy efficiencies and energy savings. The different categories are as follows:

financial support

tax incentives

mandatory measures

convenience incentives (such as special lanes for car pools)

information and education

research and development

removal of barriers to energy conservation.

12.4 1985 ENERGY DEMAND ESTIMATES

All of the scenarios analyzed for Norway assume quite high economic growth. The analysis shows that under such conditions the growth of energy demand will be of the same character as the economic growth; for example, if the rate of growth for the economy (percent per annum) is constant, then the rate of growth of the energy demand will be fairly constant also. The analysis also shows, however, that the rate of growth of the energy demand will be substantially lower than that of the economy in the period to 1985. During the period 1950 to 1972, the ratio between the rate of energy growth and the rate of GNP growth averaged approximately 1.2, and in the period 1960 to 1972 it was even higher. However, the scenarios analyzed for the period to 1985 all have ratios or "elasticities" in the range 0.7–0.8.

Figure 12.2 shows curves for energy demand for two of the most important scenarios that have been analyzed: B and C. Net energy demand is estimated to increase from about 569 PJ (14×10^6 TOE) in 1972 to about 800–900PJ ($19–21.5 \times 10^6$ TOE) in 1985, or by a factor of about 1.42 for scenario B and 1.51 for scenario C. It is important to note that the estimated energy demand includes the consumption of oil and gas used as industrial feedstocks.

Figure 12.2 also shows a curve marked "traditional development," based upon the same GNP growth as scenario C but assuming an elasticity of about 1.2. This elasticity is based upon energy delivered at the consumer end and GNP excluding extraction, refining, and primary transportation of oil and gas in order to make this development consistent with that of the period before 1972.

The difference in demand between this curve and the one for scenario C is quite significant in the long run, reflecting the reduction of the elasticity value from 1.2 to 0.7–0.8, and is due to:

1. Consumers' reactions to the higher energy prices

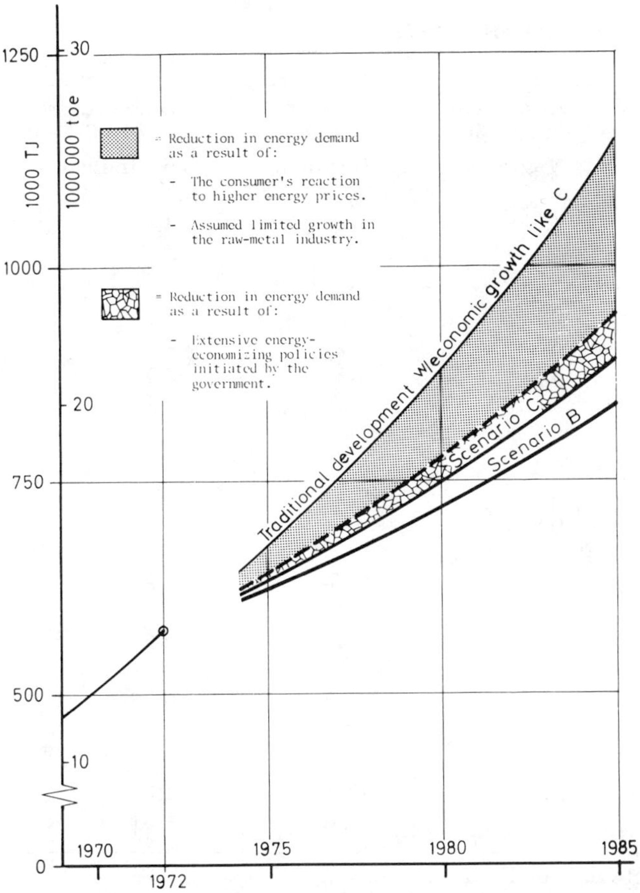

Figure 12.2 Development of energy demand at the consumer end, 1972–1985, for WAES scenarios B and C and for the case of traditional development.

that are assumed in the period to 1985.

2. Differing growth patterns, particularly in the metal-producing industries, where reduced growth has been assumed.

3. Government policies leading to improvements in energy efficiency beyond what is automatically attained by consumer reaction.

It is important to note that high and partly increasing energy prices have been assumed for the period up to 1985, whereas the energy prices were low and partly decreasing in real terms for the period 1950 to 1972.

Figure 12.2 also shows a dashed curve depicting the assumed development of energy demand if no government policies are put into operation. The other assumptions for this curve are the same as for scenario C.

Some of the assumed conservation measures can be realized only by means of research and development. Effects from such research are, of course, also reflected in the reduction of energy consumption due to the consumers' reactions to higher energy prices. Important areas of research and development are described in Section 12.5.2.

Table 12.4 presents some of the main results in the form of key figures for the scenarios analyzed. Where nothing else is stated, the following figures and conclusions related to energy demand will apply to energy delivered to the consumer. It appears from the table that the growth rate for total net energy demand is between 2.8 and 3.7 percent per annum throughout the period—about 1 percent per annum lower than the economic growth rates—for all of the scenarios that are considered realistic (A–D).

It further appears from the table that the growth in the total demand for fossil fuels is faster than that for electricity; that is, there is a shift toward oil and gas. This shift is *not* a result of an expected shift away from electricity in traditional activities and processes. Only in one area—residential heating—is an extensive shift expected, and here it is in the opposite direction. The shift that is evident from the figures is due to the fact that the sectors that depend mainly on fossil fuels expand the most. These are the transportation sector and the chemicals and allied subsector.

Table 12.4 shows that the energy demand within the transportation sector is estimated to increase much faster than the total demand. The most important reason for this is that international trade increases faster than production as a result of the continuous development toward increased international interdependence and specialization.

For the residential sector, the growth is much slower than that of total energy demand, despite the assumption of a substantial increase in the housing standard and greatly increased saturation of all presently known household appliances in 1985. The basic reasons for this are the expected low population growth and, for some of the scenarios, extensive measures aimed at achieving improved energy efficiency. Furthermore, a significant shift toward electric heating is assumed for most of the scenarios. Since an efficiency of 100 percent at the consumer end is being used for electric heating, compared to barely 70 percent on average for heating with fossil fuels, such a shift

Table 12.4
Key figures characterizing the growth of energy demand for the five WAES scenarios, 1972–1985

	A	B	C	D	E
Oil Price[1]	Increase to $17.25	Increase to $17.25	Constant at $11.50	Constant at $11.50	Decrease to $7.66
National Policy Response	Vigorous	Vigorous	Vigorous	Restrained	Restrained
National GNP Growth Rate (%/yr.)[2]	4.4	4.2	4.4	4.2	4.4
Net Growth in Energy Demand by Type of Energy (%/yr.)[3]					
Electricity	2.8	2.7	2.9	3.0	3.1
Fossil Fuels	3.3	2.9	3.5	3.4	4.1
Total	3.1	2.8	3.2	3.2	3.7
Net Growth in Energy Demand by Sector (%/yr.)[4]					
Transportation	5.1	4.5	5.1	5.0	5.6
Industrial	3.3	3.0	3.4	3.3	3.6
Residential	0.0	-0.3	0.8	1.4	2.1
Net Energy Growth/GNP Growth	0.70	0.67	0.73	0.76	0.84
Gross Growth in Total Energy Demand (%/yr.)	2.7	2.5	2.9	2.9	3.2
Gross Energy Growth/ GNP Growth	0.61	0.60	0.66	0.69	0.73
Energy Demand 1985/ Energy Demand 1972					
Net	1.48	1.42	1.51	1.51	1.60
Gross	1.42	1.38	1.49	1.46	1.51

1. WAES oil prices are in 1975 U.S. dollars per barrel of Arabian light crude oil FOB Persian Gulf.
2. Growth rates are from Table 12.1, Subtotal, recalculated for the actual period.
3. Delivered to the consumer. Includes demand for oil and gas as feedstocks.
4. Includes losses during transportation, production, and distribution.

leads to a growth rate of energy demand at the consumer end that is lower than the growth rate of utilized energy (i.e., the energy that is really contributing to residential heating).

There is no difference between gross and net energy growth in the period to 1985. The reason for this is that electricity will be produced solely by hydro power throughout the period; therefore, no significant changes in energy losses will appear.

In the remainder of this section, the results are presented in greater detail and the effects of some important factors are discussed.

12.4.1 Effect of Different Scenario Parameters

Figure 12.3 shows the energy demand for the different 1985 scenarios. A comparison of the results for the different scenarios makes it possible to study the influence of variation of the different scenario parameters. Figure 12.3 provides a basis for the following conclusions:

1. The lower economic growth rate of scenario B leads to a reduction of total energy demand by about 4 percent as compared to scenario A.

2. The lower oil price of scenario C leads to an increase in total energy demand by about 2 percent as compared to scenario A.

3. The restrained policy response and the lower oil price of scenario D lead to an increase in the total energy demand by about 6 percent as compared to scenario B.

4. The restrained policy response and the decreasing oil price of scenario E lead to an increase in total energy demand by about 8 percent as compared to scenario A.

Figure 12.4 shows the effect of a vigorous national policy response on energy demand for scenarios B and C and the distribution of the total savings

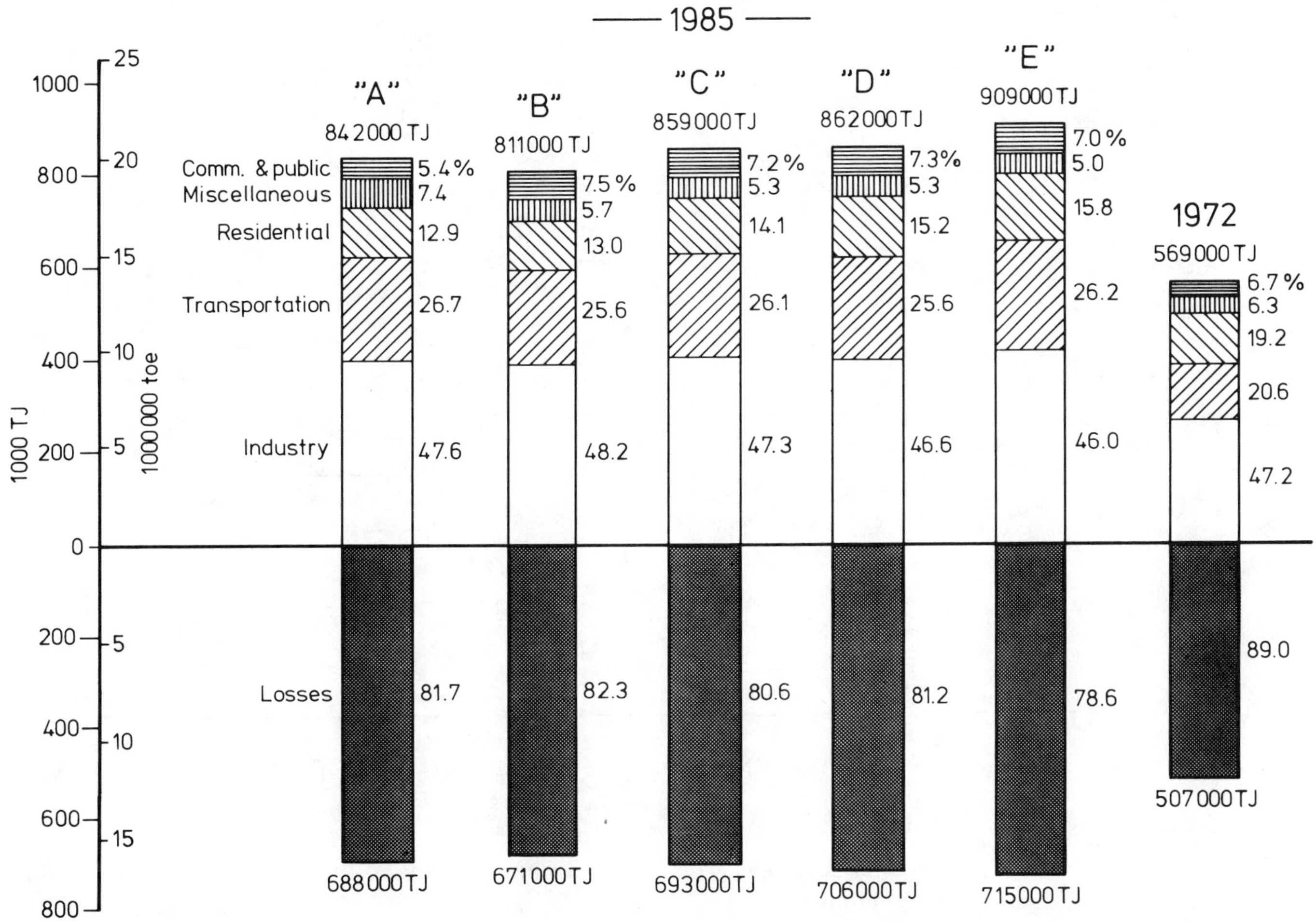

Figure 12.3 Energy demand for the five WAES scenarios, with distribution by sectors.

among the different sectors. Based on this figure, the following conclusions can be drawn:

1. The assumed vigorous policy response leads to a reduction of energy demand by about 6.5 percent for scenario B and 5 percent for scenario C. The effect is greater for scenario B due to the higher cost of energy, which makes stronger energy policies acceptable for this scenario than for scenario C.

2. The greatest energy savings are obtained within the three most energy-intensive sectors: transportation, industrial, and residential. Within the transportation sector, the main savings are due to reduction of automobile activity. In the industrial sector, the greatest savings are expected within the following groups: iron and steel, nonferrous metals, chemicals and allied, and paper and allied. Significant savings are also obtained because the energy losses are reduced in proportion to the reduction in net energy demand. In Section 12.5, the most important energy-saving policies are described, and an indication of their effect is given.

12.4.2 Energy Demand by Sector

Each sector's share of the total energy demand in 1985 is shown in Figure 12.3, from which the following conclusions can be drawn with regard to the development to 1985:

1. The share of industry, which is the largest sector, remains at about 47 percent throughout the period. This occurs despite the restrained growth in the metal-producing industries, which is balanced by the rapid expansion of the petrochemicals industry.

2. Transportation increases its share significantly from 20 percent in 1972 to about 26 percent in

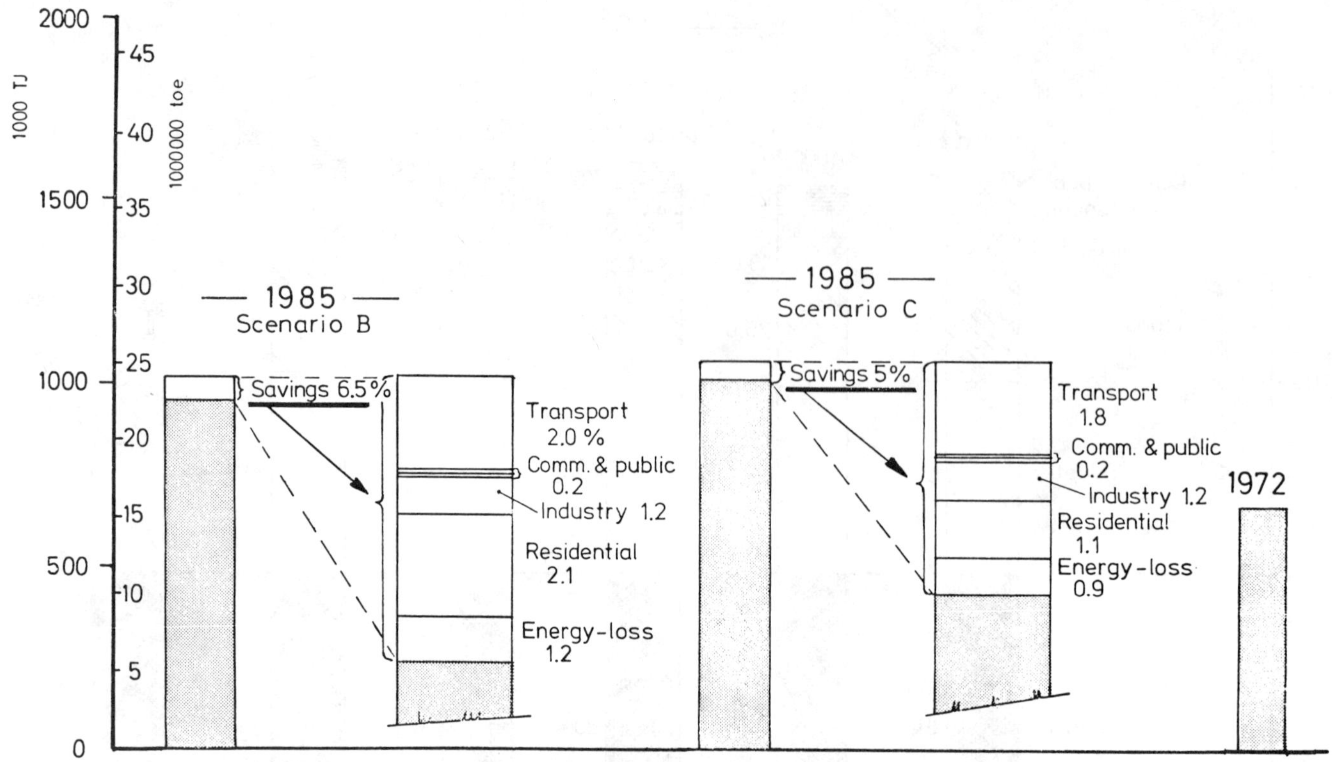

Figure 12.4 The effect of a vigorous national policy response: WAES scenarios B and C.

1985. This is a result of the continuous development toward international interdependence and specialization, which leads to a more rapid increase in freight than in production.

3. The share of the residential sector is reduced significantly from 19 percent in 1972 to about 14 percent in 1985. One reason for this is the strong shift toward more effective electric heating at the consumer end and extensive measures aimed at improving energy efficiency. The most fundamental cause is, however, that some of the most important demands within this sector are related to population size rather than economic activity level since saturation effects are becoming apparent.

4. The shares of the other sectors change very little during the period.

Figure 12.3 also shows that the constant or lower oil prices and the restrained policy response of scenarios D and E result in a much higher demand in the residential sector, giving this sector a significantly higher share in these scenarios than in the other scenarios.

12.4.3 Energy Demand by Fuel

Figure 12.5 shows the fuel mix of the energy demand for the five WAES scenarios and provides the basis for the following conclusions regarding the development of the fuel mix during the period to 1985:

1. Oil, which is the most important fuel, reduces its share somewhat from about 57 percent in 1972 to around 55 percent in 1985.

2. The share of electricity decreases slightly from 36 percent in 1972 to around 35 percent in 1985. This is basically due to the rapid expansion of oil- and gas-consuming activities such as transportation and the petrochemicals industry.

3. The share of gas increases from a negligible 0.2 percent in 1972 to 4 percent in 1985. This is due to the use of gas in the rapidly expanding petrochemicals industry.

4. The share of coal and coke decreases a little from barely 7 percent to around 6 percent in 1985

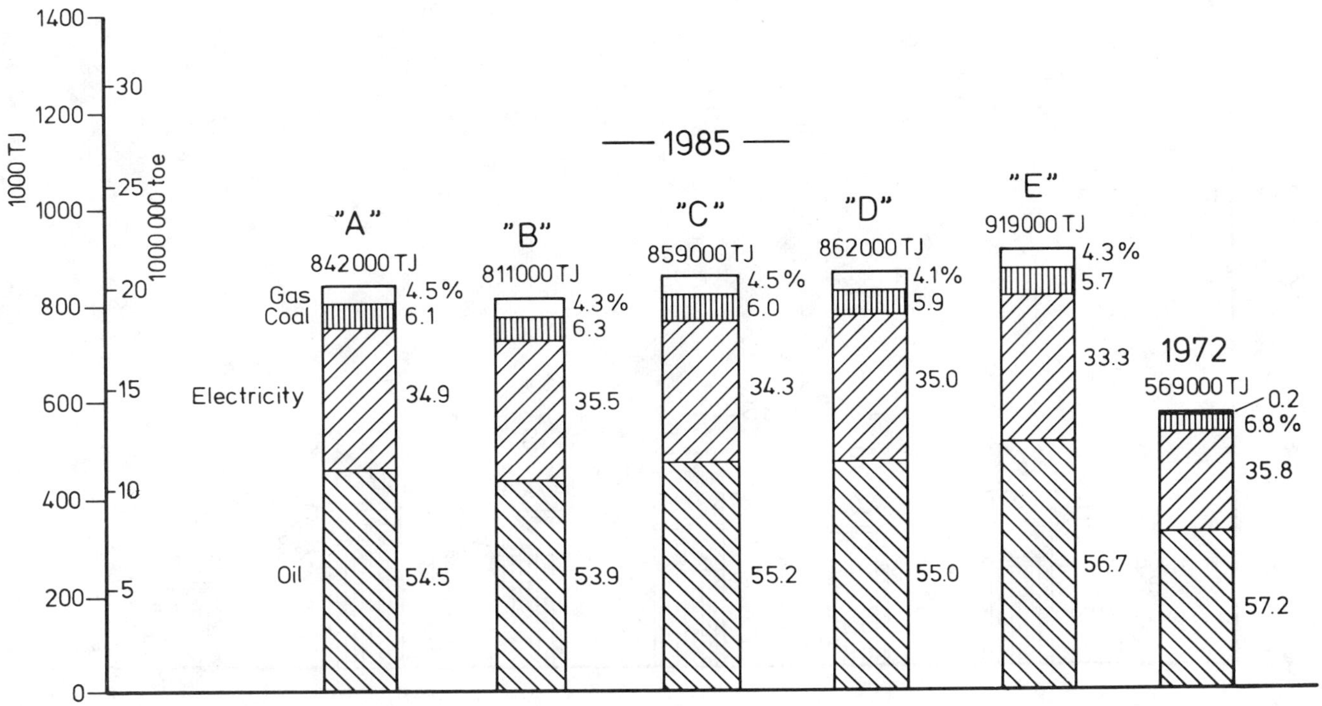

Figure 12.5 Energy demand by fuel for the five WAES scenarios.

as a result of the restrained growth assumed in the metal-producing industries.

5. On the whole, the share of fossil fuels increases slightly from around 64 percent in 1972 to about 65 percent in 1985.

6. The shift from electricity to fossil fuels is most evident in scenarios with high growth since transportation and the chemicals industry are relatively more dominant under these circumstances.

From Figure 12.5 it also appears that the decreasing oil price of scenario E leads to a larger share for oil in this scenario.

12.4.4 Energy Demand within the Transportation Sector

Figures 12.6 and 12.7 show the energy demand for 1985 within the transportation sector for passenger travel and freight, respectively. In general, it appears from these figures that freight's share of the total demand in the transportation sector increases from 56 percent in 1972 to around 61 percent for scenarios A, B, and C, to 57 percent for D, and to about 58 percent for scenario E in 1985.

Another general conclusion from this analysis is that the demand for electricity within the transportation sector will only have an insignificant share of the total in 1985. In the scenarios analyzed, it makes up less than 1 percent of the total. Electric cars are not expected to have any significance at all in the period to 1985.

Passenger Travel

Figure 12.6 provides a basis for the following conclusions concerning the development of passenger travel to 1985:

1. The energy demand for automobiles (urban plus rural) will increase by a factor of 1.6–1.9 over 1972.

2. Scenarios A and C are almost identical even though scenario A assumes a 50 percent higher crude oil price than scenario C. This occurs because the variations of this price have relatively little influence on prices for the final fuel products and particularly for gasoline, which is quite dominant.

3. Scenario B has about 50 percent lower energy demand than scenario A due to lower economic

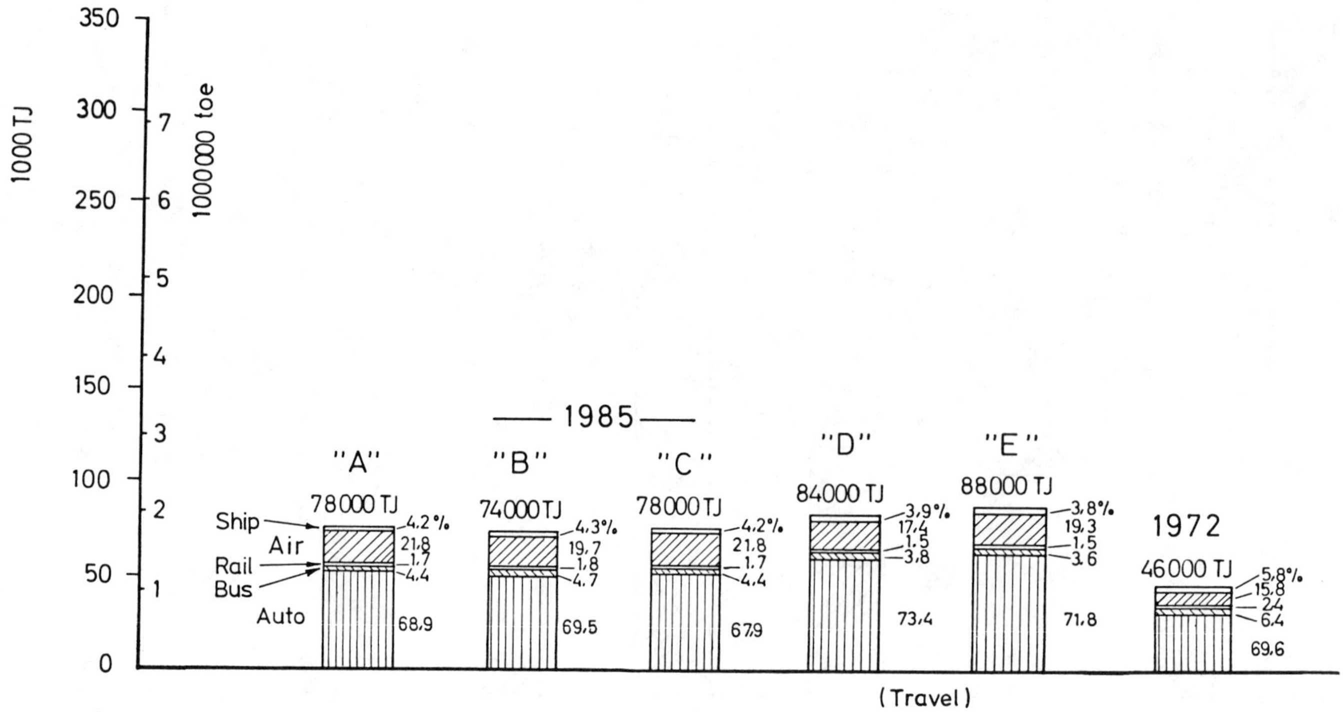

Figure 12.6 Energy demand for passenger travel for the five WAES scenarios.

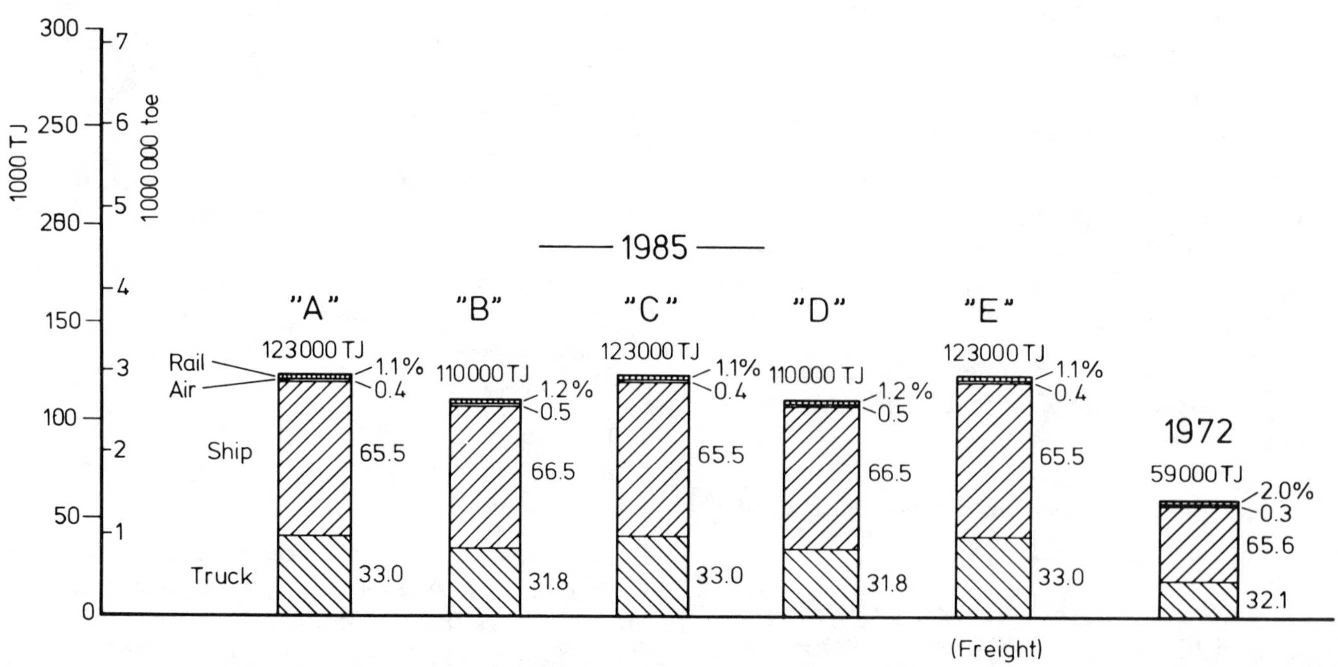

Figure 12.7 Energy demand for freight transport for the five WAES scenarios.

growth, which influences automobile activity in particular.

4. The restrained policy response for scenario D leads to around 13 percent higher energy demand than in scenario B.

5. For basically the same reason, scenario E has 13 percent higher energy demand than scenario A.

6. Air travel increases its share from barely 16 percent in 1972 to about 20 percent in 1985.

The share of the total energy demand required by automobiles stays about the same throughout the period, even if the number of autos almost doubles. Bus and rail (urban plus intercity) get reduced shares of the total despite a relative reduction in automobile commuting, which means a transfer of activity toward bus and rail. The average distance traveled per auto per year is reduced by 4–10 percent, depending on scenario, from 12,500 km in 1972. The specific fuel consumption per auto is reduced by 12–18 percent, also depending on the scenario, from around 10 km/liter (24 miles per U.S. gallon) in the base year.

Freight

For freight, the following conclusions can be drawn based on Figure 12.7:

1. The energy demand increases by a factor of 1.9–2.1, depending on the scenario, in the period 1972 to 1985.

2. Scenarios A, C, and E are almost identical with regard to both activity levels and energy demands because of their identical GNP growth rate assumptions. The differences in policy response and crude oil prices have very little influence. In the same way, scenarios B and D are nearly identical.

3. Truck (local plus intercity) and ship freight are dominant with regard to energy demand in 1985 just as in 1972. Rail and air freight still have a very small share of the total energy demand related to freight.

The specific energy consumptions have been reduced 10 to 20 percent, on average, by improved technology and better utilization of capacity.

12.4.5 Energy Demand within Industry

Figure 12.8, showing the energy demand within the industrial sector for the various 1985 scenarios, gives rise to the following conclusions:

1. The energy demand increases by a factor of 1.5 to 1.6 in the period 1972 to 1985.

2. The lower economic growth of scenario B results in a reduction of total energy demand by about 3 percent compared to scenario A.

3. The lower oil price of scenario C leads to an increase in total energy demand of only about 1 percent compared to scenario A.

4. The restrained policy response and lower oil price of scenario D cause an increase in total energy demand of around 3 percent compared to scenario B.

5. The restrained policy response and the reduction in oil price for scenario E result in an increase in total energy demand of about 4 percent compared to scenario A.

6. The chemicals and allied group shows the most rapid increase of energy demand. This is caused by the planned and expected expansion of the petrochemicals industry, especially with the production of primary plastics, which is very energy-intensive.

7. The share of the iron and steel and nonferrous metals groups decreases since it is assumed that these industries will be allocated a maximum of 2 percent more energy per year.

8. The share of the paper and allied group is reduced due to low growth and significant savings in specific energy consumption.

9. Metal manufacturing (transport equipment plus manufacturing of other metal products) is the only group other than chemicals that has an increasing share of the energy demand.

It thus appears that for such a short period as that to 1985, the price of oil has no major influence upon the energy demand of Norwegian industry. Besides the time factor, this is due to the relatively high share of electric energy and the fact that a substantial portion of the fossil fuels consumed is utilized as feedstock.

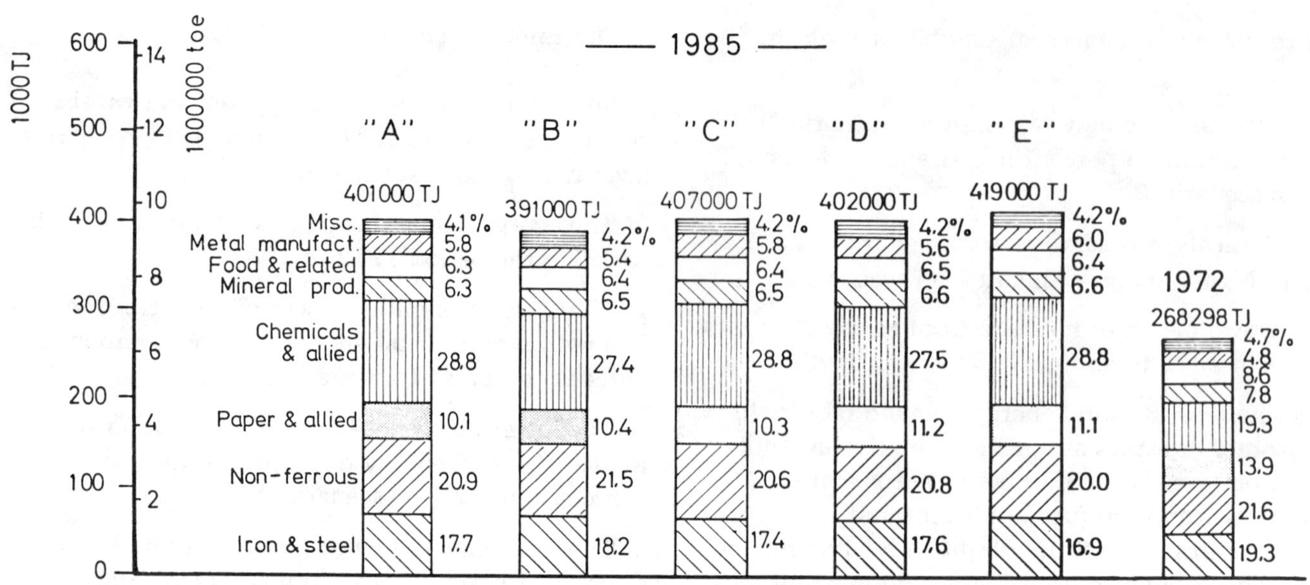

Figure 12.8 Energy demand within the industrial sector for the five WAES scenarios.

For different processes within the iron and steel group, the assumed improvements in specific energy consumption are in the area of 3–10 percent over 1972 values. No reutilization of waste heat in the production of electricity has been assumed before 1985. Improvement of the Hall process for the production of aluminum is expected to reduce the specific energy consumption by approximately 10 percent to an average of 17,250 kWh/ton in 1985. Within the paper and allied group, a 16 percent reduction in the average specific energy consumption is anticipated. The energy savings are mainly a result of better utilization of internal fuels and structural changes leading to larger production units. Finally, the substantial expansion of production of primary plastics in the petrochemicals industry causes an increase in the average specific energy consumption in the chemicals and allied group. This is because in 1985 plastics will be produced directly from natural gas, which is seven to eight times more energy-consuming than production based on the imported semiproducts that have been the raw material for the Norwegian production up to now.

12.4.6 Energy Demand within the Residential Sector

Figure 12.9 shows the energy demand in the residential sector for the different 1985 scenarios and gives rise to the following conclusions:

1. The total energy demand for scenarios A and B is virtually unchanged from 1972 to 1985.

2. The lower energy price of scenario C results in 11 percent higher energy demand than for scenario A.

3. The restrained policy response and the lower energy price of scenario D cause an increase in energy demand of about 24 percent compared to scenario B.

4. In scenario E, the restrained policy response and the decreasing oil price result in an increase in energy demand of around 32 percent compared to scenario A. An important reason for this is that scenario E has a substantially higher share of fossil fuels (oil) for heating—about the same as for the base year 1972.

5. Compared to the situation in 1972, all scenarios except E assume a significant shift toward use of electricity for space heating.

6. The share of the total energy demand related to household appliances, including water heating, increases at the expense of room heating for all scenarios. The reasons are better insulation, more widespread use of electric household appliances, and the fact that about 50 percent of the waste heat from these appliances will be contributing to the space heating.

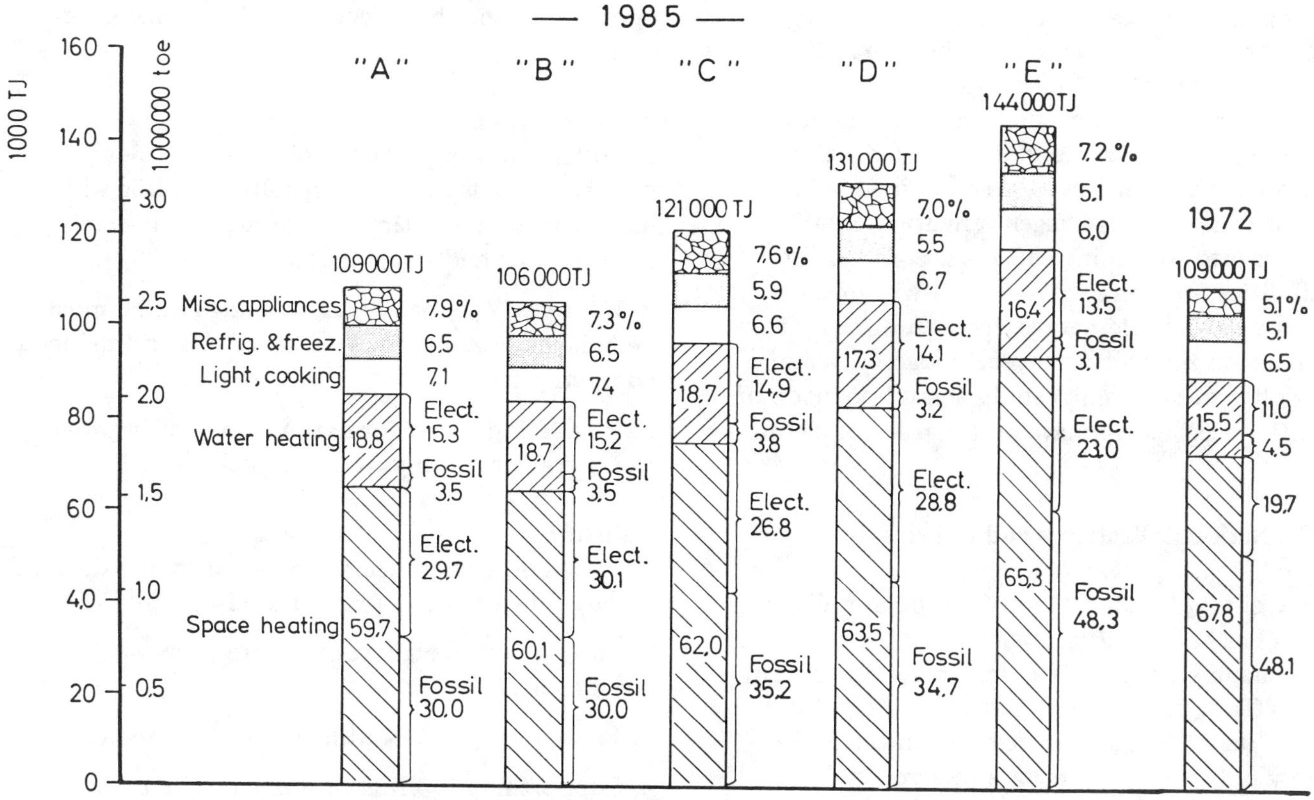

Figure 12.9 Energy demand within the residential sector for the five WAES scenarios.

It is thus apparent that the price of oil will have a significant influence on energy demand within the residential sector. This is partly because it has been assumed that the price of oil and the price of electricity will be coupled to a certain extent, so that the oil price will affect the cost and use of all energy in the residential sector. For scenario E, it is of major importance that the low oil price restricts the shift from fossil fuels to electricity in the manner assumed for the other scenarios, since at such a low oil price there is no longer the same coupling between the cost of oil and the cost of electricity. Since heating by use of fossil fuels is connected with relatively high losses—about one-third of the total consumer demand—while electric heating entails no losses at the consumer end, the same effective space heating requirement demands increased expenditure of energy if fossil fuels are used.

A total population of 1.69 million is anticipated in 1985, compared to 1.3 million in 1972. An increase of 7-10 percent in average floor area per capita is also assumed, as well as increased saturation for all important appliances. The relatively modest increase in total energy demand for the most realistic scenarios is a result of vigorous policy response, higher electricity prices, and the contribution to space heating from household appliances.

12.5 CONCLUSIONS

Changes in the energy area created by the dramatic rise in the price of oil—quadrupling during the fall of 1973 and spring of 1974—mean that many processes and a lot of equipment are no longer at a technical and economical optimum. Significant energy savings can thus be obtained in a sound, economic way by improvement of existing processes and systems, or by use of new processes and technologies that have become competitive under the new circumstances.

The new oil price has also influenced prices of other types of fuels and the price of electricity on the world market. This means that the new situation also, in an indirect way, influences electricity-consuming processes and equipment in Norway.

Government policy actions in the energy field can help to accelerate a development toward processes and equipment that are at a technical optimum in this new situation. Such a development might, in many cases, be combined with fulfillment of the more stringent environmental requirements now being enforced. Some of the possibilities for energy saving dealt with in this section will, if implemented, have an unfavorable influence on the daily and yearly load-curves of electricity production. A study should be made of the influence this will have on the price of electricity.

12.5.1 National Strategies and Policies

Energy conservation achievable through policy actions has been analyzed within the framework of this demand study. Such actions may have a significant effect, producing an approximately 6 percent reduction in total energy demand in 1985. Energy policy actions include research and development.

Our evaluation of possible policy actions has covered only the most energy-intensive sectors: transportation, industrial, and residential. The objectives of the policy actions can be described as follows: to influence energy consumers in such a way that the primary energy sources are utilized *as efficiently as possible* in order to save energy, considering constraints of the following types:

1. *Financial and economic.* As capital is usually a limiting factor, only possibilities for energy saving that give returns comparable to those of other investment possibilities are realistic.

2. *Technological.* Only technology that is commercially available can be relied upon. Lead times for entirely new technologies are long—usually one or more decades—even if large resources are used for research and development.

3. *Social and political.* Only energy-saving policies that are socially and politically acceptable can be utilized. Policies entailing effects that are unacceptable to society can be adopted only if they can be combined with countermeasures that render the combined effects acceptable.

The following strategies are the most important ones within the three sectors that we are considering.

Transportation

1. Shifting passengers and goods to less energy-intensive means of transport (although this will usually mean more time in transit, which exacts an economic and a social cost).

2. Increased occupancy and loading (which must be balanced against reduced schedules and reduced competition).

3. Regulation and information aimed at improving the energy efficiency of automobiles.

Industry

1. Increased energy prices, reducing energy demand through the effect of price elasticity.

2. Supporting changeover to more energy-efficient equipment.

3. Encouraging better utilization of waste heat.

4. Supporting recycling of waste products.

Residential

1. Affecting heat loss of new buildings through regulation.

2. Increased prices of electricity, reducing energy demand through the effect of price elasticity.

3. Supporting improvements that reduce heat loss in old buildings.

4. Providing information to people that will aid them in using energy more efficiently.

5. Encouraging by other means use of the most efficient appliances and heating systems.

6. Supporting development of new, energy-efficient systems for space heating, water heating, and ventilation.

12.5.2 Research and Development

Research and development must be regarded as a very important aspect of policy action. Much of the assumed energy savings is dependent upon the necessary technological basis being established through R&D.

Industry

The most important possibilities for energy saving in the industrial sector in Norway lie with the few groups that are dominant with regard to energy demand: iron and steel, nonferrous metals, chemicals and allied, and paper and allied. The possibilities for energy savings that are regarded as most promising in relation to R&D efforts are in the following areas:

1. Better control and improvement of existing smelting processes within the metal-producing industries.

2. Utilization of waste heat, particularly within the metal-producing industries, but also within chemicals and allied and paper and allied.

3. Better utilization of waste-type fuels within paper and allied.

It is possible to identify a relatively small number of processes that are dominant in their energy consumption and that are being used in many plants. As far as can be determined today, most of these processes will not be replaced by new ones for a long time. Concentration on R&D efforts for these processes should, therefore, make significant energy savings possible.

Residential

The best possibilities for energy savings within the residential sector are related to space heating, but significant amounts of energy may also be saved in connection with water heating. The possibilities for energy savings that are regarded as most promising in relation to R&D efforts are in the following areas:

1. Utilization of heat pump systems for space and water heating.

2. Utilization of solar energy systems for space and water heating.

3. Various measures for reduction of heat loss from buildings, such as better ventilation, better insulation, etc.

At present, residential heating in Norway accounts for about 15 percent of the total energy demand. By making use of the possibilities for energy savings mentioned above, the total energy demand within the residential sector can be stabilized at about its present level. Achievement of such an effect will require considerable R&D effort. This kind of effort may produce energy savings whose significance will increase with time. Moreover, developments with regard to residential heating might also lead to energy savings in connection with space heating in other sectors.

Transportation

The transportation sector has an increasing share of the total energy demand and is assumed to make up about one-fourth of the total energy demand in Norway in 1985. Ship freight accounts for the largest consumption of energy within this sector, even when only bunkers filled in Norway are included. Since Norway has a large merchant fleet, improved energy efficiency for ships would be of even greater importance than may appear from the national energy balance.

The following tasks, all of which aim at improved ship energy efficiency, stand out as important areas for Norwegian R&D:

1. Improvement of hull configuration.

2. Improvement of propulsion systems, especially propeller systems and possible alternatives to such systems.

3. Improvement of systems for cleaning hulls and studies of other possible methods for reducing friction.

12.5.3 Effects of Different Policy Actions

Below, in summary form, are given the results of the evaluation of policy actions for the 1985 scenarios. The effects given relate to scenario C, which assumes vigorous policy response.

Transportation

Several policies have usually been applied to each of the different modes of transportation. The most significant possibilities for energy savings are related to passenger travel. It has been taken into account that policies which influence one mode of transportation often have secondary effects on other modes of transportation.

Automobile activity (passenger-kilometers) is almost five times that of bus and rail together in 1985. Policies that apply directly to automobiles therefore result in larger energy savings than those for all the other transportation modes together.

For the automobile, the following policies have been evaluated:

1. Registration taxes that increase with weight and engine power.

2. A 50 percent increase in the price of gasoline.

3. Minimum requirements for fuel economy.

4. Miscellaneous policies such as higher parking fees.

In total, it is estimated that these policies would reduce the average driving distance per auto per year by about 7 percent, with a reduction in the average specific fuel consumption of around 10 percent.

For bus and rail, the following policies have been evaluated:

1. Subsidies on ticket prices to such an extent that prices are kept at today's level in real terms.

2. Increased information on, and better coordination of, routes and schedules.

3. Special lanes for buses.

4. Expansion of rail transport systems.

5. "Park and ride" arrangements at suburban train stations.

In total, it is estimated that these policies would increase the activity level for bus and rail by 11–17 percent and would reduce the specific energy consumption by about 8–14 percent. This last reduction is basically due to better utilization of capacity.

Industry

The evaluation has been concentrated on the industrial groups that are most energy-intensive. These groups are iron and steel, nonferrous metals, paper and allied, and chemicals and allied. The following policies have been evaluated:

1. Low-cost loans for installation of energy-saving equipment.

2. Tax credit for installation of energy-saving equipment and for recycling.

3. Increased electricity prices.

4. Financial support for research and development, especially on existing processes that are most dominant with regard to energy consumption.

In total, these policies are estimated to reduce the average specific energy consumptions as follows:

Iron and steel	1.5 percent
Nonferrous metals	3.5 percent
Paper and allied	10 percent
Chemicals and allied	3 percent.

By far the greatest reduction of the specific energy consumption is obtained for paper and allied. For the metal-producing industries as a whole, the energy savings achieved are also significant in absolute terms, since this part of the industry accounts for about 38 percent of the demand.

Residential

Within the residential sector, the following policies have been evaluated:

1. Low-cost loans for improvement of old buildings.

2. More stringent energy efficiency requirements in building codes.

3. Information about means for energy saving and selection of energy-efficient equipment.

4. Increased electricity prices.

5. Research and development.

In total, these policies are estimated to give a reduction in the average demand for heating per unit of floor area of about 7–10 percent in 1985. Moreover, some of the above-cited policies also have the effect of reducing energy demand per unit of household appliances by around 9 percent.

12.6 DEMAND STUDY WORKSHEETS

This section includes an energy use summary, an energy growth rates summary, a breakdown of sectoral percentages of total final energy demand, two sets of 1985 case comparisons, and summary and sectoral worksheets for the base year, 1972, and for each of the five WAES scenarios.

WAES NATIONAL ENERGY DEMAND STUDIES, 1972-1985
ENERGY USE SUMMARY

Country NORWAY

Scenario Assumptions

WAES Case	A		B		C		D		E			
GNP 1976-1985	High		Low		High		Low		High			
Oil Price*	$11.50 → $17.25		$11.50 → $17.25		$11.50		$11.50		$11.50 → $7.66			
Natl. Policy Response	Vigorous		Vigorous		Vigorous		Restrained		Restrained			
Natl. GNP Growth Rate 1976-1985 (%/yr.)	4.4		4.2		4.4		4.2		4.4		1972 Base Year	
Fuel Type	Fossil Fuel	Elec.	Fossil Fuel	Elec.	Fossil Fuel	Elec.	Fossil Fuel	Elec.	Fossil Fuel	Elec.	Fossil Fuel	Elec.
Units	10^{15} J/year		10^{15} J/year		10^{15} J/year		10^{15} J/year		10^{15} J/year		10^{15} J/year	
Energy-Using Sectors												
Industrial (excluding feedstocks)	175.2	184.3	171.8	180.3	180.9	184.3	180.7	182.9	190.2	187.0	116.8	127.8
Transportation	221.4	2.9	204.6	2.8	221.4	2.9	218.2	2.8	235.0	2.9	115.3	2.0
Commercial and Public	37.2	24.4	36.8	24.1	37.2	24.4	37.6	24.7	38.0	25.0	23.0	15.0
Residential	37.2	71.6	36.1	69.5	48.0	73.0	50.5	80.5	66.3	77.7	58.0	51.2
Fishing, Agriculture, Mining, Construction	35.3	10.5	35.4	10.6	35.3	10.5	35.4	10.6	35.3	10.5	28.1	8.1
Feedstocks	41.5		38.6		41.5		38.6		41.5		20.0	
Final Energy Demand (excluding processing losses)	547.8	293.7	523.3	287.4	564.3	295.1	561.0	301.5	606.3	303.0	361.2	204.1
Processing Losses	61.5	625.9	58.6	612.3	63.8	628.9	63.5	642.4	69.0	645.8	45.4	445.7
Primary Energy Input	609.3	919.6	581.9	899.7	628.1	924.0	624.5	943.9	675.3	948.8	406.6	649.8

*Oil price 1975 US$ per barrel Arabian light crude oil FOB Persian Gulf.

WAES NATIONAL ENERGY DEMAND STUDIES, 1972-1985
ENERGY GROWTH RATES SUMMARY

Country NORWAY

Scenario Assumptions										
WAES Case	A		B		C		D		E	
GNP 1976-1985	High		Low		High		Low		High	
Oil Price*	\$11.50 → \$17.25		\$11.50 → \$17.25		\$11.50		\$11.50		\$11.50 → \$7.66	
Natl. Policy Response	Vigorous		Vigorous		Vigorous		Restrained		Restrained	
Natl. GNP Growth Rate 1976-1985 (%/yr.)	4.4		4.2		4.4		4.2		4.4	
Fuel Type	Fossil Fuel	Elec.	Fossil Fuel	Elec.	Fossil Fuel	Elec.	Fossil Fuel	Elec.	Fossil Fuel	Elec.
Units (%/yr. 1972-1985)	%/yr		%/yr		%/yr		%/yr		%/yr	
Energy-Using Sectors										
Industrial (excluding feedstocks)	3.2	2.9	3.0	2.7	3.4	2.9	3.4	2.8	3.8	3.0
Transportation	5.1	2.9	4.5	2.6	5.0	2.9	5.0	2.6	5.6	2.9
Commercial and Public	3.8	3.8	3.7	3.7	3.8	3.8	3.9	3.9	3.9	4.0
Residential	(3.4)	2.6	(3.6)	2.4	(1.4)	2.8	(1.1)	3.5	1.0	3.3
Fishing, Agriculture, Mining, Construction	1.8	2.0	1.8	2.1	1.8	2.0	1.8	2.1	1.8	2.0
Feedstocks	5.8		5.2		5.8		5.2		5.8	
Final Energy Demand (excluding processing losses)	3.3	2.8	2.9	2.7	3.5	2.9	3.4	3.0	4.1	3.1
Processing Losses	2.4	2.6	2.0	2.5	2.7	2.7	2.6	2.9	3.3	2.9
Primary Energy Input	3.2	2.7	2.8	2.5	3.4	2.7	3.4	2.9	4.0	3.0

*Oil price 1975 US\$ per barrel Arabian light crude oil FOB Persian Gulf.

WAES NATIONAL ENERGY DEMAND STUDIES, 1972-1985
SECTORAL PERCENTAGES OF TOTAL

Country NORWAY

Scenario Assumptions

WAES Case	A		B		C		D		E			
GNP 1976-1985	High		Low		High		Low		High			
Oil Price*	$11.50 → $17.25		$11.50 → $17.25		$11.50		$11.50		$11.50 → $7.66			
Natl. Policy Response	Vigorous		Vigorous		Vigorous		Restrained		Restrained			
Natl. GNP Growth Rate 1976-1985 (%/yr.)	4.4		4.2		4.4		4.2		4.4		1972 Base Year	
Fuel Type	Fossil Fuel	Elec.	Fossil Fuel	Elec.	Fossil Fuel	Elec.	Fossil Fuel	Elec.	Fossil Fuel	Elec.	Fossil Fuel	Elec.
Units	%		%		%		%		%		%	
Energy-Using Sectors												
Industrial (excluding feedstocks)	32.0	62.8	32.8	62.7	32.1	62.5	32.2	60.7	31.4	61.7	32.3	62.6
Transportation	40.4	1.0	39.1	1.0	39.2	1.0	38.9	.9	38.8	1.0	31.9	1.0
Commercial and Public	6.8	8.2	7.0	8.4	6.6	8.3	6.7	8.2	6.3	8.3	6.4	7.3
Residential	6.8	24.4	6.9	24.2	8.5	24.7	9.0	26.7	10.9	25.6	16.1	25.1
Fishing, Agriculture, Mining, Construction	6.4	3.6	6.8	3.7	6.3	3.5	6.3	3.5	5.8	3.4	7.8	4.0
Feedstocks	7.6		7.4		7.3		6.9		6.8		5.5	
Final Energy Demand (excluding processing losses)	100.0	100.0	100.0	100.0	100.0	100.0	100.0	100.0	100.0	100.0	100.0	100.0
Processing Losses												
Primary Energy Input												

*Oil price 1975 US$ per barrel Arabian light crude oil FOB Persian Gulf.

WAES NATIONAL ENERGY DEMAND STUDIES
1985 CASE COMPARISONS

Country NORWAY

Scenario Cases Compared	Scenario Variables: GNP	Oil Price	National Policy Response	Energy-Use Category	Comparison (elasticity) units =
A	4.4	High	Vigorous	Transport Sector	1.69
B	4.2	High	Vigorous		
A	4.4	High	Vigorous	Industry Sector	.56
B	4.2	High	Vigorous		
A	4.4	High	Vigorous	Commercial Sector	.26
B	4.2	High	Vigorous		
A	4.4	High	Vigorous	Residential Sector	.65
B	4.2	High	Vigorous		
A	4.4	High	Vigorous	Total end use energy	.80
B	4.2	High	Vigorous		
A	4.4	High	Vigorous	Total primary energy	.68
B	4.2	High	Vigorous		

WAES NATIONAL ENERGY DEMAND STUDIES
1985 CASE COMPARISONS

Country NORWAY

Scenario Cases Compared	Scenario Variables: GNP	Oil Price	National Policy Response	Energy-Use Category	Comparison (elasticity) units =
A	High	$11.50 - 17.25	Vigorous	Transport Sector	-.000
C	High	$17.25	Vigorous		
A	High	$11.50 - 17.25	Vigorous	Industry Sector	-.063
C	High	$17.25	Vigorous		
A	High	$11.50 - 17.25	Vigorous	Commercial Sector	-.000
C	High	$17.25	Vigorous		
A	High	$11.50 - 17.25	Vigorous	Residential Sector	-.474
C	High	$17.50	Vigorous		
A	High	$11.50 - 17.25	Vigorous	Total end use energy	-.094
C	High	$17.25	Vigorous		
A	High	$11.50 - 17.25	Vigorous	Total primary energy	-.067
C	High	$17.25	Vigorous		

NATIONAL ENERGY DEMAND PROJECTIONS

Country NORWAY

Year 1972

WAES Case BASE YEAR

Summary

UNITS = TJ

Category	Oil	Gas	Coal	Electricity	Other	Total by Category (Excluding Processing Losses)
Transportation	115 311	----	----	2 027	----	117 338
Agriculture, Mining,(1) and Construction	28 118	----	----	8 110	----	36 228
Commercial	7 606	46	2 573	7 729	----	17 954
Public	12 785	----	----	7 272	----	20 057
Industrial	103 452	1 147	36 298	127 754	----	268 651
Residential	57 970	----	----	51 240	----	109 199
Total Delivered Energy by Fuel	325 242	1 193	38 871	204 127	----	
Processing Losses by Fuel	12.65%	9.6%	----	218%	----	
Gross Input by Fuel	366 385	1 308	38 871	649 786		1 076 362

Gross Energy Input (Add across bottom row)

(1) Includes forestry & fishing.

NATIONAL ENERGY DEMAND PROJECTIONS

Country NORWAY

Year 1972

WAES Case BASE YEAR

Transportation

	(A)	(B)	(C)	(D)	(E)	(F)	(G)	
	Population/Auto	Distance/Auto/Yr.(1)	Load Factor	Fraction Urban or Rural	Energy/Distance (1)	(TJ) Fossil Fuel	(TJ) Electricity	
Urban Auto	5.02	12 500 km			3.253×10^6 J/km	32 000		(1)
Rural Auto								(2)
	Activity Level(2)	% Electrified	Load Factor	Fossil Fuel/ Activity Level (3)	Electricity/Activity (3)			
Urban Bus (5)	.6			.93		550		(3)
Intercity Bus	3.27			.73		2 380		(4)
Rail (Passenger)	2.1	81		.84	.38	336	646	(5)
Air (Passenger, Domestic)	.92			4.19		3 850		(6)
Air (Passenger, International)						3 390(4)		(7)
Ship (Passenger)	.64			4.19		2 690		(8)
Rail (Freight)	2.48	67		.73	.34	597	565	(9)
Air (Freight, Domestic)	.007			31.43		220		(10)
Air (Freight, International)								(11)
Truck (Local)	4.7			4.02		18 973		(12)
Truck (Intercity)								(13)
Ship (Freight, Domestic)	9.18			1.78		16 342		(14)
Ship (Freight, International)	156.06			.14		21 848(4)		(15)
Miscellaneous						12 135	775	(16)
					Total	115 311	2 027	

(1) Estimate

(2) 10^9 passenger kilometer or 10^9 ton kilometer

(3) 10^6 J/passenger kilometer or 10^6 J/ton kilometer

(4) Fuel filled in Norway, all operators

(5) Oslo/Bergen/Trondheim.

NATIONAL ENERGY DEMAND PROJECTIONS

Country NORWAY

Year 1972

WAES Case BASE YEAR

Agriculture, Mining, and Construction

Agriculture, Forestry, and Fishing (ISIC Major Division 1)	(A) (Mill. $ ā Nkr 6.59) Value Added	(B) (J/kr) Energy/Value Added	(C) (Km^2) Area Harvested	(D) (J/km^2) Energy/Area	(E) (Tons) Physical Output	(F) (J/ton) Energy/Output	(G) (TJ) Fossil Fuel	(H) (TJ) Electricity	
Farm Machinery			7 950	$.03x10^{12}$	6 304 500	$41x10^6$	260		(17)
Greenhouses									(18)
Other Agricultural Uses	372	$23.79x10^6$			1 942 058	$4\ 564x10^6$	3 174	5 690	(19)
Forestry (ISIC 12)	110	$14.04x10^6$			4 999 354	$309x10^6$	1 544		(20)
Fishing Boats (ISIC 13)	191	$92.26x10^6$			2 906 646	$6\ 075x10^6$	17 660		(21)
Other Fishing Use (ISIC 13)									(22)
Total Fertilizer (tons)		N 78 332	P 24 365	K 58 482		Total	22 378	5 690	

Mining	(A) Value Added	(B) Energy/Value Added (J/kr)	(C) Physical Output (Tons)	(D) Energy/Physical Output (J/ton)	(E) Fossil Fuel (TJ)	(F) Electricity (TJ)	
Handling and Internal Transportation	100	$43.56x10^6$	29 346 146	$149x10^6$	1 950	2 420	(23)
Other Energy Use							(24)
				Total	1 950	2 420	

Construction	(A) Value Added	(B) Energy/Value Added	(C) Physical Output	(D) Energy/Physical Output	(E) Fossil Fuel	(F) Electricity	
Handling and Internal Transportation	841	$3.16x10^6$			2 640		(25)
Other Energy Use (1)	841	$1.38x10^6$			1 150		(26)
				Total	3 790		
				Agriculture, Mining, and Construction Total	28 118	8 110	

(1) Heating and drying of materials.

NATIONAL ENERGY DEMAND PROJECTIONS

Country NORWAY

Year 1972

WAES Case BASE YEAR

Commercial and Public

	(A)	(B)	(C)	(D)	(E)	
Commercial	Floor Area (1) (5)	Heat Gain or Loss/Area (6)	System Efficiency	Fossil Fuel (7)	Electricity (7)	
Space Heat (Fossil Fuel)	329 000	10 133 / 329x10³		10 133 (2)		(27)
Space Heat (Electricity)	329 000	7 729 / 329x10³			7 729 (3)	(28)
Air Conditioning (Fossil Fuel)						(29)
Air Conditioning (Electricity)						(30)
		Energy/Area	Electricity/Area			
Lighting						(31)
Miscellaneous	329 000	46 / 329x10³		46 (4)		(32)
Presently Unknown						(33)
			Total	10 179	7 729	
Public	Floor Area	Heat Gain or Loss/Area	System Efficiency	Fossil Fuel	Electricity	
Space Heat (Fossil Fuel)	373 000	12 785 / 373x10³		12 785 (2)		(34)
Space Heat (Electricity)	373 000	7 272 / 373x10³			7 272 (3)	(35)
Air Conditioning (Fossil Fuel)						(36)
Air Conditioning (Electricity)						(37)
		Energy/Area	Electricity/Area			
Lighting						(38)
Miscellaneous						(39)
Presently Unknown						(40)
			Total	12 785	7 272	
			Commercial and Public Total	22 964	15 001	

(1) Floor area is not available. The building statistics are based on yearly changes

(2) Includes energy for hot water supply and some unspecified uses

(3) Total value for the category, includes lighting, water heating etc.

(4) Gasworks, gas mainly for cooking and heating (restaurants etc.)

(5) Represented by number of employed

(6) Unit = TJ/person

(7) Unit = TJ.

NATIONAL ENERGY DEMAND PROJECTIONS

Country NORWAY

Year 1972

WAES Case BASE YEAR

Industrial

	(A) Value Added (1)	(B) Energy/Value Added	(C) Physical Output	(D) Energy/Output	(E) % Exported	(F) Fossil Fuel (2)	(G) Electricity	(H) Feedstocks	
	(Mill. $ â Nkr 6.59)	(J/$)	(Ton)	(J/ton)		(TJ)	(TJ)		
Iron and Steel (ISIC 371)	145	$356x10^6$	3 098 452	$16\ 674x10^6$		26 248	25 416		(41)
Nonferrous Metals (ISIC 372)	153	$326x10^6$	844 104	$68\ 604x10^6$		13 820	44 089		(42)
Paper and Allied (ISIC 34)	418	$90x10^6$	3 478 032	$10\ 721x10^6$		20 878	16 412		(43)
Chemicals and Allied (ISIC 35 minus ISIC 353)	285	$113x10^6$	2 406 916	$13\ 360x10^6$		11 177	20 981	19 670	(44)
Mineral Products (ISIC 36)	140	$149x10^6$				18 201	2 621		(45)
Food and Related (ISIC 31)	496	$46x10^6$	2 924 793	$7\ 859x10^6$		15 920	7 067		(46)
Transport Eqpt. (ISIC 384)	401	$12x10^6$				2 430	2 423		(47)
Key Industry (N.E.C.)(4)	591	$14x10^6$				3 450	4 594		(48)
Key Industry (N.E.C.)(5)	499	$19x10^6$				5 487	4 151		(49)
Misc. Manufacturing (3)						3 274		342	(50)
N.E.C. = Not Elsewhere Classified					Total	120 885	127 754	20 012	

(1) Tax not included

(2) Includes wood

(3) Residual LPG, coal and coke not distributed

(4) ISIC 38-384

(5) ISIC 32, 33 and 34.

NATIONAL ENERGY DEMAND PROJECTIONS

Country NORWAY

Year 1972

WAES Case BASE YEAR

Residential

Population 3.95×10^6 Persons/Dwelling 3.04 Dwellings 1.3×10^6

Central Space Heating		(A) Dwellings	(B)(7) Heat Loss (MJ)	(C) System Efficiency	(D)(4)	(E)(5) Fossil Fuel (TJ)	(F)(5) Electricity (TJ)	
Single Family	Fossil Fuels	(0.62)(1)	64 800	.66	.95	36 200		(51)
		762 191	(= 18 000 kWh)					
	Electricity	(0.38)(2)	(- 3 175 kWh)(3)	1.0	.95		14 820	(52)
Town or Row Houses	Fossil Fuels	(0.62)	50 400	.66	.95	10 600		(53)
		305 787	(= 14 000 kWh)					
	Electricity	(0.38)	(- 3 175 kWh)	1.0	.95		4 340	(54)
Apartments	Fossil Fuels	(0.62)	39 600	.66	.95	5 740		(55)
		228 787	(= 11 000 kWh)					
	Electricity	(0.38)	(- 3 175 kWh)	1.0	.95		2 350	(56)

	% Saturation (Fossil Fuels)	% Saturation (Electricity)(6)	Energy/Unit (Fossil Fuels)	Energy/Unit (Electricity) (kWh)			
Room Heating							(57)
Air Conditioning (Window)							(58)
Air Conditioning (Central)							(59)
Water Heating		80		4 200	4 920	12 020	(60)
Refrigerators		94		600		2 840	(61)
Lighting		100		800		4 030	(62)
Cooking		100		600		3 020	(63)
Dishwashers		5		200		15	(64)
Freezers		71		750		2 690	(65)
Television		72		110		400	(66)
Other Key Appliances		86		500		2 170	(67)
Misc. Appliances					510(8)	2 540	(68)
Potential Appliances							(69)
				Total	57 970	51 240	

(1) Share of dwellings, which are fossil heated

(2) Share of dwellings, which are elect. heated

(3) Indirect heating from appliances (50% of total consumption of electricity to such equipment)

(4) $\frac{\text{Climate factor 1972}}{\text{Climate factor normal year}}$

(5) Disaggregation of fossil fuel and electricity for space heating for each type of dwelling has been based on the average distribution for all dwellings

(6) Saturation percentages are based on the number of households which amounts to 1.4×10^6

(7) Average value for each type of dwelling, inclusive heat contribution from electrical (residential) appliances (50% of appliance energy consumption)

(8) Gasworks gas and LPG, mainly for cooking and heating.

NATIONAL ENERGY DEMAND PROJECTIONS

Country NORWAY

Year 1985

WAES Case A

UNITS = TJ

Summary

Category	Oil	Gas	Coal	Electricity	Other	Total by Category (Excluding Processing Losses)
Transportation	221 433	----	----	2 909	----	224 342
Agriculture, Mining,(1) and Construction	35 294	----	----	10 523	----	45 817
Commercial	16 985	86	----	12 955	----	30 026
Public	20 140	----	----	11 455	----	31 595
Industrial	127 386	38 286	51 025	184 256	----	400 953
Residential	37 347	----	----	71 623	----	108 870
Total Delivered Energy by Fuel	458 485	38 372	51 025	293 721	----	
Processing Losses by Fuel	12.65%	9.6%	----	213%	----	
Gross Input by Fuel	516 483	42 056	51 025	919 574	----	1 529 138

Gross Energy Input (Add across bottom row)

(1) Includes forestry & fishing.

NATIONAL ENERGY DEMAND PROJECTIONS

Country NORWAY

Year 1985

WAES Case A

Transportation

	(A)	(B)	(C)	(D)	(E)	(F)	(G)	
	Population/Auto	Distance/Auto/Yr.	Load Factor	Fraction Urban or Rural	Energy/Distance	(TJ) Fossil Fuel	(TJ) Electricity	
Urban Auto	2.65	11 375 km			2.667 x 10^6 J/km	52 630		(1)
Rural Auto								(2)
	Activity Level (1)	% Electrified	Load Factor	Fossil Fuel/ Activity Level (2)	Electricity/Activity (2)			
Urban Bus(4)	.77			.78		600		(3)
Intercity Bus	4.82			.59		2 841		(4)
Rail (Passenger)	3.3	81		.71	.52	467	844	(5)
Air (Passenger, Domestic)	2.53			3.35		8 483		(6)
Air (Passenger, International)						8 421(3)		(7)
Ship (Passenger)	.83			3.89		3 228		(8)
Rail (Freight)	3.97	67		.52	.24	680	637	(9)
Air (Freight, Domestic)	.025			24.83		607		(10)
Air (Freight, International)								(11)
Truck (Local)	11.81			3.43		40 508		(12)
Truck (Intercity)								(13)
Ship (Freight, Domestic)	22.19			1.62		35 947		(14)
Ship (Freight, International)	341.16			.13		44 351(3)		(15)
Miscellaneous						22 670	1 428	(16)
					Total	221 433	2 909	

(1) 10^9 passenger kilometer or 10^9 ton kilometer

(2) 10^6 J/passenger kilometer or 10^6 ton kilometer

(3) Fuel filled in Norway, all operators

(4) Oslo/Bergen/Trondheim.

NATIONAL ENERGY DEMAND PROJECTIONS

Country NORWAY

Year 1985

WAES Case A

Agriculture, Mining, and Construction

	(A)	(B)	(C)	(D)	(E)	(F)	(G)	(H)	
Agriculture, Forestry, and Fishing (ISIC Major Division 1)	Value Added	Energy/Value Added	(Km^2) Area Harvested	(J/km^2) Energy/ Area	(Ton) Physical Output	(J/ton) Energy/ Output	(TJ) Fossil Fuel	(TJ) Electricity	
Farm Machinery			7 451	$.04x10^{12}$	7 439 024	$41x10^6$	305		(17)
Greenhouses									(18)
Other Agricultural Uses					2 292 945	$4\ 564x10^6$	3 747	6 718	(19)
Forestry (ISIC 12)					6 789 644	$309x10^6$	2 098		(20)
Fishing Boats (ISIC 13)				-	3 177 942	$6\ 075x10^6$	19 306		(21)
Other Fishing Use (ISIC 13)									(22)
Total Fertilizer (tons)		N 92 485	P 28 767	K 69 048		Total	25 456	6 718	

	(A)	(B)	(C)	(D)	(E)	(F)	
Mining	Value Added	Energy/Value Added	Physical Output	Energy/Physical Output	Fossil Fuel	Electricity	
Handling and Internal Transportation			(Ton) 47 062 491	(J/ton) $146x10^6$	3 066	3 805	(23)
Other Energy Use							(24)
				Total	3 066	3 805	

	(A)	(B)	(C)	(D)	(E)	(F)	
Construction	Value Added	Energy/Value Added	Physical Output	Energy/Physical Output	Fossil Fuel	Electricity	
Handling and Internal Transportation	(Nkr) $9\ 815x10^6$	(J/Nkr) $.48x10^6$			4 711		(25)
Other Energy Use (1)	$9\ 815x10^6$	$.21x10^6$			2 061		(26)
				Total	6 772		
				Agriculture, Mining, and Construction Total	35 294	10 523	

(1) Heating and drying of materials.

NATIONAL ENERGY DEMAND PROJECTIONS

Country NORWAY

Year 1985

WAES Case A

Commercial and Public

	(A)	(B)	(C)	(D)	(E)	
Commercial	Floor Area (1)	Heat Gain or Loss/Area (2)	System Efficiency	Fossil Fuel (TJ)	Electricity (TJ)	
Space Heat (Fossil Fuel)	1.86	10 133 / 1.00 x .90		16 985		(27)
Space Heat (Electricity)	1.86	7 729 / 1.00 x .90			12 955	(28)
Air Conditioning (Fossil Fuel)						(29)
Air Conditioning (Electricity)						(30)
		Energy/Area	Electricity/Area			
Lighting						(31)
Miscellaneous	1.86	46 / 1.00		86 (3)		(32)
Presently Unknown						(33)
			Total	17 071	12 955	
Public	Floor Area	Heat Gain or Loss/Area	System Efficiency	Fossil Fuel	Electricity	
Space Heat (Fossil Fuel)	1.75	12 785 / 1.00 x .90		20 140		(34)
Space Heat (Electricity)	1.75	7 272 / 1.00 x .90			11 455	(35)
Air Conditioning (Fossil Fuel)						(36)
Air Conditioning (Electricity)						(37)
		Energy/Area	Electricity/Area			
Lighting						(38)
Miscellaneous						(39)
Presently Unknown						(40)
			Total	20 140	11 455	
			Commercial and Public Total	37 211	24 410	

(1) Production index; 1972: 1.00

(2) Unit = TJ/index

(3) Gas.

NATIONAL ENERGY DEMAND PROJECTIONS

Country NORWAY

Year 1985

WAES Case A

Industrial

	(A) Value Added [1]	(B) Energy/Value Added	(C) Physical Output	(D) Energy/Output	(E) % Exported	(F) Fossil Fuel [2]	(G) Electricity	(H) Feedstocks	
	(Mill. Nkr)	(J/$)	(Ton)	(J/ton)		(TJ)	(TJ)	(TJ)	
Iron and Steel (ISIC 371)			3 851 096	18 425x10^{6}		36 046	34 910		(41)
Nonferrous Metals (ISIC 372)			1 365 258	61 401x10^{6}		20 006	63 822		(42)
Paper and Allied (ISIC 34)			4 499 000	9 006x10^{6}		24 310	16 206		(43)
Chemicals and Allied (ISIC 35 minus ISIC 353)			5 009 272	14 880x10^{6}		33 542	40 996	40 960	(44)
Mineral Products (ISIC 36)	1 323	19.2x10^{6}				22 200	3 197		(45)
Food and Related (ISIC 31)			3 549 390	7 073x10^{6}		17 387	7 718		(46)
Transport Eqpt. (ISIC 384)	3 983	1.7x10^{6}				3 385	3 385		(47)
Key Industry (N.E.C.)[4]	8 147	2.0x10^{6}				6 989	9 306		(48)
Key Industry (N.E.C.)[5]	4 211	2.6x10^{6}				6 233	4 716		(49)
Misc. Manufacturing						5 083[3]		556	(50)
					Total	175 181	184 256	41 516	

N.E.C. = Not Elsewhere Classified

(1) Tax not included

(2) Includes wood

(3) Residual LPG, coal and coke not distributed

(4) ISIC 38-384

(5) ISIC 32, 33 and 34.

NATIONAL ENERGY DEMAND PROJECTIONS

Country NORWAY

Year 1985

WAES Case A

Residential

Population 4.16×10^6 Persons/Dwelling 2.53 Dwellings 1.64×10^6

		(A)	(B)	(C)	(D)	(E)	(F)	
Central Space Heating		Dwellings	Heat Loss (kWh)	System Efficiency		Fossil Fuel[3] (TJ)	Electricity (TJ)	
Single Family	Fossil Fuels	(0.40) 0.964×10^6	14 350-(3 325)[1]	0.69		23 000		(51)
	Electricity	(0.60)	13 914-(3 325)	1.00			22 870	(52)
Town or Row Houses	Fossil Fuels	(0.40) 0.387×10^6	11 159-(3 325)	0.69		6 330		(53)
	Electricity	(0.60)	10 822-(3 325)	1.00			6 270	(54)
Apartments	Fossil Fuels	(0.40) 0.289×10^6	8 768-(3 325)	0.69		3 280		(55)
	Electricity	(0.60)	8 503-(3 325)	1.00			3 230	(56)
		% Saturation (Fossil Fuels)	% Saturation[2] (Electricity)	Energy/Unit (Fossil Fuels)	Energy/Unit (Electricity) (kWh)			
Room Heating								(57)
Air Conditioning (Window)								(58)
Air Conditioning (Central)								(59)
Water Heating			90		4 200	3 837	16 643	(60)
Refrigerators			97		600		3 480	(61)
Lighting			100		800		4 280	(62)
Cooking			100		600		3 480	(63)
Dishwashers			50		200		510	(64)
Freezers			80		750		3 590	(65)
Television			95		110		650	(66)
Other Key Appliances			90		500		2 240	(67)
Misc. Appliances						800	3 980	(68)
Potential Appliances							400	(69)
					Total	37 247	71 623	

(1) Number in brackets represents heat from appliances, which can be accounted for in space heating (50% of total consumption of electricity to such equipment). This number is substracted from the heat loss when the line item total is calculated.

(2) Based on number of households

$$1972: \frac{\text{no. of dw.}}{\text{no. of ho.ho.}} = \frac{1.3 \times 10^6}{1.4 \times 10^6} = 0.92$$

1985: Assumed: 0.95

$$1985: \text{No. of households } \frac{1.64 \times 10^6}{0.95} = 1.73 \times 10^6$$

(3) Virtually all petroleum.

NATIONAL ENERGY DEMAND PROJECTIONS

Country NORWAY

Year 1985

WAES Case B

UNITS = TJ

Summary

Category	Oil	Gas	Coal	Electricity	Other	Total by Category (Excluding Processing Losses)
Transportation	204 551	----	----	2 823	----	207 374
Agriculture, Mining,(1) and Construction	35 446	----	----	10 641	----	46 087
Commercial	16 775	85	----	12 800	----	29 660
Public	19 890	----	----	11 322	----	31 212
Industrial	123 980	35 639	50 747	180 300	----	390 666
Residential	36 120	----	----	69 497	----	105 617
Total Delivered Energy by Fuel	436 762	35 724	50 747	287 383	----	
Processing Losses by Fuel	12.65%	9.6%	----	213%	----	
Gross Input by Fuel	492 012	39 154	50 747	899 731	----	1 481 644

Gross Energy Input (Add across bottom row)

(1) Includes forestry & fishing.

NATIONAL ENERGY DEMAND PROJECTIONS

Country NORWAY

Year 1985

WAES Case B

Transportation

	(A) Population/Auto	(B) Distance/Auto/Yr.	(C) Load Factor	(D) Fraction Urban or Rural	(E) Energy/Distance	(F) (TJ) Fossil Fuel	(G) (TJ) Electricity	
Urban Auto → Rural Auto	2.71	11 375 km			2.758×10^6 J/km	51 262		(1) (2)
	Activity Level(1)	% Electrified	Load Factor	Fossil Fuel/ Activity Level(2)	Electricity/Activity (2)			
Urban Bus(4)	.77			.78		600		(3)
Intercity Bus	4.82			.59		2 841		(4)
Rail (Passenger)	3.24	81		.71	.32	468	844	(5)
Air (Passenger, Domestic)	2.11			3.35		7 068		(6)
Air (Passenger, International)						(3) 7 471		(7)
Ship (Passenger)	.82			3.89		3 191		(8)
Rail (Freight)	3.86	67		.52	.24	661	620	(9)
Air (Freight, Domestic)	.019			24.83		473		(10)
Air (Freight, International)								(11)
Truck (Local) → Truck (Intercity)	10.21			3.43		35 034		(12) (13)
Ship (Freight, Domestic)	18.79			1.62		30 350		(14)
Ship (Freight, International)	329.4			.13		42 822(3)		(15)
Miscellaneous						22 670	1 359	(16)
					Total	204 551	2 823	

(1) 10^9 passenger kilometer or 10^9 ton kilometer

(2) 10^6 J/passenger kilometer or 10^6 ton kilometer

(3) Fuel filled in Norway, all operators

(4) Oslo/Bergen/Trondheim.

NATIONAL ENERGY DEMAND PROJECTIONS

Country NORWAY

Year 1985

WAES Case B

Agriculture, Mining, and Construction

Agriculture, Forestry, and Fishing (ISIC Major Division 1)	(A) Value Added	(B) Energy/Value Added	(C) (Km^2) Area Harvested	(D) (J/km^2) Energy/ Area	(E) (Ton) Physical Output	(F) (J/ton) Energy/ Output	(G) (TJ) Fossil Fuel	(H) (TJ) Electricity	
Farm Machinery			7 451	$.04x10^{12}$	7 536 585	$41x10^6$	309		(17)
Greenhouses									(18)
Other Agricultural Uses					2 316 827	$4\ 564x10^6$	3 786	6 788	(19)
Forestry (ISIC 12)					6 682 848	$309x10^6$	2 065		(20)
Fishing Boats (ISIC 13)					3 211 193	$6\ 075x10^6$	19 508		(21)
Other Fishing Use (ISIC 13)									(22)
Total Fertilizer (tons)		N 93 455	P 29 069	K 69 773		Total	25 668	6 788	

Mining	(A) Value Added	(B) Energy/Value Added	(C) Physical Output	(D) Energy/Physical Output	(E) Fossil Fuel	(F) Electricity	
Handling and Internal Transportation			(Ton) 47 655 899	(J/ton) $146x10^6$	3 105	3 853	(23)
Other Energy Use							(24)
				Total	3 105	3 853	

Construction	(A) Value Added	(B) Energy/Value Added	(C) Physical Output	(D) Energy/Physical Output	(E) Fossil Fuel	(F) Electricity	
Handling and Internal Transportation	(Nkr) $9\ 671x10^6$	(J/Nkr) $.48x10^6$			4 642		(25)
Other Energy Use(1)	$9\ 671x10^6$	$.21x10^6$			2 031		(26)
				Total	6 673		
				Agriculture, Mining, and Construction Total	35 446	10 641	

(1) Heating and drying of materials.

NATIONAL ENERGY DEMAND PROJECTIONS

Country NORWAY

Year 1985

WAES Case B

Commercial and Public

	(A)	(B)	(C)	(D)	(E)	
Commercial	Floor Area (1)	Heat Gain or Loss/Area (2)	System Efficiency	Fossil Fuel (TJ)	Electricity (TJ)	
Space Heat (Fossil Fuel)	1.84	10 133 / 1.00 x .90		16 775		(27)
Space Heat (Electricity)	1.84	7 729 / 1.00 x .90			12 800	(28)
Air Conditioning (Fossil Fuel)						(29)
Air Conditioning (Electricity)						(30)
		Energy/Area	Electricity/Area			
Lighting						(31)
Miscellaneous	1.84	46 / 1.00		85 (3)		(32)
Presently Unknown						(33)
			Total	16 860	12 800	
Public	Floor Area	Heat Gain or Loss/Area	System Efficiency	Fossil Fuel	Electricity	
Space Heat (Fossil Fuel)	1.73	12 785 / 1.00 x .90		19 890		(34)
Space Heat (Electricity)	1.73	7 272 / 1.00 x .90			11 322	(35)
Air Conditioning (Fossil Fuel)						(36)
Air Conditioning (Electricity)						(37)
		Energy/Area	Electricity/Area			
Lighting						(38)
Miscellaneous						(39)
Presently Unknown						(40)
			Total	19 890	11 322	
			Commercial and Public Total	36 750	24 122	

(1) Production index; 1972: 1.00

(2) Unit = TJ/index

(3) Gas.

NATIONAL ENERGY DEMAND PROJECTIONS

Country NORWAY

Year 1985

WAES Case B

Industrial

	(A) Value Added [1]	(B) Energy/Value Added	(C) Physical Output	(D) Energy/Output	(E) % Exported	(F) Fossil Fuel [2]	(G) Electricity	(H) Feedstocks	
	(Mill. Nkr)	(J/$)	(Ton)	(J/ton)		(TJ)	(TJ)	(TJ)	
Iron and Steel (ISIC 371)			3 851 096	$18\ 425 \times 10^6$		36 046	34 910		(41)
Nonferrous Metals (ISIC 372)			1 365 258	$61\ 401 \times 10^6$		20 006	63 822		(42)
Paper and Allied (ISIC 34)			4 499 000	$9\ 006 \times 10^6$		24 310	16 206		(43)
Chemicals and Allied (ISIC 35 minus ISIC 353)			4 652 277	$14\ 880 \times 10^6$		31 152	38 074	38 041	(44)
Mineral Products (ISIC 36)	1 323	19.2×10^6				22 200	3 197		(45)
Food and Related (ISIC 31)			3 549 390	$7\ 073 \times 10^6$		17 387	7 718		(46)
Transport Eqpt. (ISIC 384)	3 593	1.7×10^6				3 054	3 053		(47)
Key Industry (N.E.C.)[4]	7 521	2.0×10^6				6 438	8 604		(48)
Key Industry (N.E.C.)[5]	4 211	2.6×10^6				6 233	4 716		(49)
Misc. Manufacturing						4 954[3]		545	(50)
					Total	171 780	180 300	38 586	

N.E.C. = Not Elsewhere Classified

(1) Tax not included

(2) Includes wood

(3) Residual LPG, coal and coke not distributed

(4) ISIC 38-384

(5) ISIC 32, 33 and 34.

NATIONAL ENERGY DEMAND PROJECTIONS

Country NORWAY

Year 1985

WAES Case B

Residential

Population 4.16×10^6 Persons/Dwelling 2.53 Dwellings 1.64×10^6

		(A)	(B)	(C)	(D)	(E)	(F)	
Central Space Heating		Dwellings	Heat Loss	System Efficiency		Fossil Fue[3]	Electricity	
			(kWh)			(TJ)	(TJ)	
Single Family (0.964×10^6)	Fossil Fuels	(0.40)	14 183-(3 200)[1]	0.69		22 100		(51)
	Electricity	(0.60)	13 850-(3 200)	1.00			22 180	(52)
Town or Row Houses (0.387×10^6)	Fossil Fuels	(0.40)	11 032-(3 200)	0.69		6 330		(53)
	Electricity	(0.60)	10 772-(3 200)	1.00			6 330	(54)
Apartments (0.289×10^6)	Fossil Fuels	(0.40)	8 667-(3 200)	0.69		3 300		(55)
	Electricity	(0.60)	8 464-(3 200)	1.00			3 290	(56)
		% Saturation (Fossil Fuels)	% Saturation[2] (Electricity)	Energy/Unit (Fossil Fuels)	Energy/Unit (Electricity) (kWh)			
Room Heating								(57)
Air Conditioning (Window)								(58)
Air Conditioning (Central)								(59)
Water Heating			87		4 200	3 710	16 090	(60)
Refrigerators			97		600		3 480	(61)
Lighting			100		800		4 280	(62)
Cooking			100		600		3 480	(63)
Dishwashers			35		200		360	(64)
Freezers			76		750		3 400	(65)
Television			90		110		617	(66)
Other Key Appliances			90		500		2 240	(67)
Misc. Appliances						680	3 410	(68)
Potential Appliances							340	(69)
					Total	36 120	69 497	

(1) Number in brackets represents heat from appliances, which can be accounted for in space heating (50% of total consumption of electricity to such equipment). This number is substracted from the heat loss when the line item total is calculated.

(2) Based on number of households

1972: $\frac{\text{no. of dw.}}{\text{no. of ho.ho.}} = \frac{1.3 \times 10^6}{1.4 \times 10^6} = 0.92$

1985: Assumed: 0.95

1985: No. of households $\frac{1.64 \times 10^6}{0.95} = 1.73 \times 10^6$

(3) Virtually all petroleum.

NATIONAL ENERGY DEMAND PROJECTIONS

Country NORWAY

Year 1985

WAES Case C

UNITS = TJ

Summary

Category	Oil	Gas	Coal	Electricity	Other	Total by Category (Excluding Processing Losses)
Transportation	221 433	----	----	2 909	----	224 342
Agriculture, Mining,(1) and Construction	35 294	----	----	10 523	----	45 817
Commercial	16 985	86	----	12 955	----	30 026
Public	20 140	----	----	11 455	----	31 595
Industrial	133 088	38 286	51 025	184 256	----	406 655
Residential	47 990	----	----	73 030	----	121 020
Total Delivered Energy by Fuel	474 930	38 372	51 025	295 128	----	
Processing Losses by Fuel	12.65%	9.6%	----	213%	----	
Gross Input by Fuel	535 009	42 056	51 025	923 979	----	1 552 069

Gross Energy Input (Add across bottom row)

(1) Includes forestry & fishing.

NATIONAL ENERGY DEMAND PROJECTIONS

Country NORWAY

Year 1985

WAES Case C

Transportation

	(A) Population/Auto	(B) Distance/Auto/Yr.	(C) Load Factor	(D) Fraction Urban or Rural	(E) Energy/Distance	(F) (TJ) Fossil Fuel	(G) (TJ) Electricity	
Urban Auto ↗ (Rural Auto combined)	2.65	11 375 km			2.667×10^6 J/km	52 630		(1)
Rural Auto								(2)
	Activity Level(1)	% Electrified	Load Factor	Fossil Fuel/ Activity Level(2)	Electricity/Activity (2)			
Urban Bus (4)	.77			.78		600		(3)
Intercity Bus	4.82			.59		2 841		(4)
Rail (Passenger)	3.30	81		.71	.32	467	844	(5)
Air (Passenger, Domestic)	2.53			3.35		8 483		(6)
Air (Passenger, International)						8 421 (3)		(7)
Ship (Passenger)	.83			3.89		3 228		(8)
Rail (Freight)	3.97	67		.52	.24	680	637	(9)
Air (Freight, Domestic)	.025			24.83		607		(10)
Air (Freight, International)								(11)
Truck (Local) ↗ (Truck (Intercity) combined)	11.81			3.43		40 508		(12)
Truck (Intercity)								(13)
Ship (Freight, Domestic)	22.19			1.62		35 947		(14)
Ship (Freight, International)	341.16			.13		44 351 (3)		(15)
Miscellaneous						22 670	1 428	(16)
					Total	221 433	2 909	

(1) 10^9 passenger kilometer or 10^9 ton kilometer

(2) 10^6 J/passenger kilometer or 10^6 ton kilometer

(3) Fuel filled in Norway, all operators

(4) Oslo/Bergen/Trondheim.

NATIONAL ENERGY DEMAND PROJECTIONS

Country NORWAY

Year 1985

WAES Case C

Agriculture, Mining, and Construction

Agriculture, Forestry, and Fishing (ISIC Major Division 1)	(A) Value Added	(B) Energy/Value Added	(C) (Km^2) Area Harvested	(D) (J/km^2) Energy/ Area	(E) (Ton) Physical Output	(F) (J/ton) Energy/ Output	(G) (TJ) Fossil Fuel	(H) (TJ) Electricity	
Farm Machinery			7 451	$.04x10^{12}$	7 439 024	$41x10^{6}$	305		(17)
Greenhouses									(18)
Other Agricultural Uses					2 292 945	$4\ 564x10^{6}$	3 747	6 718	(19)
Forestry (ISIC 12)					6 789 644	$309x10^{6}$	2 098		(20)
Fishing Boats (ISIC 13)					3 177 942	$6\ 075x10^{6}$	19 306		(21)
Other Fishing Use (ISIC 13)									(22)
Total Fertilizer (tons)		N 92 485	P 28 767	K 69 048		Total	25 456	6 718	

Mining	(A) Value Added	(B) Energy/Value Added	(C) Physical Output	(D) Energy/Physical Output	(E) Fossil Fuel	(F) Electricity	
			(Ton)	(J/ton)	(TJ)	(TJ)	
Handling and Internal Transportation			47 062 491	$146x10^{6}$	3 066	3 805	(23)
Other Energy Use							(24)
				Total	3 066	3 805	

Construction	(A) Value Added	(B) Energy/Value Added	(C) Physical Output	(D) Energy/Physical Output	(E) Fossil Fuel	(F) Electricity	
	(Nkr)	(J/Nkr)					
Handling and Internal Transportation	$9\ 815x10^{6}$	$.48x10^{6}$			4 711	3 805	(25)
Other Energy Use(1)	$9\ 815x10^{6}$	$.21x10^{6}$			2 061		(26)
				Total	6 772		
				Agriculture, Mining, and Construction Total	35 294	10 523	

(1) Heating and drying of materials.

NATIONAL ENERGY DEMAND PROJECTIONS

Country NORWAY

Year 1985

WAES Case C

Commercial and Public

	(A)	(B)	(C)	(D)	(E)	
Commercial	Floor Area (1)	Heat Gain or Loss/Area (2)	System Efficiency	Fossil Fuel	Electricity	
Space Heat (Fossil Fuel)	1.86	$\frac{10\ 133}{1.00}$ x .90		(TJ) 16 985	(TJ)	(27)
Space Heat (Electricity)	1.86	$\frac{7\ 729}{1.00}$ x .90			12 955	(28)
Air Conditioning (Fossil Fuel)						(29)
Air Conditioning (Electricity)						(30)
		Energy/Area	Electricity/Area			
Lighting						(31)
Miscellaneous	1.86	$\frac{46}{1.00}$		86(2)		(32)
Presently Unknown						(33)
			Total	17 071	12 955	
Public	Floor Area	Heat Gain or Loss/Area	System Efficiency	Fossil Fuel	Electricity	
Space Heat (Fossil Fuel)	1.75	$\frac{12\ 785}{1.00}$ x .90		20 140		(34)
Space Heat (Electricity)	1.75	$\frac{7\ 272}{1.00}$ x .90			11 455	(35)
Air Conditioning (Fossil Fuel)						(36)
Air Conditioning (Electricity)						(37)
		Energy/Area	Electricity/Area			
Lighting						(38)
Miscellaneous						(39)
Presently Unknown						(40)
			Total	20 140	11 455	
			Commercial and Public Total	37 211	24 410	

(1) Production index; 1972: 1.00

(2) Unit = TJ/index

(3) Gas.

NATIONAL ENERGY DEMAND PROJECTIONS

Country NORWAY

Year 1985

WAES Case C

Industrial

	(A) Value Added (1)	(B) Energy/Value Added	(C) Physical Output	(D) Energy/Output	(E) % Exported	(F) Fossil Fuel (2)	(G) Electricity	(H) Feedstocks	
	(Mill. Nkr)	(J/$)	(Ton)	(J/ton)		(TJ)	(TJ)	(TJ)	
Iron and Steel (ISIC 371)			3 851 096	18 425x10^6		36 046	34 910		(41)
Nonferrous Metals (ISIC 372)			1 365 258	61 401x10^6		20 006	63 822		(42)
Paper and Allied (ISIC 34)			4 499 000	9 276x10^6		25 526	16 206		(43)
Chemicals and Allied (ISIC 35 minus ISIC 353)			5 009 272	15 215x10^6		35 219	40 996	40 960	(44)
Mineral Products (ISIC 36)	1 323	20.04x10^6				23 310	3 197		(45)
Food and Related (ISIC 31)			3 549 390	7 318x10^6		18 256	7 718		(46)
Transport Eqpt. (ISIC 384)	3 983	1.74x10^6				3 554	3 385		(47)
Key Industry (N.E.C.)(4)	8 147	2.04x10^6				7 338	9 306		(48)
Key Industry (N.E.C.)(5)	4 211	2.67x10^6				6 545	4 716		(49)
Misc. Manufacturing						5 083		556	(50)
					Total	180 883	184 256	41 516	

N.E.C. = Not Elsewhere Classified

(1) Tax not included

(2) Includes wood

(3) Residual LPG, coal and coke not distributed

(4) ISIC 38-384

(5) ISIC 32, 33 and 34.

NATIONAL ENERGY DEMAND PROJECTIONS

Country NORWAY

Year 1985

WAES Case C

Residential

Population 4.16×10^6 Persons/Dwelling 2.53 Dwellings 1.64×10^6

Central Space Heating		(A) Dwellings	(B) Heat Loss (kWh)	(C) System Efficiency	(D)	(E) Fossil Fuel(3)	(F) Electricity	
Single Family	Fossil Fuels	(0.45) 0.964×10^6	16 490–(3 440)(1)	0.69		29 540		(51)
	Electricity	(0.55)	15 270–(3 440)	1.0			22 580	(52)
Town or Row Houses	Fossil Fuels	(0.45) 0.387×10^6	12 820–(3 440)	0.69		8 520		(53)
	Electricity	(0.55)	11 880–(3 440)	1.0			6 470	(54)
Apartments	Fossil Fuels	(0.45) 0.289×10^6	10 080–(3 440)	0.69		4 510		(55)
	Electricity	(0.55)	9 330–(3 440)	1.0			3 370	(56)
		% Saturation (Fossil Fuels)	% Saturation(2) (Electricity)	Energy/Unit (Fossil Fuels)	Energy/Unit (Electricity)			
Room Heating								(57)
Air Conditioning (Window)								(58)
Air Conditioning (Central)								(59)
Water Heating			90		4 200	4 570	17 090	(60)
Refrigerators			97		600		3 520	(61)
Lighting			100		800		4 480	(62)
Cooking			100		600		3 550	(63)
Dishwashers			50		200		550	(64)
Freezers			80		750		3 630	(65)
Television			95		110		650	(66)
Other Key Appliances			90		500		2 440	(67)
Misc. Appliances						850	4 270	(68)
Potential Appliances							430	(69)
					Total	47 990	73 030	

(1) Number in brackets represents heat from appliances, which can be accounted for in space heating (50% of total consumption of electricity to such equipment). This number is substracted from the heat loss when the line item total is calculated.

(2) Based on number of households

1972: $\frac{\text{no. of dw.}}{\text{no. of ho.ho.}} = \frac{1.3 \times 10^6}{1.4 \times 10^6} = 0.92$

1985: Assumed: 0.95

1985: No. of households $\frac{1.64 \times 10^6}{0.95} = 1.73 \times 10^6$

(3) Virtually all petroleum.

NATIONAL ENERGY DEMAND PROJECTIONS

Country NORWAY

Year 1985

WAES Case D

UNITS = TJ

Summary

Category	Oil	Gas	Coal	Electricity	Other	Total by Category (Excluding Processing Losses)
Transportation	218 199	----	----	2 777	----	220 976
Agriculture, Mining,(1) and Construction	35 446	----	----	10 641	----	46 087
Commercial	17 148	85	----	13 084	----	30 317
Public	20 332	----	----	11 574	----	31 906
Industrial	132 310	36 053	50 909	182 913	----	402 185
Residential	50 497	----	----	80 505	----	131 002
Total Delivered Energy by Fuel	473 932	36 138	50 909	301 494	----	
Processing Losses by Fuel	12.65%	9.6%	----	213%	----	
Gross Input by Fuel	533 884	39 607	50 909	943 909	----	1 568 309

Gross Energy Input (Add across bottom row)

(1) Includes forestry & fishing.

NATIONAL ENERGY DEMAND PROJECTIONS

Country NORWAY

Year 1985

WAES Case D

Transportation

	(A)	(B)	(C)	(D)	(E)	(F)	(G)	
	Population/Auto	Distance/Auto/Yr.	Load Factor	Fraction Urban or Rural	Energy/Distance	(TJ) Fossil Fuel	(TJ) Electricity	
Urban Auto	2.51	12 125 km			2.873 x 10^6 J/km	61 554		(1)
Rural Auto								(2)
	Activity Level[1]	% Electrified	Load Factor	Fossil Fuel/ Activity Level[2]	Electricity/Activity[2]			
Urban Bus[4]	.75			.80		596		(3)
Intercity Bus	4.22			.61		2 572		(4)
Rail (Passenger)	2.97	81		.73	.33	441	798	(5)
Air (Passenger, Domestic)	2.16			3.31		7 160		(6)
Air (Passenger, International)						7 471[3]		(7)
Ship (Passenger)	.85			3.89		3 303		(8)
Rail (Freight)	3.86	67		.52	.24	661	620	(9)
Air (Freight, Domestic)	.019			24.83		473		(10)
Air (Freight, International)								(11)
Truck (Local)	10.21			3.43		35 034		(12)
Truck (Intercity)								(13)
Ship (Freight, Domestic)	18.79			1.62		30 350		(14)
Ship (Freight, International)	329.40			.13		42 822[3]		(15)
Miscellaneous						25 762	1 359	(16)
					Total	218 199	2 777	

(1) 10^9 passenger kilometer or 10^9 ton kilometer

(2) 10^6 J/passenger kilometer or 10^6 ton kilometer

(3) Fuel filled in Norway, all operators

(4) Oslo/Bergen/Trondheim.

NATIONAL ENERGY DEMAND PROJECTIONS

Country NORWAY

Year 1985

WAES Case D

Agriculture, Mining, and Construction

Agriculture, Forestry, and Fishing (ISIC Major Division 1)	(A) Value Added	(B) Energy/Value Added	(C) (Km^2) Area Harvested	(D) (J/km^2) Energy/ Area	(E) (Ton) Physical Output	(F) (J/ton) Energy/ Output	(G) (TJ) Fossil Fuel	(H) (TJ) Electricity	
Farm Machinery			7 451	$.04x10^{12}$	7 536 585	$41x10^6$	309		(17)
Greenhouses									(18)
Other Agricultural Uses					2 316 827	$4 564x10^6$	3 786	6 788	(19)
Forestry (ISIC 12)					6 682 848	$309x10^6$	2 065		(20)
Fishing Boats (ISIC 13)					3 211 193	$6 075x10^6$	19 508		(21)
Other Fishing Use (ISIC 13)									(22)
Total Fertilizer (tons)		N 93 455	P 29 069	K 69 773		Total	25 668	6 788	

Mining	(A) Value Added	(B) Energy/Value Added	(C) Physical Output	(D) Energy/Physical Output	(E) Fossil Fuel	(F) Electricity	
Handling and Internal Transportation			(Ton) 47 655 899	(J/km^2) $146x10^6$	(TJ) 3 105	(TJ) 3 853	(23)
Other Energy Use							(24)
				Total	3 105	3 853	

Construction	(A) Value Added	(B) Energy/Value Added	(C) Physical Output	(D) Energy/Physical Output	(E) Fossil Fuel	(F) Electricity	
Handling and Internal Transportation	(Nkr) $9 671x10^6$	(J/Nkr) $.48x10^6$			4 642		(25)
Other Energy Use(1)	$9 671x10^6$	$.21x10^6$			2 031		(26)
				Total	6 673		
				Agriculture, Mining, and Construction Total	35 446	10 641	

(1) Heating and drying of materials.

NATIONAL ENERGY DEMAND PROJECTIONS

Country NORWAY

Year 1985

WAES Case D

Commercial and Public

	(A)	(B)	(C)	(D)	(E)	
Commercial	Floor Area (1)	Heat Gain or Loss/Area (2)	System Efficiency	Fossil Fuel	Electricity	
				(TJ)	(TJ)	
Space Heat (Fossil Fuel)	1.84	$\frac{10\ 133}{1.00}$ x .92		17 148		(27)
Space Heat (Electricity)	1.84	$\frac{7\ 729}{1.00}$ x .92			13 084	(28)
Air Conditioning (Fossil Fuel)						(29)
Air Conditioning (Electricity)						(30)
		Energy/Area	Electricity/Area			
Lighting						(31)
Miscellaneous						(32)
Presently Unknown						(33)
			Total	17 233	13 084	
Public	Floor Area	Heat Gain or Loss/Area	System Efficiency	Fossil Fuel	Electricity	
Space Heat (Fossil Fuel)	1.73	$\frac{12\ 785}{1.00}$ x .92		20 332		(34)
Space Heat (Electricity)	1.73	$\frac{7\ 272}{1.00}$ x .92			11 574	(35)
Air Conditioning (Fossil Fuel)						(36)
Air Conditioning (Electricity)						(37)
		Energy/Area	Electricity/Area			
Lighting						(38)
Miscellaneous						(39)
Presently Unknown						(40)
			Total	20 332	11 574	
			Commercial and Public Total	37 565	24 658	

(1) Production index; 1972: 1.00

(2) Unit = TJ/index

(3) Gas.

NATIONAL ENERGY DEMAND PROJECTIONS

Country NORWAY

Year 1985

WAES Case D

Industrial

	(A) Value Added (1)	(B) Energy/Value Added	(C) Physical Output	(D) Energy/Output	(E) % Exported	(F) Fossil Fuel (2)	(G) Electricity	(H) Feedstocks	
	(Mill. Nkr)	(J/$)	(Ton)	(J/ton)		(TJ)	(TJ)	(TJ)	
Iron and Steel (ISIC 371)			3 799 518	$18\ 675 \times 10^6$		36 046	34 910		(41)
Nonferrous Metals (ISIC 372)			1 328 158	$63\ 116 \times 10^6$		20 006	63 822		(42)
Paper and Allied (ISIC 34)			4 499 000	$10\ 049 \times 10^6$		27 652	17 557		(43)
Chemicals and Allied (ISIC 35 minus ISIC 353)			4 652 277	$15\ 573 \times 10^6$		33 478	38 970	38 041	(44)
Mineral Products (ISIC 36)	1 323	20.04×10^6				23 310	3 197		(45)
Food and Related (ISIC 31)			3 549 390	$7\ 318 \times 10^6$		18 256	7 718		(46)
Transport Eqpt. (ISIC 384)	3 593	1.80×10^6				3 320	3 162		(47)
Key Industry (N.E.C.)(4)	7 521	2.10×10^6				6 963	8 862		(48)
Key Industry (N.E.C.)(5)	4 211	2.67×10^6				6 545	4 716		(49)
Misc. Manufacturing						5 110(3)		545	(50)
					Total	180 686	182 913	38 586	

N.E.C. = Not Elsewhere Classified

(1) Tax not included

(2) Includes wood

(3) Residual LPG, coal and coke not distributed

(4) ISIC 38-384

(5) ISIC 32, 33 and 34.

NATIONAL ENERGY DEMAND PROJECTIONS

Country NORWAY

Year 1985

WAES Case D

Residential

Population 4.16×10^6 Persons/Dwelling 2.53 Dwellings 1.64×10^6

Central Space Heating		(A) Dwellings	(B) Heat Loss (kWh)	(C) System Efficiency	(D)	(E) Fossil Fuel (3) (TJ)	(F) Electricity (TJ)	
Single Family	Fossil Fuels	(0.45) 0.964×10^6	17 340-(3 620)(1)	0.68		31 510		(51)
	Electricity	(0.55)		1.0			26 190	(52)
Town or Row Houses	Fossil Fuels	(0.45) 0.387×10^6	13 490-(3 620)	0.68		9 100		(53)
	Electricity	(0.55)		1.0			7 560	(54)
Apartments	Fossil Fuels	(0.45) 0.289×10^6	10 600-(3 620)	0.68		4 810		(55)
	Electricity	(0.55)		1.0			4 000	(56)
		% Saturation (Fossil Fuels)	% Saturation (2) (Electricity)	Energy/Unit (Fossil Fuels)	Energy/Unit (Electricity) (kWh)			
Room Heating								(57)
Air Conditioning (Window)								(58)
Air Conditioning (Central)								(59)
Water Heating			87		4 200	4 260	18 500	(60)
Refrigerators			97		600		3 630	(61)
Lighting			100		800		4 980	(62)
Cooking			100		600		3 740	(63)
Dishwashers			35		200		440	(64)
Freezers			76		750		3 550	(65)
Television			90		110		620	(66)
Other Key Appliances			90		500		2 800	(67)
Misc. Appliances						817	4 086	(68)
Potential Appliances							409	(69)
					Total	50 497	80 505	

(1) Number in brackets represents heat from appliances, which can be accounted for in space heating (50% of total consumption of electricity to such equipment). This number is substracted from the heat loss when the line item total is calculated.

(2) Based on number of households

$$1972: \quad \frac{\text{no. of dw.}}{\text{no. of ho.ho.}} = \frac{1.3 \times 10^6}{1.4 \times 10^6} = 0.92$$

1985: Assumed: 0.95

$$1985: \quad \text{No. of households } \frac{1.64 \times 10^6}{0.95} = 1.73 \times 10^6$$

(3) Virtually all petroleum.

NATIONAL ENERGY DEMAND PROJECTIONS

Country NORWAY

Year 1985

WAES Case E

UNITS = TJ

Summary

Category	Oil	Gas	Coal	Electricity	Other	Total by Category (Excluding Processing Losses)
Transportation	235 057	----	----	2 880	----	237 937
Agriculture, Mining,[1] and Construction	35 294	----	----	10 523	----	45 817
Commercial	17 362	86	----	13 243	----	30 691
Public	20 558	----	----	11 710	----	32 268
Industrial	141 780	38 731	51 195	186 986	----	418 692
Residential	66 250	----	----	77 700	----	143 950
Total Delivered Energy by Fuel	516 301	38 817	51 195	303 042	----	
Processing Losses by Fuel	12.65%	9.6%	----	213%	----	
Gross Input by Fuel	581 613	42 543	51 195	948 756	----	1 624 107

Gross Energy Input (Add across bottom row)

(1) Includes forestry & fishing.

NATIONAL ENERGY DEMAND PROJECTIONS

Country NORWAY

Year 1985

WAES Case E

Transportation

	(A)	(B)	(C)	(D)	(E)	(F)	(G)	
	Population/Auto	Distance/Auto/Yr.	Load Factor	Fraction Urban or Rural	Energy/Distance	(TJ) Fossil Fuel	(TJ) Electricity	
Urban Auto	2.45	12 125 km			2.873×10^6 J/km	63 226		(1)
Rural Auto								(2)
	Activity Level (1)	% Electrified	Load Factor	Fossil Fuel/ Activity Level (2)	Electricity/Activity (2)			
Urban Bus (4)	.75			.80		596		(3)
Intercity Bus	4.22			.61		2 572		(4)
Rail (Passenger)	3.03	81		.73	.33	449	814	(5)
Air (Passenger, Domestic)	2.59			3.31		8 585		(6)
Air (Passenger, International)						8 421 (3)		(7)
Ship (Passenger)	.89			3.89		3 353		(8)
Rail (Freight)	3.97	67		.52	.24	680	638	(9)
Air (Freight, Domestic)	.025			24.83		607		(10)
Air (Freight, International)								(11)
Truck (Local)	11.81			3.43		40 508		(12)
Truck (Intercity)								(13)
Ship (Freight, Domestic)	22.19			1.62		35 947		(14)
Ship (Freight, International)	341.16			.13		44 351 (3)		(15)
Miscellaneous						25 762	1 428	(16)
					Total	235 057	2 880	

(1) 10^9 passenger kilometer or 10^9 ton kilometer

(2) 10^6 J/passenger kilometer or 10^6 ton kilometer

(3) Fuel filled in Norway, all operators

(4) Oslo/Bergen/Trondheim.

NATIONAL ENERGY DEMAND PROJECTIONS

Country NORWAY

Year 1985

WAES Case E

Agriculture, Mining, and Construction

	(A)	(B)	(C)	(D)	(E)	(F)	(G)	(H)	
Agriculture, Forestry, and Fishing (ISIC Major Division 1)	Value Added	Energy/Value Added	(Km^2) Area Harvested	(J/km^2) Energy/ Area	(Ton) Physical Output	(J/ton) Energy/ Output	(TJ) Fossil Fuel	(TJ) Electricity	
Farm Machinery			7 451	$.04x10^{12}$	7 439 024	$41x10^6$	305		(17)
Greenhouses									(18)
Other Agricultural Uses					2 292 945	$4\ 564x10^6$	3 747	6 718	(19)
Forestry (ISIC 12)					6 789 644	$309x10^6$	2 098		(20)
Fishing Boats (ISIC 13)					3 177 942	$6\ 075x10^6$	19 306		(21)
Other Fishing Use (ISIC 13)									(22)
Total Fertilizer (tons)		N 92 485	P 28 767	K 69 048		Total	25 456	6 718	

	(A)	(B)	(C)	(D)	(E)	(F)	
Mining	Value Added	Energy/Value Added	Physical Output	Energy/Physical Output	Fossil Fuel	Electricity	
Handling and Internal Transportation			(Ton) 47 062 491	(J/ton) $146x10^6$	(TJ) 3 066	(TJ) 3 805	(23)
Other Energy Use							(24)
				Total	3 066	3 805	

	(A)	(B)	(C)	(D)	(E)	(F)	
Construction	Value Added	Energy/Value Added	Physical Output	Energy/Physical Output	Fossil Fuel	Electricity	
Handling and Internal Transportation	(Nkr) $9\ 815x10^6$	(J/Nkr) $.48x10^6$			4 711		(25)
Other Energy Use(1)	$9\ 815x10^6$	$.21x10^6$			2 061		(26)
				Total	6 772		
				Agriculture, Mining, and Construction Total	35 294	10 523	

(1) Heating and drying of materials.

NATIONAL ENERGY DEMAND PROJECTIONS

Country NORWAY

Year 1985

WAES Case E

Commercial and Public

	(A)	(B)	(C)	(D)	(E)	
Commercial	Floor Area (1)	Heat Gain or Loss/Area (2)	System Efficiency	Fossil Fuel	Electricity	
				(TJ)	(TJ)	
Space Heat (Fossil Fuel)	1.86	$\frac{10\ 133}{1.00}$ x .92		17 362		(27)
Space Heat (Electricity)	1.86	$\frac{7\ 729}{1.00}$ x .92			13 243	(28)
Air Conditioning (Fossil Fuel)						(29)
Air Conditioning (Electricity)						(30)
		Energy/Area	Electricity/Area			
Lighting						(31)
Miscellaneous	1.86	$\frac{46}{1.00}$		86 (3)		(32)
Presently Unknown						(33)
			Total	17 448	13 243	
Public	Floor Area	Heat Gain or Loss/Area	System Efficiency	Fossil Fuel	Electricity	
Space Heat (Fossil Fuel)	1.75	$\frac{12\ 785}{1.00}$ x .92		20 588		(34)
Space Heat (Electricity)	1.75	$\frac{7\ 272}{1.00}$ x .92			11 710	(35)
Air Conditioning (Fossil Fuel)						(36)
Air Conditioning (Electricity)						(37)
		Energy/Area	Electricity/Area			
Lighting						(38)
Miscellaneous						(39)
Presently Unknown						(40)
			Total	20 558	11 710	
			Commercial and Public Total	38 036	24 953	

(1) Production index; 1972: 1.00

(2) Unit = TJ/index

(3) Gas.

NATIONAL ENERGY DEMAND PROJECTIONS

Country NORWAY

Year 1985

WAES Case E

Industrial

	(A)	(B)	(C)	(D)	(E)	(F)	(G)	(H)	
	Value Added (1)	Energy/Value Added	Physical Output	Energy/Output	% Exported	Fossil Fuel (2)	Electricity	Feedstocks	
	(Mill. Nkr)	(J/$)	(Ton)	(J/ton)		(TJ)	(TJ)	(TJ)	
Iron and Steel (ISIC 371)			3 799 518	$18\ 675\text{x}10^6$		36 046	34 910		(41)
Nonferrous Metals (ISIC 372)			1 328 158	$63\ 116\text{x}10^6$		20 006	63 822		(42)
Paper and Allied (ISIC 34)			4 499 000	$10\ 342\text{x}10^6$		28 970	17 557		(43)
Chemicals and Allied (ISIC 35 minus ISIC 353)			5 009 272	$15\ 916\text{x}10^6$		37 765	41 916	40 960	(44)
Mineral Products (ISIC 36)	1 323	$20.90\text{x}10^6$				24 420	3 197		(45)
Food and Related (ISIC 31)			3 549 390	$7\ 563\text{x}10^6$		19 126	7 718		(46)
Transport Eqpt. (ISIC 384)	3 983	$1.84\text{x}10^6$				3 856	3 505		(47)
Key Industry (N.E.C.)(4)	8 147	$2.14\text{x}10^6$				7 901	9 600		(48)
Key Industry (N.E.C.)(5)	4 211	$2.74\text{x}10^6$				6 856	4 716		(49)
Misc. Manufacturing						5 244(3)		556	(50)
					Total	190 190	149 986	41 516	

N.E.C. = Not Elsewhere Classified

(1) Tax not included

(2) Includes wood

(3) Residual LPG, coal and coke not distributed

(4) ISIC 38-384

(5) ISIC 32, 33 and 34.

NATIONAL ENERGY DEMAND PROJECTIONS

Country NORWAY

Year 1985

WAES Case E

Residential

Population 4.16×10^6 Persons/Dwelling 2.53 Dwellings 1.64×10^6

		(A)	(B)	(C)	(D)	(E)	(F)	
Central Space Heating		Dwellings	Heat Loss	System Efficiency		Fossil Fuel (3) (TJ)	Electricity (TJ)	
Single Family 0.964×10^6	Fossil Fuels	(0.52)	19 640–(3 780)(1)	0.68		42 090		(51)
	Electricity	(0.48)	17 570–(3 780)	1.0			22 970	(52)
Town or Row Houses 0.387×10^6	Fossil Fuels	(0.52)	15 270–(3 780)	0.68		12 240		(53)
	Electricity	(0.48)	13 660–(3 780)	1.0			6 610	(54)
Apartments 0.289×10^6	Fossil Fuels	(0.52)	12 000–(3 780)	0.68		6 540		(55)
	Electricity	(0.48)	10 740–(3 780)	1.00			3 480	(56)
		% Saturation (Fossil Fuels)	% Saturation (2) (Electricity)	Energy/Unit (Fossil Fuels)	Energy/Unit (Electricity) (kWh)			
Room Heating								(57)
Air Conditioning (Window)								(58)
Air Conditioning (Central)								(59)
Water Heating			90		4 200	4 410	19 130	(60)
Refrigerators			97		600		3 630	(61)
Lighting			100		800		4 980	(62)
Cooking			100		600		3 740	(63)
Dishwashers			50		200		620	(64)
Freezers			80		750		3 740	(65)
Television			95		110		650	(66)
Other Key Appliances			90		500		2 810	(67)
Misc. Appliances						970	4 850	(68)
Potential Appliances							490	(69)
					Total	66 250	77 700	

(1) Number in brackets represents heat from appliances, which can be accounted for in space heating (50% of total consumption of electricity to such equipment). This number is substracted from the heat loss when the line item total is calculated.

(2) Based on number of households

$$1972: \frac{\text{no. of dw.}}{\text{no. of ho.ho.}} = \frac{1.3 \times 10^6}{1.4 \times 10^6} = 0.92$$

1985: Assumed: 0.95

$$1985: \text{No. of households } \frac{1.64 \times 10^6}{0.95} = 1.73 \times 10^6$$

(3) Virtually all petroleum.

13 Sweden

13.1 SUMMARY

Our report is based mainly on the extensive energy studies made by Swedish Government Commissions in 1973 and 1974; the Energy Forecast Commission (EPU) in particular used a methodology quite similar to that of WAES. We have, however, had to adjust their findings to account for some parameters in the WAES methodology that were not used by EPU—namely, a change in the oil price and a more vigorous government response—and the actual development of energy use in the period 1973 to 1975. We have done these adjustments by interviews with energy experts and according to recently published studies in the energy field.

The main factor influencing total energy demand is, according to our results, the economic growth rate. The effect of this factor seems to be two to three times as great as that of changes in the price of oil in a time frame of about ten years.

The combined effects of increases in oil price and vigorous but plausible government response could, however, shift the energy demand in some sectors from fossil fuels to electricity.

Vigorous government response is assumed to consist primarily of measures to increase energy conservation in the residential sector; in other sectors government response is assumed to consist of soft actions unless public opinion comes to regard it as essential that hard actions be taken to save energy.

During the 1960s, total energy input to Sweden increased by about 5 percent per year. The projections in Table 13.1 indicate much slower rates for the period to 1985.

Our forecasts indicate a continuous high growth rate for electricity in almost all sectors of society. The main area in which it would be possible to reduce the growth of electricity use is for space heat.

Due to vigorous government response, market forces, etc., quite substantial savings in fossil fuels are projected especially in space heat and the paper and allied subsector and in WAES Cases A and B, in which the oil price is increased by 50 percent and the government response is assumed to be vigorous.

Regardless of variations in the WAES parameters —economic growth, oil price, and government response—the various sectors of Swedish society will have about the same percentage of total energy demand in 1985, although the emphasis in each sector may shift between fossil fuels and electricity. However, if one compares the situation in 1985 with 1972, one finds that the transportation, commercial, public, and especially the residential sector will in 1985 have lower percentages of total primary energy input than in 1972 due to savings and the increase in processing losses. The latter is caused by the increase in electricity use. The growth rates for processing losses of electricity are, however, slightly slower than those for electricity demand as a whole due to the projected increased use of district heating, which has a much higher efficiency in converting primary energy than power plants. The possibility of utilizing nuclear power plants more efficiently by district heating is presently being discussed in Sweden.

The energy demand in the 1985 WAES cases is lower than the forecasts made by EPU due to the assumed higher oil price and more vigorous govern-

Table 13.1
Energy demand growth rates, 1972 to 1985 (% per annum)

WAES Case	A	B	C	D	E
Economic Growth	high	low	high	low	high
Oil Price	rising	rising	constant	constant	falling
Government Response	vigorous	vigorous	vigorous	restrained	restrained
Primary Energy Input[1]					
Fossil Fuels[2]	0.8	0.2	2.0	1.6	3.1
Electricity	6.3	5.6	5.9	5.2	6.9
Total	3.1	2.5	3.6	3.1	4.6

1. Including processing losses and feedstocks.
2. Excluding fossil fuel used to produce electricity.

ment response, and the actual low use of energy in 1974–1975. A detailed comparison between the EPU forecasts and other recently published forecasts can be found in Section 13.4. In general, the WAES projections are in accordance with recently published forecasts to 1980. The main difference involves specific energy demand (energy per value added) in industry for electricity. We project, compared to the developments in the 1960s, a slower but still increasing rate (+ 0.3 percent) of specific energy demand for electricity, while other studies forecast a slightly negative rate (- 0.1 percent). The difference in these rates corresponds in 1985 to a demand for electricity in the magnitude of the output from, for instance, a single, large nuclear power plant.

Even though the WAES scenarios are projections according to specific parameters and not actual forecasts, this illustrates the difficulties and the importance of making accurate forecasts, especially when considering the long lead time needed to install, for instance, a nuclear power plant. It should be pointed out, moreover, that our knowledge of electricity supply and demand is very good compared to other energy sectors, especially oil in the residential, commercial, and public sectors.

13.2 INTRODUCTION

13.2.1 National Structure

During 1973 and 1974, a number of ad hoc Government Commissions investigated different aspects of the energy problem. The two most important ones for the purposes of the WAES demand studies—the Energy Forecasting Commission (EPU) and the Energy Program Commission (EPK)—published their results during the autumn of 1974, and their findings were in large part included in the energy policy proposal presented by the Government to and approved by the Parliament in the spring of 1975.

As the methodology behind the forecasts made by EPU is very similar to that of WAES, we have to a great extent used their forecasts, revising them where necessary to suit the WAES scenario cases. In doing so, we have taken into account other energy savings studies made by the Swedish Academy of Engineering Sciences (IVA), the Industrial Institute for Economic and Social Research (IUI), etc., and also discussions with energy experts in branch institutions and individual companies.

In November 1974, a WAES Reference Group in Sweden was formed, and this has made it possible for us to utilize the experience and knowledge in the energy field in Sweden.

We gratefully acknowledge the valuable advice and assistance from the members of the Swedish Reference Groups and their Associates, especially Måns Lönnroth, Mats Höjeberg, Lars Hjorth, Åke Sjöström, Gunilla Boman, Clas Bergström, Inger Soldéus, and Pål Pentén.

Members of the Swedish Reference Group

Erik Grafström, Director General, Prime Minister's Office

Tony Hagström, Under Secretary of State, Ministry of Industry

Eric Pettersson, Director General, National Swedish Industrial Board

Jonas Norrby, Director General, National Swedish Power Administration

Lars Ingelstam, Professor, Secretariat for Future Studies

Gunnar Hambraeus, Professor, Swedish Academy of Engineering Sciences

Bo Helmerson, Director, Federation of Swedish Industries

Sam Nilsson, Secretary General, International Federation of Institutes for Advanced Study

Lars Norlin, President, Krångedegruppens Samkörning AB.

We also want to acknowledge the valuable comments from Lars Wohlin and Bo Carlsson from IUI and all other contacted experts in the energy field.

13.2.2 Statement on Data Type

The National Swedish Industrial Board is administrating and developing an Energy Forecasting Model for Sweden that will be ready for use in the summer of 1976. We hope to be able to use this model and the Brookhaven Energy System Optimization Model during the autumn of 1976 and also to apply the findings of other ongoing energy savings studies which are scheduled to be published in the summer of 1976. Until then, the forecasts presented here should be regarded as preliminary.

13.2.3 Summary of the Presentation Made by the Ministry of Industry in 1975 of the Resolution on Energy Policy Adopted by the Spring Session of the 1975 Riksdag[1]

The annual growth rate of energy consumption has been 4.5 percent in the postwar period so far. The goal is now to reduce this figure to 2 percent per annum for the period ending in 1985. Beyond that date, the aim is for consumption to be kept at a constant level from the beginning of the 1990s. At the same time, the future will bring heavy requirements for energy, which will have to be met in order to raise the universal living standard. These two aims will be met by:

deliberate conservation

comprehensive research and development

a guaranteed power supply

an active oil policy

international cooperation.

Deliberate Conservation

Planned energy utilization is an important policy instrument to encourage efficient energy saving. In Sweden, local energy planning covering the production and distribution of practically all types of energy for all consumption purposes will be used to facilitate measures including the expansion of systems of collective energy supply and the use of waste heat from industry for heating of homes and other buildings.

A considerable scope for energy conservation exists in the residential sector. Loans and grants are made available for energy conservation measures in homes. Similar financial support is given to the industrial and commercial sectors in order to promote more efficient use of energy.

Apart from economic controls, various forms of regulation are needed to achieve the aim of reducing the growth rate of energy consumption. Among other things, there are amendments to the current building legislation. A new paragraph has been inserted to the effect that a new or reconstructed building must conform to the requirements of sound heating economics.

Comprehensive Research and Development

A three-year main program of energy research with a total expenditure of Skr 336 million (U.S.$75 million) has been adopted by the Riksdag. The responsibility for the coordination and continuous observation of this program will be vested in a special Energy R&D Commission affiliated with the Ministry of Industry. The main program has been divided into the following six programs:

1. Energy utilization in industrial processes, etc.
2. Energy utilization in transport and communications
3. Energy utilization for indoor comfort
4. Retrieval of energy from goods, etc.
5. Energy production
6. General energy systems analysis.

Roughly two-thirds of the total allocation goes to the program of energy production.

A Guaranteed Power Supply

Electricity generation in Sweden is still predominantly based on domestic hydroelectric power resources. The possibilities for expansion of these resources, however, are limited due to an increasing concern for the preservation of certain areas and ecological systems for future generations. The options to consider are oil/coal-fired and nuclear-based combined power and heating plants, nuclear power stations, and hydroelectric power stations. Nuclear power offers great advantages, though these must be balanced against offsetting factors such as safety problems. It has been concluded that a continued, cautious expansion of nuclear power in Sweden is economically, logistically, and environmentally justifiable and will not entail any unacceptable risks.

An Active Oil Policy

The efforts to limit Sweden's dependence on imported oil play an important part in the nation's energy policy.

A number of direct measures of oil policy will

1 . In May 1975 the Swedish Parliament (the Riksdag) adopted a resolution on energy policy for the period up to 1985. (A summary in English titled *Energy Planning in Sweden* is available from the Ministry of Industry.)

have to be taken. One of the purposes of these measures will be to adapt our supplies to changes in the international market system. This means, for instance, establishing new channels to the oil-producing countries and also increasing state commitments within the oil trade.

International Cooperation

Energy questions are simply a subgroup of the many questions concerning raw materials and other aspects of global distribution. The police bill states that it is our duty to display international solidarity, and that we must show consideration for the needs and interests of other countries when determining the level of our own energy consumption. Ways should be sought to get a closer understanding and constructive partnership between rich and poor countries and between oil producers and oil consumers.

As a member of the International Energy Agency, Sweden takes part in activities such as allocation of emergency supplies and a long-term program of R&D in energy conservation.

13.3 METHODOLOGY

Our report is, as mentioned above, based on the extensive energy studies (involving some 70 man-years of research) by government commissions in 1973 and 1974, especially those by EPU, the Energy Forecast Commission. The methodology behind these studies is quite similar to that of WAES. EPU has four scenario-alternatives: high (1, 2) or low (3, 4) economic growth and high (1, 3) or low (2, 4) nuclear power growth. The time period is 1970 to 1985 and 1985 to 2000. The oil price is assumed to be constant at US$11.50 per barrel Arabian light crude oil FOB Persian Gulf in real terms. Government response is assumed to be more vigorous in the low growth cases than in the high growth cases, but it is not as vigorous as in the WAES definition of vigorous government response. The number of line items is some hundreds, and EPU tried to determine a trend of changes in the specific energy consumption from the 1950s in order to get a better understanding of potential changes to 1985 and 2000 coupled with probable future conservation actions. The activity levels were found through a sectoral analysis and forecasts by appropriate associations or experts. The consistency for industry was checked through an input-output analysis. The high growth cases could be affected by capital and employment constraints.

In spite of all these similarities, many figures are not suited to the WAES format and have been adjusted. WAES also uses other parameters, such as a higher and lower energy price and a more stringent concept of vigorous government response. Another difference is that WAES is making forecasts from 1976, taking into account the actual use of energy in 1973 to 1975, while EPU could not use the actual energy demand figures for these years.

WAES Case D is, however, almost identical with EPU Case 4, and WAES Case C is very close to EPU Case 2, though for a more vigorous response. We have tried to consider this as well as to adjust the EPU figures for the actual development of energy use during 1973 to 1975.

For WAES Cases A, B, and E, we have used another approach since we found it most difficult to get agreement on the effects of a higher and lower energy price on such a detailed energy demand analysis as 69 line items. We have therefore concentrated our work on the main sectors and through discussions with experts tried to get a feeling for the potential impact on the EPU forecasts—EPU alternatives 1, 3, and 2—of higher or lower energy prices and a more vigorous government response. We tried to be as detailed and specific as possible in these interviews, discussing various potential energy conservation actions or investments. The Swedish Energy Forecasting Model offers another interesting possibility for considering these parameters, and we hope to be able to use this model during the autumn of 1976.

Very little is known about the price elasticity of energy, but the general impression we got through these interviews was a feeling for a very low price elasticity at least to 1985. The result of a higher oil price would primarily be market forces that encourage further energy saving, but the effect of these forces would take a long time to be more than marginal. Moreover, problems with the balance of payments could induce the government to promote the industrial sector (increase exports) and reduce private and public consumption.

Vigorous government actions would focus mainly on increasing energy saving in space heating; elsewhere it would primarily take the form of soft actions, such as incentives, information, or subsidies, unless public opinion comes to regard it as essential that the government take hard actions to save energy.

Table 13.2, showing vigorous and restrained government policy, contains our judgments of the actual Swedish energy policy coupled with potential and politically realistic more vigorous actions. It should be noted that there could be conflicts of interest between energy-saving policies and other national policy objectives. For instance, objectives such as environmental protection and improving working conditions have priority at present over energy-saving objectives.

13.4 1985 ENERGY DEMAND—PROJECTIONS AND ANALYSIS

Summary tables for 1985 energy demand, energy growth rates, and sectoral percentages of primary energy input may be found with the other detailed worksheets in Section 13.7.

13.4.1 1985 Case Comparisons

Comparing the changes in energy demand (E) in 1985 in the various WAES scenario cases can give an indication of the impact of oil price, economic growth, and vigorous government response. One should, however, have in mind that the parameters are interrelated in a complex way, and one should be cautious to draw too firm a conclusion from the following comparisons.

Changes in Oil Price

The only difference between Cases C and A is a higher oil price in Case A. If

E_{A} and E_{C} are final energy demands (excluding processing losses and feedstocks) in Cases A and C in 1985,

E_{IA} and E_{IC} are industrial energy demands (excluding feedstocks) in Cases A and C in 1985,

E_{RA} and E_{RC} are residential energy demands in Cases A and C in 1985,

then we can calculate the percent change in energy demand in 1985 due to a 50 percent higher oil price as shown in Table 13.3.

The effect of a 50 percent higher oil price on total demand is quite small in the period 1972 to 1985, which indicates a low price elasticity. The change is much higher for residential than for industrial energy demand, which indicates a necessity for a more vigorous government response in the residential sector when the oil price is increasing.

The figures also imply a shift to electricity, especially for space heating, which is quite natural since Sweden does not have any oil and gas resources but does have quite large uranium resources.

Changes in Economic Growth

The only difference between Cases A and B is a higher economic growth rate in Case A (4 percent per year as compared to 2.5 percent). From Table 13.4 it can be seen that the effect of higher economic growth—the GNP is 20 percent higher in 1985 in Case A than in Case B—is greater than the effect of the 50 percent higher oil price and also, quite naturally, affects industrial demand more than the residential. One should note that EPU uses the same number of dwellings in 1985 regardless of high or low economic growth.

These two comparisons of scenarios, which all assume vigorous government response but differ in oil price and GNP growth rate, could therefore indicate that the effect on energy demand from changes in the economic growth rate is two to three times as great as that of changes in the price of oil over a time span of ten years.

Changes in Government Response

We do not have two WAES scenario cases which differ only in government response. From the following analysis one can, however, get an indication of the impact of a vigorous government response.

Changes in Oil Price and Government Response.

Here we compare Cases B and D, where Case B assumes a 50 percent higher oil price and vigorous response and Case D, assumes a constant price and restrained government response. We can also compare Cases C and E, where Case E assumes a 50 percent lower oil price and restrained government response, while Case C assumes a constant price and vigorous response. Table 13.5 shows these comparisons.

Table 13.2
Summary of restrained and vigorous government policy responses[1]

Policy Area	Vigorous Response	Restrained Response
Transportation		
Taxation (presently 34 Sw. öre/liter gasoline and 40 Sw. crown/m^3 gas, diesel oil, and residual fuel oil)	Doubling of existing taxes. Mileage taxation for trucks.	*Tax increases from the middle of 1975 by 5 öre/liter gasoline and 5 Sw. crowns/m^3 gas, diesel oil, and residual fuel oil bunkers. (This is to cover emergency storage costs.)*
R & D	*US$6 million for studies of alternative transport system and propulsion systems.*	
Information	Broad public information campaign regarding why and how to save energy.	
Load factor	Special lanes for car pooling; making it legal for carowners to be paid by passengers. Reducing the number of flights.	
Subsidies	Subsidize building of lighter transport equipment and more efficient engines.	
Regulations	Reduce speed limits for cars, airplanes.	
Industry		
Subsidies	*State grants of US$18 million to processes and to experimental and demonstration facilities in 1975–1976. The subsidies cover 35–50% of the costs.* More than doubling the amount of grants per year to 1985.	
R & D	*Studies of industrial energy consumption; inventory of waste heat supplies deriving from industrial processes—especially in iron and steel and forest industry—to find out what measures may have to be taken by society (US$8 million in 1975 to 1977).* Implementation of the measures proposed in these studies made by the National Board of Industry.	
Regulation		*Amendment to the Building Act making location or expansion of industry using high amounts of energy subject to governmental approval:* *• iron and steel works, metal works, ferroalloy plants* *• sawmills, mechanical pulpmills, papermills, cellulose plants* *• chemical plants for production of inorganic and organic products or other petrochemical products* *• petroleum refineries* *• nuclear power plants* *• plants for reprocessing nuclear fuel* *• steam power stations and other fossil fuel plants with an input effect exceeding 500 megawatts* *• fertilizer factories* *• cement factories.*

1. Entries in italics represent actual policy actions being implemented in the spring of 1976.

Table 13.2 (Continued)

Policy Area	Vigorous Response	Restrained Response
Retrieval of energy	*US$2 million in grants for studies of more efficient recycling of raw materials and of energy extraction through, for example, pyrolysis.* *Gradual introduction of compulsory collection of paper, which must be kept separated at source from other waste.*	
Education	*The Board of Industry will draw up training programs on energy topics relating to industrial processes, etc.*	
Residential and Commercial		
Regulation	*New legislation since January 1976 in the Building Statute to the effect that new or reconstructed buildings must conform to the requirements of sound heating economics. The National Board of Urban Planning will issue specific regulations and instructions to amplify this provision. Enforcement of the new statutory rule will be the responsibility of municipal building committees.* *The code is in principle valid for all new and rebuilt residential and other buildings. The general building codes will be supplemented during 1976 with new energy conservation paragraphs regulating building insulation and airtightness of buildings and installations. Installations should be supported by such equipment that temperature exceeding normal standards will be avoided. Larger buildings must be equipped with heat-recovery equipment for air conditioning. For water, gas, and electricity installations, individual measurements of energy will be demanded for every flat. Individual metering is also a condition to obtain state building loans.*	
Instruction	*The National Swedish Board of Public Buildings has issued instructions for maintenance, and running times are specified. In the construction of new buildings the instructions include recommendations for geometrical solutions, windows, insulation, and the installation of heating and ventilation equipment. Maximum temperatures for public buildings are also being recommended.* *In 1974, US$4.4 million was given by the government to change and rebuild existing buildings in order to save energy.*	
Information	*The Committee on Energy Conservation will continue and expand its activities such as information to the public.* *During fiscal year 1975–1976, US$0.4 million has been reserved for the education of property caretakers. The program involves conservation measures concerning subsequent servicing, such*	

Table 13.2 (Continued)

Policy Area	Vigorous Response	Restrained Response
	as adjustment of burners for efficient operation, improvement of ventilation, and so on. In all, about 4000 caretakers will attend these courses. A booklet about these questions has been distributed to all caretakers in the country.	
	A special booklet has been sent to all homeowners and caretakers on how to reduce energy consumption (e.g., improving insulation, adjusting burners).	
	A special program is being carried out to instruct homeowners and caretakers of small residences on how to improve the efficiency of their oil burners. A total of US$0.6 million was allocated for this program during 1975. During this year, 4000 homeowners and 2000 caretakers attended courses and demonstrations at 30 locations.	
	The committee on energy conservation is presently studying the conditions for making a survey of burners mandatory.	
	Continuation of these grants to 1985.	
Statistics	*The Central Bureau of Statistics has been ordered by the government to speed up and improve energy statistics, and to investigate conservation opportunities in heating, ventilation, and lighting of homes and other buildings.*	
Taxation	Implementation of energy tax if recommended by the energy taxation commission.	*A commission will be appointed to review the role of energy taxation as an instrument of energy policy.*
Grants	*US$75 million for 1975–1976 were given as loans and subsidies for improvement or replacement of heating or ventilation systems, connection of a house property to a district heating system, and improvements in the heat insulation of walls, windows, and framings of joists.*	
	For each apartment a maximum amount of 6000 Swedish crowns is given for conservation measures such as improved insulation, installation of temperature sensors, etc., of which 2000 Swedish crowns are for changing to a more efficient boiler.	
	Continuation of these grants for each year to 1985.	
Energy efficiency labeling	Mandatory energy labeling for all major consumer appliances.	The Board for Consumer Questions is looking into this area.
R&D	*US$1.5 million for prototypes and demonstration facilities to stimulate planning with new building standards, local energy planning, etc.*	
	US$12 million in 1975–1977 to study the use of energy for space heating and for household appliances.	
	A much higher amount to prototypes and demonstration facilities.	

Table 13.3
Changes in energy demand resulting from a 50 percent higher oil price

	Fossil Fuels	Electricity	Total Demand
$\frac{E_A - E_C}{E_C} \times 100$	-14.7	+7.2	-10.7
$\frac{E_{IA} - E_{IC}}{E_{IC}} \times 100$	- 4.1	-7.1	-5.0
$\frac{E_{RA} - E_{RC}}{E_{RC}} \times 100$	-44.0	+42.0	-18.0

Table 13.4
Effect of growth rate on energy demand

	Fossil Fuels	Electricity	Total Demand
$\frac{E_A - E_B}{E_B} \times 100$	+7.7	+13.7	+9.5
$\frac{E_{IA} - E_{IB}}{E_{IB}} \times 100$	+12.5	+14.0	+12.9
$\frac{E_{RA} - E_{RB}}{E_{RB}} \times 100$	- 5.0	+ 5.0	0

As compared to only a change in oil price (Table 13.3), vigorous government response seems to increase the effect of market forces on total demand only slightly, from -10.7 percent to about -12.3 percent.

Changes in Economic Growth and Government Response. By comparing the energy demand in Cases A and B and in Cases C and D, one can get another indication of the effect of vigorous government response when the economic growth rate goes from a low to a high value.

According to Table 13.6, vigorous government response can reduce the increase in demand due to an increase in economic growth from 9.5 to 6.3 percent. One should note, however, that the oil price assumption is not the same in these two case comparisons.

Sectoral Distribution of Energy Demand. Table 13.7 shows the 1972 and projected 1985 sectoral distribution of primary energy demand. It is interesting that despite the differences in oil price, economic growth, and government response assumptions among the five WAES cases, they all show the various sectors in 1985 to have about the same share of primary energy output.

The major change comes in the residential sector, whose share was about 19 percent in 1972 but is projected to account for only 10-13 percent in 1985. This change is due partly to energy saving in this sector and partly to the increasing share of processing losses, which is quite natural since electricity use will increase much more quickly for consumers than will fossil fuel use. In 1972, electricity's share of primary energy input was 35 percent, as compared to about 50 percent in 1985. It should be noted that WAES assumes a 35 percent efficiency for nuclear power, hydro power, and oil-condense. For district heating and industrial backpressure, an 85 percent efficiency has been assumed.

The high figures for processing losses indicate potential areas for increased energy efficiency—for example, by the use of nuclear power for district heating. This question has been under discussion for some time in Sweden, but the impact of a decision to use nuclear power for district heating will not be felt before 1985.

Differences in the 1985 energy demand situation also appear if one splits the demand into electricity and fossil fuel components (Table 13.8). Higher economic growth and oil price and vigorous government response will result, according to our projections, in an increased share of total primary energy input for fossil fuels in the industrial and transportation sectors and a significant reduction of the

Table 13.5
Effect of changes in both oil price and government response

Change in Oil Price	Change in Government Response	Economic Growth	% Change in Total 1985 Energy Demand
+50%	from restrained to vigorous	low	$\frac{E_B - E_D}{E_D} \times 100 = -12.3$
–50%	from vigorous to restrained	high	$\frac{E_E - E_C}{E_C} \times 100 = +12.5$

Table 13.6
Effect of changes in both economic growth and government response

Oil Price 1985	Change in Economic Growth	Change in Government Response	% Change in Total 1985 Energy Demand
US$11.50	from low to high	from restrained to vigorous	$\frac{E_C - E_D}{E_D} \times 100 = +6.3$
US$17.25	from low to high	vigorous (both cases)	$\frac{E_A - E_B}{E_B} \times 100 = +9.5$

Table 13.7
Sectoral distribution of primary energy demand, 1972 and 1985

Sector	% of Total 1972 Primary Energy Demand	% of Total 1985 Primary Energy Demand in the Five WAES Scenarios
Industrial, excl. Feedstocks	27.4	26–27
Transportation	12.9	10–11
Commercial, Public	9.6	7–8
Residential	18.6	10–13
Others	1.9	3
Feedstocks	2.7	5
Final Energy Demand	73.2	62–66
Processing Losses	26.8	34–38
Primary Energy Input	100	100

Table 13.8
Sectoral distribution of primary energy demand by source, 1972 and 1985 (%)

Sector	1972		Case A (1985)		Case D (1985)	
	Fossil Fuel	Elec.	Fossil Fuel	Elec.	Fossil Fuel	Elec.
Industrial	33	17	41	15	35	16
Transportation	19	1.0	23	0.7	20	0.8
Residential, Commercial, and Public	37	13	17	16	26	14
Others, incl. Processing Losses	11	69	19	68	16	68
Total	100	100	100	100	100	100

share of fossil fuels in the residential, commercial, and public sectors. The share of electricity consumed by processing losses is decreasing due to the increased implementation of district heating.

Energy Growth Rates

During the period 1955 to 1970, total energy input to Sweden increased by about 5.3 percent per year, as compared to about 3.0 percent per year for 1970 to 1973. This reduced rate was mainly due to the slow economic growth during 1971 and 1972 and also to the shortage of electricity in 1970 resulting from two previous dry years, which reduced the capacity for hydro power and resulted in public savings actions.

Energy consumption was reduced by 7 percent in 1974 and increased only slightly in 1975, due to savings actions and slow economic growth.

The WAES growth rate projections (Table 13.9) are averages for 1972 to 1985, but consideration has been taken of the actual use of energy in 1973 to 1975. This implies that the growth rates from 1976 to 1985 should be somewhat higher than the WAES projections for 1972 to 1985.

Compared to the development in 1955 to 1970, however, these WAES projections are much lower due to the effects of higher oil price and vigorous government response, especially for the residential

Table 13.9
Summary of growth rates for energy demand for the five WAES cases, 1972–1985 (%/yr.)

	A	B	C	D	E
Industry	+3.1	+2.2	+3.5	+2.8	+4.4
Transportation	+2.0	+0.8	+2.7	+2.1	+4.0
Commercial and Public	+0.1	+0.4	+0.9	+1.7	+2.2
Residential	−1.7	−1.7	−0.2	0	+1.2
Fishing, etc.	+7.2	+5.5	+7.7	+5.5	+8.6
Feedstocks					
Final Energy Demand	+2.0	+1.2	+2.7	+2.3	+3.8
Processing Losses	+5.7	+5.2	+5.5	+4.9	+6.2
Primary Energy Input	+3.1	+2.5	+3.6	+3.1	+4.6

and commercial and public sectors.

Even though the WAES scenarios are not forecasts but projections based on certain parameters, it could be of value to compare our projections with other forecasts in order to get a feeling for whether they are in the same range of potential developments. For example, the government's *Long Range Study 1975–1980* (SOU 1975:96), made in 1975, forecasts energy demand according to three possible developments (Table 13.10). Even though the time periods, definitions of sectors, and parameters behind the projections differ, the table indicates that WAES projections are in accordance with the latest government forecasts. It seems, however, that WAES projections of the use of fossil fuel in industry are very low. The reason for this is that there are fewer possibilities for industry to save energy in the form of electricity as compared to fuel for heating.

A higher oil price and vigorous government response will result in actions mainly to reduce fuel. The development of energy efficiency in industry will be discussed further in the next section.

Major Energy Demand Sectors

Tables 13.11 and 13.12 show the 1972 and projected 1985 energy demands for fossil fuels and electricity, respectively, for the major energy-consuming sectors. For fossil fuels, due to the quite substantial potential savings in space heat and the paper and allied subsector, these sectors have reduced their share from 58 percent in 1972 to

Table 13.10
Comparison of government and WAES projections of energy demand growth (%/yr.)

	Actual 1972–1975	Government Forecasts 1975–1980	WAES Scenarios 1972–1985
Industry			
electricity	3.3	5.5 – 6.6	4.1 – 5.7
fossil fuel	0.2	4.1 – 5.1	1.5 – 2.8
total	0.9	4.5 – 5.5	1.2 – 2.7
Transportation			
electricity	3.2	3.4	3.2
fossil fuel	1.6	2.0 – 2.5	0.9 – 2.7
total	1.7	2.1 – 2.5	0.8 – 2.7
Other Sectors			
electricity	5.0	7.4	6.0 – 8.4
fossil fuel	−1.6	−0.6	(−0.4)–(−3.8)
total	−0.4	1.2	(−0.1)–(+1.1)
Total			
electricity	4.0	6.3 – 6.8	5.2 – 6.7
fossil fuel	−0.2	1.9 – 2.4	0.1 – 1.8
total	0.5	2.7 – 3.3	1.2 – 2.7

Table 13.11
Energy demand for fossil fuels by sector, 1972 and 1985 (in TWh)[1]

	1972	1985 Case C	1985 Case D
Residential Space Heat	88	70	72
Paper and Allied[2]	58	65	60
Public and Commercial Space Heat	44	38	45
Subtotal	190	173	175
% of Total Final Energy	58%	44%	46%
Autos	30	41	39
Iron and Steel	22	50	47
Trucks	12	19	18
Subtotal	64	110	106
% of Total Final Energy	20%	28%	28%
Other Sectors	73	106	91
% of Total Final Energy	22%	27%	26%

1. Excluding feedstocks and processing losses.
2. Including internal fuel.

Table 13.12
Energy demand for electricity by sector, 1972 and 1985 (in TWh)[1]

	1972	1985 Case C	1985 Case D
Paper and Allied	13	18	17
Residential (excl. Space Heating)	12	20	18
Chemicals	5	13	12
Residential Space Heat	4	11	14
Iron and Steel	5	13	10
Public	6	15	14
Manufacturing	4	10	7
Commercial	3	8	8
Mining	2	9	5
Nonferrous Metals	3	7	8
Subtotal	57	124	113
Other Sectors	7	15	12
Total (excl. Processing Losses)	64	139	125

1. Including internally generated power.

44–47 percent in 1985. A more vigorous government response than we have assumed would perhaps look more into potential conservation policies in other sectors of society.

In almost all sectors, we project a major increase in electricity use. If a policy must be designed to reduce the use of electricity, the main target should be space heating by electricity, which in Cases C and D accounts for about 14–19 TWh. This is the main sector where substitutes might be found.

Energy Efficiencies for Industry

During the period 1960 to 1973, the specific energy demand (energy per unit of production or per value added) for industry decreased on average by 1.0 percent per year total (for fuel, –1.5 percent per year; for electricity, +0.5 percent per year).

In the government's *Long Range Forecast 1975–1980*, an annual decrease of 1.5 percent was assumed for the specific energy demand in industry.

The Industrial Institute for Economic and Social Research (IUI) has published in the spring of 1976 a forecast for specific energy demand in industry, 1973 to 1980. It is of interest to compare their forecasts with our WAES projections even though the time period and other parameters are not identical.

According to Table 13.13, the WAES projections seem to be in general agreement with the forecasts made by IUI. Higher oil price and vigorous government response lead quite naturally to a faster reduction of specific energy demand, especially for the paper and allied and mineral products subsectors.

The individual forecasts for fossil fuel and electricity specific demands are, however, quite different in the WAES projections and the forecasts of the Swedish Industrial Board (Table 13.14).

We have based our projections on published studies, mainly those of EPU, and discussions with experts in the spring of 1976. The general impression we got was that the possibilities for conserving electricity are much more limited than those for conserving fossil fuels. Conservation actions should therefore be concentrated on the reduction of fossil fuel usage. The range in Table 13.14 from +0.3 to –0.1 percent per year in specific energy demand change for electricity corresponds in 1985 to the output of, for instance, one large nuclear power station. This indicates the need for further

Table 13.13
Comparison of IUI and WAES projections of % changes per year in specific energy demand

	IUI		WAES 1972–1985				
	1973–1980		A	B	C	D	E
Growth in Production Volume or Value Added	+4.4	+5.5	+6.0	+4.1	+6.0	+4.1	+6.0
Oil Price	–	–	+50%	+50%	constant	constant	–50%
Government Response	–	–	vig.	vig.	vig.	restr.	restr.
Iron and Steel	–2.2	–1.6	–1.9	–1.7	–1.5	–1.0	–0.5
Nonferrous Metals	–0.3	–0.7	–1.0	–0.8	–0.5	–0.3	+0.1
Paper and Allied	–2.2	–3.1	–3.8	–2.8	–2.4	–2.1	–0.7
Chemicals and Allied	+1.0	+1.0	+0.2	+0.5	+0.6	+0.9	+1.1
Mineral Products	–3.4[1]	–3.6[1]	–3.8	–3.3	–3.0	–2.4	–0.9
Food and Related	–0.8	–1.2	–1.5	–1.0	–1.0	–0.6	–0.6
Manufacturing	–1.0	–2.0	–2.2	–1.4	–1.4	–0.7	–1.0
Other Sectors			–2.0	–1.2	–1.8	–0.7	–0.7
Mining	+4.7	+5.7	+5.5	+5.0	+5.8	+5.1	+6.3
Total Industry	–0.8	–1.2	–2.4	–1.6	–2.0	–1.0	–1.1

1. Cement Industry

Table 13.14
Comparison of National Swedish Industrial Board and WAES projections of % changes per year in specific energy demand

	Actual 1960–1973	Industrial 1974–1980[1]	WAES 1972–1985				
			A	B	C	D	E
Electricity	+0.5	–0.1	–0.3	+0.3	+0.3	+0.5	+0.8
Fossil Fuels	–1.5	–1.3	–3.1	–2.2	–2.8	–1.6	–1.8

1. From SOU 1975:96.

in-depth studies of the possibilities for conserving electricity.

At the same time, we project greater savings for fossil fuels, which seem reasonable when we consider the WAES parameters of higher oil price and vigorous government response.

Comparison with the Forecasts Made by EPU (The Swedish Energy Forecasting Commission)

Our forecasts are based on the four alternative forecasts made by EPU in 1972–1974. We have based WAES Case A on EPU alternative 1, Case B on EPU-3, Cases C and E on EPU-2, and Case D on EPU-4.

EPU alternatives 1 and 2 are high economic growth cases, and EPU-3 and EPU-4 are low economic growth cases. EPU alternatives 1 and 3 imply continuous expansion of nuclear power, and EPU alternatives 2 and 4 assume no more nuclear power. The oil price is constant at US$11.50 in all EPU alternatives. Government response should be considered restrained by the WAES definition even though EPU included increased R&D efforts by both the government and the private sector to stimulate energy conservation.

We have coupled Cases A and B, in which the oil price increases by 50 percent, with the two EPU nuclear alternatives, since a higher oil price would quite probably force increased nuclear power investments, and Cases C, D, and E with the EPU "no more nuclear" alternatives 2 and 4, since the oil price is the same or reduced in these WAES cases. These EPU alternatives 2 and 4 conform more closely to the actual Swedish energy policy.

The differences between the WAES cases and the EPU alternatives are due to the effects of higher oil prices, the vigorous government response, the actual use of energy in 1973 to 1975, and the information regarding energy-saving potentials that has been published after EPU or was forwarded to us in discussions with experts in different fields. As can be seen from the comparison in Table 13.15,

Table 13.15
Yearly growth rate of final energy demand, 1972–1985[1]

	EPU-2	Case C	EPU-4	Case D
Economic Growth	high	high	low	low
Oil Price, 1985	US$11.50	US$11.50	US$11.50	US$11.50
Government Response	restrained	vigorous	restrained	restrained
Growth Rate of Final Energy Demand (%/yr.)	+3.4	+2.4	+2.4	+1.7

1. Excluding feedstocks and processing losses.

Case D should be identical with EPU alternative 4. The main difference is due to the fact that in the WAES case we consider the actual energy use in 1973 to 1975.

The difference between WAES Case C and EPU-2 is also the vigorous government response, which results in even lower energy growth rates, especially in the residential sector.

In WAES Cases A and B, two additional differences from the EPU alternatives 1 and 3 are the higher oil price and the vigorous government response, in addition to the actual low use of energy in 1972 to 1975 (Table 13.16).

The only difference between WAES Case E and EPU-2 (Table 13.17) is a reduced oil price. The main reason why the total demand growth rate is almost the same and not much higher for WAES Case E is the actual low use of energy in 1974–1975 and also the fact that the impact of the so-called oil crisis of 1973–1974 will push forward energy-saving actions despite any reduction in oil price.

Table 13.18 shows the total reduction of energy demand (excluding feedstocks and processing losses) in 1985 resulting from the reduced energy growth rates in the WAES cases as compared to the EPU alternatives.

Table 13.17
Yearly growth rate of final energy demand, 1972–1985[1]

	EPU-2	Case E
Economic Growth	high	high
Oil Price, 1985	US$11.50	US$7.66
Government Response	restrained	restrained
Growth Rate of Final Energy Demand (%/yr.)	+3.4	+3.5

1. Excluding feedstocks and processing losses.

13.5 CONCLUSIONS

The need for electricity is increasing in almost all sectors due to continued automation, use of machinery and appliances, increased working and living requirements for lighting, ventilation, environmental improvements, etc., all of which is quite natural when one considers how refined a form of energy electricity is. Space heating with electricity is also increasing, and this has many advantages for attempts to encourage more efficient use of energy in heating, such as regulating the heat in a room with an inflow of sunlight.

The potential for energy saving in space heating seems most promising. More savings could probably be made in this area if the public were aware of the real need for conservation. We have, however, anticipated quite substantial savings due to the

Table 13.16
Yearly growth rate of final energy demand, 1972–1985[1]

	EPU-1	Case A	EPU-3	Case B
Economic Growth	high	high	low	low
Oil Price, 1985	US$11.50	US$17.25	US$11.50	US$17.25
Government Response	restrained	vigorous	restrained	vigorous
Growth Rate of Final Energy Demand (%/yr.)	+3.2	+1.7	+2.3	+1.0

1. Excluding feedstocks and processing losses.

Table 13.18
Difference in total 1985 energy demand (excluding feedstocks and processing losses) between the WAES cases and the respective EPU alternatives

WAES Case	A	B	C	D	E
EPU Alternative	1	3	2	4	2
Absolute Difference (TWh)	−116	−82	−79	−27	−1
% Reduction of EPU Alternative	−19.3	−15.6	−12.9	−5.1	0

higher oil price and vigorous government response. It might be questioned whether these savings will be achieved before 1985.

The same question might also be made regarding our anticipation of fossil fuel savings in the industrial sector.

Except for mining and chemicals, industry has increased its energy efficiency during the 1960s and is projected to continue this trend. Most other consumers, however, have been becoming less energy-efficient in the past few years, and this is especially true of private consumption.

More studies, especially of private consumption patterns, are needed to produce a full understanding of the driving factors—such as increased leisure time, improving material living standard, and institutional constraints—behind this decreased energy efficiency if we are to create a realistic energy conservation policy.

Regarding transportation, some factors may well result in an increase in the energy consumed by automobiles, contrary to what might be expected as energy-saving measures are adopted. Those factors include increases in per capita income and in the number of cars. Reduction of working hours per week coupled with longer vacations, an increase in the number of vacation homes, and lowering of the age of retirement can increase the use of cars and counteract energy savings.

Since automobiles use the major portion of the energy in the transportation sector, there is an important incentive for more research to study the effects of different conservation actions in that area.

The National Board of Industry is carrying out some detailed studies of industrial energy consumption. These studies include an inventory of waste heat supplies deriving from industrial processes, and of the possibility of an increased utilization of this heat for external purposes such as domestic heating. An investigation is also being made of the feasibility of expanding energy output by means of industrial backpressure. The Board is also studying industrial energy utilization in the iron and steel industry and in forest industries. All these investigations are made in collaboration with the appropriate trade and industrial associations.

Efficient use of energy in industry will come about through the adoption of many different measures. Several of these measures will evolve gradually during the next ten to twenty years as new techniques become available. Most savings till 1985, however, are expected to come essentially within the limits of today's technology. Examples of ongoing work in this field include

- recycling of waste paper in the paper and allied industry,
- continuous casting methods in the iron and steel industry,
- changing over to dry processes from wet processes in the cement industry.

The adverse environmental consequences of industrial production must be considered, and there are reasons to believe that more energy will be needed in the future to reduce such consequences in some industrial sectors, such as in the iron and steel industry. At present about 8 percent of the total electricity consumption in a typical steel plant in Sweden goes to environmental protection only. It has been suggested, however, that these requirements be made even more strict. As a result, the total energy (electricity) demand in the Swedish steel industry to keep the environment at an acceptable standard would be almost 4×10^9 kWh, or the equivalent of the total output of an 800 MWe nuclear power plant (2.5 percent of the total electricity consumption in 1985).

Finally, we would like to emphasize the following: In spite of a lot of statistics and research, one should approach the numbers in this report with caution. There still exists great uncertainty in the use of statistics that were not planned to be used in a detailed energy analysis. Because of the oil crisis, actions are now being taken to improve the statistics, but it will take time before these can be as reliable as one would want them to be.

Even if we had perfect statistics, there still exist great difficulties in the projection of energy demand. As we showed in Section 13.4, a slight difference in the expected energy efficiency development for electricity use in industry could in 1985 amount to a demand corresponding, for instance, to the output of a large nuclear power plant. More research is therefore needed to give a firm foundation regarding the potential effect of activity levels and energy conservation actions. For some sectors, we have too little knowledge to make realistic projections regarding activity levels, especially for residential, public, and commercial buildings, but also for some industrial subsectors, such as iron and steel. We have, for instance, used the same increase in the number of buildings in all WAES cases (as did EPU). Ongoing studies will in the near future yield more information and thereby allow us to refine our forecasts.

In spite of the 69 line items used by WAES, one should really make a much more detailed study to get a better picture and understanding of potential energy savings. In comparing data with other nations, a further breakdown is also necessary to allow understanding of why nations and regions may differ.

13.6 REFERENCES

Energiprognosutredningen, "Energi 1985–2000," SOU 1974:64-65, Department of Industry, 1974.

Energiprogramkommittén, "Energiforskning": SOU 1974:72–76, 1974; DsI 1974:8, 1974; DsI 1975:1, "Behov av prototyper och demonstrationsanläggningar," 1975; DsI 1975:3, "Förslag till Plan för Allmänna energisystemstudier m m," 1975. Department of Industry.

Federation of Swedish Industries, "Energy Conservation in Swedish Industry," 1975. Report from an "Energy in Industry Day" on May 2, 1974. A similar symposium was held on May 18, 1976. Proceedings from the latter symposium are in publication.

National Board of Industry, "Hushållning med energy inom industrin." Report to the Swedish Government, June 1976.

"Energihushållning m m." The Swedish Government's proposal, 1975:30, to the Riksdag on energy management.

Ångpanneföreningen, "Verksamheten," 1971–1975, Stockholm.

Ångpanneföreningen, "Spillvärmeutnyttjande" [Use of waste heat]. Report to the National Board of Industry, 1976.

Department of Finance, "Långtidsutredningen 1975," SOU 1975:89, 1975.

National Board of Industry, "Energiförsörjningen 1975–1980," SOU 1975:96, Department of Finance, 1975.

AKA-utredningen, "Använt kärnbränsle och radioaktivt avfall," SOU 1976:30, Department of Industry, 1976.

Svenska Värmeverksföreningen, "Kraftvärme 1975."

IUI:s långtidsbedömning 1976.

Kommunikationsdepartementet, "Transporter i Sverige," DsK 1975:4.

"Trafikpolitik, behov och möjligheter," SOU 1975:66.

Division of Building Technology, Royal Institute of Technology, "Energy Conservation Measures Taken in the Thermal Insulation and Installation Fields for Existing Buildings," "Designing Houses for Energy Conservation," and other papers, I. Höglund, 1974–1976.

Statens Råd för Byggnadsforskning and Professor Bo Adamson, Lunds Institute of Technology, "Yttrande angående bestämmelser i Svensk Byggnorm rörande värmehushållning m.m.," 1975.

Division of Heat and Power Engineering, "Energihushållning i tätbyggd- planeringsmetodik," L. Thörnqvist, Lund Institute of Technology, 1975.

"Energi og transport," Nordisk Komité for transportekonomisk forskning, Publikasjon nr 19, 1975.

"A Progress Report from the Future Study: Energy and Society," Secretariat for Future Studies, 1976.

"Space heating in Europe, USA and Canada," Draft Report, Nils-Eric Lindskog, 1975.

13.7 DEMAND STUDY WORKSHEETS

This section contains summary tables and WAES worksheets in the following order:

Energy Use Summary

Energy Growth Rates Summary

Sectoral Percentages of Total Primary Energy Input

Base Year (1972) worksheets (six sheets)

Case A worksheets (six sheets)

Case B worksheets (six sheets)

Case C worksheets (six sheets)

Case D worksheets (six sheets)

Case E worksheets (six sheets).

WAES NATIONAL ENERGY DEMAND STUDIES, 1972-1985
ENERGY USE SUMMARY

Country Sweden

Scenario Assumptions												
WAES Case	A		B		C		D		E			
GNP 1976-1985	High		Low		High		Low		High			
Oil Price*	$11.50 → $17.25		$11.50 → $17.25		$11.50		$11.50		$11.50 → $7.66			
Natl. Policy Response	Vigorous		Vigorous		Vigorous		Restrained		Restrained			
Natl. GNP Growth Rate 1976-1985 (%/yr.)	4		2.5		4		2.5		4		1972 Base Year	
Fuel Type	Fossil Fuel	Elec.	Fossil Fuel	Elec.	Fossil Fuel	Elec.	Fossil Fuel	Elec.	Fossil Fuel	Elec.	Fossil Fuel	Elec.
Units	Twh		Twh		Twh		Twh		Twh		Twh	
Energy-Using Sectors												
Industrial (excluding feedstocks)	162	65	144	57	169	70	157	61	191	75	118.5	33.9
Transportation	90	3	78	3	98	3	91	3	115	5	69.6	2.0
Commercial and Public	28	27	35	22	38	23	45	22	45	27	44.4	9.3
Residential	39	44	41	42	70	31	72	32	88	33	87.6	15.9
Fishing, Agriculture, Mining, Construction	17	10	14	7	18	11	15	7	19	13	7.4	3.3
Feedstocks [1)]	40	–	35	–	45	–	40	–	50	–	15.0	–
Final Energy Demand (excluding processing losses)	376	149	347	131	439	139	420	125	508	153	342.5	64.4
Processing Losses	22	285	21	267	24	277	24	256	27	306	17.3	132.7
Primary Energy Input	398	434	368	398	463	416	444	381	535	459	359.8	197.1

*Oil price 1975 US$ per barrel Arabian light crude oil FOB Persian Gulf.

1) estimates

WAES NATIONAL ENERGY DEMAND STUDIES, 1972-1985
ENERGY GROWTH RATES SUMMARY

Country Sweden

Scenario Assumptions

WAES Case	A		B		C		D		E	
GNP 1976-1985	High		Low		High		Low		High	
Oil Price*	\$11.50 → \$17.25		\$11.50 → \$17.25		\$11.50		\$11.50		\$11.50 → \$7.66	
Natl. Policy Response	Vigorous		Vigorous		Vigorous		Restrained		Restrained	
Natl. GNP Growth Rate 1976-1985 (%/yr.)	4		2.5		4		2.5		4	
Fuel Type	Fossil Fuel	Elec.	Fossil Fuel	Elec.	Fossil Fuel	Elec.	Fossil Fuel	Elec.	Fossil Fuel	Elec.
Units (%/yr. 1972-1985)	%		%		%		%		%	
Energy-Using Sectors										
Industrial (excluding feedstocks)	2.4	5.1	1.5	4.1	2.8	5.7	2.2	4.6	3.7	6.3
Transportation	2.0	3.2	0.9	3.2	2.7	3.2	2.1	3.2	3.9	7.3
Commercial and Public	-3.5	8.5	-1.8	6.9	-1.2	7.2	0.1	6.9	0.1	8.5
Residential	-6.0	8.1	-5.7	7.8	-1.7	5.3	-1.5	5.5	0.0	5.8
Fishing, Agriculture, Mining, Construction	6.6	8.9	5.0	7.0	7.1	9.7	5.6	6.0	7.5	11.1
Feedstocks	7.8	–	6.7	–	8.8	–	7.8	–	9.7	–
Final Energy Demand (excluding processing losses)	0.7	6.7	0.1	5.6	1.9	6.1	1.6	5.2	3.1	6.9
Processing Losses	1.9	6.1	1.5	5.5	2.6	5.8	2.6	5.2	3.5	6.6
Primary Energy Input	0.8	6.3	0.2	5.6	2.0	5.9	1.6	5.2	3.1	6.7

*Oil price 1975 US\$ per barrel Arabian light crude oil FOB Persian Gulf.

WAES NATIONAL ENERGY DEMAND STUDIES, 1972-1985
SECTORAL PERCENTAGES OF TOTAL

Country Sweden

Scenario Assumptions

WAES Case	A		B		C		D		E			
GNP 1976-1985	High		Low		High		Low		High			
Oil Price*	$11.50 → $17.25		$11.50 → $17.25		$11.50		$11.50		$11.50 → $7.66			
Natl. Policy Response	Vigorous		Vigorous		Vigorous		Restrained		Restrained			
Natl. GNP Growth Rate 1976-1985 (%/yr.)	4		2.5		4		2.5		4		1972 Base Year	
Fuel Type	Fossil Fuel	Elec.	Fossil Fuel	Elec.	Fossil Fuel	Elec.	Fossil Fuel	Elec.	Fossil Fuel	Elec.	Fossil Fuel	Elec.
Units	%		%		%		%		%		%	
Energy-Using Sectors												
Industrial (excluding feedstocks)	40.6	15.0	39.2	14.3	36.5	16.9	35.3	16.0	35.7	16.4	32.9	17.2
Transportation	22.6	0.7	21.2	0.7	21.2	0.7	20.4	0.8	21.5	1.1	19.3	1.0
Commercial and Public	7.0	6.2	9.5	5.5	8.2	5.6	10.1	5.8	8.4	5.9	12.3	4.7
Residential	9.9	10.1	11.1	10.5	15.1	7.5	16.2	8.4	16.5	7.2	24.4	8.1
Fishing, Agriculture, Mining, Construction	4.3	2.3	3.8	1.8	3.9	2.6	3.4	1.8	3.6	2.8	2.1	1.7
Feedstocks	10.0	–	9.5	–	9.7	–	9,0	–	9.4	–	4.2	–
Final Energy Demand (excluding processing losses)	94.4	34.3	94.3	32.9	94.8	33.4	94.5	32.8	95.0	33.4	95.2	32.7
Processing Losses	5.1	65.7	5.7	67.1	5.2	66.6	5.5	67.2	5.0	66.6	4.8	67.3
Primary Energy Input	100	100	100	100	100	100	100	100	100	100	100	100

*Oil price 1975 US$ per barrel Arabian light crude oil FOB Persian Gulf.

NATIONAL ENERGY DEMAND PROJECTIONS

Country Sweden

Year 1972

WAES Case Base year

Summary Unit Twh

Category	Oil	Gas	Coal	Electricity	Other	Total by Category (Excluding Processing Losses)
Transportation	69.6			2.0		71.6
Agriculture, Mining, and Construction	7.4			3.3		10.7
Commercial	17.0			3.3		20.3
Public	27.4			6.0		33.4
Industrial	72.1		15.9	33.9	30.5	152.4
Residential	84.6			15.9	3.0	103.5
Total Delivered Energy by Fuel	278.1		15.9	64.4	33.5	
Processing Losses by Fuel	17.3			132.7		
Gross Input by Fuel	295.4		15.9	197.1	33.5	541.7

Gross Energy Input (Add across bottom row)

NATIONAL ENERGY DEMAND PROJECTIONS

Country Sweden

Year 1972

WAES Case Base year

Transportation

	(A)	(B)	(C)	(D)	(E)	(F)	(G)	
	Population/Auto	Distance/Auto/Yr.	Load Factor	Fraction Urban or Rural	Energy/Distance	Fossil Fuel 2)	Electricity 2)	
Urban Auto	3.25	14,220 km	1.4		1 litre per 10 km	29.9		(1)
Rural Auto			2.1					(2)
	Activity Level 1)	% Electrified	Load Factor	Fossil Fuel/ Activity Level 3)	Electricity/Activity			
Urban Bus	7.0			0.20		1.5		(3)
Intercity Bus								(4)
Rail (Passenger)	4.6	100	33 4)		0.20		0.9	(5)
Air (Passenger, Domestic)	0.68			0.88		0.6		(6)
Air (Passenger, International)	8.20			0.28		2.3		(7)
Ship (Passenger)	0.15			1.7		0.25		(8)
Rail (Freight)	16.3	74	30 5)		0.10	0.6	1.1	(9)
Air (Freight, Domestic)	0.01			1.0		0.0.		(10)
Air (Freight, International)	0.13			0.77		0.1		(11)
Truck (Local)	18.2		33 5)	0.65		11.8		(12)
Truck (Intercity)								(13)
Ship (Freight, Domestic)	6.4			0.20		1.3		(14)
Ship (Freight, International)						14.9		(15)
Miscellaneous						6.4		(16)
					Total	69.6	2.0	

1) 10^9 passkm or tonkm 2) Twh 3) kWh per passkm or tonkm
4) paid seats in % of available seats 5) percentage of empty driving

NATIONAL ENERGY DEMAND PROJECTIONS

Country Sweden

Year 1972

WAES Case Base year

Agriculture, Mining, and Construction

	(A)	(B)	(C)	(D)	(E)	(F)	(G)	(H)	
Agriculture, Forestry, and Fishing (ISIC Major Division 1)	Value Added 1)	Energy/Value Added 2)	Area Harvested	Energy/ Area	Physical Output	Energy/ Output	Fossil Fuel 7)	Electricity 7)	
Farm Machinery }	2.1	1.6	2.9 ha	1.1 MWh/ha			2.6	0.7	(17)
Greenhouses }									(18)
Other Agricultural Uses }									(19)
Forestry (ISIC 12)					67.6 3)	1.2 5)	0.8		(20)
Fishing Boats (ISIC 13) }					209 4)	3.4 6)	0.7		(21)
Other Fishing Use (ISIC 13) }									(22)
Total Fertilizer (tons)		N	P	K		Total	4.1	0.7	

	(A)	(B)	(C)	(D)	(E)	(F)	
Mining	Value Added 1)	Energy/Value Added 2)	Physical Output	Energy/Physical Output 8)	Fossil Fuel	Electricity	
Handling and Internal Transportation }	0.4	12.3	34.0×10^6 ton	1.44	3.0	1.7	(23)
Other Energy Use }					0.1	0.1	(24)
				Total	3.1	1.8	

	(A)	(B)	(C)	(D)	(E)	(F)	
Construction	Value Added 1)	Energy/Value Added 2)	Physical Output	Energy/Physical Output	Fossil Fuel	Electricity	
Handling and Internal Transportation }	4.1	0.24			0.2	0.4	(25)
Other Energy Use }					–	0.4	(26)
				Total	0.2	0.8	
				Agriculture, Mining, and Construction Total	7.4	3.3	

1) 10^9 \$ 2) kWh/\$ 3) 10^6 m^3 4) 10^3 ton 5) Kwh/m^3
6) MWh/ton 7) TWh 8) kWh/ton

NATIONAL ENERGY DEMAND PROJECTIONS

Country Sweden

Year 1972

WAES Case Base year

Commercial and Public

Commercial	(A) Floor Area 1)	(B) Heat Gain or Loss/Area	(C) System Efficiency	(D) Fossil Fuel 3)	(E) Electricity	
Space Heat (Fossil Fuel)	144			17.0		(27)
Space Heat (Electricity)			1.0		0.4	(28)
Air Conditioning (Fossil Fuel)						(29)
Air Conditioning (Electricity)						(30)
		Energy/Area	Electricity/Area			
Lighting		141[2)]	20[2)]		2.9	(31)
Miscellaneous						(32)
Presently Unknown						(33)
			Total	17.0	3.3	

Public	Floor Area	Heat Gain or Loss/Area	System Efficiency	Fossil Fuel	Electricity	
Space Heat (Fossil Fuel)	232			27.4		(34)
Space Heat (Electricity)			1.0		0.5	(35)
Air Conditioning (Fossil Fuel)						(36)
Air Conditioning (Electricity)						(37)
		Energy/Area	Electricity/Area		3.6	
Lighting		143[2)]	24[2)]			(38)
Miscellaneous					1.9	(39)
Presently Unknown						(40)
			Total	27.4	6.0	
			Commercial and Public Total	44.4	9.3	

1) 10^6 m^3 2) kWh/m^3 3) TWh

NATIONAL ENERGY DEMAND PROJECTIONS

Country Sweden

Year 1972

WAES Case Base year

Industrial

	(A) Value Added 2)	(B) Energy/Value Added 3)	(C) Physical Output 4)	(D) Energy/Output 5)	(E) % Exported	(F) Fossil Fuel 6)	(G) Electricity 6)	(H) Feedstocks	
Iron and Steel (ISIC 371)	0.8	33	5.2	5.1		22.0	4.5		(41)
Nonferrous Metals (ISIC 372)	0.19	26	0.4	12.5		1.9	3.1		(42)
Paper and Allied (ISIC 34)	0.90	116	8.8	7.8		32.5	12.0		(43)
Chemicals and Allied (ISIC 35 minus ISIC 353)	1.2	7.8				4.5	4.8		(44)
Mineral Products (ISIC 36)	0.5	25.2				11.3	1.3		(45)
Food and Related (ISIC 31)	1.2	5.8				5.8	1.1		(46)
Transport Eqpt. (ISIC 384)	1.2	3.2				2.8	1.0		(47)
Key Industry (N.E.C.) 1)	4.1	2.2				6.3	2.8		(48)
Key Industry (N.E.C.)									(49)
Misc. Manufacturing	1.9	4.8				6.4	3.3		(50)
N.E.C. = Not Elsewhere Classified					**Total**	118.5	33.9		

1) Manufacturing 2) 10^9 \$ 3) kWh/\$ 4) 10^6 ton 5) MWh/ton 6) TWh

NATIONAL ENERGY DEMAND PROJECTIONS

Country Sweden

Year 1972

WAES Case Base year

Residential

Population 8.2×10^6 Persons/Dwelling 2.5 Dwellings $3.27 \times 10^6 + 0.5 \times 10^6$ vacation houses

		(A)	(B)	(C)	(D)	(E)	(F)	
Central Space Heating		Dwellings 1)	Heat Loss 2)	System Efficiency	3)	Fossil Fuel 2)	Electricity 2)	
Single Family	Fossil Fuels	1.17	19.2	58	22.6	45.5		(51)
	Electricity	0.20		100	13.5		2.7	(52)
Vacation (~~Town or Row Houses~~)	Fossil Fuels	0.2		45		2.1		(53)
	Electricity	0.3		100				(54)
Apartments	Fossil Fuels	1.87	14.0	65	13.9	40.0		(55)
	Electricity	0.03		100	13.3		0.4	(56)
		% Saturation (Fossil Fuels)	% Saturation (Electricity)	Energy/Unit (Fossil Fuels)	Energy/Unit (Electricity)			
Room Heating		93	7			87.6	4.4	(57)
Air Conditioning (Window)								(58)
Air Conditioning (Central)								(59)
Water Heating								(60)
Refrigerators			93					(61)
Lighting			99		600			(62)
Cooking		15	85		1,000			(63)
Dishwashers			5		600	3.5 3)	11.5	(64)
Freezers			45		200			(65)
Television			85		750			(66)
Other Key Appliances			38		150			(67)
Misc. Appliances								(68)
Potential Appliances								(69)
					Total	87.6	15.9	

1) 10^6 2) TWh 3) Net MWh/dwelling

NATIONAL ENERGY DEMAND PROJECTIONS

Country Sweden

Year 1985

WAES Case A

Summary TWh

Category	Oil	Gas	Coal	Electricity	Other	Total by Category (Excluding Processing Losses)
Transportation	90			3		93
Agriculture, Mining, and Construction	17			10		27
Commercial	12			10		22
Public	16			17		33
Industrial	85		40	65	37	227
Residential	39			44		83
Total Delivered Energy by Fuel	259		40	149	37	
Processing Losses by Fuel	22			285		
Gross Input by Fuel	281		40	434	37	792

Gross Energy Input (Add across bottom row)

NATIONAL ENERGY DEMAND PROJECTIONS

Country Sweden

Year 1985

WAES Case A

Transportation

	(A)	(B)	(C)	(D)	(E)	(F)	(G)	
	Population/Auto	Distance/Auto/Yr.	Load Factor	Fraction Urban or Rural	Energy/Distance	Fossil Fuel	Electricity	
Urban Auto	2.46 (Urban and Rural Auto)		1.6	75 %		39.2 (Urban and Rural Auto)		(1)
Rural Auto			2.2	25 %				(2)
	Activity Level	% Electrified	Load Factor	Fossil Fuel/ Activity Level	Electricity/Activity			
Urban Bus	12.0[1] (Urban and Intercity Bus)			0.15[3] (Urban and Intercity Bus)		1.8 (Urban and Intercity Bus)		(3)
Intercity Bus								(4)
Rail (Passenger)	7.2[1]				0.13[3]		0.9	(5)
Air (Passenger, Domestic)	2.35[1]			0.50[3]		1.2		(6)
Air (Passenger, International)	25.7[1]			0.20[3]		5.1		(7)
Ship (Passenger)						0.3		(8)
Rail (Freight)	34.8[2]				0.07[4]		2.4	(9)
Air (Freight, Domestic)	0.03[2]			0.75[4]		0.02		(10)
Air (Freight, International)	0.37[2]			0.60[4]		0.22		(11)
Truck (Local)	48.0[2] (Local and Intercity Truck)			0.35 (Local and Intercity Truck)		16.8		(12)
Truck (Intercity)								(13)
Ship (Freight, Domestic)	20.0[2]			0.14		2.8		(14)
Ship (Freight, International)						16.0		(15)
Miscellaneous						7.5		(16)
					Total	90.4	3.3	

1) 10^9 passkm 2) 10^9 tonkm 3) kWh/passkam 4) kWh/tonkm

NATIONAL ENERGY DEMAND PROJECTIONS

Country Sweden

Year 1985

WAES Case A

Agriculture, Mining, and Construction

	(A)	(B)	(C)	(D)	(E)	(F)	(G)	(H)	
Agriculture, Forestry, and Fishing (ISIC Major Division 1)	Value Added	Energy/Value Added	Area Harvested	Energy/ Area	Physical Output	Energy/ Output	Fossil Fuel	Electricity	
Farm Machinery	}								(17)
Greenhouses	4.0[1)]	1.3 kWh/$	2.9x10^6 ha	1.8 MWh/ha			3.8	1.4	(18)
Other Agricultural Uses									(19)
Forestry (ISIC 12)					90x10^6 m^3	1.2 kWh/m^3	1.1		(20)
Fishing Boats (ISIC 13)							}		(21)
Other Fishing Use (ISIC 13)							0.8		(22)
Total Fertilizer (tons)		N ____	P ____	K ____		Total	5.8	1.4	

	(A)	(B)	(C)	(D)	(E)	(F)	
Mining	Value Added	Energy/Value Added	Physical Output	Energy/Physical Output	Fossil Fuel	Electricity	
Handling and Internal Transportation	0.8 [1)]		66.5x10^6 ton	2.9 kWh/ton	8.9	8.2	(23)
Other Energy Use					0.4	0.5	(24)
				Total	11.3	8.0	

	(A)	(B)	(C)	(D)	(E)	(F)	
Construction	Value Added	Energy/Value Added	Physical Output	Energy/Physical Output	Fossil Fuel	Electricity	
Handling and Internal Transportation	6.0[1)]	0.22 kWh/$			0.3	0.5	(25)
Other Energy Use						0.5	(26)
				Total	0.3	1.0	
				Agriculture, Mining, and Construction Total	17.4	10.4	

1) 10^9 $

NATIONAL ENERGY DEMAND PROJECTIONS

Country: Sweden

Year: 1985

WAES Case: A

Commercial and Public

Commercial	(A) Floor Area 1)	(B) Heat Gain or Loss/Area	(C) System Efficiency	(D) Fossil Fuel 3)	(E) Electricity 3)	
Space Heat (Fossil Fuel)	168			12.0		(27)
Space Heat (Electricity)					4.0	(28)
Air Conditioning (Fossil Fuel)						(29)
Air Conditioning (Electricity)						(30)
		Energy/Area 2)	Electricity/Area 2)			
Lighting		130	35		5.8	(31)
Miscellaneous						(32)
Presently Unknown						(33)
			Total	12.0	9.8	

Public	Floor Area	Heat Gain or Loss/Area	System Efficiency	Fossil Fuel	Electricity	
Space Heat (Fossil Fuel)	252			15.8		(34)
Space Heat (Electricity)					5.0	(35)
Air Conditioning (Fossil Fuel)						(36)
Air Conditioning (Electricity)						(37)
		Energy/Area 2)	Electricity/Area 2)			
Lighting		130	48		12.0	(38)
Miscellaneous						(39)
Presently Unknown						(40)
			Total	15.8	17.0	
			Commercial and Public Total	27.8	26.8	

1) 10^6 m^3 2) kWh/m^3 3) TWh

NATIONAL ENERGY DEMAND PROJECTIONS

Country Sweden

Year 1985

WAES Case A

Industrial

	(A) Value Added 1)	(B) Energy/Value Added 2)	(C) Physical Output 3)	(D) Energy/Output 4)	(E) % Exported	(F) Fossil Fuel 5)	(G) Electricity 5)	(H) Feedstocks	
Iron and Steel (ISIC 371)			14.7	4.0		47.8	11.0		(41)
Nonferrous Metals (ISIC 372)			1.1	11.1		3.6	8.5		(42)
Paper and Allied (ISIC 34)			16.6	4.7		61.6	16.5		(43)
Chemicals and Allied (ISIC 35 minus ISIC 353)	3.0	8.0				11.4	12.6		(44)
Mineral Products (ISIC 36)	0.61	15.2				8.2	1.1		(45)
Food and Related (ISIC 31)	1.7	4.8				6.5	1.7		(46)
Transport Eqpt. + (ISIC 384) Manufacturing	12.5	1.8				14.8	9.3		(47)
Key Industry (N.E.C.)									(48)
Key Industry (N.E.C.)									(49)
Misc. Manufacturing	3.3	3.7				7.8	4.3		(50)
					Total	161.7	65.1		

N.E.C. = Not Elsewhere Classified

1) 10^9\$ 2) kWh/\$ 3) 10^6 ton 4) MWh/ton 5) TWh

NATIONAL ENERGY DEMAND PROJECTIONS

Country Sweden

Year 1985

WAES Case A

Residential

Population 8.6×10^6 Persons/Dwelling 2.3 Dwellings $3.7 \times 10^6 + 0.8 \times 10^6$ vacation houses

		(A)	(B)	(C)	(D)	(E)	(F)	
Central Space Heating		Dwellings 1)	Heat Loss	System Efficiency	2)	Fossil Fuel 3)	Electricity 3)	
Single Family	Fossil Fuels	0.45		0.74	20.0	12.1		(51)
	Electricity	1.20		1.00	12.5		15.0	(52)
Vacation ~~Town or Row~~ Houses	Fossil Fuels	0.2				1.0		(53)
	Electricity	0.6					5.0	(54)
Apartments	Fossil Fuels	1.65		0.80	12.5	25.8		(55)
	Electricity	1.40		1.00	12.5		5.0	(56)
		% Saturation (Fossil Fuels)	% Saturation (Electricity)	Energy/Unit (Fossil Fuels)	Energy/Unit (Electricity)			
Room Heating						38.9	25.0	(57)
Air Conditioning (Window)								(58)
Air Conditioning (Central)								(59)
Water Heating								(60)
Refrigerators					}			(61)
Lighting					}			(62)
Cooking					}			(63)
Dishwashers					} 5.0 4)		18.5	(64)
Freezers					}			(65)
Television					}			(66)
Other Key Appliances					}			(67)
Misc. Appliances					}			(68)
Potential Appliances					}			(69)
					Total	38.9	43.5	

1) 10^6 2) Net MWh/dwelling and year 3) TWh 4) MWh/dwelling and year

NATIONAL ENERGY DEMAND PROJECTIONS

Country Sweden

Year 1985

WAES Case B

Summary TWh

Category	Oil	Gas	Coal	Electricity	Other	Total by Category (Excluding Processing Losses)
Transportation	78			3		81
Agriculture, Mining, and Construction	14			7		21
Commercial	14			9		23
Public	21			13		34
Industrial	72		36	57	36	201
Residential	41			42		83
Total Delivered Energy by Fuel	240		36	131	36	
Processing Losses by Fuel	21			267		
Gross Input by Fuel	261		36	398	36	731

Gross Energy Input (Add across bottom row)

NATIONAL ENERGY DEMAND PROJECTIONS

Country Sweden

Year 1985

WAES Case B

Transportation

	(A)	(B)	(C)	(D)	(E)	(F)	(G)	
	Population/Auto	Distance/Auto/Yr.	Load Factor	Fraction Urban or Rural	Energy/Distance	Fossil Fuel 3)	Electricity 3)	
Urban Auto	2.66		1.5	0.75		35.0		(1)
Rural Auto			2.1	0.25				(2)
	Activity Level 1)	% Electrified	Load Factor	Fossil Fuel/ Activity Level 2)	Electricity/Activity 2)			
Urban Bus	8.6			0.17		1.5		(3)
Intercity Bus								(4)
Rail (Passenger)	5.1				0.15		0.8	(5)
Air (Passenger, Domestic)	1.35			0.53		0.7		(6)
Air (Passenger, International)	14.8			0.22		3.3		(7)
Ship (Passenger)						0.3		(8)
Rail (Freight)	23.4				0.08		1.8	(9)
Air (Freight, Domestic)	0.02			0.80		0.02		(10)
Air (Freight, International)	0.26			0.63		0.16		(11)
Truck (Local)	31.7			0.39		12.4		(12)
Truck (Intercity)								(13)
Ship (Freight, Domestic)	11.5			0.15		1.7		(14)
Ship (Freight, International)						16.0		(15)
Miscellaneous						6.5		(16)
					Total	77.6	2.6	

1) 10^9 passkm or tonkm 2) kWh/passkm or tonkm 3) TWh

NATIONAL ENERGY DEMAND PROJECTIONS

Country: Sweden

Year: 1985

WAES Case: B

Agriculture, Mining, and Construction

Agriculture, Forestry, and Fishing (ISIC Major Division 1)	(A) Value Added 1)	(B) Energy/Value Added 2)	(C) Area Harvested	(D) Energy/ Area	(E) Physical 3) Output	(F) Energy/ 5) Output	(G) Fossil 7) Fuel	(H) Electricity 7)	
Farm Machinery	3.4 (Farm Machinery, Greenhouses, Other Agricultural Uses)	1.4							(17)
Greenhouses							3.4	1.4	(18)
Other Agricultural Uses					90	1.2	1.1		(19)
Forestry (ISIC 12)							0.6 (Forestry and Fishing Boats)		(20)
Fishing Boats (ISIC 13)									(21)
Other Fishing Use (ISIC 13)									(22)
Total Fertilizer (tons)		N	P	K		Total	5.1	1.4	

Mining	(A) Value Added 1)	(B) Energy/Value Added 2)	(C) Physical Output 4)	(D) Energy/Physical Output 6)	(E) Fossil Fuel	(F) Electricity	
Handling and Internal Transportation	0.6 (Handling and Other Energy Use)		50.6	2.70	8.0	4.9	(23)
Other Energy Use					0.3	0.5	(24)
				Total	8.3	5.4	

Construction	(A) Value Added 1)	(B) Energy/Value Added 2)	(C) Physical Output	(D) Energy/Physical Output	(E) Fossil Fuel	(F) Electricity	
Handling and Internal Transportation	5.0 (Handling and Other Energy Use)	0.22			0.3	0.4	(25)
Other Energy Use					–	0.2	(26)
				Total	0.3	0.6	
				Agriculture, Mining, and Construction Total	13.7	7.4	

1) 10^9 \$ 2) kWh/\$ 3) 10^6 m^3 4) 10^6 ton 5) kWh/m^3
6) kWh/ton 7) TWh

NATIONAL ENERGY DEMAND PROJECTIONS

Country Sweden

Year 1985

WAES Case B

Commercial and Public

	(A)	(B)	(C)	(D)	(E)	
Commercial	Floor Area 1)	Heat Gain or Loss/Area	System Efficiency	Fossil Fuel 3)	Electricity 3)	
Space Heat (Fossil Fuel)	168			14.2		(27)
Space Heat (Electricity)					3.0	(28)
Air Conditioning (Fossil Fuel)						(29)
Air Conditioning (Electricity)						(30)
		Energy/Area 2)	Electricity/Area 2)			
Lighting		135	33		5.5	(31)
Miscellaneous						(32)
Presently Unknown						(33)
			Total	14.2	8.5	

Public	Floor Area 1)	Heat Gain or Loss/Area	System Efficiency	Fossil Fuel 3)	Electricity 3)	
Space Heat (Fossil Fuel)	252			21.0		(34)
Space Heat (Electricity)					3.0	(35)
Air Conditioning (Fossil Fuel)						(36)
Air Conditioning (Electricity)						(37)
		Energy/Area 2)	Electricity/Area 2)			
Lighting		135	40		10.0	(38)
Miscellaneous						(39)
Presently Unknown						(40)
			Total	21.0	13.0	
			Commercial and Public Total	35.2	21.5	

1) 10^6 m^3 2) kWh/m^3 3) TWh

NATIONAL ENERGY DEMAND PROJECTIONS

Country Sweden

Year 1985

WAES Case B

Industrial

	(A) Value Added 1)	(B) Energy/Value Added 2)	(C) Physical Output 3)	(D) Energy/Output 4)	(E) % Exported	(F) Fossil Fuel 5)	(G) Electricity 5)	(H) Feedstocks	
Iron and Steel (ISIC 371)			12.7	4.1		43.2	9.0		(41)
Nonferrous Metals (ISIC 372)			1.0	11.3		3.5	7.8		(42)
Paper and Allied (ISIC 34)			13.0	5.4		55.6	15.1		(43)
Chemicals and Allied (ISIC 35 minus ISIC 353)	2.4	8.3				9.0	11.0		(44)
Mineral Products (ISIC 36)	0.5	16.2				7.4	0.7		(45)
Food and Related (ISIC 31)	1.4	5.1				5.7	1.4		(46)
Transport Eqpt.+ (ISIC 384) Manufacturing	9.0	2.0				11.7	7.1		(47)
Key Industry (N.E.C.)									(48)
Key Industry (N.E.C.)									(49)
Misc. Manufacturing	2.8	4.1				7.6	4.4		(50)
					Total	143.7	56.5		

N.E.C. = Not Elsewhere Classified

1) 10^9 \$ 2) kWh/\$ 3) 10^6 ton 4) MWh/ton 5/TWh

NATIONAL ENERGY DEMAND PROJECTIONS

Country Sweden

Year 1985

WAES Case B

Residential

Population 8.6×10^6 Persons/Dwelling 2.3 Dwellings $3.7 \times 10^6 + 0.8 \times 10^6$ vacation houses

		(A)	(B)	(C)	(D)	(E)	(F)	
Central Space Heating		Dwellings 1)	Heat Loss	System Efficiency	2)	Fossil Fuel 3)	Electricity 3)	
Single Family	Fossil Fuels	0.35		67	21.0	11.0		(51)
	Electricity	1.30		100	13.0		16.9	(52)
Vacation ~~Town or Row~~ Houses	Fossil Fuels	0.2				1.1		(53)
	Electricity	0.6					3.0	(54)
Apartments	Fossil Fuels	1.62		72	13.0	29.3		(55)
	Electricity	0.43		100	13.0		5.6	(56)
		% Saturation (Fossil Fuels)	% Saturation (Electricity)	Energy/Unit (Fossil Fuels)	Energy/Unit (Electricity)			
Room Heating						41.4	25.5	(57)
Air Conditioning (Window)								(58)
Air Conditioning (Central)								(59)
Water Heating								(60)
Refrigerators								(61)
Lighting					4.4 4)			(62)
Cooking							16.3	(63)
Dishwashers								(64)
Freezers								(65)
Television								(66)
Other Key Appliances								(67)
Misc. Appliances								(68)
Potential Appliances								(69)
					Total	41,4	41.8	

Note: the values 4.4 4) and 16.3 are bracketed to cover rows (61) through (69).

1) 10^6 2) Net MWh/dwelling and year 3) TWh 4) MWh/dwelling and year

NATIONAL ENERGY DEMAND PROJECTIONS

Country Sweden

Year 1985

WAES Case C

Summary TWh

Category	Oil	Gas	Coal	Electricity	Other	Total by Category (Excluding Processing Losses)
Transportation	98			3		101
Agriculture, Mining, and Construction	19			11		30
Commercial	16			8		24
Public	22			15		37
Industrial	86		44	70	39	239
Residential	70			31		101
Total Delivered Energy by Fuel	311		44	139	39	
Processing Losses by Fuel	24			277		
Gross Input by Fuel	355		44	416	39	854

Gross Energy Input
(Add across bottom row)

NATIONAL ENERGY DEMAND PROJECTIONS

Country Sweden

Year 1985

WAES Case C

Transportation

	(A) Population/Auto	(B) Distance/Auto/Yr.	(C) Load Factor	(D) Fraction Urban or Rural	(E) Energy/Distance	(F) Fossil Fuel 3)	(G) Electricity 3)	
Urban Auto	2.46 (Urban and Rural Auto)			0.75		41.3 (Urban and Rural Auto)	-	(1)
Rural Auto				0.25				(2)
	Activity Level 1)	% Electrified	Load Factor	Fossil Fuel/ Activity Level 2)	Electricity/Activity 2)			
Urban Bus	12.0 (Urban and Intercity Bus)			0.18 (Urban and Intercity Bus)		2.2 (Urban and Intercity Bus)		(3)
Intercity Bus								(4)
Rail (Passenger)	7.2	100			0.15	-	1.1	(5)
Air (Passenger, Domestic)	2.35			0.54		1.3		(6)
Air (Passenger, International)	25.7			0.23		5.9		(7)
Ship (Passenger)	0.2			1.5		0.3		(8)
Rail (Freight)	34.8	100			0.07		2.4	(9)
Air (Freight, Domestic)	0.03			0.80		0.02		(10)
Air (Freight, International)	0.37			0.62		0.23		(11)
Truck (Local)	48.0 (Local and Intercity Truck)			0.40 (Local and Intercity Truck)		19.2 (Local and Intercity Truck)		(12)
Truck (Intercity)								(13)
Ship (Freight, Domestic)	20.0			0.15		3.0		(14)
Ship (Freight, International)						16.3		(15)
Miscellaneous						8.0		(16)
					Total	97.9	3.4	

1) 10^9 passkm or 10^9 tonkm 2) kWh per passkm or tonkm 3) TWh

NATIONAL ENERGY DEMAND PROJECTIONS

Country Sweden

Year 1985

WAES Case C

Agriculture, Mining, and Construction

Agriculture, Forestry, and Fishing (ISIC Major Division 1)	(A) Value Added 1)	(B) Energy/Value Added 2)	(C) Area Harvested 3)	(D) Energy/Area 4)	(E) Physical Output	(F) Energy/Output	(G) Fossil Fuel 9)	(H) Electricity 9)	
Farm Machinery									(17)
Greenhouses	4.0	1.5	2.9	2.1			4.7	1.4	(18)
Other Agricultural Uses									(19)
Forestry (ISIC 12)					90[5]	1.3[6]	1.2		(20)
Fishing Boats (ISIC 13)							0.9		(21)
Other Fishing Use (ISIC 13)									(22)
Total Fertilizer (tons)		N ____	P ____	K ____		Total	6.8	1.4	

Mining	(A) Value Added 1)	(B) Energy/Value Added	(C) Physical Output 7)	(D) Energy/Physical Output 8)	(E) Fossil Fuel	(F) Electricity	
Handling and Internal Transportation	0.8		66.5	3.0	11.1	7.9	(23)
Other Energy Use					0.4	0.6	(24)
				Total	11.5	8.5	

Construction	(A) Value Added 1)	(B) Energy/Value Added 2)	(C) Physical Output	(D) Energy/Physical Output	(E) Fossil Fuel	(F) Electricity	
Handling and Internal Transportation	6.0	0.23			0.3	0.6	(25)
Other Energy Use					–	0.5	(26)
				Total	0.3	1.1	
				Agriculture, Mining, and Construction Total	18.6	11.0	

1) 10^9 \$ 2) kWh/\$ 3) 10^6 ha 4) MWh/ha
5) 10^6 m^3 6) kWh/m^3 7) 10^6 ton 8) kWh/ton 9) TWh

NATIONAL ENERGY DEMAND PROJECTIONS

Country Sweden

Year 1985

WAES Case C

Commercial and Public

Commercial	(A) Floor Area 1)	(B) Heat Gain or Loss/Area	(C) System Efficiency	(D) Fossil Fuel 3)	(E) Electricity 3)	
Space Heat (Fossil Fuel)	168			16.4		(27)
Space Heat (Electricity)			1.0		2.0	(28)
Air Conditioning (Fossil Fuel)						(29)
Air Conditioning (Electricity)						(30)
		Energy/Area 2)	Electricity/Area 2)			
Lighting		145	36		6.0	(31)
Miscellaneous						(32)
Presently Unknown						(33)
			Total	16.4	8.0	

Public	Floor Area	Heat Gain or Loss/Area	System Efficiency	Fossil Fuel	Electricity	
Space Heat (Fossil Fuel)	252			21.5		(34)
Space Heat (Electricity)					2.5	(35)
Air Conditioning (Fossil Fuel)						(36)
Air Conditioning (Electricity)						(37)
		Energy/Area 2)	Electricity/Area 2)			
Lighting		145	50		12.5	(38)
Miscellaneous						(39)
Presently Unknown						(40)
			Total	21.5	15.0	
			Commercial and Public Total	37.9	23.0	

1) 10^6 m^3 2) kWh/m^3 3) TWh

NATIONAL ENERGY DEMAND PROJECTIONS

Country Sweden

Year 1985

WAES Case C

Industrial

	(A) Value Added 1)	(B) Energy/Value Added 2)	(C) Physical Output 3)	(D) Energy/Output 4)	(E) % Exported	(F) Fossil Fuel 5)	(G) Electricity 5)	(H) Feedstocks	
Iron and Steel (ISIC 371)	(2.3)		14.7	4.2		49.5	12.6		(41)
Nonferrous Metals (ISIC 372)	(0.5)		1.1	11.7		4.0	8.9		(42)
Paper and Allied (ISIC 34)	(1.7)		16.6	5.0		65.0	17.6		(43)
Chemicals and Allied (ISIC 35 minus ISIC 353)	3.0	9.0				11.8	13.4		(44)
Mineral Products (ISIC 36)	0.6	17.0				9.1	1.2		(45)
Food and Related (ISIC 31)	1.7	5.1				6.9	1.8		(46)
Transport Eqpt. + (ISIC 384) Manufacturing	12.5	2.0				15.0	10.0		(47)
Key Industry (N.E.C.)									(48)
Key Industry (N.E.C.)									(49)
Misc. Manufacturing	3.3	3.8				8.0	4.5		(50)
					Total	169.3	70.0		

N.E.C. = Not Elsewhere Classified

1) 10^9 \$ 2) kWh/\$ 3) 10^6 ton 4) MWh/ton 5) TWh

NATIONAL ENERGY DEMAND PROJECTIONS

Country: Sweden

Year: 1985

WAES Case: C

Residential

Population: 8.6×10^6 Persons/Dwelling: 2.3 Dwellings: $3.7 \times 10^6 + 0.8 \times 10^6$ vacation houses

Central Space Heating		(A) Dwellings 1)	(B) Heat Loss	(C) System Efficiency	(D) 2)	(E) Fossil Fuel 3)	(F) Electricity 3)	
Single Family	Fossil Fuels	1.25	10.2	0.72	21.0	36.4		(51)
	Electricity	0.40		1.0	13.0		5.2	(52)
Town or Row Houses	Fossil Fuels	0.2				1.4		(53)
	Electricity	0.6					4.6	(54)
Apartments	Fossil Fuels	1.97	7.6	0.77	13.0	33.2		(55)
	Electricity	0.08		1.00	13.0		1.0	(56)

	% Saturation (Fossil Fuels)	% Saturation (Electricity)	Energy/Unit (Fossil Fuels)	Energy/Unit (Electricity)			
Room Heating					70.0	10.8	(57)
Air Conditioning (Window)							(58)
Air Conditioning (Central)							(59)
Water Heating							(60)
Refrigerators		100		}		}	(61)
Lighting		100		}		}	(62)
Cooking		100		} 5.5 4)		} 20.4	(63)
Dishwashers		60		}		}	(64)
Freezers		90		}		}	(65)
~~Television~~ Washing machine		85		}		}	(66)
Other Key Appliances Vacuum cleaner		120		}		}	(67)
Misc. Appliances				}		}	(68)
Potential Appliances				}		}	(69)
				Total	70.0	31.2	

1) 10^6 2) Net MWh/dwelling and year 3) TWh 4) MWh per dwelling and year

NATIONAL ENERGY DEMAND PROJECTIONS

Country Sweden

Year 1985

WAES Case D

Summary Unit TWh

Category	Oil	Gas	Coal	Electricity	Other	Total by Category (Excluding Processing Losses)
Transportation	91			3		94
Agriculture, Mining, and Construction	15			7		22
Commercial	19			8		27
Public	26			14		40
Industrial	82		38	61	37	218
Residential	72			32		104
Total Delivered Energy by Fuel	305		38	125	37	
Processing Losses by Fuel	24			256		
Gross Input by Fuel	329		38	381	37	785

Gross Energy Input (Add across bottom row)

NATIONAL ENERGY DEMAND PROJECTIONS

Country Sweden

Year 1985

WAES Case D

Transportation

	(A)	(B)	(C)	(D)	(E)	(F)	(G)	
	Population/Auto	Distance/Auto/Yr.	Load Factor	Fraction Urban or Rural	Energy/Distance	Fossil Fuel 3)	Electricity 3)	
Urban Auto	2.64		1.4	0.75		38.9		(1)
Rural Auto			2.0	0.25				(2)
	Activity Level 1)	% Electrified	Load Factor	Fossil Fuel/ Activity Level 2)	Electricity/Activity 2)			
Urban Bus	8.6			0.20		1.8		(3)
Intercity Bus								(4)
Rail (Passenger)	5.1				0.19	-	1.0	(5)
Air (Passenger, Domestic)	1.35			0.76		1.0		(6)
Air (Passenger, International)	14.8			0.27		3.9		(7)
Ship (Passenger)						0.3		(8)
Rail (Freight)	23.4				0.08	-	1.8	(9)
Air (Freight, Domestic)	0.02			0.9		0.02		(10)
Air (Freight, International)	0.26			0.69		0.18		(11)
Truck (Local)	31.7			0.60		19.0		(12)
Truck (Intercity)								(13)
Ship (Freight, Domestic)	11.5			0.18		2.1		(14)
Ship (Freight, International)						16.3		(15)
Miscellaneous						7.0		(16)
					Total	90.5	2.8	

1) 10^9 passkm or tonkm 2) kWh per passkm or tonkm 3) TWh

NATIONAL ENERGY DEMAND PROJECTIONS

Country Sweden

Year 1985

WAES Case D

Agriculture, Mining, and Construction

Agriculture, Forestry, and Fishing (ISIC Major Division 1)	(A) Value Added 1)	(B) Energy/Value Added 2)	(C) Area Harvested 3)	(D) Energy/Area 4)	(E) Physical Output 5)	(F) Energy/Output 6)	(G) Fossil Fuel 7)	(H) Electricity 7)	
Farm Machinery									(17)
Greenhouses	3.4	1.6	2.9	1.9			4.0	1.4	(18)
Other Agricultural Uses									(19)
Forestry (ISIC 12)					90	1.3	1.2		(20)
Fishing Boats (ISIC 13)									(21)
Other Fishing Use (ISIC 13)							0.7		(22)
Total Fertilizer (tons)		N	P	K		Total	5.9	1.4	

Mining	(A) Value Added	(B) Energy/Value Added	(C) Physical Output 8)	(D) Energy/Physical Output 9)	(E) Fossil Fuel	(F) Electricity	
Handling and Internal Transportation	0.6		50.6	2.75	8.7	4.5	(23)
Other Energy Use					0.3	0.5	(24)
				Total	8.9	5.0	

Construction	(A) Value Added	(B) Energy/Value Added 2)	(C) Physical Output	(D) Energy/Physical Output	(E) Fossil Fuel	(F) Electricity	
Handling and Internal Transportation	5.0	0.24			0.3	0.5	(25)
Other Energy Use						0.4	(26)
				Total	0.3	0.9	
				Agriculture, Mining, and Construction Total	15.1	7.3	

1) 10^9 \$ 2) kWh/\$ 3) 10^6 ha 4) MWh/ha 5) 10^6 m^3 6) kWh/m^3
7) TWh 8) 10^6 ton 9) kWh/ton

NATIONAL ENERGY DEMAND PROJECTIONS

Country Sweden

Year 1985

WAES Case D

Commercial and Public

	(A)	(B)	(C)	(D)	(E)	
Commercial	Floor Area 1)	Heat Gain or Loss/Area	System Efficiency	Fossil Fuel 3)	Electricity 3)	
Space Heat (Fossil Fuel)	168			18.5		(27)
Space Heat (Electricity)					2.0	(28)
Air Conditioning (Fossil Fuel)						(29)
Air Conditioning (Electricity)						(30)
		Energy/Area 2)	Electricity/Area 2)			
Lighting		157	35		5.9	(31)
Miscellaneous						(32)
Presently Unknown						(33)
			Total	18.5	7.9	
Public	Floor Area 1)	Heat Gain or Loss/Area	System Efficiency	Fossil Fuel	Electricity	
Space Heat (Fossil Fuel)	252			25.7		(34)
Space Heat (Electricity)					2.5	(35)
Air Conditioning (Fossil Fuel)						(36)
Air Conditioning (Electricity)						(37)
		Energy/Area 2)	Electricity/Area 2)			
Lighting		158	45		11.4	(38)
Miscellaneous						(39)
Presently Unknown						(40)
			Total	25.7	13.9	
			Commercial and Public Total	44.2	21.8	

1) 10^6 m^3 2) kWh/m^3 3) TWh

NATIONAL ENERGY DEMAND PROJECTIONS

Country Sweden

Year 1985

WAES Case D

Industrial

	(A) Value Added 1)	(B) Energy/Value Added 2)	(C) Physical Output 3)	(D) Energy/Output 4)	(E) % Exported	(F) Fossil Fuel 5)	(G) Electricity 5)	(H) Feedstocks	
Iron and Steel (ISIC 371)	(2.0)		12.7	4.5		47.2	10.4		(41)
Nonferrous Metals (ISIC 372)	(0.5)		1.0	12.1		4.0	8.1		(42)
Paper and Allied (ISIC 34)	(1.3)		13.0	5.6		60.3	16.7		(43)
Chemicals and Allied (ISIC 35 minus ISIC 353)	2.4	8.8				9.5	11.5		(44)
Mineral Products (ISIC 36)	0.5	18.4				8.3	0.9		(45)
Food and Related (ISIC 31)	1.5	5.4				6.7	1.4		(46)
Transport Eqpt. + (ISIC 384) Manufacturing	9.3	2.2				13.2	7.1		(47)
Key Industry (N.E.C.)									(48)
Key Industry (N.E.C.)									(49)
Misc. Manufacturing	2.8	4.4				7.6	4.7		(50)
					Total	156.8	60.8		

N.E.C. = Not Elsewhere Classified

1) 10^9 \$ 2) kWh/\$ 3) 10^6 ton 4) MWh/ton 5) TWh

NATIONAL ENERGY DEMAND PROJECTIONS

Country: Sweden

Year: 1985

WAES Case: D

Residential

Population: 8.6×10^6 Persons/Dwelling: 2.3 Dwellings: $3.7 \times 10^6 + 0.8 \times 10^6$ vacation houses

		(A)	(B)	(C)	(D)	(E)	(F)	
Central Space Heating		Dwellings 1)	Heat Loss 2)	System Efficiency	3)	Fossil Fuel 2)	Electricity 2)	
Single Family	Fossil Fuels	0.91	12.7	0.6	21.0	31.8		(51)
	Electricity	0.74		1.0	13.0		9.6	(52)
Vacation ~~Town or Row Houses~~	Fossil Fuels	0.2				1.6		(53)
	Electricity	0.6					3.0	(54)
Apartments	Fossil Fuels	1.97	12.5	0.68	13.5	39.0		(55)
	Electricity	0.08		1.0	13.0		1.0	(56)
		% Saturation (Fossil Fuels)	% Saturation (Electricity)	Energy/Unit (Fossil Fuels)	Energy/Unit (Electricity)			
Room Heating						72.4	13.6	(57)
Air Conditioning (Window)								(58)
Air Conditioning (Central)								(59)
Water Heating								(60)
Refrigerators			100					(61)
Lighting			100					(62)
Cooking			100					(63)
Dishwashers			30		5.0 4)		18.5	(64)
Freezers			80					(65)
~~Television~~ Washing machine			74					(66)
~~Other Key Appliances~~ Vacuum cleaner			115					(67)
Misc. Appliances								(68)
Potential Appliances								(69)
					Total	72.4	32.1	

(The values 5.0 4) and 18.5 are bracketed to cover rows (61)–(69).)

1) 10^6 2) TWh 3) Net MWh/dwelling and year 4) MWh/dwelling and year

NATIONAL ENERGY DEMAND PROJECTIONS

Country Sweden

Year 1985

WAES Case E

Summary

Category	Oil	Gas	Coal	Electricity	Other	Total by Category (Excluding Processing Losses)
Transportation	115			5		120
Agriculture, Mining, and Construction	19			13		32
Commercial	19			10		29
Public	26			17		43
Industrial	104		50	75	37	266
Residential	88			33		121
Total Delivered Energy by Fuel	371		50	153	37	
Processing Losses by Fuel	27			306		
Gross Input by Fuel	398		50	459	37	944

Gross Energy Input (Add across bottom row)

NATIONAL ENERGY DEMAND PROJECTIONS

Country Sweden

Year 1985

WAES Case E

Transportation

	(A)	(B)	(C)	(D)	(E)	(F)	(G)	
	Population/Auto	Distance/Auto/Yr.	Load Factor	Fraction Urban or Rural	Energy/Distance	Fossil Fuel 3)	Electricity 3)	
Urban Auto }	2.46					50.0		(1)
Rural Auto }								(2)
	Activity Level 1)	% Electrified	Load Factor	Fossil Fuel/ Activity Level 2)	Electricity/Activity			
Urban Bus }	12.0			0.20		2.4		(3)
Intercity Bus }								(4)
Rail (Passenger)	7.2				0.2		1.4	(5)
Air (Passenger, Domestic)	2.75			0.75		1.8		(6)
Air (Passenger, International)	25.7			0.26		6.7		(7)
Ship (Passenger)	0.2			1.6		0.3		(8)
Rail (Freight)	34.8				0.09		3.1	(9)
Air (Freight, Domestic)	0.03			1.0		0.03		(10)
Air (Freight, International)	0.37			0.75		0.28		(11)
Truck (Local) }	48.0			0.50		24.0		(12)
Truck (Intercity) }								(13)
Ship (Freight, Domestic)	20.0			0.19		3.8		(14)
Ship (Freight, International)						17.0		(15)
Miscellaneous						9.0		(16)
					Total	115.3	4.5	

1) 10^9 passkm or tonkm 2) kWh per passkm or tonkm 3) TWh

NATIONAL ENERGY DEMAND PROJECTIONS

Country Sweden

Year 1985

WAES Case E

Agriculture, Mining, and Construction

Agriculture, Forestry, and Fishing (ISIC Major Division 1)	(A) Value Added 1)	(B) Energy/Value Added 2)	(C) Area Harvested 3)	(D) Energy/Area 4)	(E) Physical Output	(F) Energy/Output	(G) Fossil Fuel 9)	(H) Electricity 9)	
Farm Machinery	} 4.0	1.6					} 4.9	} 1.5	(17)
Greenhouses									(18)
Other Agricultural Uses									(19)
Forestry (ISIC 12)					90	1.4	1.3		(20)
Fishing Boats (ISIC 13)							} 0.9		(21)
Other Fishing Use (ISIC 13)									(22)
Total Fertilizer (tons)		N	P	K		Total	7.1	1.5	

Mining	(A) Value Added 1)	(B) Energy/Value Added	(C) Physical Output 7)	(D) Energy/Physical Output 8)	(E) Fossil Fuel	(F) Electricity	
Handling and Internal Transportation	} 0.8			3.2	10.9	9.4	(23)
Other Energy Use					0.3	0.6	(24)
				Total	11.2	10.0	

Construction	(A) Value Added 1)	(B) Energy/Value Added 2)	(C) Physical Output	(D) Energy/Physical Output	(E) Fossil Fuel	(F) Electricity	
Handling and Internal Transportation	6.0	0.24			0.3	0.6	(25)
Other Energy Use					–	0.5	(26)
				Total	0.3	1.1	
				Agriculture, Mining, and Construction Total	18.6	12.6	

1) 10^9 \$ 2) kWh/\$ 3) 10^6 ha 4) MWh/ha
5) 10^6 m^3 6) kWh/m^3 7) 10^6 ton 8) kWh/ton 9) TWh

NATIONAL ENERGY DEMAND PROJECTIONS

Country Sweden

Year 1985

WAES Case E

Commercial and Public

Commercial	(A) Floor Area 1)	(B) Heat Gain or Loss/Area	(C) System Efficiency	(D) Fossil Fuel 3)	(E) Electricity 3)	
Space Heat (Fossil Fuel)	168			19.0		(27)
Space Heat (Electricity)					2.0	(28)
Air Conditioning (Fossil Fuel)						(29)
Air Conditioning (Electricity)						(30)
		Energy/Area 2)	Electricity/Area 2)			
Lighting		170	45		8.2	(31)
Miscellaneous						(32)
Presently Unknown						(33)
			Total	19.0	10.2	

Public	Floor Area	Heat Gain or Loss/Area	System Efficiency	Fossil Fuel	Electricity	
Space Heat (Fossil Fuel)				26.4		(34)
Space Heat (Electricity)	252				2.5	(35)
Air Conditioning (Fossil Fuel)						(36)
Air Conditioning (Electricity)						(37)
		Energy/Area 2)	Electricity/Area 2)			
Lighting		170	55		14.3	(38)
Miscellaneous						(39)
Presently Unknown						(40)
			Total	26.4	16.8	
			Commercial and Public Total	45.4	27.0	

1) 10^6 m^3 2) kWh/m^3 3) TWh

NATIONAL ENERGY DEMAND PROJECTIONS

Country Sweden

Year 1985

WAES Case E

Industrial

	(A) Value Added 1)	(B) Energy/Value Added 2)	(C) Physical Output 3)	(D) Energy/Output 4)	(E) % Exported	(F) Fossil Fuel 5)	(G) Electricity 5)	(H) Feedstocks	
Iron and Steel (ISIC 371)			14.7	4.8		57.5	13.5		(41)
Nonferrous Metals (ISIC 372)			1.1	12.6		4.5	9.4		(42)
Paper and Allied (ISIC 34)			16.6	7.1		72.0	18.5		(43)
Chemicals and Allied (ISIC 35 minus ISIC 353)	3.0	9.0				14.0	13.0		(44)
Mineral Products (ISIC 36)	0.61	21.0				12.3	1.4		(45)
Food and Related (ISIC 31)	1.7	5.4				7.2	2.0		(46)
Transport Eqpt. + (ISIC 384) Manufacturing	12.5	2.1				14.7	11.5		(47)
Key Industry (N.E.C.)									(48)
Key Industry (N.E.C.)									(49)
Misc. Manufacturing	3.3	4.4				8.5	6.0		(50)
N.E.C. = Not Elsewhere Classified					**Total**	190.7	75.3		

1) 10^9 \$ 2) kWh/\$ 3) 10^6 ton 4) MWh/ton 5) TWh

NATIONAL ENERGY DEMAND PROJECTIONS

Country Sweden

Year 1985

WAES Case E

Residential

Population 8.6×10^6 Persons/Dwelling 2.3 Dwellings $3.7 \times 10^6 + 0.8 \times 10^6$ vacation houses

		(A)	(B)	(C)	(D)	(E)	(F)	
Central Space Heating		Dwellings 1)	Heat Loss 2)	System Efficiency	3)	Fossil Fuel 2)	Electricity 2)	
Single Family	Fossil Fuels	1.25		0.60	22.5	46.9		(51)
	Electricity	0.40		1.00	13.5		5.4	(52)
Vacation ~~Town or Row~~ Houses	Fossil Fuels	0.2				2.1		(53)
	Electricity	0.6					4.2	(54)
Apartments	Fossil Fuels	1.97		0.68	13.5	39.1		(55)
	Electricity	0.08		1.00	13.5		1.1	(56)
		% Saturation (Fossil Fuels)	% Saturation (Electricity)	Energy/Unit (Fossil Fuels)	Energy/Unit (Electricity)			
Room Heating						88.1	10.7	(57)
Air Conditioning (Window)								(58)
Air Conditioning (Central)								(59)
Water Heating								(60)
Refrigerators					}		}	(61)
Lighting					}		}	(62)
Cooking					} 6.0 4)		} 22.2	(63)
Dishwashers					}		}	(64)
Freezers					}		}	(65)
Television					}		}	(66)
Other Key Appliances					}		}	(67)
Misc. Appliances					}		}	(68)
Potential Appliances					}		}	(69)
					Total	88.1	32.9	

1) 10^6 2) TWh 3) Net MWh/dwelling, year 4) MWh/dwelling, year

United Kingdom

14.1 INTRODUCTION AND SUMMARY

In the United Kingdom, energy consumption has been, and will be, strongly influenced by the availability of indigenous fuels, and it is therefore not possible to present a study of demand without also discussing supply. In 1972 the UK imported more than half the energy consumed in that year, but by 1980 it is likely that energy imports will be balanced by energy exports. If a moderately optimistic view is taken on North Sea oil potential, the UK could be a net exporter of oil by 1985, and in that year nearly 20 percent of the energy consumed could be provided by indigenous natural gas. However, even with moderate optimism about the North Sea, it seems likely that the UK will again become a net importer of energy before the year 2000.

A discussion of UK energy demand and policy implications to 1985 would therefore be both incomplete and misleading if it did not also give some consideration to energy supply and to forward prospects to the year 2000. Whilst concentrating on the detailed numerical discussion of energy demand within the WAES scenario cases for 1985, this report therefore also summarises projections of supply and demand to the year 2000. Since these projections lead to requirements for energy imports to the UK which seem likely to increase quite rapidly around the year 2000, it is necessary to take a view on the world energy situation at that time.

An examination of the world energy situation to the year 2000 forms one of the central objectives of the WAES program and it would not be appropriate for the results to be anticipated here. However, a view on the world energy situation is essential to an analysis of UK energy prospects, since it determines the degree of concern that might be expressed about an energy policy to 1985 that led to increasing import requirements in the period 1990 to 2000. If the world was expected to encounter an energy scarcity in this period, then the UK policy implications would be different from those if no world energy scarcity was anticipated. In our analysis of the UK situation we have taken the view that a policy that assumes there is a risk of a world energy scarcity is more prudent than one that ignores this risk. The penalty for error in the latter case could be more severe than the penalty for error in the former.

The first stage of an analysis of energy policy that takes account of an uncertain future is the investigation of some of the possibilities or scenarios. These are not forecasts, but are intended to be internally consistent pictures associated with specified assumptions such as economic growth. Given these assumptions, one seeks to evaluate energy demand and supply in a consistent manner. In this report we outline the scenario assumptions and describe the corresponding results on energy demand and supply for the UK in 1985. These results cannot be unique consequences of our scenario assumptions since our projection methods are not infallible or precise even in a "surprise free" situation. However, they do indicate some of the problems that require consideration, particularly when placed in the context of our year 2000 projections for the UK which are also summarised in this report.

The report presents as its "conclusions" some of these problems and the policy alternatives that require further consideration. Their identification represents the second stage of policy analysis. The third stage would be concerned with the assessment of optimal policies under conditions of uncertainty, and the fourth with the methods of decision making by which an optimal policy could be achieved. It is hoped that some limited aspects of these further stages will be considered in a later report, together with a more detailed assessment of the UK towards the year 2000 in the context of the WAES global studies.

14.2 BACKGROUND DISCUSSION AND SCENARIO DEFINITIONS

The UK energy projections have been made using the background provided by the common scenario assumptions agreed upon by WAES. These assumptions include the specification of world economic growth to 1985 at either a high level of about 5.0 percent per annum, or a low level of about 3.5 percent, which may be compared with the historical average annual growth rate of 5.0 percent for the period 1960 to 1972. The corresponding annual figures for 1977 to 1985 for the UK have been taken to be 3.1 percent and 1.8 percent, respectively, for real growth in GDP excluding the North Sea element. These growth assumptions represent possible trends, and no account is taken of business cycles, or other fluctuations, except indirectly by our use of historical data in the analysis, where trend figures result from a succession of cycles. They are intended as 'typical' high and low choices, and are not intended as the highest or lowest plausible possibilities. For the period 1972 to 1977 a combination of actual, preliminary and forecast growth rates are used for economic growth and energy demand. These are summarised in Table 14.1, and they suggest that in 1977 GDP may be about 5 percent higher than in 1972, but the total energy demand may be about the same.

The second key scenario variable specified by the WAES collaboration is the world price of oil, which is to be taken at one of three levels. The first assumes that oil prices remain constant in real terms until 1985, the second assumes an increase by a factor of 3/2 by 1985, and the third assumes a decrease by a factor of 2/3 by 1985.

It is *assumed* in the WAES scenarios that economic growth and energy prices are *not* coupled. The logic behind this assumption is that the collaboration aims to study the dependence of future energy demand on each of the key scenario variables taken separately. If individual countries in WAES were to change their economic growth assumptions when the oil price assumption is changed, this would necessarily introduce the unwanted coupling. This requirement does make it difficult for the UK projections, since a higher oil price would certainly have significant economic effects, and even if the GDP is considered with the element from North Sea oil removed, it is likely to be affected by the world oil price. However, it is not inconsistent to assume that other economic factors (which we do not specify) are also different when the oil price is different, in such a way that we can compare three scenarios for the UK that have a given GDP and three alternatives for the oil price.

We have therefore adopted the WAES scenario conventions on economic growth and oil price. These lead to the five scenarios shown in Table 14.2 to be studied for the period to 1985.

The third key variable specified for the WAES collaboration is the government policy response on energy. The two situations envisaged are:

1. "vigorous but plausible" government intervention to encourage energy conservation and to stimulate energy supply development;

2. "restrained" government intervention.

The latter is taken to mean no significant intervention beyond that supported by short to medium term market expectations as indicated by the energy price assumptions. Although we would certainly advocate a vigorous policy, we are aware of some of the problems that limit the extent and potential effects of government intervention in the UK. In this context, we have therefore adopted an "expected" government response assumption for the UK, instead of using the WAES terminology. We believe that this lies between the WAES choices of "vigorous" and "moderate", and it has the consequence that more energy savings could be feasible, over and above those included in our scenario projections for future energy demand.

It was observed in Section 14.1 that an adequate discussion of policy in relation to UK energy demand to 1985 requires an examination of the national picture on supply and demand to the year 2000. Although the WAES scenarios for the period 1985 to 2000 have not been finally selected at the time of writing, it is expected that they will include alternatives of high and low world economic growth, together with a specification of the major energy supply source that is chosen for replacing potential increases in oil demand during this period. The latter has been suggested as *either* coal *or* nuclear. We view a "low coal" scenario for the UK

Table 14.1
Average annual percentage growth rates to 1977

AAPG	1972–1973 (actual)	1973–1974 (actual)	1974–1975 (preliminary)	1975–1976 (forecast)	1976–1977 (forecast)
GDP	5	−0.5	−2	+0.5	+1.5
Energy	4	−4	−2	+0.5	+1

Table 14.2
Scenario definitions, 1977–1985

Scenario	Average Annual Economic Growth (percent)		World Oil Price[b]	Government Policy Response	
	World	UK[a]		WAES	UK
A	high	3.1	$11.50 rising to $17.25	vigorous	"expected"
B	low	1.8	$11.50 rising to $17.25	vigorous	"expected"
C	high	3.1	$11.50 constant	vigorous	"expected"
D	low	1.8	$11.50 constant	restrained	"expected"
E	high	3.1	$11.50 falling to $7.66	restrained	"expected"

a. The growth rates for the UK represent the GDP after subtracting out the element corresponding to North Sea oil revenue. The resulting "GDP" is thought to have a closer relation to energy demand than if oil revenue was included.
b. Prices are in constant 1975 U.S. dollars per barrel of light Arabian crude oil FOB Persian Gulf.

in the year 2000 with some disfavour since it seems doubtful as a policy choice, but the purpose of WAES is to study alternatives and we have therefore considered for the UK the WAES scenario choices indicated in Table 14.3.

The projections, described in Sections 14.4 and 14.6, are based on modifications of past trends in sectors of the UK economy, taking account of the higher energy prices specified in the WAES scenarios for oil and implied for other fuels and electricity, and also of the changing availability of different fuels in the UK. They are, of course, "surprise free" projections based on the economic trends and assumptions in the scenarios. The relation to the scenarios is judgemental rather than mechanical, since the UK response to higher energy prices under growth conditions cannot be rigorously deduced from either its earlier behaviour when energy prices were low, or its more recent behaviour during the recession. The methodology is described in more detail in Section 14.3.

The projections and the report have been prepared in Cambridge, but we wish to note the valuable input to the study from the WAES Participants and Associates in Shell International and British Petroleum, and also the very useful comments on the projections and advice received from the WAES (UK) Energy Consultative Group. Further details of the WAES organisation in the UK are given in Section 14.6.

14.3 METHODOLOGY

Each national team in WAES is permitted to adapt the general Workshop methodology (see Chapter 1) to conform to the special circumstances of their country. Thus the approach of the UK study differs somewhat from what is described as the "WAES methodology".

The UK projections for energy demand under the WAES scenario assumptions are based on a 'four-fuel' breakdown of energy consumption in each of ten economic sectors. Subsidiary studies have been carried out in more detail in some of these sectors but are not presented here since they are of a preliminary nature designed as exploratory surveys with a view to future research. The 'fuel' and sector breakdown are shown in Table 14.4.

Historical data for the fuels and energy used in the ten sectors was analysed both in terms of net

Table 14.3
Scenario definitions, 1985–2000

Scenario[b]	Average Annual Economic Growth (percent)		World Oil Price[a]	Principal Replacement Fuel
	World	UK		
C1	high	3.0	$17.25	coal
C2	high	3.0	$17.25	nuclear
D7	low	1.6	$11.50	coal
D8	low	1.6	$11.50	nuclear

a. Prices are in constant 1975 U.S. dollars per barrel of light Arabian crude oil FOB Persian Gulf.
b. Scenarios C1 and C2 (1985 to 2000) are coupled to scenario C (1977 to 1985); scenarios D7 and D8 to scenario D.

Table 14.4
'Fuels' and sectors for UK energy projections

Economic Sectors	Energy Carrier or Fuel
(1) Steel and Iron	(1) Coal
(2) Other Industry	(2) Oil
(3) Transport: Road	(3) Gas
(4) Transport: Rail	(4) Electricity
(5) Transport: Water	
(6) Transport: Air	
(7) Public Administration	
(8) Residential	
(9) Commercial and Agriculture	
(10) Feedstocks and Non-Energy Uses of Fuels	

energy supplied and in terms of useful energy supplied. Useful energy is defined for each fuel and in each economic sector by means of a fixed ratio between useful and net energy that takes account of the average efficiency with which the fuel, or electricity, is used in the sector. In the four sectors listed in Table 14.5, the extent of fuel substitution in the past leads to growth rates for useful energy that differ significantly from those for net energy, and in these cases the modified trends for future demand were based on useful energy. For the others, net energy was used as the basis for projections.

Base year data were obtained for 1972 for economic or physical activities and net fuel and energy used in each of the main economic sectors listed in Table 14.4, and for useful energy for those listed in Table 14.5. Historical data were examined in greater detail in some sectors as indicated below. Some of the more detailed information is given in the WAES worksheets for the base year 1972 which are included in Section 14.6. These base year worksheets are designed to facilitate comparative studies between different countries, and it is for this reason that they are included in this report. However, we do not have adequate time series data on many of the items listed on the worksheets and so could not use them as a basis for our projections in general, although they were used in particular sectors.

Table 14.5
Conventions for ratios of useful to net energy

	Coal	Oil	Gas	Electricity
Rail	0.10	0.20	–	0.90
Agriculture and Commercial	0.60	0.52	0.65	0.90
Public Administration	0.60	0.52	0.65	0.90
Residential	0.35	0.76	0.65	0.90

The projections for energy demand in 1985 for each of the scenarios defined in Table 14.2 were made on a judgemental basis, outlined for each sector in the following paragraphs. These projections take the following factors into account:

1. Estimates of supply availability of different fuels

2. Estimates of activity in different sectors and subsectors of the economy

3. Historical trends in energy demand by sector and their relation to economic activity and to physical activity (where relevant and assessable)

4. The effects of fuel substitution on the efficiency with which energy supplied can be used in some sectors

5. Estimates of the effects of probable energy conservation measures and other responses to higher prices

6. Some preliminary and rather general judgement about the kind of economic policy and energy policy that might be followed in the UK, with regard to such items as consumer expenditure, investment, and the mix of energy supply (these aspects are mentioned below in relation to certain sectors of the economy)

7. Initial advice from WAES Associates and others in the UK, and subsequent advice and comments from members of the WAES (UK) Energy Advisory Group.

The judgemental nature of our approach to projections of energy demand is considered to be necessary not only because of economic uncertainties, but also because these are reflected in government policy on energy which can have a significant effect on many of the factors noted above. Our projections are thought to be plausible and reasonably self-consistent, but the disaggregated figures by sector and particularly by the type of fuel are likely to be more uncertain than energy demand within the context of each scenario.

In assessing UK demand by type of fuel, we have been strongly influenced by the changes in the availability of indigenous fuel supplies. The assumptions on future supplies are given in more detail in Sections 14.4 and 14.6 but we summarise them here. The decline in coal production and demand that took place in the 1960s is assumed to be halted, giving approximately constant production at around 125 million tonnes to 1985. Thereafter, to the year 2000, two of the WAES scenarios are interpreted to assume a fall in coal production (low coal, high nuclear), and two others to assume a slight increase (high coal, low nuclear). Indigenous oil or petroleum supply in 1985 is assumed to match or exceed demand, and in general this is taken to be the residual fuel. The use of oil as a residual continues to hold also for projections to the year 2000, though indigenous supply is not expected to exceed demand in that year (within the context of the scenario assumptions). Natural gas production is assumed to match consumption and could be in the range 60 to 90 million tonnes of coal equivalent in the 1980s. Its actual level will be influenced by government policy on rates of depletion, by expectations about resources and reserves, by long term contracts for the purchase of North Sea gas, by government policy on the price of gas, and by the price of alternative fuels. The growth of electricity demand to 1985 is assessed by sector, taking into account both historic trends and our judgement of the effects of economic growth, price changes and energy conservation measures.

The resulting growth rates for energy consumption in the UK economic sectors listed in Table 14.4 are listed in Table 14.6 for two of the five scenarios studied to 1985 and for two of the four extensions to the year 2000. For the period 1972 to 1977 figures for actual demand and short term forecasts were used, so that the growth rates for this period include a large energy saving as a consequence of the energy crisis (see also Table 14.1 for the effect on total energy demand). The 1972 to 1977 growth rates thus appear small in relation to both historic and projected values.

The figures for growth rates given in Table 14.6 were derived by means of the list of relevant factors noted above. A few special points require additional notes which are given here:

1. Energy demand for steel was calculated from estimates of future steel production (related to GDP growth) and specific energy use per tonne of steel, taking note of past improvements in specific energy requirements and further gains resulting from the investment program of the steel industry and the higher energy prices, as given by the scenario assumptions.

2. "Other industry" was initially analysed by subsectors, but the detailed assessment required for this to be useful is not feasible in the time-scale of the WAES program. The growth of energy consumption was therefore estimated from GDP growth coupled with modifications of past trends.

3. Road transport energy demand was initially calculated by an analysis of activity and associated energy requirements, but was subsequently modified as a result of discussions with WAES (UK) advisors.

4. Rail transport energy demand was related to expectations about the size of the network in active use, and about continuing expansion of road transport.

5. Water transport is affected significantly by activity in the North Sea, but this is assumed to level off and remain nearly constant in the future.

6. Air transport includes international demand in the UK. It is assumed that the boom in air transport of the 1960s and early 1970s has come to an end, to be replaced by a period in which the growth rate lies between UK and world GDP growth rates.

7. Growth in energy consumption for public administration is expected to be curtailed during the period to 1985, partly due to government policy to invest more in manufacturing industry, and partly due to an ongoing successful program for energy conservation in public buildings.

8. The total consumption of energy in the residential sector in the UK has remained almost constant on a 'heat supplied' or 'net energy' basis for the period 1960 to 1972, although standards of heating have improved and the number of dwellings has increased. The principal reason for this is the gain in efficiency that has taken place as a result of fuel substitution from coal, to oil, gas and electricity. The substitution has involved the purchase of new and more efficient equipment, which can be represented by a gain in useful energy using the

Table 14.6
Summary of assumptions about sectoral growth (AAPG) in energy demand

	Historical Average, 1960–1962 to 1970–1972	1972 –1977	Case C 1977–1985	Case C1 1985–2000	Case D 1977–1985	Case D7 1985–2000
GDP Growth (% p.a.)			3.1	3.0	1.8	1.6
Steel (net energy)	-0.2	-0.2	2.4	2.0	1.8	1.2
Other Industry (net energy)	1.9	0.0	2.0	2.0	1.5	1.2
Transport: Road (net energy)	6.1	0.5	2.7	2.0	1.8	1.3
Transport: Rail (useful energy)	-6.4	0.0	0.4	2.0	-0.1	1.3
Transport: Water (net energy)	-1.9	7.0	1.0	0.0	1.0	0.0
Transport: Air (net energy)	6.4	0.5	3.9	2.0	2.9	1.3
Public Administration (net energy)	3.5	0.2	1.6	2.0	0.8	1.3
Residential (useful energy)	2.8	1.0	3.2	2.0	2.2	1.3
Commercial and Agriculture (net energy)	1.0	0.0	2.0	2.0	1.2	1.3
Feedstocks and Non-energy Uses of Fuels	6.5	0.5	5.8	3.5	4.0	2.5

AAPG = average annual percentage growth

coefficients given in Table 14.5. For the residential sector these coefficients are:

$R_1 = 0.9$ (electricity)

$R_2 = 0.35$ (coal and other solid fuels)

$R_3 = 0.76$ (oil fuel)

$R_4 = 0.65$ (gas fuel).

The total useful energy is given by

$$E \text{ (useful)} = R_1 D_1 + R_2 E_2 + R_3 E_3 + R_4 E_4,$$

where E_j denotes the net energy supplied by electricity or fuel type j ($j = 1, 2, 3, 4$). In the residential sector the total useful energy increased at an average rate of 2.6 percent per annum from 1960 to 1972, which is almost the same as GDP growth in that period. The useful energy for the period 1972 to 1977 was calculated from actual and forecast fuel use, and for 1977 to 1985 it was initially related to GDP growth, and then perturbed downwards to allow for the higher energy prices and energy conservation in the scenarios studied. The total useful energy for 1985 is converted into useful energy by fuel type and electricity using modified trends, and converted to net energy demand by dividing by the coefficients R_j noted above. The rate of fuel substitution, and equipment replacement and tendency to saturation, are then checked for plausibility and discussed with WAES advisors after which further revisions are made. This leads to the growth rates indicated in Table 14.6, in which the results of a similar procedure for 1985 to 2000 are also included. The effective result is an increase also in the net energy supplied since the benefits of fuel substitution to give gains in efficiency and useful energy have nearly been exhausted.

9. The consumption of net energy by the commercial and agricultural sector given in Table 14.6 is influenced by expectations of somewhat slower growth and restrictions on further growth in electricity demand which is already high in this sector.

10. The growth rate in the use of petroleum for chemical feedstocks has been as high as 9 percent per annum in recent years. This rate is expected to

slow down, and the rate for other non-energy uses (bitumen for roads, etc.) is taken to be similar to the growth of GDP. The resulting growth rates are given in Table 14.6.

The energy industries' own use of fuel and the losses from converting primary energy input to the net energy supplied to the ten economic sectors were assumed to amount to 2 percent for the coal industry, 7.5 percent for petroleum and 10 percent for the gas industry. A 15 percent allowance is made for the electricity industry's own use and losses between the busbar and the consumer. Hence, the generated electricity (G) equals (electricity demand)/(0.85). This is generated in one of five ways: from coal, oil, gas, hydro or nuclear. The equivalent primary energy, which we nominally attribute to hydro and nuclear production, assumes the same conversion efficiency, f, as the average of fossil fuel stations. Thus, if the nominal thermal input so derived for hydro is H and for nuclear is N, the electricity required from fossil fuel is F_e, given by

$$F_e = G - fH - fN,$$

and hence the fossil fuel input required will be

$$(1/f)F_e.$$

The distribution of this requirement between coal, oil and gas was based on assumptions about energy policy, taking account of coal availability and constraints on future generating plants given the pattern of existing plants.

The energy coefficient is defined as the ratio of the AAPG (average annual percentage growth) in energy consumption to the AAPG of gross domestic product (GDP). In the UK, the energy coefficient has risen from 0.6 in the early 1960s to about 0.9 in the early 1970s before the energy crisis. This suggested a value of about 1.0 might have been expected in the future if the world price of oil had not risen by a factor of four. Further arguments for an expected increase in the energy coefficient come from the expected increase in energy consumption by the domestic sector as a result of approaching the end of the gains in efficiency that came from fuel substitution and new appliances. The higher energy prices may be expected to moderate and possibly reverse the increase in the energy coefficient. This intuitive argument is consistent with our calculations of primary energy demand from the more detailed analysis of economic sectors, from which we obtain the results given in Table 14.8. The right hand column of Table 14.8, giving the energy coefficients for 1972 to 1985, makes use also of Table 14.1. The figures in this column contain the initial energy conservation effects implied by Table 14.1. The oil income element is excluded from GDP in our discussion since North Sea oil production uses only a small amount of energy in relation to the value added, and it would give misleading changes in the energy coefficient if it was included.

Energy supply prospects play a vital role in our methodology, as has been noted earlier. This is because we expect government policy and internal pricing to ensure that all indigenous coal production is used, though there may be periods in which stocks are accumulated. We also expect all natural gas production to be used, but to be directed towards premium uses, except possibly for some load levelling uses and some associated gas production. We expect some restraint on natural gas production but find it difficult either to assess the peak level or to determine criteria on which any such constraint should be based. The problem of possible constraints on North Sea oil production as a measure of resource conservation is even more difficult to assess. We are not convinced that the conventional arguments based on discount rates are valid in the special case of North Sea oil, neither do we accept that market forces would necessarily produce the best solution. Our doubts arise partly from the size of the resource, which must surely require more sophisticated economic modelling than is implied by the discount rate method, and partly from the difficulty of assessing the response of the market in a period that may precede a scarcity of oil on the world market. Finally, the depletion policy should take account of risks and penalties associated with various possible outcomes, taking both the UK and the world situation into account. The first objective of WAES scenario studies is to examine possible outcomes, and only after this has been achieved can risks and

Table 14.7
Assumed (busbar) efficiency of fossil fuel power stations

	1972	1985	2000
f	0.31	0.35	0.375

Table 14.8
Energy coefficients

Scenario	AAPG of Primary Energy Consumption (a)	AAPG of GDP Excluding Oil Element (b)	Energy Coefficient (a/b)	Energy Coefficient
1977–1985				1972–1985
A	2.4	3.1	0.77	0.64
B	1.5	1.8	0.83	0.62
C	2.6	3.1	0.85	0.69
D	1.7	1.8	0.94	0.70
E	2.9	3.1	0.94	0.78
1985–2000				
C1	2.1	3.0	0.7	
C2	2.1	3.0	0.7	
D7	1.3	1.6	0.8	
D8	1.3	1.6	0.8	

AAPG = average annual percentage growth.

penalties and optimal policies be discussed. Therefore, our assumptions on energy supply for the UK are necessarily provisional, and we would expect to revise them in later reports, as our understanding of the problems and options progresses.

Policy on energy supply is uncertain not only because of difficulties of analysis but also because the extent of reserves and resources is not known. UK indigenous energy reserves and resources are summarised in Tables 14.9 to 14.11.

Our assumptions about UK fossil fuel reserves are summarised in Table 14.12, and the related supplies of fuel for our projections are given below in Table 14.14.

14.4 SUMMARY OF RESULTS

Energy demand projections are summarised in Table 14.13. These projections are related to indigenous supply estimates in Table 14.14 for scenarios C and D in 1985 and for alternative projections of each to the year 2000. Further details for the projections of energy demand and primary energy input are given in the form of WAES demand study worksheets in Section 14.6.5.

Coal supply has been taken to balance demand (approximately) in 1985 and is about 130 million tonnes a year. This should be achieved on the current UK plan for coal. Production figures after this date depend on policy decisions yet to be taken. The WAES scenarios C1 and D7 assume high coal production and low nuclear, and we have somewhat arbitrarily selected coal production for the

Table 14.9
Coal reserves and resources

Category of Reserves	Amount (10^9 tonnes)
1. 'Classified Reserves' (all coal accessible from existing collieries and currently economically extractable)	3.5
2. 'Unclassified Reserves' (all coal accessible from existing collieries that could become economically extractable with further proving or with beneficial changes in mining methods and/or marketing requirements)	2.5
3. Potential Classified Reserves (Classified Reserves at new mine prospects)	0.5
4. All Other 'Physically Workable' Coal (⩾ 2ft thick, ⩽ 4000ft deep) in Proved Coalfields	92[a]

a. In addition to the cumulative total of ~ 98 x 10^9 tonnes above, it is thought that additional resources in unproved coalfields could make the cumulative total of physically workable coal up to ~ 160 x 10^9 tonnes.

Table 14.10
Petroleum reserves (Department of Energy 1975 estimates)

Category of Reserves[a]	Amount (10^9 tonnes)
1. Proven (technically and economically producible)	1.35
2. Probable (greater than 50% chance of being technically and economically producible)	0.96
3. Possible (less than 50% chance of being technically and economically producible)	0.88

a. These estimates refer to all areas of the UK Continental Shelf that have been licensed. The possible cumulative total of 3.05 thousand million tonnes could be raised to 4.5 thousand million tonnes when unlicensed, but already designated, areas are taken into account.

Table 14.11
Gas reserves (Department of Energy estimates for gas remaining in known discoveries at 31 December 1975)

Category of Reserves[a]	Amount (10^{12} ft^3)	Amount (10^9 tonnes of oil equivalent)
Proven	28.7	0.69
Probable	11.5	0.28
Possible	10.3	0.25
Total	50.5	1.22

a. These figures exclude any gas associated with undiscovered oil, and also only include the UK sector of the Frigg field.

year 2000 at the levels of 140 and 123 million tonnes for these scenarios (high and low economic growth, respectively). In both cases substantial coal imports would be required to balance our projections of demand. For WAES scenarios C2 and D8 we assume low coal production and high nuclear, and our coal production assumptions are 90 and 82 million tonnes, respectively, in the year 2000. The coal import requirements are still substantial but are lower than is required by our 'low nuclear' scenarios C1 and D7.

Table 14.12
Assumptions about effective reserves

Fuel	Assumptions
Coal	No effective limit due to reserves. Production is limited by investment policy, manpower estimates and lead times for new developments.
Oil	For the projections, reserves of 4000 million tonnes are assumed to have been proved by the late 1980s. This assumption is considered to be somewhat optimistic.
Gas	For the projections, we assume a reserves figure of 1300 million tonnes of oil equivalent.

Oil production depends on the success of future exploration, on prices and on depletion policy. At the time of writing a reserves figure of 2.0 billion tonnes seems reasonably assured, and this might be increased to 3.0 billion within the North Sea areas already under investigation. Beyond this figure there is much less certainty, and we may be somewhat optimistic to have chosen a reserves figure of 4.0 billion tonnes for our calculations. Given this choice, if it had all been proved by 1985, the annual production could be as high as 250 million tonnes per year, or 11 x 10^{18} J/yr. However, it is unlikely that the full reserves figure would have been proved by that date, and there may be some policy constraint on development, so we have therefore chosen, somewhat arbitrarily, a single illustrative production figure of 7 x 10^{18}J/yr for 1985 for use in our projections. The production in the year 2000 is obtained from calculation corresponding to an eventual reserves-to-production ratio of ten years.

Gas production prospects, like those for oil, depend on assumptions about future exploration successes, on energy prices and on depletion policy. Associated gas production has a technical dependence on the production of the oil with which it is associated. We assume a gas supply from the UK sector of the North Sea in the region of 1.7 x 10^{18}J/yr for 1985 (67 million tonnes of coal equivalent). To this is added a figure of about 0.4 x 10^{18}J/yr from the Norwegian sector of the Frigg field. Both of these are included in our tables under "indigenous supply" since if it was included under "imports" this would imply participation in a world natural gas market. The latter is not the case since supplies are already secured, and they will have the same effect on UK gas demand as if the production was indigenous. The total gas supply assumed for 1985 is equivalent to 82 million tonnes of coal. For the year 2000, the supplies are calculated from our assumed figure for reserves (Table 14.12) using an eventual reserves-to-production ratio of ten years.

Electricity production is determined by our projections of future demand using the WAES scenario assumptions. These projections take into account the high cost of electricity relative to other fuels when used to supply low grade heat such as for

Table 14.13
Summary of UK energy demand projections

Year	Scenario	Units	Final Energy Demand (excluding processing losses) Coal	Oil	Gas	Electricity	Primary Energy Input Fossil (direct)	Fuel for Electricity	Total Primary Energy Input
1960		10^{18} J	3.34	1.31	0.28	0.36	6.00	1.52	7.52
		%	63	25	5	7	80	20	100
1972		10^{18} J	1.39	3.49	0.97	0.74	6.50	2.81	9.31
		%	21	53	15	11	70	30	100
1985	A	10^{18} J	1.05	3.92	1.86	1.11	7.54	3.74	11.28
		%	13	49	23	14	67	33	100
	B	10^{18} J	1.03	3.73	1.74	1.00	7.15	3.37	10.53
		%	14	49	23	13	68	32	100
	C	10^{18} J	1.06	4.00	1.87	1.16	7.66	3.89	11.55
		%	13	49	23	14	66	34	100
	D	10^{18} J	1.03	3.81	1.73	1.02	7.23	3.44	10.67
		%	14	50	23	13	68	32	100
	E	10^{18} J	1.05	4.04	1.89	1.19	7.72	3.99	11.71
		%	13	49	23	15	66	34	100
2000	C1	10^{18} J	1.68	6.05	1.47	1.80	10.24	5.65	15.89
		%	15	55	13	16	64	36	100
	C2	10^{18} J	1.34	6.19	1.47	1.85	9.98	5.81	15.79
		%	12	57	14	17	63	37	100
	D7	10^{18} J	1.40	5.18	1.35	1.32	8.77	4.13	12.90
		%	15	56	15	14	68	32	100
	D8	10^{18} J	1.17	5.23	1.35	1.35	8.58	4.22	12.80
		%	13	57	15	15	67	33	100

space heating. They also take into account the lead times for constructing power stations. Nuclear power capacity in 1985, for the UK, is almost entirely determined by current plans, and we assume for that date: present nuclear capacity plus capacity under construction, together with 4 x 660 MWe from SGHWR plants, and two further 1000 MWe nuclear stations, running at an average load factor of 60 percent. (Note that we use the convention for primary energy input for hydro and nuclear stations that it is the equivalent fossil fuel that would be required, at average efficiencies for 1972, at 35 percent efficiency in 1985 and at 37.5 percent in 2000.) Our projected electricity output capacities in 1985 depend slightly on economic growth assumptions but are given approximately in Table 14.15; these figures may be reduced slightly by retirements. These figures imply average load factors of about 40 percent for coal-fired stations, 60 percent for nuclear, and 20 to 40 percent for oil depending on the scenario (oil varies considerably since it is taken to be the marginal fuel for electricity production).

Our projections for electricity demand for the year 2000 follow criteria similar to those used for 1985, except that in the case of the 'high nuclear' scenarios C2 and D8 we assume a faster growth for electricity demand in the industrial sector. In all scenarios there is assumed to be a strong dependence of electricity demand on economic growth. The assumptions for nuclear power generation begin from calculations for Case C2, the high growth, high nuclear case. For this we assume only 1 GWe is ordered annually until 1983 when the first SGHWR should be commissioned. We then assume annual orders of 2 GWe until 1986, by which date the SGHWRs should be fully proved in commercial operation. Thereafter, 3 then 4 then 5 GWe are assumed to be ordered annually, though the last increase is delayed to 1991 due to our assumption that capital constraints will limit the rate of building nuclear power stations. Assuming a time of six years from ordering to commissioning, we obtain 58 GWe in the year 2000 for our high nuclear figure (scenario C2). In this case residual electricity demand requires also the construction of further coal-fired power stations; this applies to a greater extent in Case C1, which is a

Table 14.14
Summary of projections for UK primary energy consumption and production in units of 10^{18} joules[a]

Year and Scenario	Category	Coal	Oil	Gas	Hydro	Nuclear	Total
1972	Consumption	3.15	4.68	1.19	0.04	0.25	9.31
	Production	3.02	0.00	1.15	0.04	0.25	4.47
	Imports	0.13	4.68	0.03	0.00	0.00	4.84
1985	Consumption	3.30	5.31	2.10	0.04	0.81	11.55
C	Production	3.30	7.00	2.10[b]	0.04	0.81	13.25
	Exports	0.00	1.69	0.00	0.00	0.00	1.69
1985	Consumption	3.30	4.62	1.94	0.04	0.77	10.67
D	Production	3.30	7.00	1.94[b]	0.04	0.77	13.05
	Exports	0.00	2.38	0.00	0.00	0.00	2.38
2000	Consumption	4.89	7.66	1.63	0.04	1.67	15.89
C1	Production	3.56	3.39	0.97	0.04	1.67	9.63
	Imports	1.33	4.26	0.66	0.00	0.00	6.26
2000	Consumption	3.62	7.57	1.63	0.04	2.93	15.79
C2	Production	2.30	3.39	0.97	0.04	2.93	9.63
	Imports	1.32	4.18	0.66	0.00	0.00	6.26
2000	Consumption	3.68	6.22	1.50	0.04	1.46	12.90
D7	Production	3.14	3.39	1.10	0.04	1.46	9.13
	Imports	0.54	2.83	0.40	0.00	0.00	3.77
2000	Consumption	2.48	6.28	1.50	0.04	2.51	12.80
D8	Production	2.09	3.39	1.10	0.04	2.51	9.13
	Imports	0.38	2.89	0.40	0.00	0.00	3.67

a. Approximate fuel equivalents: 10^6 tonne coal = 25.5×10^{15} joules; 10^6 tonne oil = 44.0×10^{15} joules.
b. Gas production includes about 0.4×10^{18} J from the Norwegian sector of the Frigg field.

high coal scenario. In the latter case, the nuclear capacity is taken to be 33 GWe. In the low economic growth cases D7 and D8, the figures are 29 GWe and 50 GWe respectively.

14.5 DISCUSSION AND CONCLUSIONS

Our projections for UK energy consumption under the WAES scenario assumptions to 1985 have been presented in the context of the UK energy supply pattern, taking into account the longer-term energy prospects through the year 2000. A discussion of energy demand in the UK in 1985 would be incomplete without also considering supply policy, since this affects relative pricing and the fuel mix, which (for example) bears directly on the useful energy component for key sectors such as the residential sector. Policy on energy conservation to 1985, and policy on energy supply, should be influenced by expectations both for UK indigenous supplies in later years and for the world energy supply. The WAES program is directed towards a study of world energy prospects, but its initial estimates for energy demand will necessarily arise from the preliminary estimates of individual countries. In order to avoid anticipating conclusions that may later be deduced on a global basis, it is initially necessary for national projections to be made using WAES scenarios, assuming that national growth will not be constrained by world energy supply availability, except to the extent that this is reflected in the initial WAES scenario assumptions about energy prices. If this leads in due course to a total of all national energy demand estimates that exceeds the estimates for world energy supplies in some future year, this would mean that the initial WAES scenario assumptions are inconsistent. Revised scenarios could change one or more of the assumptions about economic growth, energy prices or the level of energy conservation effort. The UK

Table 14.15
Estimated output capacities for electric power in 1985

Source	Estimated 1985 Output Capacity (GWe)
Coal and Dual-Fired	52
Oil	19
Nuclear	15
Hydro (excluding pumped storage)	1.3
Other (gas turbine, etc.)	3.3
Total	90

projections presented in this report assume continued UK economic growth through to the year 2000 in the context of the WAES scenarios. If, for example, the WAES global scenario for high economic growth turns out to be inconsistent, the UK high growth projections would need to be revised with new scenario assumptions. For the 1985 projections, this seems unlikely, but for the year 2000 we doubt whether the WAES high growth scenario will be consistent, unless other national projections have a much greater slowdown in demand than we have assumed for the UK. This possibility of a worldwide energy scarcity before the year 2000 for a high growth scenario is an essential ingredient to our discussion of the UK to 1985. The choice of a UK energy policy in this intermediate period is likely to be influenced by the possible problems in the longer-term future.

The indigenous supply of oil in the UK is expected to exceed the inland demand by 1980. Thereafter, the level and timing for net exports or imports will depend on future discoveries and on policy decisions concerning resource depletion that will be made within the next few years. Our projections for 1985 have assumed a net export of oil for all WAES scenarios, but even with our moderately optimistic assumptions about reserves, they lead to net imports for oil for all scenarios by the year 2000 that are well in excess of the present total oil demand. Alternative projections maintaining a net oil deficit would permit indigenous supplies to remain at a high level beyond the year 2000.

The indigenous supply of natural gas in the UK will be increased by the addition of associated gas when the oil flow increases, and this is expected to be adequate to meet demand throughout the 1980s. In our longer-term projections there will be a shortfall in indigenous natural gas supplies before the year 2000, and they would need to be supplemented either by imports of LNG, or by synthetic gas made from oil or from coal. The optimum choice would depend on a more precise assessment of the world picture in the year 2000 than we have available at the present time. These factors would also influence depletion policy during the 1980s.

Our projections to 1985 all assume that indigenous coal production will be maintained at about 130 million tonnes a year until 1985. We have had some difficulty in producing an energy demand picture that is plausible and that uses all the indigenous coal production. The view we have taken is that all the indigenous coal production at this level, or higher, could be used in the year 2000, and that the maintenance of the coal industry in the intermediate period is likely to be given high priority. This imposes a constraint on our demand projections for 1985 which requires the use of 130 million tonnes of coal annually. Some increase in the coal burn for electricity power generation was therefore assumed, but this in turn is influenced by our choice for the growth rate for electricity demand.

Our projections to the year 2000 are made on the WAES scenarios for *either* high coal *or* high nuclear development. The massive energy imports that would be required for the UK in the year 2000 suggest that a policy for *both* high indigenous coal production *and* high nuclear development might be preferred. The scenario projections that allow indigenous coal production to decline between 1985 and 2000 would seem unlikely as a policy choice, and would contradict the present policy that is reflected in our scenarios to 1985, where coal production is maintained.

Nuclear power for 1985 is essentially determined by existing plant and construction programs. Beyond 1985 to the year 2000 our projections suggest that nuclear power is very unlikely to increase fast enough to avoid large imports of energy to the UK by the latter date. This conclusion implies a need to maintain indigenous coal production, which then produces the coal demand constraint for 1985 discussed above.

Energy conservation estimates have a significant influence on our energy demand projections to 1985. They can be interpreted as preventing a rise in the energy coefficient (the ratio of energy AAPG to economic [GDP] AAPG). A potential rise from its earlier historic value of 0.7 was already indicated by a rise in its average values for the late 1960s and early 1970s. The expected increase in residential energy demand, which would follow from the ending of the benefits to energy efficiency caused by fuel substitution away from solid fuel in recent years, would also tend to increase the growth rate of energy demand relative to that for GDP. The compensating influence of higher energy prices and energy conservation is expected to keep the energy coefficient below unity. Its value depends on the scenario, and the energy demand–energy price elasticity to 1985 is about -0.04 for our projections.

For the period to 1985, energy conservation will tend to reduce the growth in demand for electricity as well as for fuels supplied for direct use. The increase in nuclear power for base load will magnify the effect that energy conservation has on coal demand through its effect on electricity. There is therefore a conflict between two of the central policy requirements, suggested by our long-term projections and our assumption of a potential scarcity of world energy supplies before the year 2000. These two policy requirements are the maintenance of indigenous coal production and consumption, and the maintenance of a strong program for energy conservation. The lead times for the expansion of coal production and for the achievement of effective results from energy conservation require that both the coal and conservation programs be actively developed throughout the intermediate period. If these requirements are satisfied, our projections for 1985 could be too high both for total energy requirements, and for electricity growth and hence for coal demand in the electricity sector. The resulting discrepancy between coal supply and demand in 1985 could be resolved only by an increase in coal demand by other sectors, of which the industrial, public and commercial sectors seem to be the only serious candidates. Policy instruments that are available to effect such a change include the relative pricing of coal, limitations on natural gas availability, regulations and incentives. A substantial effort on research and development in coal utilisation, distribution and storage, coupled with a major marketing and incentive program, could also make a significant contribution, and in the long run could be the most important of any package of measures to stimulate the use of coal.

14.6 REFERENCES AND TECHNICAL INFORMATION

14.6.1 Authorship

This demand study has been prepared by Edmund Crouch, Richard Eden and David Pullin of the Energy Research Group, Cavendish Laboratory, Cambridge, England.

14.6.2 Acknowledgements

The authors wish to acknowledge the help of their WAES collaborators in the UK, namely, James Davidson, Gareth Price, Hans DuMoulin and Aart Beijdorff of Shell International, Robert Belgrave, Michael Clegg and Andrew Flower of British Petroleum, and Professor Sir William Hawthorne, of Churchill College. They are also indebted to other members of the Energy Research Group in the Cavendish Laboratory for their assistance. The authors take final responsibility for the assumptions, projections and comments given in the Report. However, they are grateful for advice from the WAES (UK) Energy Consultative Group which includes representatives from the following organisations:

British Gas Corporation

British Petroleum

British Steel Corporation

Central Electricity Generating Board

Department of Energy

Electricity Council

Esso Petroleum (UK)

Imperial Chemicals Industries

National Coal Board

Shell International

Shell (UK)

UK Atomic Energy Authority.

14.6.3 Publications Used to Compile Data

The Digest of UK Energy Statistics (Department of Energy, annual)

Energy Trends (Department of Energy, monthly)

Digest of Monthly Statistics (HMSO)

Annual Abstract of Statistics (HMSO)

Highway Statistics (HMSO, annual)

Passenger Transport in GB (HMSO, annual)

Forecasts of Vehicles and Traffic in GB (Transport and Road Research Laboratory)

British Railways Board Statistics

British Railways Board Annual Report and Accounts

UK Petroleum Industry Statistics: Consumption and Refinery Production (Institute of Petroleum)

Development of the Oil and Gas Resources in the UK (HMSO)

British Gas Annual Report and Accounts

Handbook of Electricity Supply Statistics

The Electricity Council Statement of Accounts and Statistics

Electricity Council UNIPEDE Statistics (unpublished)

Family Expenditure Survey (HMSO, annual)

Agriculture Statistics (HMSO, annual)

Central Statistical Office Input-Output Tables for the UK.

14.6.4 Conversion Factors and Conventions

The conversion factors used in this report are shown in Table 14.16. It should be noted that the thermal content of UK coal is somewhat lower than the conventional figure of 28.8 GJ per tonne that is often used internationally.

14.6.5 Demand Study Worksheets

In this section will be found the following WAES demand study worksheets:

Energy Use Summary

Energy Growth Rates Summary

Sectoral Percentages of Total Final Energy Demand

Sectoral Percentages of Total Primary Energy Input

1985 Case Comparisons (two sheets)

Base Year (1972) worksheets.

Table 14.16
Conversion factors

Energy	10^{18} joules = 10^9 GJ = 947 x 10^6 Btu = 278 TWh = 22.7 million tonnes of oil equivalent = 39.2 million tonnes of coal equivalent = 25.0 thousand million metres cubed of natural gas
Power	10^{18} joules/year = 32 GW = 32 x 10^9 Watts = 0.45 million barrels per day oil equivalent
Calorific Values	1 tonne oil equivalent = 44.0 GJ 1 tonne coal equivalent = 25.5 GJ 10^3 standard m^3 natural gas (at 0°C 760mm dry) = 37.9 x 10^3 standard ft^3 (at 60°F 30″ wet) = 40.0 GJ 10^6 B/d.O.E. = 2.2 x 10^9 GJ per annum = 2.2 x 10^{18} J per annum
Electricity	A 1000 MW (= 1 GW) power station at 60 percent load factor and at 35 percent conversion efficiency (assumed for 1985) requires annually a primary energy input of 54 x 10^{15} joules, or 1.23 million tonnes of oil equivalent, or 2.12 tonnes of coal equivalent. At 37.5 percent efficiency (for year 2000) an input of 50.5 x 10^{15} joules is assumed to be required for a 1 GWe nuclear station.

WAES NATIONAL ENERGY DEMAND STUDIES, 1972-1985
ENERGY USE SUMMARY

Country United Kingdom

Scenario Assumptions

WAES Case	A		B		C		D		E			
GNP 1976-1985	High		Low		High		Low		High			
Oil Price*	$11.50 → $17.25		$11.50 → $17.25		$11.50		$11.50		$11.50 → $7.66			
Natl. Policy Response	Expected		Expected		Expected		Expected		Expected			
Natl. GDP Growth Rate 1977-1985 (%/yr.) **	3.1		1.8		3.1		1.8		3.1		1972 Base Year	
Fuel Type	Fossil Fuel	Elec.	Fossil Fuel	Elec.	Fossil Fuel	Elec.	Fossil Fuel	Elec.	Fossil Fuel	Elec.	Fossil Fuel	Elec.
Units	10^{18}J PER ANNUM											
Energy-Using Sectors												
Industrial (excluding feedstocks)	2.40	0.40	2.37	0.37	2.44	0.42	2.36	0.38	2.47	0.43	2.16	0.26
Transportation	1.57	0.01	1.46	0.01	1.59	0.01	1.49	0.01	1.61	0.01	1.24	0.01
Public	0.33	0.07	0.30	0.06	0.32	0.08	0.31	0.06	0.32	0.08	0.32	0.05
Residential	1.31	0.46	1.24	0.43	1.33	0.48	1.24	0.44	1.35	0.49	1.21	0.31
Fishing, Agriculture, and Commercial	0.25	0.17	0.28	0.14	0.24	0.18	0.28	0.14	0.25	0.18	0.29	0.11
Feedstocks	0.97	0.00	0.84	0.00	1.00	0.00	0.87	0.00	0.98	0.00	0.62	0.00
Final Energy Demand (excluding processing losses)	6.83	1.11	6.50	1.00	6.93	1.16	6.57	1.02	6.98	1.19	5.85	0.74
Processing Losses	0.71	2.63	0.65	2.37	0.73	2.73	0.66	2.42	0.74	2.80	0.65	2.07
Primary Energy Input	11.28		10.53		11.55		10.67		11.71		9.30	

*Oil price 1975 US$ per barrel Arabian light crude oil FOB Persian Gulf.

** excluding oil

WAES NATIONAL ENERGY DEMAND STUDIES, 1972-1985
ENERGY GROWTH RATES SUMMARY

Country United Kingdom

Scenario Assumptions										
WAES Case	A		B		C		D		E	
GNP 1976-1985	High		Low		High		Low		High	
Oil Price*	$11.50 → $17.25		$11.50 → $17.25		$11.50		$11.50		$11.50 → $7.66	
Natl. Policy Response	Expected		Expected		Expected		Expected		Expected	
Natl. GDP Growth Rate 1977 1985 (%/yr.) **	3.1		1.8		3.1		1.8		3.1	
Fuel Type	Fossil Fuel	Elec.	Fossil Fuel	Elec.	Fossil Fuel	Elec.	Fossil Fuel	Elec.	Fossil Fuel	Elec.
Units (%/yr. 1972-1985)	%	%	%	%	%	%	%	%	%	%
Energy-Using Sectors										
Industrial (excluding feedstocks)	0.8	3.3	0.7	2.6	0.9	3.6	0.7	2.8	1.1	3.8
Transportation	1.8	1.7	1.2	1.2	1.9	1.7	1.4	1.2	2.0	1.7
Public	0.0	3.7	-0.4	1.8	0.0	4.0	-0.1	1.6	0.1	4.2
Residential	0.7	3.0	0.2	2.5	0.8	3.3	0.2	2.7	0.9	3.5
Fishing, Agriculture, and Commercial	-1.5	3.7	-0.4	2.0	-1.4	4.0	-0.2	2.0	-1.4	4.2
Feedstocks	3.5	0.0	2.3	0.0	3.7	0.0	2.6	0.0	3.5	0.0
Final Energy Demand (excluding processing losses)	1.2	3.2	0.8	2.4	1.3	3.5	0.9	2.6	1.4	3.7
Processing Losses	0.7	1.9	0.1	1.1	0.9	2.2	0.2	1.2	1.0	2.4
Primary Energy Input	1.5		1.0		1.7		1.1		1.8	

*Oil price 1975 US$ per barrel Arabian light crude oil FOB Persian Gulf.

** excluding oil

WAES NATIONAL ENERGY DEMAND STUDIES, 1972-1985
SECTORAL PERCENTAGES OF TOTAL FINAL ENERGY DEMAND

Country United Kingdom

Scenario Assumptions

WAES Case	A		B		C		D		E		1972 Base Year	
GNP 1976-1985	High		Low		High		Low		High			
Oil Price*	$11.50 → $17.25		$11.50 → $17.25		$11.50		$11.50		$11.50 → $7.66			
Natl. Policy Response	Expected		Expected		Expected		Expected		Expected			
Natl. GDP Growth Rate 1977-1985 (%/yr.) **	3.1		1.8		3.1		1.8		3.1			
Fuel Type	Fossil Fuel	Elec.	Fossil Fuel	Elec.	Fossil Fuel	Elec.	Fossil Fuel	Elec.	Fossil Fuel	Elec.	Fossil Fuel	Elec.
Units	%	%	%	%	%	%	%	%	%	%	%	%
Energy-Using Sectors												
Industrial (excluding feedstocks)	35.1	36.0	36.5	37.0	35.2	36.2	35.9	37.3	35.4	36.1	36.9	35.1
Transportation	23.0	0.9	22.5	1.0	22.9	0.9	22.7	1.0	23.1	0.8	21.2	1.4
Public	4.8	6.3	4.6	6.0	4.6	6.9	4.7	5.9	4.6	6.7	5.5	6.8
Residential	19.2	41.4	19.1	43.0	19.2	41.4	18.9	43.1	19.3	41.2	20.7	41.9
Fishing, Agriculture, and Commercial	3.7	15.3	4.3	14.0	3.5	15.5	4.3	13.7	3.6	15.1	5.0	14.9
Feedstocks	14.2	0.0	12.9	0.0	14.4	0.0	13.2	0.0	14.0	0.0	10.6	0.0
Final Energy Demand (excluding processing losses)	100.0	100.0	100.0	100.0	100.0	100.0	100.0	100.0	100.0	100.0	100.0	100.0
Processing Losses												
Primary Energy Input												

*Oil price 1975 US$ per barrel Arabian light crude oil FOB Persian Gulf.

** excluding oil

WAES NATIONAL ENERGY DEMAND STUDIES, 1972-1985
SECTORAL PERCENTAGES OF TOTAL PRIMARY ENERGY INPUT

Country United Kingdom

Scenario Assumptions

WAES Case	A		B		C		D		E			
GNP 1976-1985	High		Low		High		Low		High			
Oil Price*	$11.50 → $17.25		$11.50 → $17.25		$11.50		$11.50		$11.50 → $7.66			
Natl. Policy Response	Expected		Expected		Expected		Expected		Expected			
Natl. GDP Growth Rate 1977-1985 (%/yr.) **	3.1		1.8		3.1		1.8		3.1		1972 Base Year	
Fuel Type	Fossil Fuel	Elec.	Fossil Fuel	Elec.	Fossil Fuel	Elec.	Fossil Fuel	Elec.	Fossil Fuel	Elec.	Fossil Fuel	Elec.
Units	%	%	%	%	%	%	%	%	%	%	%	%
Energy-Using Sectors												
Industrial (excluding feedstocks)	21.3	3.5	22.5	3.5	21.1	3.6	22.1	3.6	21.1	3.7	23.2	2.8
Transportation	13.9	0.1	13.9	0.1	13.8	0.1	14.0	0.1	13.7	0.1	13.3	0.1
Public	2.9	0.6	2.8	0.6	2.8	0.7	2.9	0.6	2.7	0.7	3.4	0.5
Residential	11.6	4.1	11.8	4.1	11.5	4.2	11.6	4.1	11.5	4.2	13.0	3.3
Fishing, Agriculture, and Commercial	2.2	1.5	2.7	1.3	2.1	1.6	2.6	1.3	2.1	1.5	3.1	1.2
Feedstocks	8.6	0.0	8.0	0.0	8.7	0.0	8.2	0.0	8.4	0.0	6.7	0.0
Final Energy Demand (excluding processing losses)	60.5	9.8	61.7	9.5	60.0	10.0	61.6	9.6	59.6	10.2	62.9	8.0
Processing Losses	6.3	23.3	6.2	22.5	6.3	23.6	6.2	22.7	6.3	23.9	7.0	22.3
Primary Energy Input	100.0		100.0		100.0		100.0		100.0		100.0	

*Oil price 1975 US$ per barrel Arabian light crude oil FOB Persian Gulf.
** excluding oil

WAES NATIONAL ENERGY DEMAND STUDIES
1985 CASE COMPARISONS

Country United Kingdom

Scenario Cases Compared	Scenario Variables: GDP (1)	Oil Price	National Policy Response	Energy-Use Category	Comparison GDP (elasticity) (2) units =
A	3.1%	$17.25		Primary Energy Input	0.68
B	1.8%	$17.25			
C	3.1%	$11.50		Primary Energy Input	0.78
D	1.8%	$11.50			

(1) GDP excluding oil

(2) Definition

$$\frac{\left(\frac{E_x - E_y}{E_x + E_y}\right)}{\left(\frac{Y_x - Y_y}{Y_x + Y_y}\right)}$$

(E_x = Energy use in Case x

Y_x = GDP in Case x)

WAES NATIONAL ENERGY DEMAND STUDIES
1985 CASE COMPARISONS

Country United Kingdom

Scenario Cases Compared	Scenario Variables: GDP[1]	Oil Price	National Policy Response	Energy-Use Category	Price Comparison (~~elasticity~~)[2] units =
A	3.1%	$17.25		Primary Energy Input	-0.059
C	3.1%	$11.50			
C	3.1%	$11.50		Primary Energy Input	-0.034
E	3.1%	$11.50			
B	1.8%	$17.25		Primary Energy Input	-0.033
D	1.8%	$11.50			

(1) GDP excluding oil

(2) Definition

$$\frac{\left(\frac{E_x - E_y}{E_x + E_y}\right)}{\left(\frac{P_x - P_y}{P_x + P_y}\right)}$$

(E_x = Energy use in Case x

P_x = Oil price in Case x)

NATIONAL ENERGY DEMAND PROJECTIONS

Country United Kingdom

Year 1972

WAES Case

Transportation

	(A)	(B)	(C)	(D)	(E)	(F)	(G)	
	Population/Auto	Distance/Auto/Yr.	Load Factor	Fraction Urban or Rural	Energy/Distance	Fossil Fuel	Electricity	
Urban Auto	4.29 (Urban and Rural)	14,320 km				622 (Urban and Rural)	Autos and Motorcycles	(1)
Rural Auto								(2)
	Activity Level	% Electrified	Load Factor	Fossil Fuel/ Activity Level	Electricity/Activity			
Urban Bus						48 (Urban and Intercity)	Including Taxis	(3)
Intercity Bus								(4)
Rail (Passenger)	13.0 x 10^9 passenger km	42.9 ≠	* 96.58				8.1	(5)
Air (Passenger, Domestic)	+ 2.168					49 (Domestic and International)	Including Freight	(6)
Air (Passenger, International)	+ 20.00							(7)
Ship (Passenger)								(8)
Rail (Freight)	21025	9.5 ≠				See under passengers		(9)
Air (Freight, Domestic)	+ 21.76					See under passengers (Domestic and International)		(10)
Air (Freight, International)	+ 658.7							(11)
Truck (Local)	88600 (Local and Intercity)					315 (Local and Intercity)		(12)
Truck (Intercity)								(13)
Ship (Freight, Domestic) $						19.4		(14)
Ship (Freight, International) $						232		(15)
Miscellaneous inc Military and Government						13.2		(16)
					Total	1488	8.1	

* Passenger km per loaded train km
\+ Scheduled Air Services
UNITS: 10^9 passenger km (Travel)
10^6 tonne km (Freight)
$ Freight and passengers
≠ By passenger or tonne km

NATIONAL ENERGY DEMAND PROJECTIONS

Country United Kingdom

Year 1972

WAES Case

Agriculture, Mining, and Construction

	(A)	(B)	(C)	(D)	(E)	(F)	(G)	(H)	
Agriculture, Forestry, and Fishing (ISIC Major Division 1)	Value Added	Energy/Value Added	Area Harvested	Energy/ Area	Physical Output	Energy/ Output	Fossil Fuel	Electricity	
Farm Machinery							**	**	(17)
Greenhouses	1450*	0.06					0	0	(18)
Other Agricultural Uses			72.14				79	13.8	(19)
Forestry (ISIC 12)			Included in Agriculture						(20)
Fishing Boats (ISIC 13)	130*	0.18			0.942		23.5		(21)
Other Fishing Use (ISIC 13)									(22)
Total Fertilizer (thousand tonnes)		N 909.2	P 441.8	K 404.0		Total	103	13.8	

	(A)	(B)	(C)	(D)	(E)	(F)	
Mining	Value Added	Energy/Value Added	Physical Output	Energy/Physical Output	Fossil Fuel	Electricity	
Handling and Internal Transportation	868	0.05	334	7.8	**	**	(23)
Other Energy Use					38.1	* 4.9	(24)
				Total	38.1	4.9	

	(A)	(B)	(C)	(D)	(E)	(F)	
Construction	Value Added	Energy/Value Added	Physical Output	Energy/Physical Output	Fossil Fuel	Electricity	
Handling and Internal Transportation	3541	0.02			**	**	(25)
Other Energy Use					59.3	1.73	(26)
				Total	59.3	1.73	
				Agriculture, Mining, and Construction Total	200	20	

** Gasoline & diesel included in transport
* Estimate

UNITS: 10^6 GJ, £(1972) (value added), 10^6 tonnes

NATIONAL ENERGY DEMAND PROJECTIONS

Country United Kingdom

Year 1972

WAES Case

Commercial and Public

Commercial	(A) Floor Area	(B) Heat Gain or Loss/Area	(C) System Efficiency	(D) Fossil Fuel	(E) Electricity	
Space Heat (Fossil Fuel)						(27)
Space Heat (Electricity)						(28)
Air Conditioning (Fossil Fuel)						(29)
Air Conditioning (Electricity)						(30)
		Energy/Area	Electricity/Area	219	93.7	
Lighting						(31)
Miscellaneous						(32)
Presently Unknown						(33)
			Total	219	93.7	

Public	Floor Area	Heat Gain or Loss/Area	System Efficiency	Fossil Fuel	Electricity	
Space Heat (Fossil Fuel)						(34)
Space Heat (Electricity)					45.5	(35)
Air Conditioning (Fossil Fuel)						(36)
Air Conditioning (Electricity)						(37)
		Energy/Area	Electricity/Area	332		
Lighting					6.7	(38)
Miscellaneous					Included in space heat above	(39)
Presently Unknown						(40)
			Total	332	52.2	
			Commercial and Public Total	551	146	

UNITS: 10^6 GJ

NATIONAL ENERGY DEMAND PROJECTIONS

Country United Kingdom

Year 1972

WAES Case

Industrial

	(A) Value Added	(B) Energy/Value Added	(C) Physical Output	(D) Energy/Output	(E) % Exported	++ (F) Fossil Fuel	(G) Electricity	(H) Feedstocks	
Iron and Steel (ISIC 371)	787*	.85	19.35	34.5	15.3%	632.6	35.3		(41)
Nonferrous Metals (ISIC 372)	289*	.20	1.59	36.0	18.9%	40.8	16.4		(42)
Paper and Allied (ISIC 34)	1460	.11	4.443	34.5	8.1%	142.8	10.5		(43)
Chemicals and Allied (ISIC 35 minus ISIC 353)	1300	.31	*12	33.4	26.2%	352.9	47.8	410	(44)
Mineral Products (ISIC 36)	167	.81			16.2%	125.7	9.71		(45)
Food and Related (ISIC 31)	1958	.11			8.8%	192.4	19.9		(46)
Transport Eqpt. (ISIC 384)	1687+	.07	$6x10^5$ vehicles	190GJ/vehicle	32.4%	*95.1	20.4		(47)
Cement	661**	.19	18.05	7.1	4.8%	120.2	7.49		(48)
Engineering	4501	.05			25.9%	180.3	*38.7		(49)
Misc. Manufacturing	3831	.09			15.1%	290.0	36.2	10.0	(50)
					Total	2172	242	420	

* Estimated

** Including bricks and pottery

\+ Excluding shipbuilding which is included in engineering

++ 1971 figures based on financial data

UNITS: £(1972)million, GJ/£(1972), 10^6 tonnes, GJ/tonne, %, 10^6 GJ

NATIONAL ENERGY DEMAND PROJECTIONS

Country United Kingdom

Year 1972

WAES Case

Residential

Population 55.9×10^6 Persons/Dwelling 2.85 Dwellings 19.64×10^6

Central Space Heating		(A) Dwellings	(B) Heat Loss	(C) System Efficiency	(D)	(E) Fossil Fuel	(F) Electricity	
Single Family	Fossil Fuels							(51)
	Electricity							(52)
Town or Row Houses	Fossil Fuels							(53)
	Electricity							(54)
Apartments	Fossil Fuels							(55)
	Electricity							(56)

		% Saturation (Fossil Fuels)	% Saturation (Electricity) *	Energy/Unit (Fossil Fuels)	Energy/Unit (Electricity)			
Room Heating			75**					(57)
Air Conditioning (Window)			} < 0.5					(58)
Air Conditioning (Central)								(59)
Water Heating			63					(60)
Refrigerators			69					(61)
Lighting			100					(62)
Cooking			41					(63)
Dishwashers			< 2					(64)
Freezers			6					(65)
Television			96					(66)
Other Key Appliances	– washing machine		64					(67)
Misc. Appliances	– kettle		60					(68)
Potential Appliances								(69)
					Total	1207	312.8	

* Per cent of domestic electrical consumers owning appliances

** Includes portable electric heaters

15.1 SUMMARY AND CONCLUSIONS

In recent years there have been a great many published projections of U.S. energy demand. It is not the intent of this report to increase that collection by one. Rather, the intent is to highlight specific changes in energy use in response to explicit sets of assumptions, or scenario cases, chosen for use in an international workshop.

Total energy demand in the United States in 1985, given the set of explicit workshop assumptions, is estimated to range from 80.8 to 106.9 Quads (1 Quad = 10^{15} British thermal units). Within this 25 percent variation in energy demand for the WAES cases (nearly 12 million barrels per day of oil equivalent), a few substantial shifts in the division of energy use among major economic sectors can be seen. For example, industrial energy use, as a percentage of the total, is projected to increase from some 27 percent in 1972 to about 30–32 percent in the 1985 cases—due to relatively greater efficiency improvements in other sectors as well as to higher growth rates in the industrial sector. Transportation energy use, on the other hand, declines from 32 percent of the total in 1972 to about 28 percent in the 1985 cases—reflecting the projected large improvements in energy efficiency in this sector in all 1985 cases.

Energy demand grows at average annual rates (1972 to 1985) of between 1.0 and 3.2 percent for the five WAES scenario cases. These figures can be compared with the recent growth of energy demand, from 1962 to 1972, of 4.0 percent per annum.

Energy use in some subsectors is projected to grow at significantly higher or lower rates than the total energy demand. For example, chemicals, iron and steel, petrochemical feedstocks, and residential electricity use are expected to be high energy growth sectors. In contrast, energy use by automobiles will decline from 1972 to 1985 because of the substantial improvements projected in mileage efficiency.

This report provides estimates of energy demand based on specific economic growth, oil price, and national energy policy assumptions. The estimates show both the economic activity and the energy efficiency associated with each of nearly 70 energy-use categories, as outlined in Chapter 1.

Figure 15.1 summarizes the sectoral totals for final energy demand in Quads. Case B—low economic growth coupled with a rising oil price and vigorous national energy policy assumptions—results in the lowest final energy demand, while Case E—high economic growth coupled with a falling oil price and restrained national energy policy assumptions—results in the largest final energy demand. There is a difference of 26 Quads between the highest and lowest 1985 energy demand estimates shown in Figure 15.1. (These same data are exhibited in tabular form in the worksheets at the end of this chapter.)

In this figure and in general throughout this report, processing losses are fuels used to transport, refine, and convert energy. The figures used are very approximate and are provided merely to show gross totals. Losses of 10 percent are assumed for fossil fuels and 70.25 percent for electricity. The latter figure results from an assumed average 35 percent generating efficiency and 85 percent transmission and distribution efficiency.

In addition to energy demand projections for WAES cases A through E, this report contains estimates for a Case C′—which differs from Case C only in its assumptions about national energy policy. Thus, the C-C′ pair allows for careful analysis of the effects of energy policies. Figure 15.2 shows the energy-use totals for Cases C and C′.

The vigorous energy policies used in Case C save 9.9 Quads, or about 10 percent of the energy used in the restrained energy policy case C′. Much of this saving occurs in the residential and commercial sectors. Further substantial energy reduction occurs in the transportation sector, even though both restrained and vigorous cases assume the current mandated, fairly stringent automobile efficiency standards.

The overall projections in this report are comparable to those of other recent studies of U.S. energy demand, even though the input assumptions may be quite different. In particular, the energy-use totals of the constant oil price and high economic growth cases in this range (Cases C′ and C) are fairly consistent with other recently published pro-

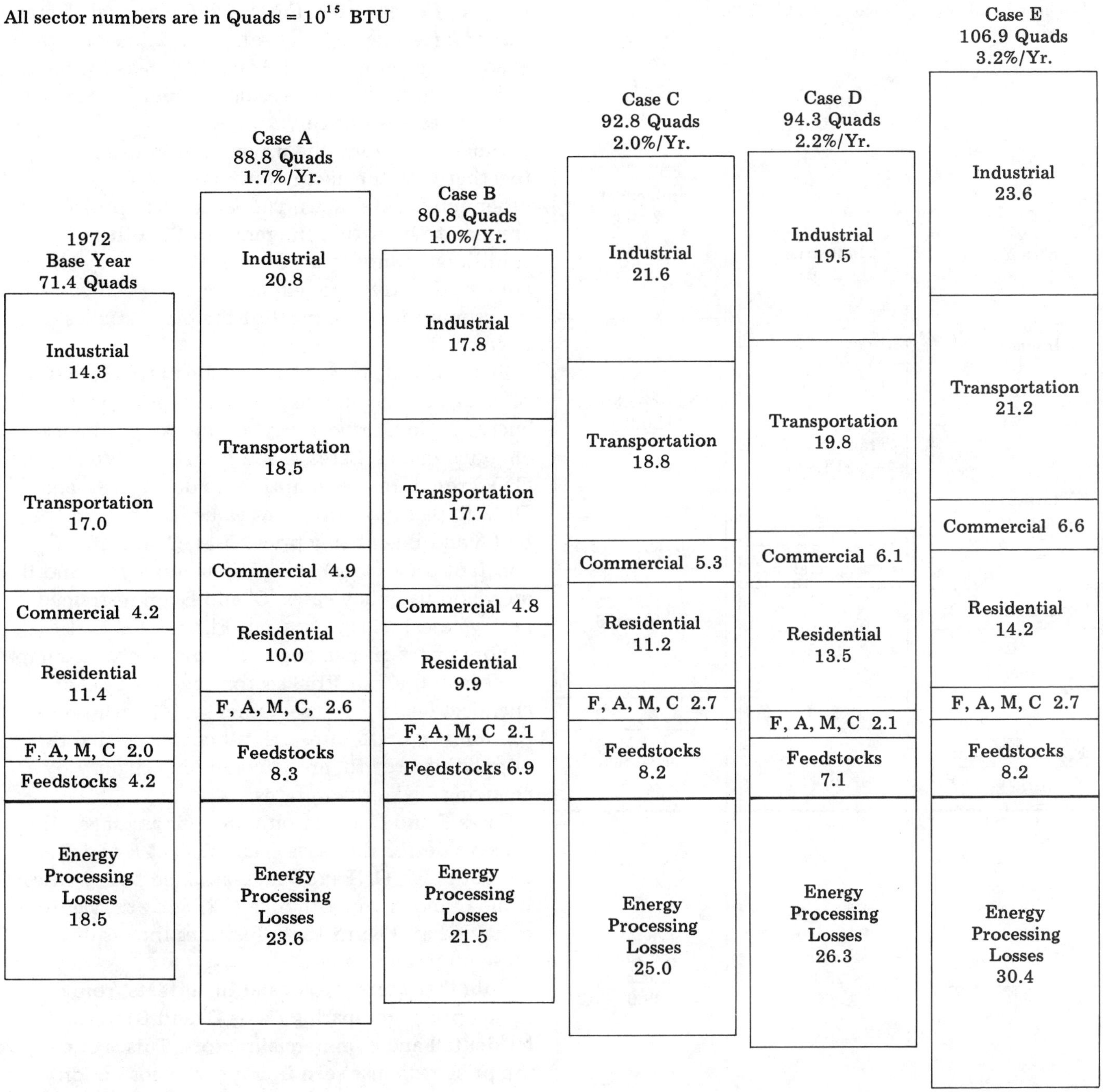

Commercial includes the public sector
F, A, M, C: fishing, agriculture, mining and construction

Figure 15.1 Summary of U.S. energy demand 1972 and Case A–E projections for 1985. All sector numbers are in Quads (1 Quad = 10^{15} Btu). Energy growth rates are averages, in percent per year, 1972–1985.

All sector numbers are in Quads = 10^{15} BTU

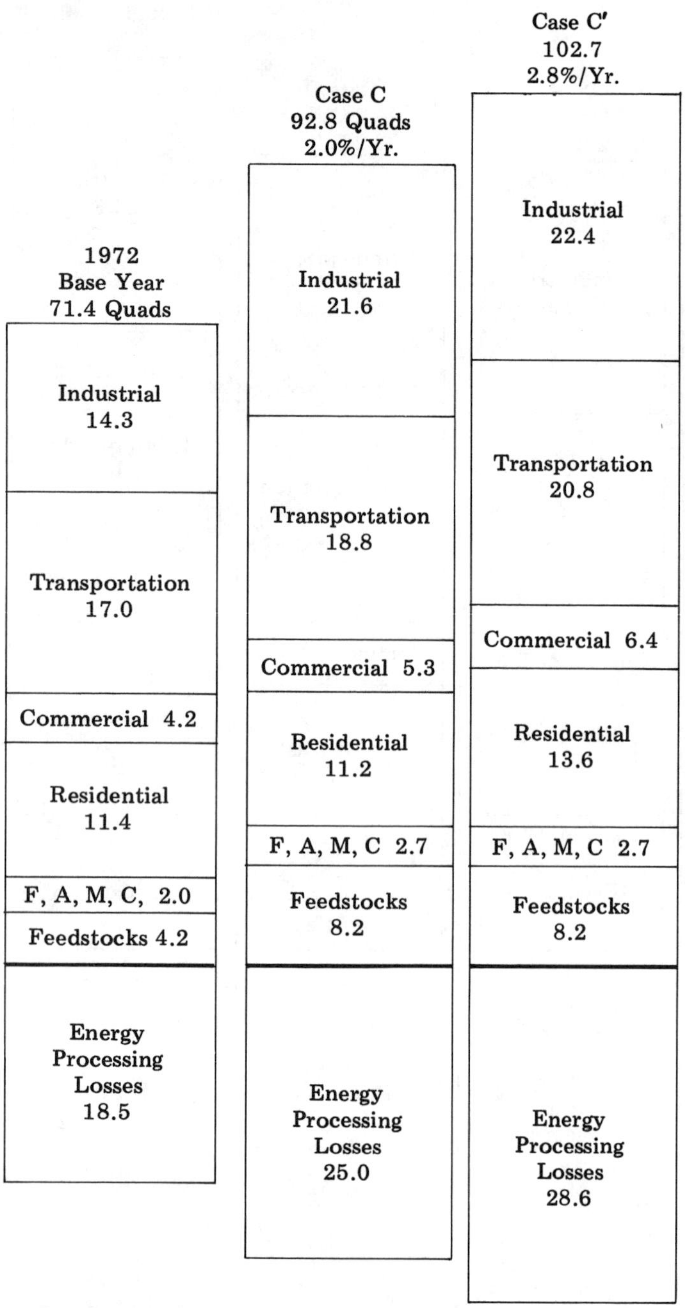

Commercial includes the public sector
F, A, M, C: fishing, agriculture, mining and construction

Figure 15.2 Case C′ projection of U.S. energy demand in 1985, as compared to 1972 data and the Case C projection. Sector numbers are in Quads, energy growth rates in %/yr., 1972–1985.

jections. For example, FEA's *1976 National Energy Outlook* (reference 9 in Section 15.2.2) estimates energy consumption of 99 and 93 Quads for the \$13 per barrel "business as usual" and "conservation" cases, respectively.

Table 15.1 shows the projections in this report together with the energy estimates from several other published studies. The WAES case projections are seen to lie within the range of the others. Additionally, note that the WAES case projections are significantly more explicit in the specification of assumptions than most of the other studies cited.

Figures 15.3, 15.4, and 15.5 illustrate the results of changes in economic growth, oil price, and energy policy, respectively, on the total primary energy demand. Cases A and B differ only in the GNP growth rate assumption, as do Cases C′ and D. Each pair has a different value for the price of oil (A and B are rising price cases, C′ and D are constant price cases) and national policy (A and B are vigorous policy cases, C′ and D are restrained policy cases). It is clear from Figure 15.3 that the major impact of changing economic growth assumptions is in the industrial sector, including petrochemical feedstocks use. Other sectors, whose economic growth rates are rather closely tied to population growth, are affected only slightly by changing GNP growth rates.

Cases E and C′ differ only in their assumed oil price values, as do Cases C and A. Yet E and C′ are both high GNP growth, restrained policy cases, while C and A are both high GNP growth, vigorous policy cases. Figure 15.4 illustrates the result of these oil price changes.

Note that there are substantial effects from a falling price (comparing Cases C′ and E) in the residential and commercial sectors. This is due to the price response seen in improving insulation and in the purchase of more efficient appliances. Other sectors show a lesser response to a decline in energy prices.

Finally, Cases C and C′ differ in their assumptions about national policy. Their comparison is discussed in Section 15.4.6, and Figure 15.5 shows the differences between the two cases.

It is important to note that no single study can exhaust all the important issues that affect energy demand. Even though several important topics are not considered in this report, the

Table 15.1
Comparison of United States 1985 energy demand projections

Study	Total Energy Demand (Quads)	1985 Oil Price (per barrel)[a]	Economic Growth Rate (%/yr.)	National Policy Response
Ford Foundation EPP				
Technical Fix	91.3	not explicit	not explicit	conservation
Historical Growth	116.1	not explicit	not explicit	historical
OECD				
$9	100.1	$9 (1972 $)	4.2[b]	not explicit
$6 (historical)	105.8	$6 (1972 $)	4.2[b]	not explicit
FEA National Energy Outlook				
Conservation	92.7	1975 real price	4.8	conservation
Business-As-Usual	98.9	1975 real price	4.8	business-as-usual
Historical	107.7	$8	4.8	historical
WAES				
A	88.8	$17.25	high (4.4)	vigorous
B	80.8	$17.25	low (3.2)	vigorous
C	92.8	1975 real price	high (4.4)	vigorous
C′	102.7	1975 real price	high (4.4)	restrained
D	94.3	1975 real price	low (3.2)	restrained
E	106.9	$7.66	high (4.4)	restrained

a. WAES and FEA oil prices are in 1975 $. WAES and OECD prices are FOB Persian Gulf. FEA prices are CIF United States.
b. 1971 to 1985.

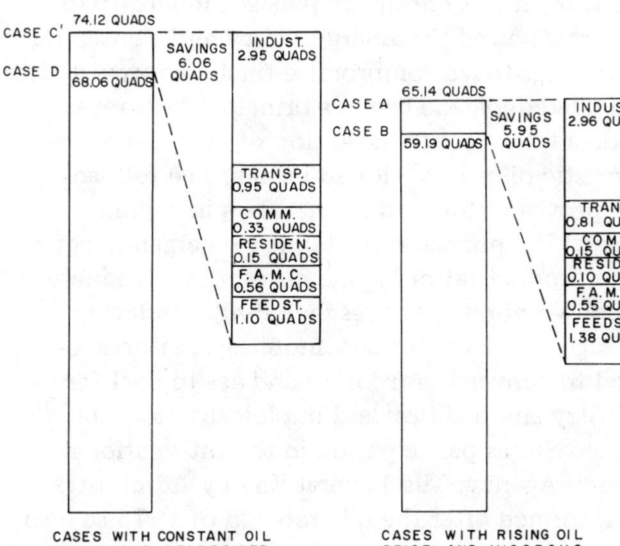

Figure 15.3 Results of changes in economic growth for final energy demand.

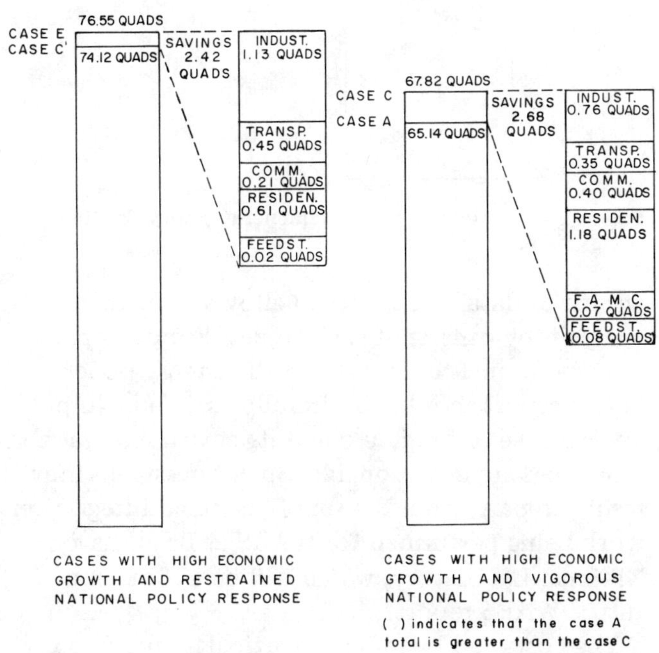

Figure 15.4 Results of changes in oil price for final energy demand.

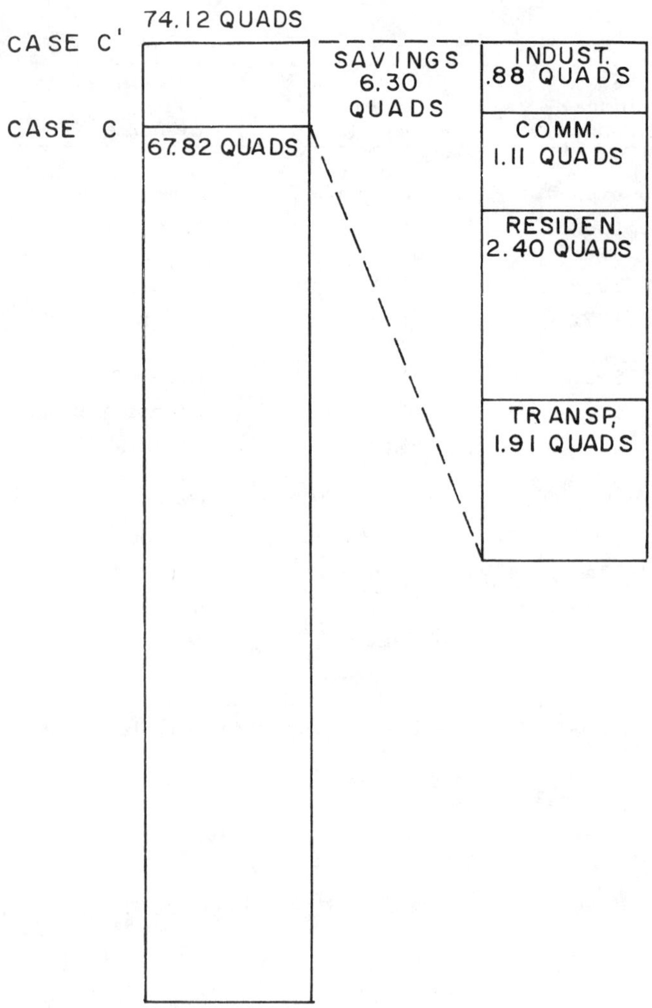

Figure 15.5 Results of changes in energy policy for final energy demand.

wealth of data herein should allow for further analysis, by others, of such topics. For example, the environmental impacts of the energy policies in this report should be carefully assessed. No policy is a wise or feasible one if its environmental consequences are unreasonable. Such assessment may result, in part, from the supply-demand integration work being performed for WAES at Brookhaven National Laboratory, which will be a topic in a future WAES report.

The costs and benefits of particular energy policies should be carefully weighed. It is possible that some policies designed to reduce energy demand may be economically or socially unacceptable. The potentials for new energy-utilizing technologies should also be assessed with an eye to the important issue of national priorities for research and development programs for energy conservation. Such an assessment could also include the many possible energy demand influences of increased solar usage—whether decentralized or centralized, whether for heating and hot water or for electricity generation.

These and other issues remain vital ones for careful study using a consistent, clear, explicit set of the basic relevant data. This report hopes to provide those basic data, along with an analysis of some of the primary determinants of energy demand in the next decade.

15.2 INTRODUCTION

15.2.2 The United States Energy Situation

The energy policy climate in the United States since 1973 has been marked by confusion. There has been little increase in widespread understanding of the long-term energy supply picture or of the long-term policy and technology options that are available. After nearly three years of debate, there is no general consensus regarding the adoption of long-term policy directions.

Although this confusion persists, in late 1975 Congress passed the Energy Policy and Conservation Act—a legislative compromise finally approved after extended debate. The act's principal features include a forty-month extension of controls over domestic oil prices with an initial price rollback and provision for gradual increases in ceiling prices;[1] the provision of standby emergency petroleum product rationing measures; the introduction of conservation measures including mandatory mileage standards for automobiles; measures related to conversion from oil and gas to coal for industry and utilities; and implementation of United States participation in the International Energy Agency. The Federal Energy Administration, formed after the oil embargo of 1973 to oversee emergency petroleum allocation procedures and oil price controls, will administer this new act.

With no major interests fully satisfied with the outcome of this legislation, it seems certain to be no more than a starting point for the evolution of national policies over the next few years. Today, several major problems persist, including: a slowdown in domestic oil and gas production; the con-

1. Controlled domestic oil prices are currently well below world oil prices.

tinuing debate over natural gas and oil pricing and their potential impact on future supplies and demands; a shortage of uranium enrichment and nuclear fuel reprocessing capacity; and indecision on the balance to be struck between energy and environmental requirements. Delays in the resolution of these and other matters have resulted in increased oil imports, exceeding domestic production for the first time, and during a period when total energy demand has been relatively flat.

This year, 1976, is the first full year of operation for the new Energy Research and Development Administration (ERDA), which began operation in 1975 with the consolidation of energy technology programs from several federal agencies. Already there has been notable progress toward the integration and balancing of an energy research, development, and demonstration program, and also toward an improved balance between support for new supply technologies and support for technologies that improve end-use efficiency. The most recent national research, development, and demonstration plan—ERDA 76-1 (reference 21 in Section 15.2.2)—gives high priority to conservation measures. This plan outlines the long-term problem of accomplishing a transition from oil and gas to the more abundant resources of coal, nuclear fuel, and renewable energy forms. ERDA proposes the development of a diverse set of technologies, providing a wide range of options to choose from as the energy system passes through this transition period.

Also operating in its first full year during 1976 is the Nuclear Regulatory Commission, which continues the regulatory functions of the former Atomic Energy Commission. This commission plays a central role in deliberations on safety and environmental protection—issues that will strongly affect the future of nuclear power in the United States.

In view of the expected resource and environmental constraints on United States energy supply development in the coming decade and the resulting fact that a substantial part of incremental demand will have to be met from oil imports, estimates of future energy demand assume critical importance for analysts and decision makers dealing with conservation policies, energy supplies, and the international oil market. Given the recent economic recovery in the industrialized nations, including the United States, it seems most likely that the demand for energy will increase in the next few years—although perhaps at a lower growth rate than prior to the embargo due to increased prices. The energy demand projections in this report represent detailed estimates of possible growth rates in the next decade under alternative, explicit assumptions. The relationships among price, policy, and capital turnover in energy-consuming equipment are outlined in a degree of detail not usually found in demand projections—a fact which makes this study particularly useful for determining the impacts of particular conservation policies on overall energy demands.

In a future report, these calculations of energy required at the point of delivery to the consumer will be integrated with the WAES estimates of possible United States energy supplies, and with other national energy demand and supply studies, into global assessments of feasible energy futures. In addition, the United States demand and supply projections will be integrated,[2] taking account of the global assessment (e.g., of import availabilities and export requirements) to incorporate detailed analyses of the United States energy processing and supply infrastructure. Such analyses include calculations of energy used in the transport, conversion, processing, and distribution of energy. (In this demand report, such energy processing losses are calculated only in a very approximate way, so that rough totals of primary energy demand may be compared.) The global and national supply-demand integration processes are described briefly in Chapter 1.

15.2.2 The Data

The many recent energy studies in the United States have proved valuable in the preparation of this report of energy consumption in 1972 and projected energy demands in 1985. A listing of the major sources of data for this report is given in this section.

Judgmental estimates were required in a few instances where reliable data were unavailable. The projections herein are a result of the best assessments of the Associates, together with related sup-

2. As an aid in its national integrations of supply and demand, the United States WAES members will be using the Brookhaven Energy System Optimization Model.

porting information from published sources and discussions with experts.

In order to minimize gross errors of judgment, many consistency checks were incorporated into the study. Each demand sector has been reviewed to ensure consistency in each scenario case. This review made use of published data, individual expertise, and a second econometric model.

The estimates in this study are of final *end-use* energy demand—energy required at the point of use by the consumer. The study focuses on efficiencies of energy used directly for goods and services: industrial production, transportation, home heating, air conditioning, etc.

The major sources of data for this report of United States energy demand are as follows:

1. A. D. Little, Inc. "An Impact Assessment of ASHRAE Standard 90-75 Energy Conservation in New Building Design." Report #C-78309 (December 1975).

2. William Ahern et al. "Energy Alternatives for California: Paths to the Future," Rand #R-1793-CSA/RF (December 1975).

3. M. Beller, ed. *Sourcebook for Energy Assessment.* Prepared for ERDA by the National Center for Analysis of Energy Systems, Brookhaven National Laboratory Associated Universities, Inc. (December 1975).

4. S. Berman et al. "California Electrical Energy Consumption." UCID #3487.

5. David Birch et al. *America's Housing Needs: 1970 to 1980*, Joint Center for Urban Studies, MIT and Harvard University (December 1973).

6. The Center for Auto Safety. *Mobile Homes* (New York: Grossman, 1975), pp. 106–107.

7. John G. Myers et al. *Energy Consumption in Manufacturing* (Cambridge, Mass.: Ballinger Publishing Co., 1974).

8. Otto Eckstein and Sara Johnson. "The Many Sources of Instability and the New DRI Long-Term Projections." In *Data Resources, U.S. Long-Term Review—Summer 1975* (Data Resources, Inc.).

9. Federal Energy Administration. *1976 National Energy Outlook.* FEA-N-75/713 (February 1976).

10. Ford Foundation Energy Policy Project. *A Time to Choose* (Cambridge, Mass.: Ballinger Publishing Co., 1974), Appendix A.

11. E. C. Hise. "Seasonal Fuel Utilization Efficiency of Residential Heating Systems." ORNL-NSF-E p. 82.

12. The National Petroleum Council. *Potential for Energy Conservation in the U.S. 1974–1978.* Library of Congress Catalog Card No. 74-79107 (September 10, 1974).

13. *Project Independence*, Vol. 1: *Residential and Commercial Energy Use Patterns 1970–1990* (November 1974).

14. *Project Independence*, Vol. 2: *Blueprint and Energy Conservation: Transportation Sectors* (November 1974).

15. *Project Independence*, Vol. 3: *Energy Conservation in the Manufacturing Sector 1954–1990* (November 1974).

16. R. G. Salter et al. "Energy Conservation in Non-Residential Buildings." Rand #R-1623-NSF (March 1975).

17. U.S. Bureau of Mines. *Minerals Yearbook* (1973).

18. U.S. Bureau of the Census. *Census of Manufactures.* MC72(SR)-6 (1972).

19. U.S. Department of Commerce. *Construction Review.* Issues of January 1974, December 1975.

20. U.S. Department of Commerce. *Statistical Abstract of the U.S. 1975.*

21. U.S. Energy Research and Development Administration. *A National Plan for Energy Research, Development and Demonstration: Creating Energy Choices for the Future.* Washington, D.C. (April 1976).

15.2.3 The Contributors

Mr. Paul S. Basile, a staff member of the WAES Secretariat, coordinated the writing and editing of the overall report, with major support in developing projected energy efficiency improvements from research assistants Mr. Richard Cheston and Mr. Miles Harbur. Workshop Associates from the Atlantic Richfield Company together with Dr. A. J. Finizza and Ms. M. S. Crist from the

Atlantic Richfield Company and Dr. Gerald Adams of Wharton Economic Forecasting Associates developed the economic activity levels for the 1985 cases. Mr. Robert Corkhill of Allied Chemical Corporation assisted in the review of the 1985 activity levels. Several U.S. Associates with special expertise reviewed and contributed substantially to many sections of the report.

In addition, Dr. Sam Berman of Lawrence Berkeley Laboratory and Stanford University contributed to the review of projections in the residential and commercial sectors. Dr. Robert Socolow of the Center for Environmental Studies at Princeton University assisted in the development of the base year information for the residential sector. Mr. Jon Gardner of Allied Chemical Corporation contributed to the review of the base year data and the industrial sector projected demands and efficiencies. Mr. Steven C. Carhart, of the Brookhaven National Laboratory (formerly a consultant to the Secretariat) provided invaluable direction and counseling during the initial stages of this study.

Not every contributor to this study has had the opportunity to approve every portion of this report, although they are all supportive of the approach. The study remains the responsibility of the United States Associates in WAES and the Secretariat, with the general approval and support of the United States Participants in the Workshop.

15.3 METHODOLOGY OF THE STUDY

The general methodology adopted by the Workshop for energy demand studies is described in Chapter 1. Here a description is given of the process by which the global scenario assumptions of Chapter 1 are translated into specific national assumptions for the United States.

Economic activity levels for the five WAES scenario cases are developed using the Wharton Annual and Industry Forecasting Model.[3] Such a long-range econometric model of the United States economy is useful for this purpose because its explicit mathematical structure captures the essence of economic relationships (i.e., what affects what and by how much), because its high level of detail permits analysis of specific alternatives, and because it explicitly specifies input assumptions.

An existing base case of the Wharton Model is modified to fit the alternative WAES scenarios. This base case assumes four years of above-average growth (1977 to 1980), a return to a full-employment economy in the early 1980s (with a delicate balance during 1977 to 1979 among real growth, low unemployment, and inflation), and an average annual growth rate of 3.6 percent in real GNP for 1977 to 1983.

The results of modifying this base case to fit the WAES cases are presented in Section 15.4.1.

15.3.1 Economic Growth Assumptions

The growth rates of real gross national product (GNP) used for the United States energy demand study were derived from the initial assumptions of "high" and "low" real world economic growth rates described in Chapter 1. In the United States, these global rates were translated into "target" national rates of 4.5 and 3.0 percent per annum, respectively, between 1977 and 1985. The target rates of 4.5 and 3.0 percent were achieved by modifying the base case run of the Wharton Model, which had a growth rate of 3.6 percent per annum from 1977 to 1983.

Modifying the Wharton base case was an iterative process. To raise the base 3.6 percent annual GNP growth rate to fit the WAES high growth cases, the steps taken included:

- Increasing government transfer payments
- Increasing growth of world trade
- Instituting a ten billion dollar tax cut in 1982
- Instituting a twenty billion dollar tax cut in 1983.

These changes in economic policy assumptions increased the GNP growth rate to the 4.5 percent target for the 1977 to 1985 period.

To lower the Wharton base case 3.6 percent annual growth rate to fit the WAES low economic growth cases, the changes in economic policy assumptions included:

- Instituting a very tight money policy in 1977–1979, with continued tight money policy thereafter

3. One description of the model is provided in R. S. Preston, "The Wharton Long Term Model: Input-Output Within the Context of a Macro Forecasting Model." *International Economic Review*, Vol. 16, No. 1 (February 1975).

- Reducing growth of world trade
- Reinstating the original growth in transfer payments
- Instituting no tax cuts in 1982 and 1983.

These steps succeeded in reducing the Wharton base case to the low growth target of 3.0 percent per annum for the 1977 to 1985 period.

15.3.2 Energy Price Assumptions

The WAES scenario case world oil prices of $11.50 constant, $11.50 rising to $17.25 by 1985, and $11.50 falling to $7.66 by 1985 (in constant 1975 United States dollars per barrel of Arabian light crude oil) were translated into United States energy prices for this study.

The Wharton Model has three energy price variables: (1) the price of imported oil, (2) domestic energy prices, and (3) the deflator for consumer gasoline and oil purchases. The first two variables are exogenous—their values are determined by the model user. The third variable is endogenous and depends only partially on the other two energy prices. Consequently, it does not change much from the high to the low oil price cases, but instead tracks the pattern of projected inflation in the overall economy. Table 15.2 shows the 1985 values for these variables in the three alternative WAES energy price cases.

In the Wharton Model, the input assumption for the price of imported oil was handled through a real price index variable for crude, nonfood imports, of which oil represents 85 percent. This index is in real terms (it excludes inflation) and therefore shows only the effect of WAES input assumptions.

Domestic energy prices, although not specified by WAES, were assumed to move with the price of imported oil. The model variable representing domestic energy prices is the price deflator for the mining sector, 68 percent of which is oil and gas, 14 percent coal, and the remainder metal mining and stone quarrying. In all WAES cases this deflator moved exactly with the price of imported oil.

These changes in energy price assumptions had a significant effect on the economy in the model, with higher energy prices reducing the 4.5 percent target real GNP growth rate. Consequently, monetary policy adjustments were used to retarget real growth to a 4.4 percent average annual growth rate in real GNP for the United States, 1977–1985, for the three WAES high growth scenario cases (A, C, and E).

In the low growth scenarios, the energy price assumptions also affected the economy in the model, causing it to diverge from the 3.0 percent growth target. Monetary policy adjustments retargeted growth to a 3.2 percent average annual growth rate in real GNP for the United States,

Table 15.2
Wharton Model price variables for three oil price cases

		Results in 1985[1]		
	1976 "Actual" Starting Point	Falling Price Case	Constant Price Case	Rising Price Case
Oil Price Input Assumptions (1975 constant dollars)	$11.50	$7.66	$11.50	$17.25
% Change from 1976		–33.4%	0%	50.0%
Price Index of Imported Oil[2]	305.0	203.0	305.0	457.0
% Change from 1976		–33.4%	0%	49.8%
Mining Sector Deflator	169.0	178.0	262.0	386.0
% Change from 1976		5.3%	55.0%	128.4%
Gasoline and Oil Consumption Deflator	211.4	314.2	319.4	327.0
% Change from 1976		48.6%	50.8%	54.8%
GNP Price Deflator	195.9	282.6	288.7	299.5
% Change from 1976		44.3%	47.4%	52.9%

1. The results for 1985 shown are for the high economic growth case. The input assumptions for the price of imported oil and the mining sector deflator are exogenous and thus are identical in the high and low economic growth cases. The other variables vary only slightly from the low economic growth cases.
2. This index is in real terms and thus excludes inflation.

1977–1985, for the two low growth cases (B and D).

15.3.3 Energy Policy Assumption

National energy policy is an important element in the assessments of changes in energy efficiencies for the WAES demand projections. Workshop scenario cases specify "vigorous" and "restrained" categories of national energy policy. In the United States study these labels have been interpreted so that existing practices and measures embodied in existing legislation such as the Energy Policy and Conservation Act (PL 94-163) of December 1975 are contained in the "restrained" policy cases. Practices and measures *in addition to* existing policies, designed to improve energy efficiencies, are contained in the "vigorous" policy cases.

The national policy response scenario variable does not, as a first approximation, affect economic activity levels. Therefore, the impact of policy, as discussed here, is on energy efficiencies alone.

In the analysis, alternative policies are explicitly separated into price and nonprice effects. That is, actions that would be induced by rising energy prices are included in price response calculations, while actions that would *not* be induced by rising energy prices are included in policy response calculations. Thus, all "vigorous" energy demand policies contained herein are nonprice policies designed to improve energy efficiencies.

In order to assess policy response independently of the other scenario variables, a Case C′ was defined and tested by the United States group. Case C′ differs from Case C only in that it is a "restrained" policy case, while Case C is a "vigorous" policy case. Both cases assume high economic growth and energy prices constant in real terms.

The specific "vigorous" and "restrained" policies selected in this study, and their calculated energy savings, will be discussed in detail in Section 15.4.

15.3.4 Economic Activity Levels and Technical Coefficients

Each category of energy demand is described by both an economic activity level and an efficiency of energy use or "technical coefficient." Energy demand is projected by separately estimating the values for the economic activity and technical efficiency of each category and then multiplying them. The estimates for each item depend directly on the scenario assumptions. Changing economic growth rates and oil prices affect economic activity levels; changing oil prices and energy policies affect technical coefficients.[4]

Scenario-dependent economic activity measures for the major energy-consuming categories are derived from the Wharton Model results. These categories include:

Total auto vehicle miles

Residential housing starts

Nonresidential construction

Farm production

Agricultural services, forestry, and fishing

Metal mining

Mining and quarrying of nonmetallic minerals

Contract construction

Industries: iron and steel
- nonferrous metals
- motor vehicles
- stone, clay, and glass
- food and beverages
- paper
- total manufacturing

Economic activity for those categories that cannot be derived from the Wharton Model are developed from published data, modified to ensure consistency with overall Wharton results. Most of these estimates are based on the 1974 Project Independence Report and other reports listed in Section 15.2.2. Wherever possible, alternative sources have been checked as an additional verification of activity level projections.[5]

The energy efficiencies—energy consumed per unit

4. It is postulated here that economic activity levels are not affected by national energy policies, and technical coefficients are not affected by economic growth rates. In reality, of course, one cannot completely separate energy efficiencies from economic growth, nor economic activity levels from national energy policy actions. In this study, however, such interactions are assumed to be second-order effects which do not alter the results markedly.

5. See, in particular, the review of the commercial and residential sectors, as cited in Sections 15.4.3 and 15.4.4.

of activity—are developed by making extensive use of existing literature projections, historical trends, and discussions with experts. Wherever possible, estimates are made of efficiencies of a specific process, so that each efficiency measure has some engineering meaning. For example, in order to determine the energy efficiency in the iron and steel industry, one should know the efficiencies—and potential improvements—in different types of blast furnaces and steel mills.

Technical coefficients are calculated for a base or reference case for 1985, Case C′, which closely approximates recent historical trends and is quite similar to the base case of several published U.S. energy studies. These coefficients represent the basic relationship between energy use and some measure of economic activity (Btu/lb of output, Btu/dollar of value added, Btu/sq. ft. of area heated, etc.). Technical coefficients are changed from case to case on the basis of changes in oil price and national energy policy assumptions.

The effects of price changes are included in the technical coefficients through consideration of new, more energy-efficient technologies which might be induced by higher energy prices. In some sectors, where technologies are not specifically identifiable, changes in coefficients (such as energy per dollar of value added) are correlated with past performance of such coefficients under price changes. This more technical approach differs from conventional analyses of price elasticity as a percentage change in energy purchased relative to percentage changes in price.

Both economic activity levels and energy efficiencies are affected by—in addition to economic growth, oil price, and energy policy—the rate of population growth. Population directly affects, for example, the total number of passenger-miles traveled, the number of housing starts, and growth in the service (commercial and public) sector. In this report, the United States Census Series II projections are used, resulting in a population of 234 million in 1985 (a 0.9 percent per annum rate of growth) for all cases.

15.4 1985 ENERGY DEMAND ESTIMATES

This section presents the overall results and analyses of the projections of U.S. energy demand for six different scenario cases.[6] In the sections that follow, the five WAES cases are compared and contrasted in several dimensions: economics, energy efficiencies, and energy consumption rates and totals in each sector (transportation, commercial, residential, and industrial). In addition, the effects of U.S. national energy policy are presented and discussed by use of a sixth scenario case C′ described above.

15.4.1 Economic Analysis

As described in Section 15.3, the base case of the Wharton Model was modified to achieve the high and low target growth rates for the United States. A summary of the results for the high growth scenarios is presented in Table 15.3.

A number of implications can be drawn from these scenarios. First, a 4.5 percent annual growth rate would be very hard for the United States to sustain for eight years regardless of energy price level; it must be viewed in reference to the historical average of 3.6 percent. Conventionally, long-run real growth is a function of growth in the labor force, in productivity, and in hours worked; but hours worked are declining, and population growth is assumed low. Thus the only sources of a high growth rate are high productivity and low unemployment. The outlook for improved productivity is not good, but even assuming a return to normal productivity gains, unemployment must drop to 3 percent to generate 4.5 percent growth in the 1980s. This is a very risky assumption.

In addition, although high growth rates are compatible with high oil prices, there are costs. Among these are higher money supply growth (which may portend high future inflation), lower real consumption, shifts away from durable goods consumption, and possible balance of payments problems.

Another possible consequence of high growth is the risk of failing to achieve the required low unemployment rate, with the result of further inflation and, for the high oil price case, economic

6. For a comprehensive documentation of sources, logic, and calculations contained in these energy projections, see "1985 U.S. Energy Demand Projections: Worksheets and Supporting Documentation," available for $10.00 per copy ($13.00 outside North America) from the WAES Secretariat, E40-167, Massachusetts Institute of Technology, Cambridge, Massachusetts, 02139, United States of America.

Table 15.3
Results to 1985 in the high economic growth scenarios

WAES Case	Case E	Case C	Case A
Oil Price ($/bbl)	$7.66	$11.50	$17.25
Real GNP (10^9 1958 $)	1207.5	1207.0	1207.9
Annual % Change, 1976-1985	4.4%	4.4%	4.4%
GNP Inflation, Index	282.6	288.7	299.5
Annual % Change, 1976-1985	4.2%	4.2%	4.8%
Money Supply (M1)	569.1	611.1	661.4
Annual % Change, 1976-1985	6.1%	7.0%	7.9%
Unemployment Rate	3.3%	3.0%	3.1%
Annual % Change, 1976-1985	4.5%	4.6%	4.6%

infeasibility.

A summary of the economic results for the low growth scenarios is presented in Table 15.4. They indicate that if money is tight in 1977 to 1979 and interest rates rise accordingly, the impact on inflation will be modest, while the loss of real growth and the increase in unemployment will be high.

Several major output indicators from Wharton Model runs for WAES cases are used directly in the energy demand analysis. These major indicators are shown for all cases in Table 15.5.

15.4.2 Transportation Sector

In 1972, the transportation sector represented 35 percent of final energy demand. The scenarios developed here project that proportion to decline to around 28 percent in 1985. The primary reason for the expected decline is the absolute decline in automobile gasoline consumption that will result from mandated fuel economy improvements. Other modes of transport also exhibit significant energy efficiency improvements.

Figure 15.6 shows the 1972 base year and 1985 energy demand projections for the total transportation sector and for automobiles, air travel, and freight movement.

Table 15.4
Results to 1985 in the low economic growth scenarios

WAES Case	Case D	Case B
Oil Price ($/bbl)	$11.50	$17.25
Real GNP (10^9 1958 $)	1083.0	1081.6
Annual % Change, 1976-1985	3.2%	3.2%
GNP Inflation, Index	287.4	298.4
Annual % Change, 1976-1985	4.35%	4.8%
Money Supply (M1)	437.0	479.9
Annual % Change, 1976-1985	3.0%	4.1%
Unemployment Rate	6.3%	6.7%
Annual % Change, 1976-1985	6.65%	6.6%

As shown in the figure, the proportion of total transportation sector energy used by automobiles is expected to decline from 54 percent in 1972 to around 45 percent in 1985—or from 9.13 Quads in 1972 to 8.0-9.3 Quads in 1985. This is due to a projected negative growth rate in automobile gasoline consumption for all cases except the falling oil price case. Energy demand for air travel, both domestic and international, is expected to increase from 8 percent in 1972 to about 12 percent of the 1985 transportation sector energy use. Even though the derived energy growth rates for freight movements are lower than GNP growth rates in all cases, energy used in moving freight increases from 17 percent in 1972 to 20 percent of the transportation sector energy use in 1985.

Figure 15.7 illustrates the results of energy policy, economic growth, and oil price changes on energy demand in the transportation sector. Note that the vigorous energy policies described here have a somewhat larger impact on energy demand than WAES case changes in oil price or economic growth.

Automobiles

The economic activity levels and the energy efficiency estimates for automobiles are shown in Table 15.6.

The measure of economic activity for autos—billions of vehicle miles traveled (VMT) for all cars on the road—was calculated by using the estimates of gasoline and oil expenditures (in constant dollars) available from the Wharton results. It should be noted that there are only small differences in VMT resulting from the energy price assumptions; that is, there is only a reduction of 14 billion miles (less than 1 percent) between Cases E and A even

Table 15.5
Major economic indicators used in the energy demand analysis: Growth rate per year of output from 1972 to 1985

Sector	Growth rate 1972–1985 (%/yr.)[1]				
	Case A	Case B	Case C	Case D	Case E
GNP	3.3	2.4	3.3	2.4	3.3
Nonresidential Construction (10^9 sq. ft./year)	1.45	1.28	1.40	1.15	1.41
Residential Construction: Private Housing Starts (10^6 dwelling units/year)	1.96	1.82	1.99	1.86	1.98
Autos and Parts	3.4	3.0	3.6	3.1	3.5
Gasoline and Oil	3.0	2.7	3.1	2.8	3.1
Contract Construction	3.3	(0.5)	2.4	(2.0)	2.3
Manufacturing	3.8	2.8	3.7	2.7	3.7
Iron and Steel	5.6	3.6	5.5	3.5	5.4
Nonferrous Metals	5.9	4.6	5.8	4.4	5.8
Motor Vehicles and Nonauto Transport Equipment and Ordnance	3.4	2.7	3.4	2.7	3.4
Stone, Clay, and Glass	3.9	2.4	3.6	2.1	3.5
Food and Beverage	2.3	2.4	2.5	2.6	2.5
Paper	3.2	2.5	3.2	2.5	3.2
Chemicals	3.2	2.5	3.1	2.4	3.1
Petroleum	3.4	2.7	3.1	2.4	2.9

Parentheses indicate negative growth rates.
1. All figures are in %/yr. except those for residential and nonresidential construction.

though energy prices are assumed to differ by 225 percent by 1985 between the two cases. While gasoline prices would not be expected to show the same relative change as crude oil prices—because of refining and distribution costs—the changes might be greater than those derived in the Wharton Model and used here. For example, the 1976 report by the United States Federal Energy Administration (reference 9 in Section 15.2.2) estimates an increase of 10 percent in real gasoline prices due to a 23 percent increase in domestic crude oil prices.

Another important assumption behind the estimated VMT figures in Table 15.6 is that this activity level is independent of energy policy. This assumption is not entirely valid since it is generally recognized that total vehicle miles traveled is affected by automobile efficiency (miles per gallon or Btu per mile) as well as by gasoline prices, disposable income, etc.

Even though the WAES case assumptions are usually different from and more explicit than the assumptions in other studies, the estimates of VMT shown in Table 15.6 are in general agreement with other estimates of 1985 VMT. For example, the Federal Energy Administration's *1976 National Energy Outlook* derived an estimate of 1443 billion VMT for their conservation scenario using assumptions similar to those of Case C, which has an estimate of 1445 billion VMT.

The energy efficiency estimates shown in Table 15.6 are calculated by estimating new car sales from the Wharton Model output of real expenditures for autos and auto parts and by assuming that new cars will meet the fuel economy standards required by the U.S. Energy Policy and Conservation Act of 1975 (P.L. 94-163). This law requires that all manufacturers meet a production-weighted standard of 18 miles per gallon (mpg) in 1978, 19 mpg in 1979, 20 mpg in 1980, and 27.5 mpg (4545 Btu per mile) in 1985. Using annual VMT and new car sales estimates from the Wharton Model, the fuel economy standards for new cars, and historical auto scrappage rates,[7] the average energy efficiency of all cars in use in 1985 was calculated as 5774 Btu per mile (21.7 mpg) for Case C,

7. Historical data and an analysis of this are provided in: Wayne A. Daniel, "An Estimation and Characterization of Annual Passenger-Car Populations," Research Laboratories Publication GMR-1898, General Motors Corporation, Warren, Michigan, May 30, 1975.

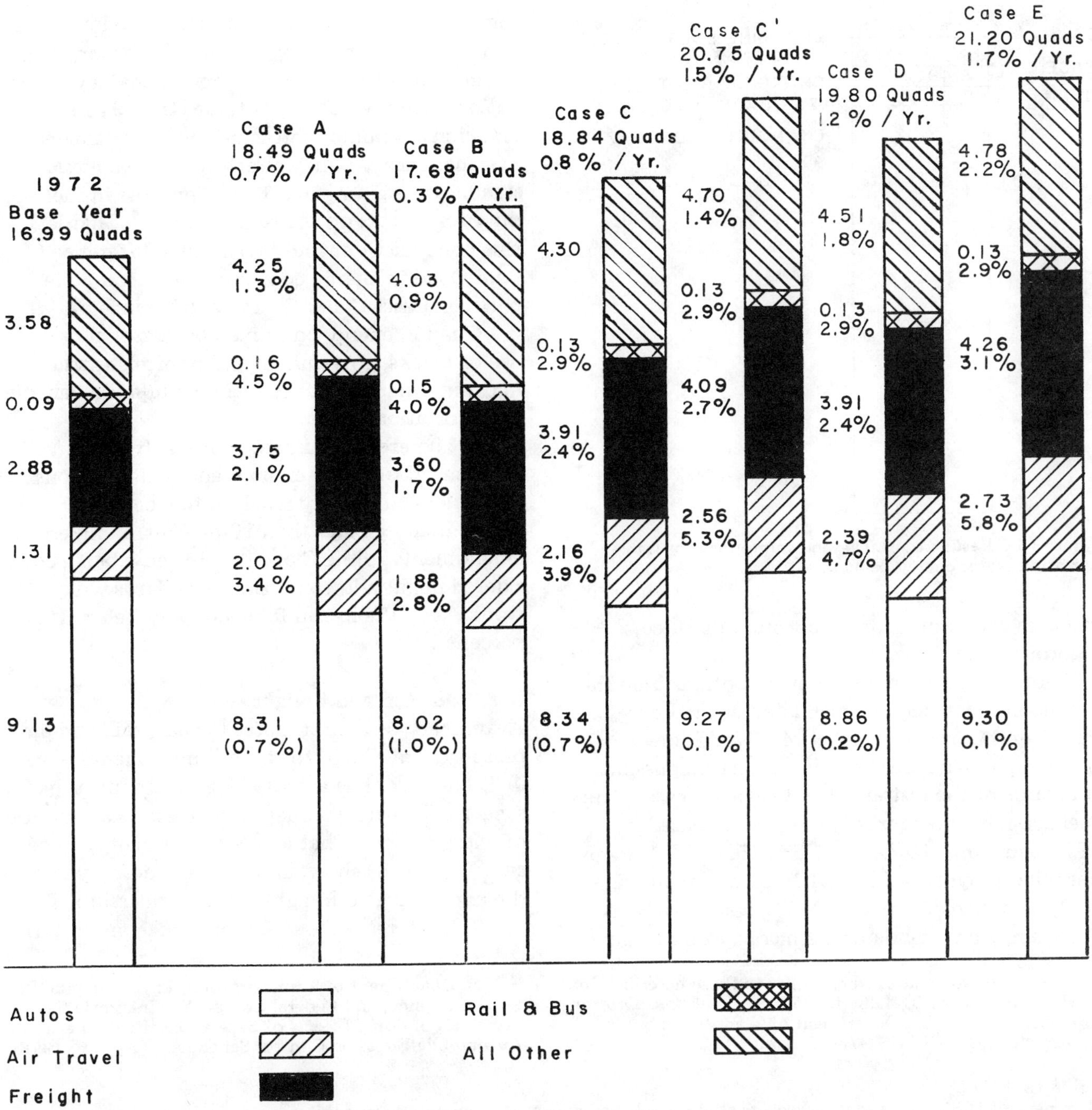

Figure 15.6 Transportation sector energy demand, 1972 and 1985, and energy growth rates, 1972–1985. The first number in each subsector is total demand in Quads; the second is the growth rate in %/yr. Parentheses indicate a negative growth rate.

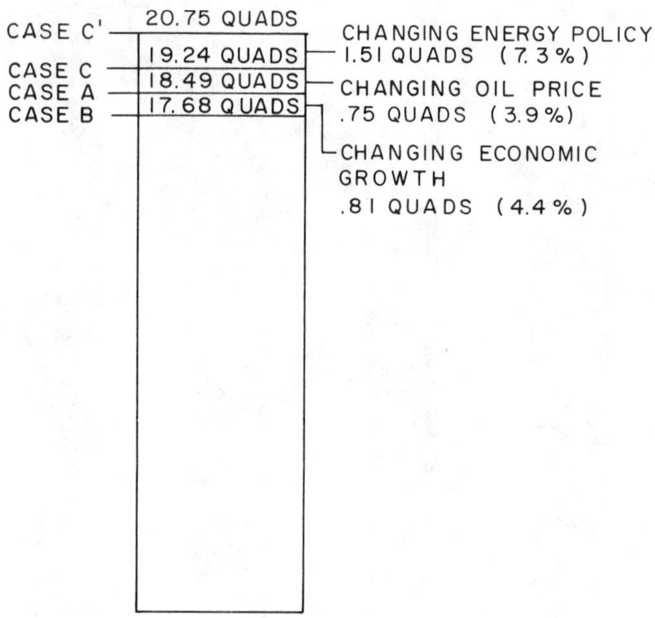

Figure 15.7 Results of changes in assumptions for transportation sector energy demand.

the high economic growth, constant oil price, vigorous policy case.[8]

Vigorous and restrained policies both assume the fuel economy standards mandated in the Energy Policy and Conservation Act of 1975. There is, however, a difference between the auto emissions standards of the restrained and vigorous cases. The fuel economy standards in the Act are based upon 1976 emissions standards of 1.5, 15, and 3.1 grams per mile of hydrocarbons (HC), carbon monoxide (CO), and nitrogen oxides (NO_x), respectively. As these emissions standards are made more stringent, they will reduce fuel economy, and the law therefore provides for an adjustment in this standard.

Thus the assumption used here is that a vigorous national energy policy continues the 1976 auto emissions standards through 1985. A restrained national energy policy strengthens the emissions standards, resulting in a 10 percent loss in fuel economy for all cars on the road in 1985.[9] That is, this policy action would result in a difference of 0.93 Quad in energy consumed by automobiles between Cases C and C′, and a difference of 0.4 Quad in energy consumed by personal use of trucks, government and nonfreight trucking, and other categories included in miscellaneous transportation.

The 10 percent estimate is probably conservative since the existing federally mandated emissions schedule would likely result in fuel economy losses substantially greater than 10 percent. Congressional amendments to the Clean Air Act now being considered might allow emissions standards which would keep the loss in fuel economy below 10 percent.

Bus, Rail, Air, and Freight

A summary of the estimates for bus, rail, and air passenger travel and for freight movements is shown in Table 15.7. The activity levels (billions of passenger-miles) are assumed to increase as a function of energy prices. That is, the higher energy price cases (A and B) show an increase in activity due to the expected gain in public transit ridership. This results in increased load factors which are in turn

8. Calculation procedures were similar to those described in: H. H. Gould and A. C. Malliaris, *Highway Fuel Consumption Computer Model*, U.S. Department of Transportation, Report No. DOT-TSC-OST-73-43, April 1974.

9. U.S. Department of Transportation, Environmental Protection Agency, and Federal Energy Administration. "Analysis of Some Effects of Several Specified Alternative Automobile Emission Control Schedules" (April 6, 1976).

Table 15.6
Summary of 1985 transportation sector activity levels and energy efficiencies: Automobiles[1]

WAES Case	A	B	C	D	E	1972
Oil Price	Rising to $17.25	Rising to $17.25	Constant at $11.50	Constant at $11.50	Falling to $7.66	
National Policy Response	Vigorous	Vigorous	Vigorous	Restrained	Restrained	
National GNP Growth Rate (%/yr.)	4.4	3.2	4.4	3.2	4.4	
Activity Level (10^9 vehicle miles)	1435	1383	1445	1392	1449	986
Energy Efficiency (Btu/mile)	5790	5798	5774	6365	6417	9260

1. Estimates of energy demand for the personal use of trucks are included in the miscellaneous category of the worksheets.

Table 15.7
Summary of 1985 transportation sector activity levels and energy efficiencies: Air travel, freight, rail travel, and bus[1]

WAES Case	A	B	C	D	E	1972 Base Year
Oil Price	Increase to $17.25	Increase to $17.25	Constant at $11.50	Constant at $11.50	Decrease to $7.66	
National Policy Response	Vigorous	Vigorous	Vigorous	Restrained	Restrained	
National GNP Growth Rate (%/yr.)	4.4	3.2	4.4	3.2	4.4	
Activity Level (10^9 passenger-miles or 10^9 ton-miles)						
Air Travel	387	360	411	382	416	169
Freight	2547	2445	2556	2455	2613	1880
Bus and Rail	107	100	83	80	83	57.5
Energy Efficiency (Btu per passenger-mile or per ton-mile)						
Air Travel	5222	5221	5258	6242	6555	7751
Freight	1473	1473	1530	1591	1630	1532
Bus and Rail	1536	1537	1599	1550	1611	1696

1. The transportation sector worksheets show separate estimates for domestic and international air travel; for rail, local and intercity truck, air, and ship freight; for urban bus, intercity bus, and passenger rail.

reflected in the energy efficiency estimates. The basic source of data for Table 15.7 and the other tables in this section is the 1974 Project Independence Report (reference 14 in Section 15.2.2).

Air travel is expected to increase sharply over the next few years, and the increased load factors are expected to result in gains in energy efficiency. Moreover, a vigorous energy policy could involve encouraging the Civil Aeronautics Board to remove certain restrictions so that the number of flights could be reduced. It is estimated that this change in regulatory policy would save about 0.4 Quad of energy, as indicated in Table 15.8.

Freight modes are summarized in Table 15.7 on the basis of ton-miles. Energy price increases are expected to increase rail freight shipments. This modal shift is described in the detailed worksheets in Section 15.5. The "vigorous" energy policy assumption is that the Interstate Commerce Commission would ease its restrictions on railroad cars crossing interstate boundaries. Such deregulation would save about 0.18 Quad.

15.4.3 Commercial and Public Sector

The potential for improved energy efficiency in

Table 15.8
Effect of national policy response: Transportation sector

Policy Type	Vigorous Response Measure (Case C)	Restrained Response Measure (Case C′)	End-Use Energy Savings (Case C′ – Case C)
Auto emissions standards	1976 standards	More stringent standards	0.93 Quad (autos) 0.40 Quad (other vehicles)
Government regulation of airline load factors	C.A.B. regulation to reduce the number of flights	No additional regulation	0.40 Quad
Government regulation of railroads	I.C.C. to ease its restrictions on railcars crossing interstate boundaries	No change in present regulatory policy	0.18 Quad
			Total transportation energy savings: 1.91 Quads[1]

1. This represents 19 percent of the overall C′ – C savings.

commercial buildings is high. The most effective of the various imaginable energy-saving measures are most easily implemented in new buildings. This is significant because the commercial sector is, in these projections, a high growth sector; almost half of the total 1985 commercial floor area will have been constructed after 1972. Government energy policies can be quite effective in setting standards for new buildings, for new heating, ventilating, and air conditioning (HVAC) systems, and for new appliances and office equipment. Such policies for new buildings are included in the vigorous policy cases here, and their potential impact on energy use assessed.

Reductions in energy use in existing buildings are more difficult to achieve. They depend greatly upon many individual actions to improve efficiency through the retrofitting of old buildings. There is little that policy can do in this area other than public encouragement of energy-saving practices (through advertising, etc.) and tax incentives.

Figure 15.8 shows the energy demand and energy growth rates for the major activities in the commercial sector. Fossil fuel space heating is, and will continue to be, by far the largest component of commercial sector energy use.

Figure 15.9 illustrates the results of energy policy, economic growth, and oil price changes on commercial sector energy demand. Note that energy policies can have a fairly significant impact on energy demand in this sector (given the other case assumptions), while the variations in oil price and economic growth used here have relatively little effect.

In a careful review of this sector and the residential sector, the numbers in this report were compared with several recent published estimates such as the FEA's *1976 National Energy Outlook* (ref-

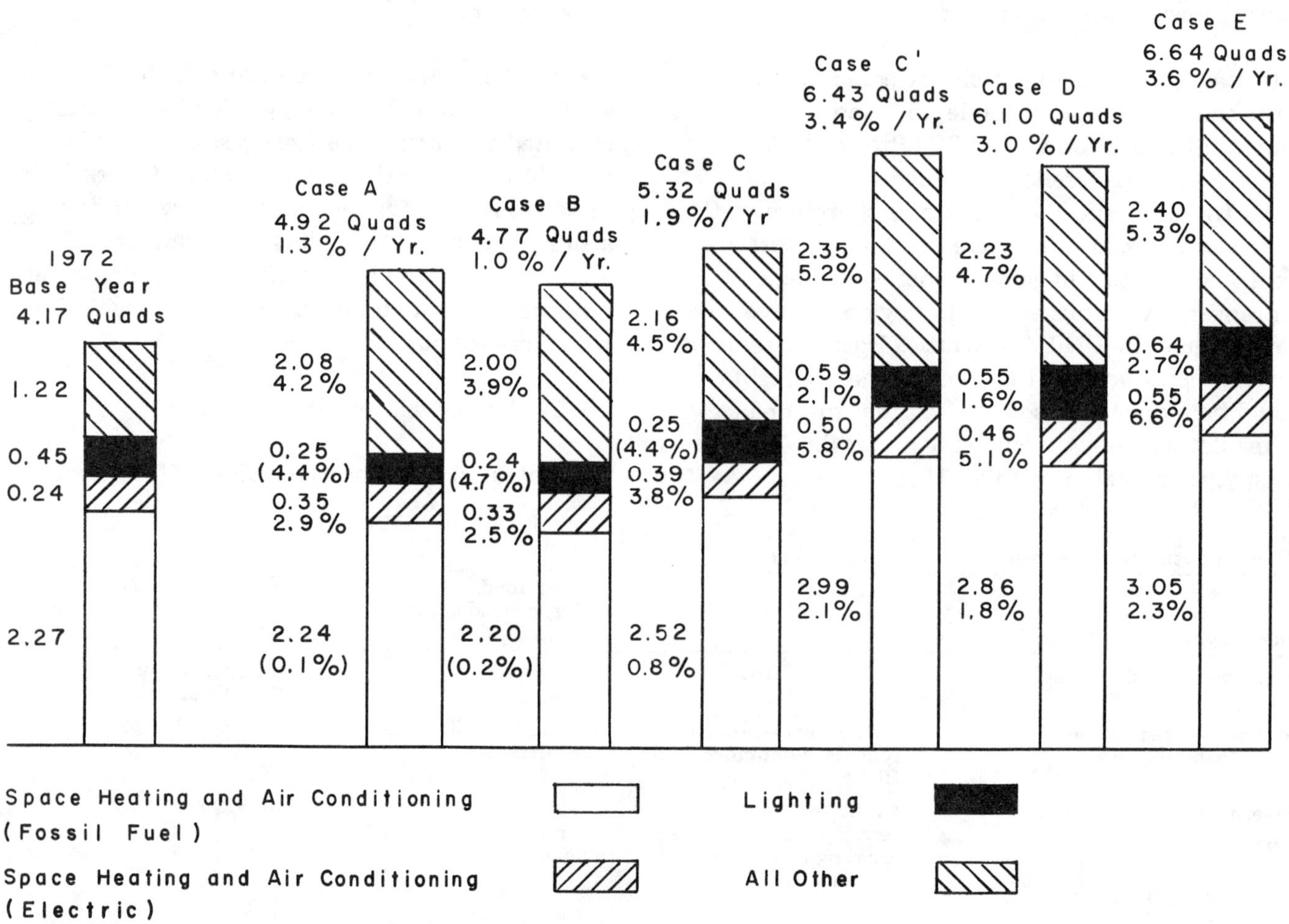

Figure 15.8 Commercial sector energy demand, 1972 and 1985, and energy growth rates, 1972–1985. The first number in each subsector is total demand in Quads; the second is the growth rate in %/yr. Parentheses indicate a negative growth rate.

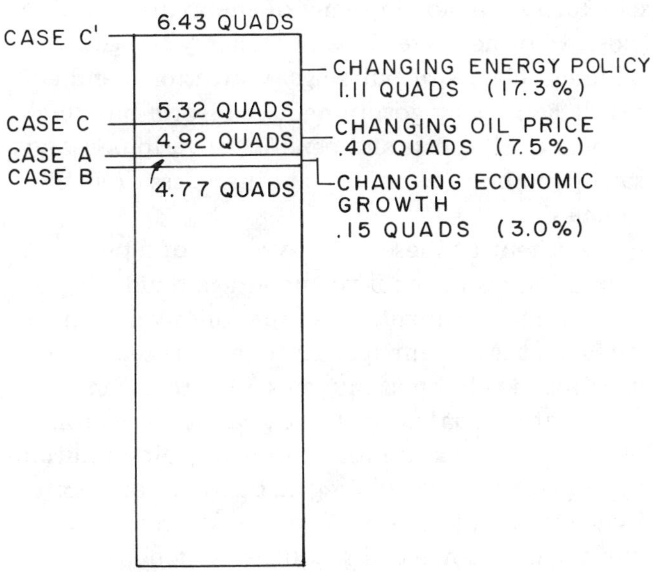

Figure 15.9 Results of changes in assumptions for commercial sector energy demand.

erence 9). The results showed that the high growth case activity levels were just slightly (3 to 4 percent) higher than the mean value of other projections. Technical coefficients in this study were found to be in substantial agreement with published estimates. Energy consumption in the restrained cases is a bit higher (about 5 percent) than some of the recent estimates, while energy consumption in the vigorous cases is somewhat lower.

Table 15.9 contains the major activity levels and a set of energy efficiency estimates for the commercial sector. Note that there is some evidence that the commercial sector grows rapidly in the high price cases to compensate for the lower growth in energy-intensive activities. That is, floor area for Case A (high price) is slightly greater than floor area for Case C (constant price), though both are high growth, vigorous policy cases.

In both the commercial and residential sectors, total dwellings are disaggregated into categories of "old" (built before 1972) and "new" (built after 1972). The activity level assessment in each sector defines the *rate* at which new, presumably more efficient, buildings are built, resulting in changing percentages of old and new buildings existing in 1985. As the GNP growth rate increases, the stock turnover rate is higher and a greater percentage of new, more efficient stock appears in the market. The result is that the *average* energy efficiency of all stock (old and new) is higher in high growth cases than in low growth cases.

Identical commercial buildings in the same city can show large differences in energy consumption due to differing styles of operation and types of

Table 15.9
Summary of 1985 commercial sector activity levels and energy efficiencies

WAES Case	A	B	C	D	E	1972 Base Year
Oil Price	Rising to $17.25	Rising to $17.25	Constant at $11.50	Constant at $11.50	Falling to $7.66	
National Policy Response	Vigorous	Vigorous	Vigorous	Restrained	Restrained	
National GNP Growth Rate (%/yr.)	4.4	3.2	4.4	3.2	4.4	
Activity Level (10⁹ sq. ft. of floor area)	36.6	35.1	36.1	33.9	36.2	23.4
Energy Efficiency						
Space Heat (FF)						
Heat Loss[1]	4.30-5.68	4.30-5.68	4.95-6.40	6.18-7.10	6.46-7.10	7.10
System Efficiency	0.70-0.80	0.70-0.80	0.70-0.80	0.70-0.75	0.70-0.75	0.70
Space Heat (Electric)						
Heat Loss[1]	4.30	4.30	4.95	6.18	6.46	7.10
System Efficiency	1.0	1.0	1.0	1.0	1.0	1.0
Air Conditioning (FF) System Efficiency (COP)[2]	2.2	2.2	2.2	2.0	1.8	1.8
Air Conditioning (Electric) System Efficiency (COP)[2]	2.5	2.5	2.5	2.2	2.0	1.8
Lighting[1]	0.69	0.69	0.69	1.63	1.77	1.92

1. 10^4 Btu/sq. ft.
2. Coefficient of performance.

occupants. There is, however, great potential for reduced energy usage through policies designed to improve building operation. For example, one vigorous energy policy proposed here requires the installation of automated thermostat controls to regulate building temperature: the building temperature would be automatically maintained or adjusted to predetermined temperature settings by a mechanical device or a computer rather than by manual temperature adjustments. Another vigorous policy results in reduced average lighting levels through mandated levels of power expenditure per unit area. A third area of vigorous energy policy for new construction would be to set higher efficiency and insulation standards for ancillary equipment such as for water heating, cooking, and refrigeration. These policies, and their potential energy savings, are outlined in Table 15.10.

Government policies affecting both new and old commercial buildings focus on encouraging changes in building use and operating procedures that may be cost-effective only over a long period of time. There are a number of economic incentive policies available to government for use in the commercial sector, such as sales taxes on energy, revised rate schedules, peak-load pricing of energy, or taxes (positive or negative) based on energy budgets for different classes of buildings. Educational and advisory actions by government to inform building owners of the costs and benefits of various energy-saving methods constitute another class of viable policies.

The intent of these incentives and of a public information campaign is to encourage building owners and occupants to adopt conservation measures in at least four specific areas: the reduction of lighting, ideally by as much as 50 percent; internal temperature control with the goal of moderating building temperature for heating and air conditioning; operational scheduling of building temperature by automated temperature controls; and equipment maintenance and feedback control.

In all, the vigorous policies cited here could reduce energy by approximately 1.1 Quads, or by almost 20 percent.

15.4.4 Residential Sector

Energy use in the residential sector is quite responsive to higher oil prices. Higher fuel and electric

Table 15.10
Effect of national policy response: Commercial sector

Policy Type	Vigorous Response Measure (Case C)	Restrained Response Measure (Case C′)	End-Use Energy Savings (Case C –Case C′)
Space Heating and Air Conditioning			
Relating to heat loss (gain)	Require automatic thermostat controls for new construction	No regulation standards	0.51 Quad for space heating and 0.07 Quad for air conditioning
	Improved insulating glass standards for new construction	Maintain existing standards	
	Government incentives to moderate temperature settings for heating and cooling	Maintain voluntary guidelines	
Relating to system efficiency	Government standards for equipment maintenance and feedback control	Maintain voluntary guidelines	
Lighting	Mandatory standards of proper levels of illumination through education and inspection (both old and new buildings)	Maintain voluntary guidelines	0.34 Quad
Miscellaneous	Standards for efficiency and insulation of new refrigerators, freezers, ovens, water heaters	No policy standards	0.19 Quad
	Require electric ignition units to replace pilot lights for new units		Total commercial energy savings: 1.11 Quads[1]

1. This represents 11 percent of the overall C′–C savings.

bills are expected to encourage householders to invest in improved insulation and to make purchasers more cost-conscious when they buy appliances.

Residential electricity use has grown at a 7 to 7.5 percent rate over the past several years. In 1974, almost 50 percent of all new household construction had electric space heating.[10] One of the important factors accounting for this new trend is the continuing curtailment of natural gas hookups. For the period 1972 to 1985, it is estimated that 45 percent of the new housing starts will have electric space heating. Overall, some 20 percent of all occupied dwellings are projected to have electric space heating in 1985.

Figure 15.10 shows residential sector energy use and energy growth rates. Space heating in single family homes will continue to be the dominant user of energy. Central air conditioning is seen to grow quite rapidly, reflecting the expected shift to electricity use in residences.

Figure 15.11 shows the results of changes in energy policy, oil price, and economic growth on residential sector energy demand. Note that energy

10. "26th Annual Electrical Industry Forecast," *Electrical World* (New York: McGraw-Hill), September 15, 1975.

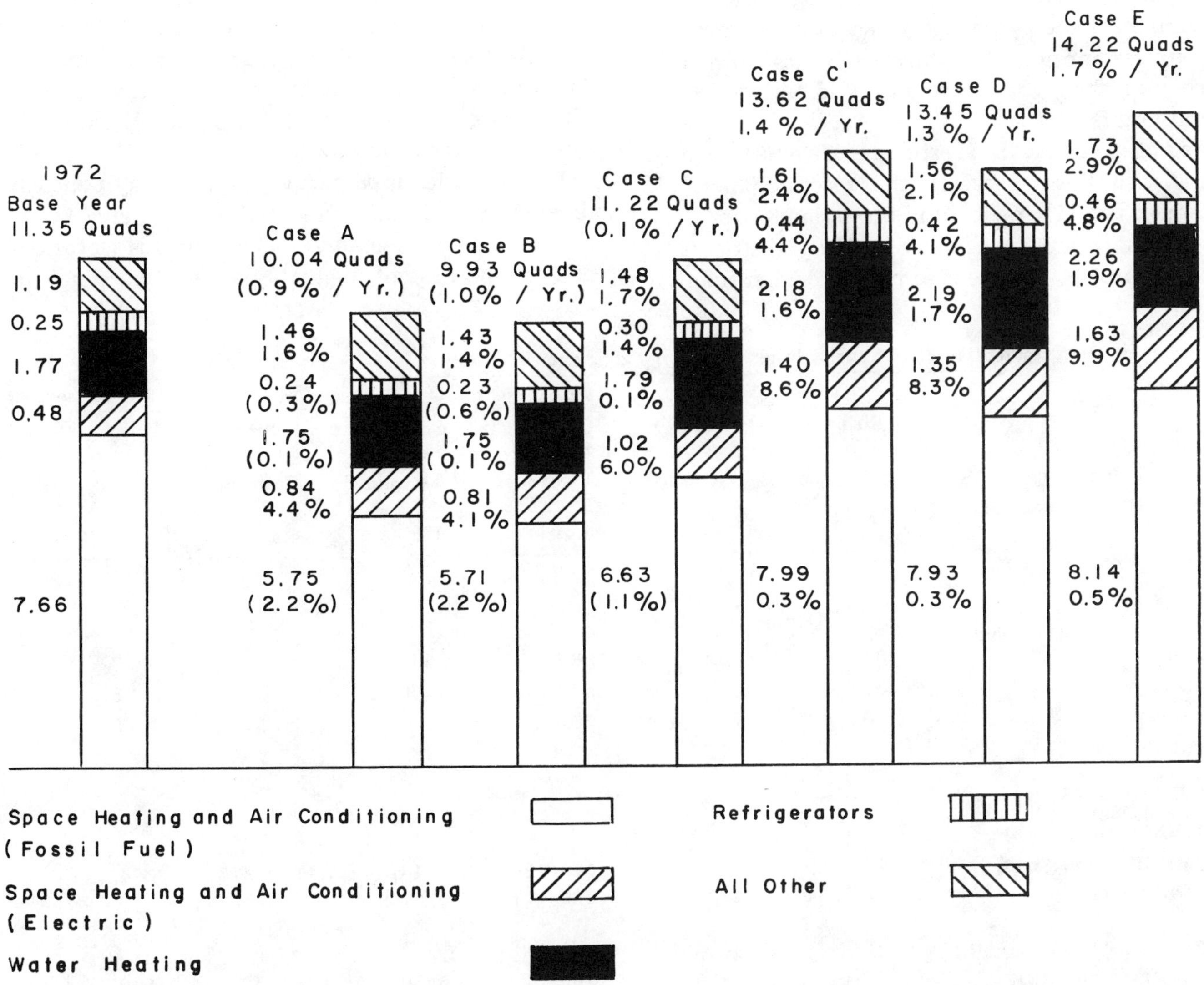

Figure 15.10 Residential sector energy demand, 1972 and 1985, and energy growth rates, 1972–1985. The first number in each subsector is total demand in Quads; the second is the growth rate in %/yr. Parentheses indicate a negative growth rate.

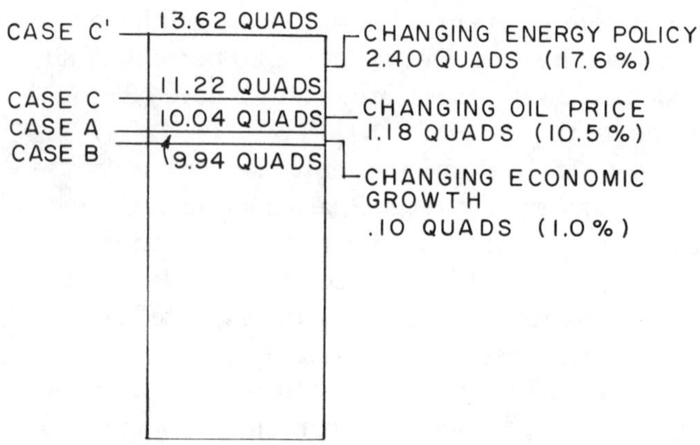

Figure 15.11 Results of changes in assumptions for residential sector energy demand.

policy and oil price changes provide significant potential for energy savings. Changes of economic growth assumptions from high to low, as defined here, offer insignificant reductions in energy demand in this sector.

As noted earlier, these estimates have been compared with those of several other recent studies. There was good agreement in the estimates for single-family and multifamily housing. The overall housing projection here is somewhat higher than that of other sources because of the larger projections for mobile homes in this report.

Residential dwellings are divided into three categories: single-family dwellings, mobile homes, and apartments. The number of housing starts each year is given by the Wharton Model for each scenario case. An annual removal rate (encompassing demolitions plus conversions to other types of dwellings) of 38 percent of new housing starts, or 1 percent of the total housing stock, would lead to the number of dwellings shown in Table 15.11 for each case. Table 15.11 also contains energy efficiency estimates.

Economic activities for appliances are measured by saturation levels. The saturation levels for appliances range to a high of 165 percent for televisions (i.e., two-thirds of all occupied dwellings in 1985 will have two television sets). Growth rates for clothes dryers, dishwashers, and central air conditioners are projected to increase rapidly as families' purchasing power expands.

The potential impact of vigorous energy policies on energy use in the residential sector is great; but identifying viable energy policies for this sector is difficult because of the highly decentralized nature

Table 15.11
Summary of 1985 residential sector activity levels and energy efficiencies

WAES Case	A	B	C	D	E	1972 Base Year
Oil Price	Rising to $17.25	Rising to $17.25	Constant at $11.50	Constant at $11.50	Falling to $7.66	
National Policy Response	Vigorous	Vigorous	Vigorous	Restrained	Restrained	
National GNP Growth Rate (%/yr.)	4.4	3.2	4.4	3.2	4.4	
Activity Level (millions of units)						
Total Number of Occupied Dwellings	86.4	85.2	86.6	85.6	86.5	66.7
Single-Family Homes	52.4	51.7	52.5	51.9	52.5	44.8
Apartments	26.7	26.3	26.8	26.4	26.7	19.3
Mobile Homes	7.3	7.2	7.3	7.2	7.3	2.7
Energy Efficiency Space Heating (heat loss in therms/yr.)						
Single-Family Homes	723–482	723–482	786–524	874–743	874–813	874
Apartments	478–384	478–384	520–417	578–491	578–509	578
Water Heating						
Fossil Fuel (10^6 Btu/unit)	22.4	22.4	22.9	28.5	29.87	31.5
Electric (kWh/unit)	4710	4710	4905	5450	5514	5627
Air Conditioning (kWh/unit)	1830	1830	1840	2200	3162	3546
Refrigerators (kWh/unit)	711	711	889	1290	1316	1084

of much of the potential conservation in residences. The most significant conservation measures involve decisions made by millions of individual householders.

Table 15.12 summarizes the restrained and vigorous policies selected. Most of these policies are directed toward new construction and new appliances. There are no policy levers to guarantee thermostat setbacks or improvements in the energy efficiencies of existing dwellings. The key to household behavior with regard to a given measure is the full price and cost effectiveness of that measure (i.e., the relationship of initial cost to future savings). Otherwise, residential policies include public information campaigns to alert and instruct consumers regarding conservation measures and their potential, and tax incentives to encourage those measures that involve capital outlays for changes in building components.

These vigorous residential conservation policies were selected for their high *potential* to reduce consumption. It should be remembered, however, that many other measures might be implemented whose *actual* reduction of energy use could be more significant than that of the policies cited. Energy conservation is more likely to be achieved through many small savings of energy than through a few big ones.

The efficiency of fossil fuel heating systems can be improved by as much as 60 percent by better load matching (i.e., reducing the heating system capacity in order to conform with the size of the dwelling unit); improved burner efficiency, including the replacement of pilot lights with electric ignition; and proper maintenance of the system. The efficiency of electric heating systems can be

Table 15.12
Effect of national policy response: Residential sector

Policy Type	Vigorous Response Measure (Case C)	Restrained Response Measure (Case C′)	End-Use Energy Savings (Case C′–Case C)
Heating System Efficiency Standards			
Fossil Fuel System	Require load matching of system with dwelling unit needs	No definitive FHA standards	1.66 Quads total savings for space heating
	Require electric ignition for all fossil fuel systems		
	Publicity for guidelines for system maintenance		
Electric System	Require minimum standards of efficiency based on heat pump system efficiencies	Rely on the marketplace to raise heating unit efficiencies	
Dwelling Unit Insulation Standards	Require standards such as R-19 for ceilings and R-11 for walls	Existing minimum FHA standards	
	For improved insulation standards, require weather stripping and caulking		
Air Conditioning System Standards	Require minimum standards of efficiency based on heat pump system efficiencies	Rely on the marketplace to raise air conditioning unit efficiency	0.08 Quad
Water Heating System Standards	Set insulation standards for heater unit and pipes	No changes from P.L. 94-163	0.39 Quad
	Set temperature-setting standards		
Household Appliances	Set improved minimum efficiency and insulation standards	No changes from P.L. 94-163	0.27 Quad
			Total residential energy savings: 2.40 Quads[1]

1. This represents 24 percent of the overall C′–C savings.

substantially improved by the adoption of heat pump systems to replace electrical resistance and other less efficient heating systems.

Air conditioning energy requirements are expected to fall due to improved building insulation. Further improvement will result from higher coefficients of performance for air conditioning systems. Air conditioning efficiency improvements could range as high as 50 percent.

Water heating systems will be made more efficient by improvements in the insulation of the heater jacket and pipes and lowered temperature settings. The improvement in efficiency could reach as high as 40 percent depending on the degree of application of these procedures.

15.4.5 Industrial Sector

The industrial sector is projected to be the fastest-growing sector in the period 1972 to 1985. Total end-use energy demand in the industrial sector (excluding petrochemical feedstocks) is projected to be approximately 32 percent of the total 1985 end-use demand, compared to 27 percent in 1972. If petrochemical feedstocks are also included, the share rises to around 40 percent.

Figure 15.12 shows energy demand and energy growth rates in the industrial sector for six major industries.

Note that energy use in the industrial sector varies considerably with changes in economic growth: the high growth cases (A, C, C′, and E) show significantly greater energy use than low growth cases (B and D). Also, higher oil prices somewhat discourage growth in energy consumption: energy use in Case A is less than that in Case C which is less than that in Case E.

Figure 15.13, which shows the results of changes in the values of the scenario variables, demonstrates the large impact of changing economic growth on energy demand in this sector.

The chemicals and iron and steel industries are major energy users that show particular sensitivity to economic growth. Iron and steel energy use grows faster than overall industrial energy use in the high growth cases and slower than overall industrial energy use in the low growth cases.

The activity levels for the principal energy-consuming industries are summarized in Table 15.13. (The chemicals industry numbers do not include petrochemical feedstocks.) Note that energy price has little projected impact on economic activity; that is, activity levels for Cases A, C, C′, and E (high growth but with different oil prices) and for Cases B and D (low growth but with different oil prices) are nearly the same in the industrial sector.

The chemicals industry activity level was the only one not derived from the Wharton Model. Rather, following the suggestion of industry experts, the Federal Reserve Board Production Indices for the chemicals industry were derived for the WAES scenario cases.[11] The growth rates for these indices were then used in the Wharton Model to define exogenously the chemicals industry growth rate. Since no significant change in the economics of the cases resulted, the FRB production indices are used for the chemicals industry in all cases.

The figures in Table 15.14 represent a measure of the basic relationship between energy use and value added and show the expected future improvement in that relationship as a function of economic growth, oil price, and government policy. Notice that rather substantial reductions in the ratio of energy use to value added, on the basis of historical trends, are expected in the iron and steel, nonferrous metals, paper and allied, and chemicals industries.

Energy use is calculated from these data according to the following relationship:

$$E85 = E72 \times \frac{VA85}{VA72} \times \frac{E85/VA85}{E72/VA72},$$

where

$E85$ = energy used in 1985

$E72$ = energy used in 1972

$VA85$ = value added in 1985

$VA72$ = value added in 1972.

11. This was done using a Data Resources Inc. econometric model, as described below.

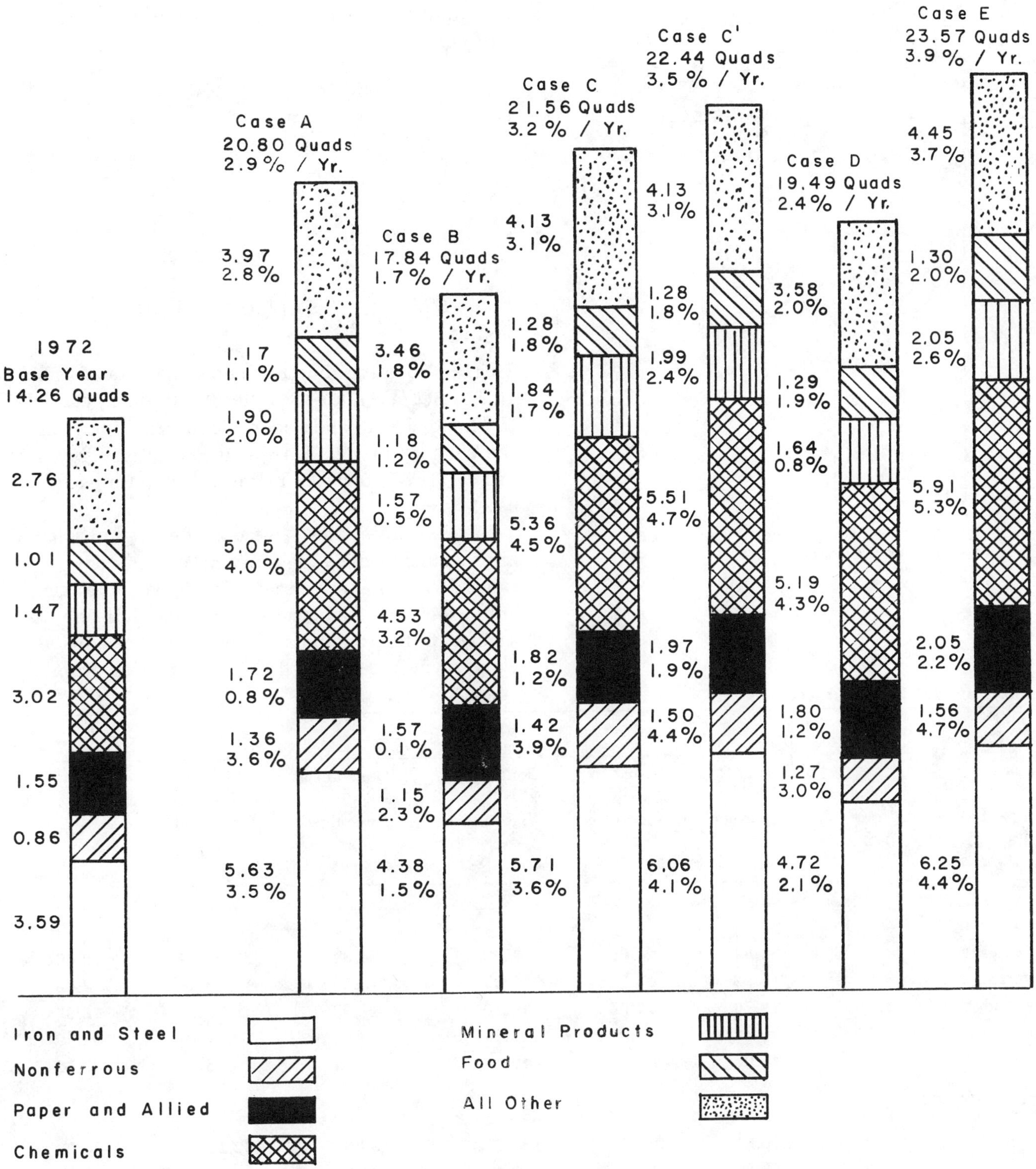

Figure 15.12 Industrial sector energy demand, 1972 and 1985 (excluding petrochemical feedstocks), and energy growth rates, 1972–1985. The first number in each subsector is the total demand in Quads; the second is the growth rate in %/yr. Parentheses indicate a negative growth rate.

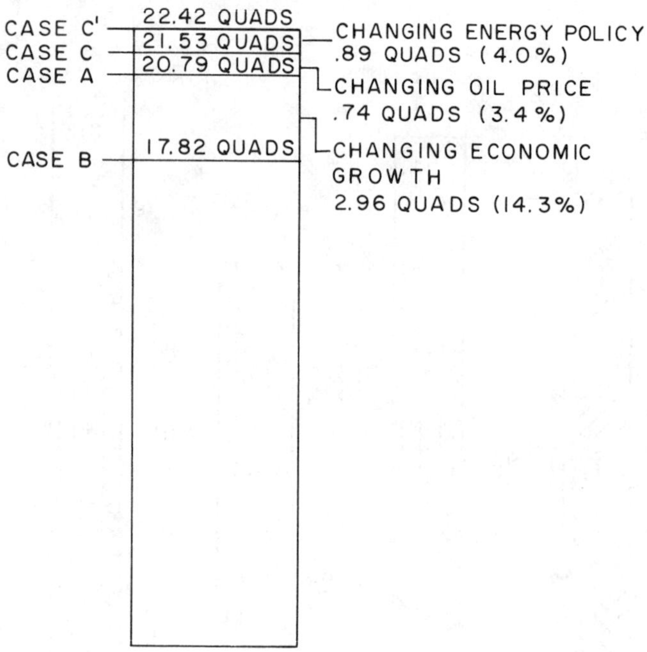

Figure 15.13 Results of changes in assumptions for industrial sector energy demand.

The numbers in Table 15.14 are (in units of percent per year):

$$\Delta\,(E/VA) = \frac{(E_2/VA_2) - (E_1/VA_1)}{(E_1/VA_1)}$$

$$= \frac{E_2/VA_2}{E_1/VA_1} - 1.$$

Thus

$$E85 = E72 \times \frac{VA85}{VA72} \times [1 + \Delta\,(E/VA)]^{13}.$$

This last equation illustrates how the data of Table 15.14, $\Delta(E/VA)$, are used along with base year data ($E72$ and $VA72$) and economic information from the Wharton Model ($VA85$) to calculate energy use in 1985 ($E85$) for each major industry for each case.

In these projections, five sets of energy efficiency improvement policy measures have been considered. The first set, usage of waste steam to generate electricity, is broadly applicable, but only the savings in the iron and steel industry is calculated. Total savings stemming from this practice, from

Table 15.13
Summary of 1985 industrial sector activity levels and energy efficiencies

WAES Case	A	B	C	D	E	1972 Base Year
Oil Price	Rising to $17.25	Rising to $17.25	Constant at $11.50	Constant at $11.50	Falling to $7.66	
National Policy Response	Vigorous	Vigorous	Vigorous	Restrained	Restrained	
National GNP Growth Rate (%/yr.)	4.4	3.2	4.4	3.2	4.4	
Activity Level (value added, 10^9 1972 $)						
Iron and Steel	31.84	24.80	31.23	24.34	30.92	15.60
Nonferrous Metals	16.22	13.71	15.89	13.38	15.89	7.66
Paper and Allied	19.70	17.98	19.70	17.98	19.70	13.06
Mineral Products	20.62	17.07	19.94	16.39	19.77	12.59
Food and Related	48.06	48.69	49.46	49.77	49.15	35.62
Energy Efficiency (10^3 Btu/$ value added)						
Iron and Steel	176.8	176.6	182.8	193.9	202.1	230.0
Nonferrous Metals	83.8	83.9	89.4	94.9	98.2	111.7
Paper and Allied	91.4	91.8	96.4	104.6	108.6	123.7
Mineral Products	92.1	92.0	92.3	100.1	103.7	108.3
Food and Related	24.3	24.2	25.9	25.9	26.4	28.4

Table 15.14
Industrial sector energy efficiencies: Percent change per year in energy use per dollar value added[1]

WAES Case	A	B	C	C′	D	E
Iron and Steel	2.0	2.0	1.75	1.3	1.3	1.0
Nonferrous Metals	2.2	2.2	1.75	1.3	1.3	1.0
Paper and Allied	2.3	2.3	1.9	1.3	1.3	1.0
Chemicals[2]	2.0	2.0	1.7	1.5	1.5	1.0
Mineral Products	1.8	1.8	1.5	1.2	1.2	0.9
Food and Related	1.2	1.2	0.7	0.7	0.7	0.5
Miscellaneous	1.0	1.0	0.6	0.6	0.6	0
Chemical Feedstocks	0	0	0	0	0	0

1. The data of this table have been compiled using trend data from *Energy Consumption in Manufacturing* and from *Project Independence* (references 7 and 15 in Section 15.2.2).

2. Chemicals industry energy efficiencies are rates of change of the ratio of energy use to unit output.

new technologies, and from recycling are estimated to conserve 0.35 Quad in 1985, as shown in Table 15.15.

A second set of policy measures was assumed for the aluminum industry. Adoption of the Alcoa process in new plants and increased recycling to 13.6 percent recovery were estimated to yield a 0.08 Quad reduction. A third policy set would upgrade the energy efficiency of the cement and glass container industries, where expensive labor and low-cost fuel have set a historical precedent for inefficient energy use. Kiln improvements and changeovers in cement production, and use of the submerged combustion process in container manufacturing can be expected to yield a total energy saving of 0.15 Quad.

The other two policy sets, for the paper and chemicals industries, rely on tax incentives to encourage industry to adopt energy-efficient processes in spite of high investment costs.

It is easier to estimate the potential energy savings from various industrial conservation measures than to specify the nonprice policies by which these measures may be implemented. Government could, theoretically, mandate the recycling of certain materials and require more efficient processes in particular industries, but such an approach might be ideologically or politically unacceptable as well as administratively impossible. The Rand Study, "Energy Alternatives for California" (reference 2 in Section 15.2.2), finds that expectations of future prices and availabilities of fuel are the fundamental motivators in industrial energy consumption patterns. Capital changeovers to enable the use of new fuels and technologies are very expensive, and even short-run disruptions in supply can be disastrous to an industry. The result is that industrial adoption of more energy-efficient but capital-intensive measures may not be justified without strong expectations of higher future prices and relatively stable sources of supply.

Alternative Projections

A review of these estimates of industrial sector energy demand has been undertaken by industrial experts within the WAES-U.S. group. This review involved use of the Data Resource Inc. (DRI) Long Term Model, with fiscal and monetary changes incorporated to produce 1985 economic scenarios comparable to those produced by the Wharton Model. The resulting industrial sector energy demands are given in Table 15.16.

The industrial energy demand totals from the two methods are quite similar, as can be seen by comparing Table 15.16 with Figure 15.12. The difference in industrial sector energy totals is not greater than 5 percent in any case. Thus the separate analysis using the DRI econometric model, with somewhat different assumptions, and with different units, fundamentally substantiates the WAES projections.

There are, of course, differences in the energy projections for specific industries. Note that the DRI-based analysis—which measures economic activity in Federal Reserve Board Production Indices (shown in Table 15.17) rather than in dollars of value added—shows somewhat lower energy demands for the iron and steel and nonferrous metals industries, while projecting larger energy demands for the paper and allied group. These individual industry differences, which in some cases are substantial, can be traced to a number of factors such as the uncertainties inherent in any projection or the different economic relationships embodied in the two econometric models.[12] More importantly, perhaps, the differences may be due to the energy efficiencies assumed. From a review of historical trends and recent industrial conservation progress,

12. To avoid such potential inconsistencies, the energy demand figures actually used for all energy-using activities (except the chemicals industry) are based on Wharton Model economic indicators. The DRI-based numbers are presented here for comparative purposes only.

Table 15.15
Effect of national policy response: Industrial sector

Policy Type	Vigorous Response Measure (Case C)	Restrained Response Measure (Case C′)	End-Use Energy Savings (Case C - Case C′)
Iron and Steel			
Tax incentives to encourage adoption of more efficient technologies in the steel industry	Investment tax credit to encourage investment in higher efficiency foundaries	Reliance on natural replacement of old plants with newer, more efficient ones	0.35 Quad
Nonferrous Metals			
Research and development to improve the Alcoa process for primary aluminum production	Federal allocation of money for practical research	Reliance on aluminum companies' research and development efforts	0.08 Quad
Tax incentives and publicity to encourage the recycling of aluminum	Federal encouragement of recycling through publicity and education	Reliance on higher prices for recycled aluminum to increase supply	
Paper and Allied			
Tax incentives to encourage adoption of more efficient technologies in the paper board industry	Investment tax credit to encourage investment in improved paper-forming technologies that reduce the water content of the paper	Reliance on natural replacement of old plants with newer, more efficient ones	0.15 Quad
Chemicals			
Tax incentives to encourage adoption of more efficient technologies in the chemicals industry	Investment tax credit to chemicals companies to encourage investment in higher efficiency plants	Reliance on natural replacement of old plants with newer, more efficient ones	0.15 Quad
Mineral Products			
Tax incentives to encourage adoption of the dry process technology for cement	Investment tax credit to cement companies to encourage investment in higher efficiency plants	Reliance on natural replacement of old plants with newer, more efficient ones	0.15 Quad
			Total industrial energy savings: 0.88 Quad[1]

1. This represents 9 percent of the overall C′–C savings.

energy efficiencies were developed in terms of an average annual reduction in energy consumption per unit of physical output; these energy efficiencies are summarized in Table 15.18.

15.4.6 Energy Policy Summary

"Vigorous" energy policies, as defined in this report, save some 9.9 Quads—or 10 percent—of the end-use energy that would be demanded in 1985 if "restrained" policies were implemented. This saving is calculated at an oil price level of $11.50 (constant from 1972 to 1985) and at a "high" real economic growth rate of 4.4 percent per year. At other price levels or growth rates, the potential savings from vigorous policies would presumably be different.

Figure 15.5 illustrates the sectoral breakdown of the total energy savings from all policy measures. The major vigorous policies that produce such savings are: relaxed auto emissions standards, regulation of airline load factors, strengthened residential and commercial heating system efficiency and insulation standards for buildings, and various tax incentives for better energy-use efficiencies in industries.

It should be noted that the policies adopted here are intended to be neither a comprehensive nor a recommended set. Rather, they were chosen to be representative of the type and scale of possible energy policies that might be pursued during the next decade.

Table 15.16
Industrial sector energy demand, 1972 and 1985, and energy growth rates, 1972-1985, resulting from a DRI-based review of the industrial sector[1]

WAES Case	A		B		C		C′		D		E		1972 Base Year
GNP 1976-1985	High		Low		High		High		Low		High		
Oil Price[2]	$11.50→$17.25		$11.50→$17.25		$11.50		$11.50		$11.50		$11.50→$7.66		
Natl. Policy Response	Vigorous		Vigorous		Vigorous		Restrained		Restrained		Restrained		
Natl. GNP Growth Rate 1976-1985 (%/yr.)	4.4		3.2		4.4		4.4		3.2		4.4		
	Total (Quads)	%/Yr.	Total (Quads)	%/Yr.	Total (Quads)	%/Yr.	Total (Quads)	%/Yr.	Total (Quads)	%/Yr.	Total (Quads)	%/Yr.	Total (Quads)
Major Energy-Use Categories													
Iron and Steel	4.68	1.9	3.97	0.6	4.89	2.3	4.96	2.4	4.34	1.3	5.09	2.6	3.65
Nonferrous Metals	1.21	2.2	1.05	1.1	1.25	2.5	1.27	2.6	1.11	1.5	1.29	2.7	0.91
Paper and Allied	2.20	2.6	1.97	1.7	2.30	2.9	2.33	3.1	2.14	2.4	2.41	3.3	1.57
Chemicals[3]	5.05	4.0	4.53	3.2	5.36	4.5	5.51	4.7	5.19	4.2	5.91	5.3	3.02
Mineral Products	1.74	1.8	1.59	1.1	1.85	2.3	1.90	2.5	1.72	1.7	2.00	2.9	1.38
Food	1.24	1.3	1.18	0.9	1.29	1.6	1.33	1.8	1.29	1.6	1.41	2.3	1.05
All Other	4.18	3.7	3.07	1.2	4.36	4.0	4.54	4.3	4.26	3.8	4.78	4.7	2.62
Total Fuel	20.31	2.8	17.33	1.5	21.27	3.2	21.79	3.3	20.04	2.7	22.85	3.7	14.21
Chemical Feedstocks	6.37	6.2	5.71	5.3	6.50	6.4	6.50	6.4	6.12	5.9	6.52	6.4	2.93
Total	26.68	3.5	23.04	2.3	27.77	3.8	28.29	3.9	26.16	3.3	29.37	4.2	17.14

1. The industrial sector here excludes the petroleum and coal products industries (USSIC 29). Energy used in these industries is estimated in the supply-demand integration procedure.
2. Oil price 1975 US$ per barrel Arabian light crude oil FOB Persian Gulf.
3. Does not include petrochemical feedstocks.

Table 15.17
Summary of 1985 industrial sector activity levels: DRI-based comparison using industrial production indices

WAES Case	A	B	C	D	E	1972 Base Year
Oil Price	Rising to $17.25	Rising to $17.25	Constant at $11.50	Constant at $11.50	Falling to $7.66	
National Policy Response	Vigorous	Vigorous	Vigorous	Restrained	Restrained	
GNP Growth Rate (%/yr.)	4.4	3.2	4.4	3.2	4.4	
Activity Level (Industrial Production Index)						
Iron and Steel	1.629	1.379	1.658	1.451	1.638	1.071
Nonferrous Metals	1.944	1.680	1.959	1.717	1.912	1.236
Paper and Allied	2.073	1.853	2.104	1.959	2.096	1.282
Chemicals and Allied	3.033	2.720	3.094	2.915	3.106	1.396
Mineral Products	1.814	1.628	1.824	1.658	1.791	1.186
Food and Related	1.706	1.626	1.707	1.664	1.699	1.186

Table 15.18
Industrial sector energy efficiencies (average percent per annum reduction in energy consumption per unit output), 1972–1985

WAES CASE	A	B	C	C′	D	E
Iron and Steel	1.3	1.3	1.1	1.0	1.0	0.7
Nonferrous Metals	1.3	1.3	1.1	1.0	1.0	0.7
Paper and Allied	1.1	1.1	0.9	0.8	0.8	0.5
Chemicals and Allied	2.0	2.0	1.7	1.5	1.5	1.0
Clay, Glass, and Stone	1.5	1.5	1.2	1.0	1.0	0.5
Food and Related	1.5	1.5	1.2	1.0	1.0	0.5
Transport Equipment	0.5	0.5	0.2	0.0	0.0	−0.7
Total Manufacturing	1.5	1.5	1.2	1.0	1.0	0.5
Petroleum Products	2.0	2.0	1.5	1.2	1.2	0.5

15.5 SUMMARY TABLES AND WORKSHEETS

The following pages contain tables summarizing the overall results of the U.S. energy demand projections to 1985, plus detailed worksheets showing all of the economic and technical estimates made in these projections. The data sheets are presented in the following sequence:

Energy Use Summary 1972 and 1985 for Cases A-E

Energy Use Summary 1972 and 1985 for Cases C and C′

Energy Growth Rates Summary 1972–1985 for Cases A-E

Energy Growth Rates Summary 1972–1985 for Cases C and C′

Sectoral Percentages of Total Final Energy Demand 1972 and 1985 for Cases A-E

Energy Demand and Energy Growth Rates Summaries for the Transportation, Commercial, Residential, and Industrial Sectors (four sheets)

National Demand Study Worksheets:

- 1972 Base Year (six sheets, including a summary)
- Case A (five sheets)
- Case B (five sheets)
- Case C (five sheets)
- Case D (five sheets)
- Case E (five sheets)
- Case C′ (five sheets).

WAES NATIONAL ENERGY DEMAND STUDIES, 1972-1985
ENERGY USE SUMMARY

Country USA

Scenario Assumptions

WAES Case	A		B		C		D		E			
GNP 1976-1985	High		Low		High		Low		High			
Oil Price*	\$11.50 → \$17.25		\$11.50 → \$17.25		\$11.50		\$11.50		\$11.50 → \$7.66			
Natl. Policy Response	Vigorous		Vigorous		Vigorous		Restrained		Restrained			
Natl. GNP Growth Rate 1976-1985 (%/yr.)	4.4		3.2		4.4		3.2		4.4		1972 Base Year	
Fuel Type	Fossil Fuel	Elec.	Fossil Fuel	Elec.	Fossil Fuel	Elec.	Fossil Fuel	Elec.	Fossil Fuel	Elec.	Fossil Fuel	Elec.
Units [1]	Quad	Quad	Quad	Quad	Quad	Quad	Quad	Quad	Quad	Quad	Quad	Quad
Energy-Using Sectors												
Industrial (excluding feedstocks)	18.13	2.67	15.52	2.32	18.76	2.80	16.96	2.53	20.54	3.03	12.45	1.81
Transportation	18.47	.02	17.66	.02	18.82	.02	19.78	.02	21.18	.02	16.98	.01
Commercial and Public	2.94	1.98	2.89	1.88	3.26	2.06	3.65	2.45	3.89	2.75	2.75	1.42
Residential	7.33	2.71	7.37	2.57	8.23	2.99	9.97	3.48	10.19	4.04	9.47	1.88
Fishing, Agriculture, Mining, Construction	2.44	.17	1.92	.14	2.50	.18	1.96	.16	2.50	.19	1.90	.13
Feedstocks	8.28		6.90		8.20		7.10		8.22		4.19	
Final Energy Demand (excluding processing losses)	57.59	7.55	52.26	6.93	59.77	8.05	59.42	8.64	66.52	10.03	47.74	5.25
Processing Losses [2]	5.76	17.83	5.23	16.36	5.98	19.01	5.94	20.40	6.65	23.68	4.96	13.47
Primary Energy Input	63.35	25.38	57.49	23.29	65.75	27.06	65.36	29.04	73.17	33.71	52.70	18.72

*Oil price 1975 US\$ per barrel Arabian light crude oil FOB Persian Gulf.

1 Quad = 10^{15} BTU

2 Processing losses are fuels used to transport, refine and convert energy. The figures here are very approximate, and are provided merely to show gross totals. Losses of 10 percent are assumed for fossil fuels and 70.25 percent for electricity. The latter figure results from an assumed 35 percent generating efficiency and 85 percent transmission and distribution efficiency.

WAES NATIONAL ENERGY DEMAND STUDIES, 1972-1985
ENERGY USE SUMMARY : CASES C AND C'

Country USA

Scenario Assumptions						
WAES Case	C		C'			
GNP 1976-1985	High		High			
Oil Price*	$11.50		$11.50			
Natl. Policy Response	Vigorous		Restrained			
Natl. GNP Growth Rate 1976-1985 (%/yr.)	4.4		4.4		1972 Base Year	
Fuel Type	Fossil Fuel	Elec.	Fossil Fuel	Elec.	Fossil Fuel	Elec.
Units	Quad	Quad	Quad	Quad	Quad	Quad
Energy-Using Sectors						
Industrial (excluding feedstocks)	18.76	2.80	19.57	2.87	12.45	1.81
Transportation	18.82	0.02	20.73	0.02	16.98	0.01
Commercial and Public	3.26	2.06	3.82	2.61	2.75	1.42
Residential	8.23	2.99	9.93	3.69	9.47	1.88
Fishing, Agriculture, Mining, Construction	2.50	0.18	2.50	0.18	1.90	0.13
Feedstocks	8.20		8.20		4.19	
Final Energy Demand (excluding processing losses)	59.77	8.05	64.75	9.37	47.74	5.25
Processing Losses	5.98	19.01	6.48	22.13	4.96	13.47
Primary Energy Input	65.75	27.06	71.23	31.50	52.70	18.72

*Oil price 1975 US$ per barrel Arabian light crude oil FOB Persian Gulf.

WAES NATIONAL ENERGY DEMAND STUDIES, 1972-1985
ENERGY GROWTH RATES SUMMARY

Country USA

Scenario Assumptions

WAES Case	A		B		C		D		E	
GNP 1976-1985	High		Low		High		Low		High	
Oil Price*	$11.50 → $17.25		$11.50 → $17.25		$11.50		$11.50		$11.50 → $7.66	
Natl. Policy Response	Vigorous		Vigorous		Vigorous		Restrained		Restrained	
Natl. GNP Growth Rate 1976-1985 (%/yr.)										
Fuel Type	Fossil Fuel	Elec.	Fossil Fuel	Elec.	Fossil Fuel	Elec.	Fossil Fuel	Elec.	Fossil Fuel	Elec.
Units (%/yr. 1972-1985)										
Energy-Using Sectors										
Industrial (excluding feedstocks)	2.93	3.04	1.70	1.89	3.20	3.38	2.41	2.61	3.92	4.02
Transportation	.65	4.49	.30	4.49	.79	4.49	1.18	4.49	1.71	4.83
Commercial and Public	.51	2.61	.37	2.20	1.31	2.93	2.19	4.31	2.70	5.24
Residential	(1.95)	2.85	(1.91)	2.43	(1.08)	3.63	.39	4.85	.56	6.06
Fishing, Agriculture, Mining, Construction	1.94	2.09	.08	.57	2.13	2.53	.24	1.61	2.13	2.96
Feedstocks	5.37		3.91		5.29		4.13		5.31	
Final Energy Demand (excluding processing losses)	1.45	2.83	.70	2.15	1.74	3.33	1.70	3.91	2.58	5.10
Processing Losses										
Primary Energy Input										

*Oil price 1975 US$ per barrel Arabian light crude oil FOB Persian Gulf.

Note: ( ) indicates a negative growth rate

WAES NATIONAL ENERGY DEMAND STUDIES, 1972-1985
ENERGY GROWTH RATES SUMMARY: CASES C AND C'

Country USA

Scenario Assumptions

WAES Case	C		C′	
GNP 1976-1985	High		High	
Oil Price*	$11.50		$11.50	
Natl. Policy Response	Vigorous		Restrained	
Natl. GNP Growth Rate 1976-1985 (%/yr.)	4.4		4.4	
Fuel Type	Fossil Fuel	Elec.	Fossil Fuel	Elec.
Units (%/yr. 1972-1985)				
Energy-Using Sectors				
Industrial (excluding feedstocks)	3.20	3.38	3.54	3.58
Transportation	0.79	4.49	1.55	4.49
Commercial and Public	1.31	2.93	2.55	4.82
Residential	(1.08)	3.63	0.36	5.32
Fishing, Agriculture, Mining, Construction	2.13	2.53	2.13	2.53
Feedstocks	5.29		5.29	
Final Energy Demand (excluding processing losses)	1.74	3.33	2.37	4.55
Processing Losses				
Primary Energy Input				

*Oil price 1975 US$ per barrel Arabian light crude oil FOB Persian Gulf.

Note: ( ) indicates a negative growth rate.

WAES NATIONAL ENERGY DEMAND STUDIES, 1972-1985
SECTORAL PERCENTAGES OF TOTAL

Country USA

Scenario Assumptions												
WAES Case	A		B		C		D		E			
GNP 1976-1985	High		Low		High		Low		High			
Oil Price*	$11.50 → $17.25		$11.50 → $17.25		$11.50		$11.50		$11.50 → $7.66			
Natl. Policy Response	Vigorous		Vigorous		Vigorous		Restrained		Restrained			
Natl. GNP Growth Rate 1976-1985 (%/yr.)	4.4		3.2		4.4		3.2		4.4		1972 Base Year	
Fuel Type	Fossil Fuel	Elec.	Fossil Fuel	Elec.	Fossil Fuel	Elec.	Fossil Fuel	Elec.	Fossil Fuel	Elec.	Fossil Fuel	Elec.
Units	%	%	%	%	%	%	%	%	%	%	%	%
Energy-Using Sectors												
Industrial (excluding feedstocks)	31.5	35.4	29.7	33.4	31.4	34.7	28.5	29.3	30.9	30.1	26.1	34.5
Transportation	32.1	0.3	33.8	0.3	31.5	0.3	33.3	0.3	31.8	0.2	35.6	0.2
Commercial and Public	5.1	26.2	5.5	27.2	5.5	25.6	6.1	28.4	5.8	27.4	5.8	27.0
Residential	12.7	35.9	14.1	37.1	13.8	37.2	16.8	40.3	15.3	40.3	19.8	35.8
Fishing, Agriculture, Mining, Construction	4.2	2.3	3.7	2.0	4.2	2.2	3.3	1.9	3.8	1.9	4.0	2.5
Feedstocks	14.4		13.2		13.7		11.9		12.4		8.8	
Final Energy Demand (excluding processing losses)	100	100	100	100	100	100	100	100	100	100	100	100
Processing Losses												
Primary Energy Input												

*Oil price 1975 US$ per barrel Arabian light crude oil FOB Persian Gulf.

TRANSPORTATION SECTOR ENERGY DEMAND AND ENERGY GROWTH RATES 1972-1985

Country USA

Scenario Assumptions													
WAES Case	A		B		C		C'		D		E		
GNP 1976-1985	High		Low		High		High		Low		High		
Oil Price*	$11.50 → $17.25		$11.50 → $17.25		$11.50		$11.50		$11.50		$11.50 → $7.66		
Natl. Policy Response	Vigorous		Vigorous		Vigorous		Restrained		Restrained		Restrained		1972 Base Year
Natl. GNP Growth Rate 1976-1985 (%/yr.)	4.4		3.2		4.4		4.4		3.2		4.4		
	Total	%/Yr.	Total	%/Yr.	Total	%/Yr.	Total	%/Yr.	Total	%/Yr.	Total	%/Yr.	Total
Units	Quad		Quad		Quad		Quad		Quad		Quad		Quad
Major Energy-Use Categories													
Autos	8.31	(0.7)	8.02	(1.0)	8.34	(0.7)	9.27	0.1	8.86	(0.2)	9.30	0.1	9.13
Air Travel	2.02	3.4	1.88	2.8	2.16	3.9	2.56	5.3	2.39	4.7	2.73	5.8	1.31
Freight	3.75	2.1	3.60	1.7	3.91	2.4	4.09	2.7	3.91	2.4	4.26	3.1	2.88
Rail and Bus	0.16	4.5	0.15	4.0	0.13	2.9	0.13	2.9	0.13	2.9	0.13	2.9	0.09
All Other	4.25	1.3	4.03	0.9	4.30	1.4	4.70	2.1	4.51	1.8	4.78	2.2	3.58
Total	18.49	0.7	17.68	0.3	18.84	0.8	20.75	1.5	19.80	1.2	21.20	1.8	16.99

*Oil price 1975 US$ per barrel Arabian light crude oil FOB Persian Gulf.

Note: ( ) indicates a negative growth rate.

COMMERCIAL SECTOR ENERGY DEMAND AND
ENERGY GROWTH RATES 1972-1985

Country USA

Scenario Assumptions

WAES Case	A		B		C		C'		D		E		
GNP 1976-1985	High		Low		High		High		Low		High		
Oil Price*	$11.50 → $17.25		$11.50 → $17.25		$11.50		$11.50		$11.50		$11.50 → $7.66		
Natl. Policy Response	Vigorous		Vigorous		Vigorous		Restrained		Restrained		Restrained		
Natl. GNP Growth Rate 1976-1985 (%/yr.)	4.4		3.2		4.4		4.4		3.2		4.4		1972 Base Year
	Total	%/Yr.	Total	%/Yr.	Total	%/Yr.	Total	%/Yr.	Total	%/Yr.	Total	%/Yr.	Total
Units	Quad		Quad		Quad		Quad		Quad		Quad		Quad
Major Energy-Use Categories													
Space Heat (FF)	2.23	(0.1)	2.19	(0.2)	2.51	0.8	2.97	2.1	2.85	1.8	3.03	2.3	2.26
Space Heat (Elec.)	0.19	6.9	0.17	6.0	0.21	7.7	0.26	9.5	0.23	8.5	0.28	10.1	0.08
Air Conditioning	0.17	0	0.17	0	0.19	0.9	0.26	3.3	0.24	2.7	0.29	4.2	0.17
Lighting	0.25	(4.4)	0.24	(4.7)	0.25	(4.4)	0.59	2.1	0.55	1.6	0.64	2.8	0.45
All Other	2.08	4.2	2.00	3.9	2.16	4.5	2.35	5.2	2.23	4.7	2.40	5.3	1.22
Total	4.92	1.3	4.77	1.0	5.32	1.9	6.43	3.4	6.10	3.0	6.64	3.7	4.17

*Oil price 1975 US$ per barrel Arabian light crude oil FOB Persian Gulf.

Note: ( ) indicates a negative growth rate.

RESIDENTIAL SECTOR ENERGY DEMAND AND ENERGY GROWTH RATES 1972-1985

Country USA

Scenario Assumptions

WAES Case	A		B		C		C'		D		E		
GNP 1976-1985	High		Low		High		High		Low		High		
Oil Price*	$11.50 → $17.25		$11.50 → $17.25		$11.50		$11.50		$11.50		$11.50 → $7.66		
Natl. Policy Response	Vigorous		Vigorous		Vigorous		Restrained		Restrained		Restrained		
Natl. GNP Growth Rate 1976-1985 (%/yr.)	4.4		3.2		4.4		4.4		3.2		4.4		1972 Base Year
	Total	%/Yr.	Total	%/Yr.	Total	%/Yr.	Total	%/Yr.	Total	%/Yr.	Total	%/Yr.	Total
Units	Quad		Quad		Quad		Quad		Quad		Quad		Quad
Major Energy-Use Categories													
Space Heat (FF)	5.75	(2.2)	5.71	(2.2)	6.63	(1.1)	7.99	0.3	7.93	0.3	8.14	0.5	7.66
Space Heat (Elec.)	0.48	4.5	0.47	4.4	0.66	7.1	0.96	10.2	0.94	10.1	1.02	10.8	0.27
Water Heating	1.75	(0.1)	1.75	(0.1)	1.79	0.1	2.18	1.6	2.19	1.7	2.26	1.9	1.77
Air Cond.-Central	0.27	8.8	0.25	8.2	0.27	8.8	0.33	10.5	0.30	9.7	0.49	13.9	0.09
Refrigerators	0.24	(0.3)	0.23	(0.6)	0.30	1.4	0.44	4.4	0.42	4.1	0.46	4.8	0.25
All Other	1.55	1.3	1.52	1.2	1.57	1.4	1.72	2.1	1.67	1.9	1.85	2.7	1.31
Total	10.04	(0.9)	9.93	(1.0)	11.22	(0.1)	13.62	1.4	13.45	1.3	14.22	1.8	11.35

*Oil price 1975 US$ per barrel Arabian light crude oil FOB Persian Gulf.

Note ( ) indicates a negative growth rate.

INDUSTRIAL SECTOR[1] ENERGY DEMAND AND ENERGY GROWTH RATES 1972 - 1985

Country USA

Scenario Assumptions

WAES Case	A		B		C		C′		D		E		
GNP 1976-1985	High		Low		High		High		Low		High		
Oil Price*	\$11.50 → \$17.25		\$11.50 → \$17.25		\$11.50		\$11.50		\$11.50		\$11.50 → \$7.66		
Natl. Policy Response	Vigorous		Vigorous		Vigorous		Restrained		Restrained		Restrained		
Natl. GNP Growth Rate 1976-1985 (%/yr.)	4.4		3.2		4.4		4.4		3.2		4.4		1972 Base Year
	Total	%/Yr.	Total	%/Yr.	Total	%/Yr.	Total	%/Yr.	Total	%/Yr.	Total	%/Yr.	Total
Units	Quad		Quad		Quad		Quad		Quad		Quad		Quad
Major Energy-Use Categories													
Iron and Steel	5.63	3.5	4.38	1.6	5.71	3.6	6.06	4.1	4.72	2.1	6.25	4.4	3.59
Nonferrous	1.36	3.6	1.15	2.3	1.42	3.9	1.50	4.4	1.27	3.0	1.56	9.7	0.86
Paper and Allied	1.72	0.8	1.57	0.1	1.82	1.2	1.97	1.9	1.80	1.2	2.05	2.2	1.55
Chemicals[2]	5.05	4.0	4.53	3.2	5.36	4.5	5.51	4.7	5.19	4.3	5.91	5.3	3.02
Mineral Products	1.90	2.0	1.57	0.5	1.84	1.7	1.99	2.4	1.64	0.9	2.05	2.6	1.47
Food	1.17	1.1	1.18	1.2	1.28	1.8	1.28	1.8	1.29	1.9	1.30	2.0	1.01
All Other	3.97	2.9	3.46	1.7	4.13	3.1	4.13	3.1	3.58	2.3	4.45	3.7	2.76
Total Fuel	20.80	2.9	17.84	1.7	21.56	3.2	22.44	3.5	19.49	2.4	23.57	3.9	14.26
Chemical Feedstocks[2]	6.36	6.1	5.71	5.3	6.49	6.3	6.49	6.3	6.12	5.8	6.52	6.3	2.93
Total	27.16	3.6	23.55	2.4	28.05	3.8	28.93	4.1	25.61	3.1	30.09	4.4	17.19

*Oil price 1975 US\$ per barrel Arabian light crude oil FOB Persian Gulf.

1. The industrial sector here excludes the petroleum and coal products industries (SIC 29). Energy used in these industries is estimated in the supply-demand integration procedure.
2. Growth rate based on the Federal Reserve Board Production Indices.

NATIONAL ENERGY DEMAND PROJECTIONS

Country USA

Year Base Year, 1972

WAES Case

Summary [1]

Category	Oil	Gas	Coal	Electricity	Other	Total by Category (Excluding Processing Losses)
Transportation	16.98	----	----	.01	----	16.99
Agriculture, Mining, and Construction	1.90	----	----	.13	----	2.03
Commercial [2]	1.20	1.37	.18	1.42	----	4.17
Public						
Industrial	4.96	7.63	4.05	1.81	----	18.45
Residential	3.77	5.70	----	1.88	----	11.35
Total Delivered Energy by Fuel	28.81	14.70	4.23	5.25	----	
Processing Losses by Fuel	3.17	1.62	.17	13.47	----	
Gross Input by Fuel	31.98	16.32	4.40	18.72	----	71.42

Gross Energy Input (Add across bottom row)

[1] All numbers are in Quads = 10^{15} BTU.

[2] Public is included in the commercial data

NATIONAL ENERGY DEMAND PROJECTIONS

Country USA

Year Base Year, 1972

WAES Case

Transportation

	(A)	(B)	(C)	(D)	(E)	(F)	(G)	
	Population/Auto	Distance/Auto/Yr. [1]	Load Factor	Fraction Urban or Rural	Energy/Distance	Fossil Fuel [2]	Electricity [2]	
Urban Auto	2.28	986.4 x 10^9	1.4	.73	9300 BTU/mile	6.70		(1)
Rural Auto				.27	9117 BTU/mile	2.43		(2)
	Activity Level [3]	% Electrified	Load Factor [4]	Fossil Fuel/ Activity Level [5]	Electricity/Activity			
Urban Bus	23.3		15	1720		.04		(3)
Intercity Bus	25.6		22	1170		.03		(4)
Rail (Passenger)	8.6			2620	349	.0225	.005	(5)
Air (Passenger, Domestic)	119			8990		1.07		(6)
Air (Passenger, International)	50			4800		.24		(7)
Ship (Passenger)								(8)
Rail (Freight)	784			710		.556		(9)
Air (Freight, Domestic)	3.9			28200		.11		(10)
Air (Freight, International)						.10		(11)
Truck (Local)	69.5		3	7140		.496		(12)
Truck (Intercity)	392		10	2600		1.02		(13)
Ship (Freight, Domestic)	631			475		.3		(14)
Ship (Freight, International)						.3		(15)
Miscellaneous						3.566	.008	(16)
					Total	16.981	.013	

1 Total Vehicle Miles

2 Energy Totals in Quads = 10^{15} BTU

3 10^9 Passenger Miles or 10^9 Ton Miles

4 $\frac{\text{Passenger Miles}}{\text{Vehicle Miles}}$ or $\frac{\text{Ton Miles}}{\text{Vehicle Miles}}$

5 BTU/Passenger Mile or BTU/Ton Mile

NATIONAL ENERGY DEMAND PROJECTIONS

Country USA

Year Base Year, 1972

WAES Case ______

Agriculture, Mining, and Construction

	(A)	(B)	(C)	(D)	(E)	(F)	(G)	(H)	
Agriculture, Forestry, and Fishing (ISIC Major Division 1)	Value Added	Energy/Value Added	Area Harvested	Energy/ Area	Physical Output	Energy/ Output	Fossil Fuel [1]	Electricity [1]	
Farm Machinery	$\$28.0 \times 10^9$	20,000 BTU/$	293×10^6 Acre	1.91×10^6 BTU/Acre			.56		(17)
Greenhouses									(18)
Other Agricultural Uses	$\$28.0 \times 10^9$	18,571 BTU/$	293×10^6 Acre	1.77×10^6 BTU/Acre	$\$65.6 \times 10^9$ [2]		.43	.09	(19)
Forestry (ISIC 12)					$12.2 \times 10^9 ft^3$				(20)
Fishing Boats (ISIC 13)									(21)
Other Fishing Use (ISIC 13)					4.71×10^9 lbs.				(22)
Total Fertilizer (tons)		N 10.4×10^6	P 9.1×10^6	K 7.2×10^6		Total	.99	.09	

	(A)	(B)	(C)	(D)	(E)	(F)	
Mining	Value Added	Energy/Value Added	Physical Output	Energy/Physical Output	Fossil Fuel	Electricity	
Handling and Internal Transportation							(23)
Other Energy Use	$\$ 5.105 \times 10^9$		$\$ 7.167 \times 10^9$ [2]		.216	.044	(24)
				Total			

	(A)	(B)	(C)	(D)	(E)	(F)	
Construction	Value Added	Energy/Value Added	Physical Output	Energy/Physical Output	Fossil Fuel	Electricity	
Handling and Internal Transportation	$\$68.2 \times 10^9$		$\$124. \times 10^9$ [2]		.69		(25)
Other Energy Use							(26)
				Total			
				Agriculture, Mining, and Construction Total	1.90	.13	

[1] Energy Totals in Quads = 10^{15} BTU.

[2] Dollar Value of Shipments.

NATIONAL ENERGY DEMAND PROJECTIONS

Country USA

Year Base Year, 1972

WAES Case

Commercial and Public [1]

Commercial	(A) Floor Area [2]	(B) Heat Gain or Loss/Area	(C) System Efficiency	(D) Fossil Fuel [3]	(E) Electricity [3]	
Space Heat (Fossil Fuel)	(10^9 Sq Ft) 22.23	7.10×10^4 BTU/Sq Ft/Yr	.7	2.255		(27)
Space Heat (Electricity)	1.17	7.10×10^4 BTU/Sq Ft/Yr	1.00		.083	(28)
Air Conditioning (Fossil Fuel)	.82	2.558×10^4 BTU/Sq Ft/Yr	1.80 (COP)	.012		(29)
Air Conditioning (Electricity)	10.93	2.558×10^4 BTU/Sq Ft/Yr	1.80 (COP)		.155	(30)
		Energy/Area	Electricity/Area			
Lighting	23.4		1.92×10^4 BTU/Sq Ft/Yr		.449	(31)
Miscellaneous				.486	.729	(32)
Presently Unknown						(33)
			Total	2.753	1.416	

Public	Floor Area	Heat Gain or Loss/Area	System Efficiency	Fossil Fuel	Electricity	
Space Heat (Fossil Fuel)						(34)
Space Heat (Electricity)						(35)
Air Conditioning (Fossil Fuel)						(36)
Air Conditioning (Electricity)						(37)
		Energy/Area	Electricity/Area			
Lighting						(38)
Miscellaneous						(39)
Presently Unknown						(40)
			Total			
			Commercial and Public Total			

[1] Public is included in the commercial data.

[2] Total Floor Space = 23.4×10^9 Sq. Ft.

[3] Energy Totals in Quads = 10^{15} BTU.

NATIONAL ENERGY DEMAND PROJECTIONS

Country USA

Year Base Year, 1972

WAES Case

Industrial

	(A) Value Added	(B) Energy/Value Added	(C) Physical Output	(D) Energy/Output	(E) % Exported	(F) Fossil Fuel [1]	(G) Electricity [1]	(H) Feedstocks	
	(10^9 $)	(MBTU/$)							
Iron and Steel (ISIC 371)	15.60	230.0			8.2	3.382	.206		(41)
Nonferrous Metals (ISIC 372)	7.66	111.7			8.0	.580	.276		(42)
Paper and Allied (ISIC 34)	13.06	118.5			4.96	1.419	.128		(43)
Chemicals and Allied (ISIC 35 minus ISIC 353)	32.41	93.3			7.2	2.654	.370	2.929	(44)
Mineral Products (ISIC 36)	12.59	108.3			.12	1.376	.096		(45)
Food and Related (ISIC 31)	35.62	28.4			2.69	.892	.119		(46)
Transport Eqpt. (ISIC 384)	39.80	10.7			4.7	.323	.102		(47)
Key Industry (N.E.C.)[2]								1.264	(48)
Key Industry (N.E.C.)									(49)
Misc. Manufacturing	191.52	12.2				1.821	.516		(50)
N.E.C. = Not Elsewhere Classified					**Total**	12.447	1.813	4.193	

[1] Energy Totals in Quads = 10^{15} BTU.

[2] Asphalt and Road Tar Feedstocks.

NATIONAL ENERGY DEMAND PROJECTIONS

Country USA

Year Base Year, 1972

WAES Case

Residential

Population 209×10^6 Persons/Dwelling 3.13 Dwellings 66.7×10^6

		(A)	(B)	(C)	(D)	(E)	(F)	
Central Space Heating		Dwellings (10^6 Units)	Heat Loss[1]	System Efficiency		Fossil Fuel[2]	Electricity[2]	
Single Family	Fossil Fuels	41.65	874	.63		5.778		(51)
	Electricity	3.13	700	1.0			.219	(52)
Town or Row Houses[4]	Fossil Fuels	2.60	495	.63		.204		(53)
	Electricity	.06	495	1.0			.003	(54)
Apartments	Fossil Fuels	18.30	578	.63		1.679		(55)
	Electricity	.96	463	1.0			.044	(56)
		% Saturation (Fossil Fuels)	% Saturation (Electricity)	Energy/Unit (Fossil Fuels)	Energy/Unit[3] (Electricity)			
Room Heating[5]		12	32	8.8×10^6 BTU	938	.070	.068	(57)
Air Conditioning (Window)			27		1986		.122	(58)
Air Conditioning (Central)		0.3	11	17.4×10^6 BTU	3546	.003	.089	(59)
Water Heating		66	30	31.5×10^6 BTU	5627	1.387	.384	(60)
Refrigerators			100		1084		.247	(61)
Lighting			100		762		.173	(62)
Cooking		56	43	9.4×10^6 BTU	1042	.351	.102	(63)
Dishwashers			22		352		.018	(64)
Freezers			31		1348		.095	(65)
Television			149		413		.140	(66)
Other Key Appliances[6]			71		88		.014	(67)
Misc. Appliances			100		712		.162	(68)
Potential Appliances								(69)
					Total	9.472	1.880	

[1] Heat loss is in 10^5 BTU/Year.

[2] Energy Totals in Quads = 10^{15} BTU.

[3] Energy/Unit (Electricity) is in kilowatt-hours=3412 BTU

[4] Mobile Homes

[5] Clothes Dryer

[6] Clothes Washer

NATIONAL ENERGY DEMAND PROJECTIONS

Country USA

Year 1985

WAES Case Case A

Transportation

	(A)	(B)	(C)	(D)	(E)	(F)	(G)	
	Population/Auto	Distance/Auto/Yr.[1]	Load Factor	Fraction Urban or Rural	Energy/Distance	Fossil Fuel[2]	Electricity[2]	
Urban Auto		$1{,}435.1 \times 10^9$			5790 BTU/Mile	8.31		(1)
Rural Auto					(21.6 m.p.g.)			(2)
	Activity Level[3]	% Electrified	Load Factor[4]	Fossil Fuel/ Activity Level[5]	Electricity/Activity			
Urban Bus	40.3		17.9	1540		.06		(3)
Intercity Bus	49.0		20.0	1160		.06		(4)
Rail (Passenger)	17.5			2400		.04	.003	(5)
Air (Passenger, Domestic)	246.0		67.5	6020		1.48		(6)
Air (Passenger, International)	141.0			3830		.54		(7)
Ship (Passenger)								(8)
Rail (Freight)	1178.0			630		.74		(9)
Air (Freight, Domestic)	7.3			23600		.17		(10)
Air (Freight, International)				19300		.18		(11)
Truck (Local)	114.2			6740		.77		(12)
Truck (Intercity)	510.0			2150		1.10		(13)
Ship (Freight, Domestic)	737.0			490		.36		(14)
Ship (Freight, International)						.43		(15)
Miscellaneous						4.23	.02	(16)
					Total	18.47	.023	

[1] Total Vehicle Miles

[2] Energy Totals in Quads = 10^{15} BTU

[3] 10^9 Passenger Miles or 10^9 Ton Miles

[4] $\frac{\text{Passenger Miles}}{\text{Vehicle Miles}}$ or $\frac{\text{Ton Miles}}{\text{Vehicle Miles}}$

[5] BTU/Passenger Mile or BTU/Ton Mile

NATIONAL ENERGY DEMAND PROJECTIONS

Country USA

Year 1985

WAES Case Case A

Agriculture, Mining, and Construction

	(A)	(B)	(C)	(D)	(E)	(F)	(G)	(H)	
Agriculture, Forestry, and Fishing (ISIC Major Division 1)	Value Added	Energy/Value Added	Area Harvested	Energy/Area	Physical Output	Energy/Output	Fossil Fuel [1]	Electricity [1]	
Farm Machinery			345×10^6 Acres	1.83×10^6 BTU/Acre			.63		(17)
Greenhouses									(18)
Other Agricultural Uses	$\$42.6 \times 10^9$	17,000 BTU/$ V.A.	345×10^6 Acres	2.14×10^6 BTU/Acre	$\$99.9 \times 10^9$ [2]		.61	.13	(19)
Forestry (ISIC 12)					18.9×10^9 ft^3				(20)
Fishing Boats (ISIC 13)									(21)
Other Fishing Use (ISIC 13)					7.3×10^9 lbs				(22)
Total Fertilizer (tons)		N ______	P ______	K ______		Total	1.24	.13	

	(A)	(B)	(C)	(D)	(E)	(F)	
Mining	Value Added	Energy/Value Added	Physical Output	Energy/Physical Output	Fossil Fuel	Electricity	
Handling and Internal Transportation							(23)
Other Energy Use	$\$5.2 \times 10^9$		$\$7.30 \times 10^9$ [2]		.21	.04	(24)
				Total			

	(A)	(B)	(C)	(D)	(E)	(F)	
Construction	Value Added	Energy/Value Added	Physical Output	Energy/Physical Output	Fossil Fuel	Electricity	
Handling and Internal Transportation	$\$103.6 \times 10^9$		$\$188.4 \times 10^9$ [2]		.99		(25)
Other Energy Use							(26)
				Total			
				Agriculture, Mining, and Construction Total	2.44	.17	

[1] Energy Totals in Quads = 10^{15} BTU

[2] Dollar Value of Shipments

NATIONAL ENERGY DEMAND PROJECTIONS

Country USA

Year 1985

WAES Case Case A

Commercial and Public [1]

Commercial	(A) Floor Area [2]	(B) Heat Gain or Loss/Area[3]	(C) System Efficiency	(D) Fossil Fuel [4]	(E) Electricity [4]	
	(10^9 Sq Ft)					
Space Heat (Fossil Fuel)	32.24	5.68 to 4.30 x 10^4 BTU/Sq Ft	.70 to .80	2.23		(27)
Space Heat (Electricity)	4.4	4.30 x 10^4 BTU/Sq Ft	1.00		.19	(28)
Air Conditioning (Fossil Fuel)	1.28	1.78 x 10^4 BTU/Sq Ft	2.2 (COP)	.01		(29)
Air Conditioning (Electricity)	22.89	1.78 x 10^4 BTU/Sq Ft	2.5 (COP)		.16	(30)
		Energy/Area	Electricity/Area			
Lighting	36.64		.69 x 10^4 BTU/Sq Ft		.25	(31)
Miscellaneous				.70	1.38	(32)
Presently Unknown						(33)
			Total	2.94	1.98	

Public	Floor Area	Heat Gain or Loss/Area	System Efficiency	Fossil Fuel	Electricity	
Space Heat (Fossil Fuel)						(34)
Space Heat (Electricity)						(35)
Air Conditioning (Fossil Fuel)						(36)
Air Conditioning (Electricity)						(37)
		Energy/Area	Electricity/Area			
Lighting						(38)
Miscellaneous						(39)
Presently Unknown						(40)
			Total			
			Commercial and Public Total			

[1] Public is included in the commercial data.

[2] Total Floor Space = 36.64 x 10^9 Sq. Ft.

[3] The first number is heat gain or loss/square foot for old commercial units, and the second number is for new commercial units (i.e., buildings built after 1972).

[4] Energy Totals in Quads = 10^{15} BTU.

NATIONAL ENERGY DEMAND PROJECTIONS

Country USA

Year 1985

WAES Case Case A

Industrial

	(A)	(B)	(C)	(D)	(E)	(F)	(G)	(H)	
	Value Added	Energy/Value Added	Physical Output	Energy/Output	% Exported	Fossil Fuel [1]	Electricity[1]	Feedstocks	
Iron and Steel (ISIC 371)	(10^9 \$) 31.84	(MBTU/\$) 176.8				5.31	.32		(41)
Nonferrous Metals (ISIC 372)	16.22	83.8				.92	.44		(42)
Paper and Allied (ISIC 34)	19.70	87.6				1.58	.14		(43)
Chemicals and Allied (ISIC 35 minus ISIC 353)						4.43	.62	6.36	(44)
Mineral Products (ISIC 36)	20.62	92.1				1.78	.12		(45)
Food and Related (ISIC 31)	48.06	24.3				1.03	.14		(46)
Transport Eqpt. (ISIC 384)	61.89	9.4				.44	.14		(47)
Key Industry (N.E.C.)[2]								1.92	(48)
Key Industry (N.E.C.)									(49)
Misc. Manufacturing	316.54	10.7				2.64	.75		(50)
N.E.C. = Not Elsewhere Classified					**Total**	18.13	2.67	8.28	

[1] Energy Totals in Quads = 10^{15} BTU

[2] Asphalt and Road Tar Feedstocks

NATIONAL ENERGY DEMAND PROJECTIONS

Country USA

Year 1985

WAES Case Case A

Residential

Population 234×10^6 Persons/Dwelling 2.71 Dwellings 86.44×10^6

		(A)	(B)	(C)	(D)	(E)	(F)	
Central Space Heating		Dwellings	Heat Loss[1]	System Efficiency		Fossil Fuel [2]	Electricity[2]	
Single Family	Fossil Fuels	(10^6 Units) 43.76	723 484	.70 (old units) .80 (new units)		4.20		(51)
	Electricity	8.60	621 424	1.5 (COP)			.28	(52)
Town or Row Houses [4]	Fossil Fuels	4.71	414 364	.70 (old units) .80 (new units)		.24		(53)
	Electricity	2.63	364	1.5 (COP)			.06	(54)
Apartments	Fossil Fuels	21.01	478 384	.70 (old units) .80 (new units)		1.31		(55)
	Electricity	5.73	400 372	1.5 (COP)			.14	(56)
		% Saturation (Fossil Fuels)	% Saturation (Electricity)	Energy/Unit (Fossil Fuels)	Energy/Unit [3] (Electricity)			
Room Heating [5]		12	57	7.3×10^6 BTU	846	.08	.14	(57)
Air Conditioning (Window)			33		920		.09	(58)
Air Conditioning (Central)			50		1830		.27	(59)
Water Heating		66	34	22.4×10^6 BTU	4710	1.28	.47	(60)
Refrigerators			115		711		.24	(61)
Lighting			100		648		.19	(62)
Cooking		30	65	8.3×10^6 BTU	1093	.22	.21	(63)
Dishwashers			60		290		.05	(64)
Freezers			38		807		.09	(65)
Television			165		300		.15	(66)
Other Key Appliances [6]			78		117		.03	(67)
Misc. Appliances			100		1020		.30	(68)
Potential Appliances								(69)
					Total	7.33	2.71	

[1] Heat loss is in 10^5 BTU/Year. The first number is the heat loss for old dwelling units, and the second number is for new dwelling units (i.e., those built after 1972).

[2] Energy Totals in Quads = 10^{15} BTU.

[3] Energy/Unit (Electricity) is in kilowatt-hours = 3412 BTU

[4] Mobile Homes

[5] Clothes Dryer

[6] Clothes Washer

NATIONAL ENERGY DEMAND PROJECTIONS

Country USA

Year 1985

WAES Case Case B

Transportation

	(A)	(B)	(C)	(D)	(E)	(F)	(G)	
	Population/Auto	Distance/Auto/Yr.[1]	Load Factor	Fraction Urban or Rural	Energy/Distance	Fossil Fuel[2]	Electricity[2]	
Urban Auto		$1,382.8 \times 10^9$			5798 BTU/Mi	8.02		(1)
Rural Auto					(21.6 m.p.g.)			(2)
	Activity Level[3]	% Electrified	Load Factor[4]	Fossil Fuel/ Activity Level[5]	Electricity/Activity			
Urban Bus	38.3		17.9	1540		.06		(3)
Intercity Bus	45.6		20.0	1160		.05		(4)
Rail (Passenger)	16.3			2390		.04	.003	(5)
Air (Passenger, Domestic)	228.8		67.5	6020		1.38		(6)
Air (Passenger, International)	131.1			3830		.50		(7)
Ship (Passenger)								(8)
Rail (Freight)	1119.1			630		.70		(9)
Air (Freight, Domestic)	6.8			23500		.16		(10)
Air (Freight, International)				19300		.17		(11)
Truck (Local)	109.6			6740		.74		(12)
Truck (Intercity)	494.7			2150		1.06		(13)
Ship (Freight, Domestic)	714.9			490		.35		(14)
Ship (Freight, International)						.42		(15)
Miscellaneous						4.01	.02	(16)
					Total	17.66	.023	

[1] Total Vehicle Miles

[2] Energy Totals in Quads = 10^{15} BTU

[3] 10^9 Passenger Miles or 10^9 Ton Miles

[4] $\frac{\text{Passenger Miles}}{\text{Vehicle Miles}}$ or $\frac{\text{Ton Miles}}{\text{Vehicle Miles}}$

[5] BTU/Passenger Mile or BTU/Ton Mile

NATIONAL ENERGY DEMAND PROJECTIONS

Country USA

Year 1985

WAES Case Case B

Agriculture, Mining, and Construction

Agriculture, Forestry, and Fishing (ISIC Major Division 1)	(A) Value Added	(B) Energy/Value Added	(C) Area Harvested	(D) Energy/Area	(E) Physical Output	(F) Energy/Output	(G) Fossil Fuel[1]	(H) Electricity[1]	
Farm Machinery			345×10^6 acres	1.83×10^6 BTU/acre			.63		(17)
Greenhouses									(18)
Other Agricultural Uses	$37.8 x 10^9	17,000 BTU/$V.A.	345×10^6 acres	1.86×10^6 BTU/acre	$88.5 x 10^9 [2]		.53	.11	(19)
Forestry (ISIC 12)					$17.1 \times 10^9 ft^3$				(20)
Fishing Boats (ISIC 13)									(21)
Other Fishing Use (ISIC 13)					6.6×10^9 lbs		1.16	.11	(22)
Total Fertilizer (tons)		N	P	K		Total			

Mining	(A) Value Added	(B) Energy/Value Added	(C) Physical Output	(D) Energy/Physical Output	(E) Fossil Fuel	(F) Electricity	
Handling and Internal Transportation							(23)
Other Energy Use	$3.82 x 10^9		$5.35 x 10^9 [2]		.15	.031	(24)
				Total			

Construction	(A) Value Added	(B) Energy/Value Added	(C) Physical Output	(D) Energy/Physical Output	(E) Fossil Fuel	(F) Electricity	
Handling and Internal Transportation	$64.0 x 10^9		$116.4 x 10^9 [2]		.61		(25)
Other Energy Use							(26)
				Total			
				Agriculture, Mining, and Construction Total	1.92	.14	

[1] Energy Totals in Quads = 10^{15} BTU

[2] Dollar Value of Shipments

NATIONAL ENERGY DEMAND PROJECTIONS

Country USA

Year 1985

WAES Case Case B

Commercial and Public[1]

Commercial	(A) Floor Area[2]	(B) Heat Gain or Loss/Area[3]	(C) System Efficiency	(D) Fossil Fuel[4]	(E) Electricity[4]	
Space Heat (Fossil Fuel)	31.1	5.68 to 4.30 x 10^4 BTU/Sq Ft	.70 to .80	2.19		(27)
Space Heat (Electricity)	4.0	4.30 x 10^4 BTU/Sq Ft	1.0		.17	(28)
Air Conditioning (Fossil Fuel)	1.23	1.78 x 10^4 BTU/Sq Ft	2.2 (COP)	.01		(29)
Air Conditioning (Electricity)	21.79	1.78 x 10^4 BTU/Sq Ft	2.5 (COP)		.16	(30)
		Energy/Area	Electricity/Area			
Lighting	35.1		.69 x 10^4 BTU/Sq Ft		.24	(31)
Miscellaneous				.69	1.31	(32)
Presently Unknown						(33)
			Total	2.89	1.88	

Public	Floor Area	Heat Gain or Loss/Area	System Efficiency	Fossil Fuel	Electricity	
Space Heat (Fossil Fuel)						(34)
Space Heat (Electricity)						(35)
Air Conditioning (Fossil Fuel)						(36)
Air Conditioning (Electricity)						(37)
		Energy/Area	Electricity/Area			
Lighting						(38)
Miscellaneous						(39)
Presently Unknown						(40)
			Total			
			Commercial and Public Total			

[1] Public is included in the commercial data.

[2] Total Floor Space = 35.10 x 10^9 Sq. Ft.

[3] The first number is heat gain or loss/square foot for old commercial units, and the second number is for new commercial units (i.e., buildings built after 1972).

[4] Energy Totals in Quads = 10^{15} BTU.

NATIONAL ENERGY DEMAND PROJECTIONS

Country USA

Year 1985

WAES Case Case B

Industrial

	(A) Value Added	(B) Energy/Value Added	(C) Physical Output	(D) Energy/Output	(E) % Exported	(F) Fossil Fuel [1]	(G) Electricity[1]	(H) Feedstocks	
Iron and Steel (ISIC 371)	(10^9 \$) 24.80	(MBTU/\$) 176.6				4.13	.25		(41)
Nonferrous Metals (ISIC 372)	13.71	83.9				.78	.37		(42)
Paper and Allied (ISIC 34)	17.98	87.5				1.44	.13		(43)
Chemicals and Allied (ISIC 35 minus ISIC 353)						3.98	.55	5.71	(44)
Mineral Products (ISIC 36)	17.07	92.0				1.47	.10		(45)
Food and Related (ISIC 31)	48.69	24.2				1.04	.14		(46)
Transport Eqpt. (ISIC 384)	56.48	9.4				.40	.13		(47)
Key Industry (N.E.C.)[2]								1.19	(48)
Key Industry (N.E.C.)									(49)
Misc. Manufacturing	272.85	10.7				2.28	.65		(50)
N.E.C. = Not Elsewhere Classified					**Total**	15.52	2.32	6.90	

[1] Energy Totals in Quads = 10^{15} BTU.

[2] Asphalt and Road Tar Feedstocks.

NATIONAL ENERGY DEMAND PROJECTIONS

Country USA

Year 1985

WAES Case Case B

Residential

Population 234×10^6 Persons/Dwelling 2.75 Dwellings 85.22×10^6

		(A)	(B)	(C)	(D)	(E)	(F)	
Central Space Heating		Dwellings	Heat Loss[1]	System Efficiency		Fossil Fuel[2]	Electricity[2]	
		(10^6 Units)						
Single Family	Fossil Fuels	43.37	723 482	.70 (old units) .80 (new units)		4.18		(51)
	Electricity	8.34	621 424	1.5 (COP)			.27	(52)
Town or Row Houses[4]	Fossil Fuels	4.66	414 364	.70 (old units) .80 (new units)		.23		(53)
	Electricity	2.58	364	1.5 (COP)			.06	(54)
Apartments	Fossil Fuels	20.75	478 384	.70 (old units) .80 (new units)		1.30		(55)
	Electricity	5.52	400 372	1.5 (COP)			.14	(56)
		% Saturation (Fossil Fuels)	% Saturation (Electricity)	Energy/Unit (Fossil Fuels)	Energy/Unit[3] (Electricity)			
Room Heating[5]		12	53	7.3×10^6 BTU	846	.08	.13	(57)
Air Conditioning (Window)			33		920		.09	(58)
Air Conditioning (Central)			46		1830		.25	(59)
Water Heating		70	30	22.4×10^6 BTU	4710	1.34	.41	(60)
Refrigerators			112		711		.23	(61)
Lighting			100		648		.19	(62)
Cooking		34	61	8.3×10^6 BTU	1093	.24	.19	(63)
Dishwashers			56		290		.05	(64)
Freezers			38		807		.09	(65)
Television			160		300		.14	(66)
Other Key Appliances[6]			78		117		.03	(67)
Misc. Appliances			100		1020		.30	(68)
Potential Appliances								(69)
					Total	7.37	2.57	

[1] Heat loss is in 10^5 BTU/Year. The first number is the heat loss for old dwelling units, and the second number is for new dwelling units (i.e., those built after 1972).

[2] Energy Totals in Quads = 10^{15} BTU.

[3] Energy/Unit (Electricity) is in kilowatt-hours=3412 BTU

[4] Mobile Homes

[5] Clothes Dryer

[6] Clothes Washer

NATIONAL ENERGY DEMAND PROJECTIONS

Country USA

Year 1985

WAES Case Case C

Transportation

	(A)	(B)	(C)	(D)	(E)	(F)	(G)	
	Population/Auto	Distance/Auto/Yr.[1]	Load Factor	Fraction Urban or Rural	Energy/Distance	Fossil Fuel[2]	Electricity[2]	
Urban Auto		1,444.6 x 10^9			5774 BTU/Mi	8.34		(1)
Rural Auto					(21.7 m.p.g.)			(2)
	Activity Level[3]	% Electrified	Load Factor[4]	Fossil Fuel/ Activity Level[5]	Electricity/Activity			
Urban Bus	37.8		17.9	1590		.06		(3)
Intercity Bus	34.4		20.0	1160		.04		(4)
Rail (Passenger)	11.0			2730		.03	.003	(5)
Air (Passenger, Domestic)	268.8		67.5	6040		1.62		(6)
Air (Passenger, International)	142.0			3800		.54		(7)
Ship (Passenger)								(8)
Rail (Freight)	1163.2			630		.73		(9)
Air (Freight, Domestic)	11.3			26500		.30		(10)
Air (Freight, International)				22300		.18		(11)
Truck (Local)	114.2			6740		.77		(12)
Truck (Intercity)	530.0			2150		1.14		(13)
Ship (Freight, Domestic)	737.0			490		.36		(14)
Ship (Freight, International)						.43		(15)
Miscellaneous						4.28	.02	(16)
					Total	18.82	.023	

1 Total Vehicle Miles

2 Energy Totals in Quads = 10^{15} BTU

3 10^9 Passenger Miles or 10^9 Ton Miles

4 $\frac{\text{Passenger Miles}}{\text{Vehicle Miles}}$ or $\frac{\text{Ton Miles}}{\text{Vehicle Miles}}$

5 BTU/Passenger Mile or BTU/Ton Mile

NATIONAL ENERGY DEMAND PROJECTIONS

Country USA

Year 1985

WAES Case Case C

Agriculture, Mining, and Construction

	(A)	(B)	(C)	(D)	(E)	(F)	(G)	(H)	
Agriculture, Forestry, and Fishing (ISIC Major Division 1)	Value Added	Energy/Value Added	Area Harvested	Energy/ Area	Physical Output	Energy/ Output	Fossil Fuel [1]	Electricity [1]	
Farm Machinery			345×10^6 BTU/Acre	1.91×10^6 BTU/Acre			.66		(17)
Greenhouses									(18)
Other Agricultural Uses	$\$41.7 \times 10^9$	18,000 BTU/$	345×10^6 Acre	2.2×10^6 BTU/Acre	$\$97.7 \times 10^9$ [2]		.64	.13	(19)
Forestry (ISIC 12)					$18.9 \times 10^9 ft^3$				(20)
Fishing Boats (ISIC 13)									(21)
Other Fishing Use (ISIC 13)					7.3×10^9 lbs				(22)
Total Fertilizer (tons)		N ______	P ______	K ______		Total	1.30	.13	

	(A)	(B)	(C)	(D)	(E)	(F)	
Mining	Value Added	Energy/Value Added	Physical Output	Energy/Physical Output	Fossil Fuel	Electricity	
Handling and Internal Transportation							(23)
Other Energy Use	$\$6.09 \times 10^9$		$\$8.55 \times 10^9$ [2]		.26	.05	(24)
				Total			

	(A)	(B)	(C)	(D)	(E)	(F)	
Construction	Value Added	Energy/Value Added	Physical Output	Energy/Physical Output	Fossil Fuel	Electricity	
Handling and Internal Transportation	$\$93.0 \times 10^9$		$\$169.1 \times 10^9$ [2]		.94		(25)
Other Energy Use							(26)
				Total			
				Agriculture, Mining, and Construction Total	2.50	.18	

[1] Energy Totals in Quads = 10^{15} BTU.

[2] Dollar Value of Shipments.

NATIONAL ENERGY DEMAND PROJECTIONS

Country USA

Year 1985

WAES Case Case C

Commercial and Public [1]

	(A)	(B)	(C)	(D)	(E)	
Commercial	Floor Area [2]	Heat Gain or Loss/Area [3]	System Efficiency	Fossil Fuel [4]	Electricity [4]	
Space Heat (Fossil Fuel)	(10^9 Sq Ft) 31.86	6.40 to 4.95 x 10^4 BTU/Sq Ft	.70 to .80	2.51		(27)
Space Heat (Electricity)	4.25	4.95 x 10^4 BTU/Sq Ft	1.00		.21	(28)
Air Conditioning (Fossil Fuel)	1.26	2.02 x 10^4 BTU/Sq Ft	2.2 (COP)	.01		(29)
Air Conditioning (Electricity)	22.56	2.02 x 10^4 BTU/Sq Ft	2.5 (COP)		.18	(30)
		Energy/Area	Electricity/Area			
Lighting	36.11		.69 x 10^4 BTU/Sq Ft		.25	(31)
Miscellaneous				.74	1.42	(32)
Presently Unknown						(33)
			Total	3.26	2.06	
Public	Floor Area	Heat Gain or Loss/Area	System Efficiency	Fossil Fuel	Electricity	
Space Heat (Fossil Fuel)						(34)
Space Heat (Electricity)						(35)
Air Conditioning (Fossil Fuel)						(36)
Air Conditioning (Electricity)						(37)
		Energy/Area	Electricity/Area			
Lighting						(38)
Miscellaneous						(39)
Presently Unknown						(40)
			Total			
			Commercial and Public Total			

[1] Public is included in the commercial data.

[2] Total Floor Space = 36.11 x 10^9 Sq. Ft.

[3] The first number is heat gain or loss/square foot for old commercial units, and the second number is for new commercial units (i.e., buildings built after 1972).

[4] Energy Totals in Quads = 10^{15} BTU.

NATIONAL ENERGY DEMAND PROJECTIONS

Country USA

Year 1985

WAES Case Case C

Industrial

	(A) Value Added	(B) Energy/Value Added	(C) Physical Output	(D) Energy/Output	(E) % Exported	(F) Fossil Fuel [1]	(G) Electricity	(H) Feedstocks	
	(10^9 $)	(MBTU/$)							
Iron and Steel (ISIC 371)	31.23	182.8				5.38	.33		(41)
Nonferrous Metals (ISIC 372)	15.89	89.4				.96	.46		(42)
Paper and Allied (ISIC 34)	19.70	92.2				1.67	.15		(43)
Chemicals and Allied (ISIC 35 minus ISIC 353)						4.70	.66	6.49	(44)
Mineral Products (ISIC 36)	19.94	92.3				1.72	.12		(45)
Food and Related (ISIC 31)	49.46	25.9				1.13	.15		(46)
Transport Eqpt. (ISIC 384)	61.66	9.9				.46	.15		(47)
Key Industry (N.E.C.) [2]								1.71	(48)
Key Industry (N.E.C.)									(49)
Misc. Manufacturing	311.42	11.3				2.74	.78		(50)
					Total	18.76	2.80	8.20	

N.E.C. = Not Elsewhere Classified

[1] Energy Totals in Quads = 10^{15} BTU.

[2] Asphalt and Road Tar Feedstocks.

NATIONAL ENERGY DEMAND PROJECTIONS

Country USA

Year 1985

WAES Case Case C

Residential

Population 234×10^6 Persons/Dwelling 2.70 Dwellings

Central Space Heating		(A) Dwellings	(B) Heat Loss[1]		(C) System Efficiency	(D)	(E) Fossil Fuel[2]	(F) Electricity[2]	
Single Family	Fossil Fuels	(10^6 Units) 43.82	786	524	.66 (old units) .75 (new units)		4.85		(51)
	Electricity	8.64	658	461	1.2 (COP)			.37	(52)
Town or Row Houses[4]	Fossil Fuels	4.71	436	396	.66 (old units) .75 (new units)		.27		(53)
	Electricity	2.63		396	1.2 (COP)			.09	(54)
Apartments	Fossil Fuels	21.05	520	417	.66 (old units) .75 (new units)		1.51		(55)
	Electricity	5.75	435	405	1.2 (COP)			.20	(56)

	% Saturation (Fossil Fuels)	% Saturation (Electricity)	Energy/Unit (Fossil Fuels)	Energy/Unit[3] (Electricity)			
Room Heating[5]	12	57	7.3×10^6 BTU	846	.08	.14	(57)
Air Conditioning (Window)		33		929		.09	(58)
Air Conditioning (Central)		50		1840		.27	(59)
Water Heating	66	34	22.9×10^6 BTU	4905	1.30	.49	(60)
Refrigerators		115		889		.30	(61)
Lighting		100		716		.21	(62)
Cooking	30	65	8.3×10^6 BTU	1093	.22	.21	(63)
Dishwashers		60		290		.05	(64)
Freezers		38		807		.09	(65)
Television		165		300		.15	(66)
Other Key Appliances[6]		78		117		.03	(67)
Misc. Appliances		100		1020		.30	(68)
Potential Appliances							(69)
				Total	8.23	2.99	

[1] Heat loss is in 10^5 BTU/Year. The first number is the heat loss for old dwelling units, and the second number is for new dwelling units (i.e., those built after 1972).

[2] Energy Totals in Quads = 10^{15} BTU.

[3] Energy/Unit (Electricity) is in kilowat-hours=3412 BTU

[4] Mobile Homes

[5] Clothes Dryer

[6] Clothes Washer

NATIONAL ENERGY DEMAND PROJECTIONS

Country USA

Year 1985

WAES Case Case D

Transportation

	(A)	(B)	(C)	(D)	(E)	(F)	(G)	
	Population/Auto	Distance/Auto/Yr.[1]	Load Factor	Fraction Urban or Rural	Energy/Distance	Fossil Fuel[2]	Electricity[2]	
Urban Auto		$1,392.3 \times 10^9$			6365 BTU/Mi	8.86		(1)
Rural Auto					(19.6 m.p.g.)			(2)
	Activity Level[3]	% Electrified	Load Factor[4]	Fossil Fuel/ Activity Level[5]	Electricity/Activity			
Urban Bus	35.9		17.9	1560		.06		(3)
Intercity Bus	33.4		20.0	1170		.04		(4)
Rail (Passenger)	10.7			2430		.03	.003	(5)
Air (Passenger, Domestic)	250.0			7520		1.88		(6)
Air (Passenger, International)	132.1			3830		.51		(7)
Ship (Passenger)								(8)
Rail (Freight)	1002.2			630		.63		(9)
Air (Freight, Domestic)	10.5			26500		.28		(10)
Air (Freight, International)						.17		(11)
Truck (Local)	108.5			6750		.73		(12)
Truck (Intercity)	618.5			2150		1.33		(13)
Ship (Freight, Domestic)	714.9			490		.35		(14)
Ship (Freight, International)						.42		(15)
Miscellaneous						4.49	.020	(16)
					Total	19.78	.023	

[1] Total Vehicle Miles

[2] Energy Totals in Quads = 10^{15} BTU

[3] 10^9 Passenger Miles or 10^9 Ton Miles

[4] $\frac{\text{Passenger Miles}}{\text{Vehicle Miles}}$ or $\frac{\text{Ton Miles}}{\text{Vehicle Miles}}$

[5] BTU/Passenger Mile or BTU/Ton Mile

NATIONAL ENERGY DEMAND PROJECTIONS

Country USA

Year 1985

WAES Case Case D

Agriculture, Mining, and Construction

	(A)	(B)	(C)	(D)	(E)	(F)	(G)	(H)	
Agriculture, Forestry, and Fishing (ISIC Major Division 1)	Value Added	Energy/Value Added	Area Harvested	Energy/Area	Physical Output	Energy/Output	Fossil Fuel [1]	Electricity [1]	
Farm Machinery			345 x 10^6 Acre	1.91 x 10^6 BTU/Acre			.66		(17)
Greenhouses									(18)
Other Agricultural Uses	$36.9 x 10^9	19,000 BTU/$	345 x 10^6 Acre	2.00 BTU/Acre	$86.5 x 10^9 [2]		.57	.12	(19)
Forestry (ISIC 12)					17.1 x 10^9Ft				(20)
Fishing Boats (ISIC 13)									(21)
Other Fishing Use (ISIC 13)					6.6 x 10^9lbs				(22)
Total Fertilizer (tons)		N ____	P ____	K ____		Total	1.23	.12	

	(A)	(B)	(C)	(D)	(E)	(F)	
Mining	Value Added	Energy/Value Added	Physical Output	Energy/Physical Output	Fossil Fuel	Electricity	
Handling and Internal Transportation							(23)
Other Energy Use	$4.69 x 10^9		$6.58 x 10^9 [2]		.20	.04	(24)
				Total			

	(A)	(B)	(C)	(D)	(E)	(F)	
Construction	Value Added	Energy/Value Added	Physical Output	Energy/Physical Output	Fossil Fuel	Electricity	
Handling and Internal Transportation	$52.8 x 10^9		$94.2 x 10^9 [2]		.53		(25)
Other Energy Use							(26)
				Total			
				Agriculture, Mining, and Construction Total	1.96	.16	

[1] Energy Totals in Quads = 10^{15} BTU.

[2] Dollar Value of Shipments.

NATIONAL ENERGY DEMAND PROJECTIONS

Country USA

Year 1985

WAES Case Case D

Commercial and Public [1]

Commercial	(A) Floor Area [2]	(B) Heat Gain or Loss/Area [3]	(C) System Efficiency	(D) Fossil Fuel [4]	(E) Electricity [4]	
Space Heat (Fossil Fuel)	(10^9 Sq Ft) 30.19	7.10 to 6.18 x 10^4 BTU/Sq Ft	.70 to .75	2.85		(27)
Space Heat (Electricity)	3.7	6.18 x 10^4 BTU/Sq Ft	1.0		.23	(28)
Air Conditioning (Fossil Fuel)	1.19	2.379 x 10^4 BTU/Sq Ft	2.0 (COP)	.01		(29)
Air Conditioning (Electricity)	20.81	2.379 x 10^4 BTU/Sq Ft	2.2 (COP)		.23	(30)
		Energy/Area	Electricity/Area			
Lighting	33.89		1.63 x 10^4 BTU/Sq Ft		.55	(31)
Miscellaneous				.79	1.44	(32)
Presently Unknown						(33)
			Total	3.65	2.45	

Public	Floor Area	Heat Gain or Loss/Area	System Efficiency	Fossil Fuel	Electricity	
Space Heat (Fossil Fuel)						(34)
Space Heat (Electricity)						(35)
Air Conditioning (Fossil Fuel)						(36)
Air Conditioning (Electricity)						(37)
		Energy/Area	Electricity/Area			
Lighting						(38)
Miscellaneous						(39)
Presently Unknown						(40)
			Total			
			Commercial and Public Total			

1 Public is included in the commercial data.

2 Total Floor Space = 33.89 x 10^9 Sq. Ft.

3 The first number is heat gain or loss/square foot for old commercial units, and the second number is for new commercial units (i.e., buildings built after 1972).

4 Energy Totals in Quads = 10^{15} BTU.

NATIONAL ENERGY DEMAND PROJECTIONS

Country USA

Year 1985

WAES Case Case D

Industrial

	(A) Value Added	(B) Energy/Value Added	(C) Physical Output	(D) Energy/Output	(E) % Exported	(F) Fossil Fuel[1]	(G) Electricity[1]	(H) Feedstocks	
	(10^9 $)	(MBTU/$)							
Iron and Steel (ISIC 371)	24.34	193.9				4.45	.27		(41)
Nonferrous Metals (ISIC 372)	13.38	94.9				.86	.41		(42)
Paper and Allied (ISIC 34)	17.98	100.0				1.65	.15		(43)
Chemicals and Allied (ISIC 35 minus ISIC 353)						4.55	.64	6.12	(44)
Mineral Products (ISIC 36)	16.39	100.1				1.53	.11		(45)
Food and Related (ISIC 31)	49.77	25.9				1.14	.15		(46)
Transport Eqpt. (ISIC 384)	56.59	9.7				.42	.13		(47)
Key Industry (N.E.C.)[2]								.98	(48)
Key Industry (N.E.C.)									(49)
Misc. Manufacturing	268.82	11.3				2.36	.67		(50)
N.E.C. = Not Elsewhere Classified					**Total**	16.96	2.53	7.10	

[1] Energy Totals in Quads = 10^{15} BTU.

[2] Asphalt and Road Tar Feedstocks.

NATIONAL ENERGY DEMAND PROJECTIONS

Country USA

Year 1985

WAES Case Case D

Residential

Population 234×10^6 Persons/Dwelling 2.74 Dwellings 85.57×10^6

		(A)	(B)		(C)	(D)	(E)	(F)	
Central Space Heating		Dwellings	Heat Loss[1]		System Efficiency		Fossil Fuel[2]	Electricity[2]	
Single Family	Fossil Fuels	(10^6 Units) 43.46	874	743	.63 (old units) .7 (new units)		5.80		(51)
	Electricity	8.45	700	658	1.0			.57	(52)
Town or Row Houses[4]	Fossil Fuels	4.66	495		.63 (old units) .7 (new units)		.34		(53)
	Electricity	2.58	495		1.0			.13	(54)
Apartments	Fossil Fuels	20.85	578	491	.63 (old units) .7 (new units)		1.79		(55)
	Electricity	5.57	463	435	1.0			.24	(56)

	% Saturation (Fossil Fuels)	% Saturation (Electricity)	Energy/Unit (Fossil Fuels)	Energy/Unit[3] (Electricity)			
Room Heating[5]	12	53	8.1×10^6 BTU	900	.08	.14	(57)
Air Conditioning (Window)		33		1100		.11	(58)
Air Conditioning (Central)		46		2200		.30	(59)
Water Heating	70	30	28.5×10^6 BTU	5450	1.71	.48	(60)
Refrigerators		112		1290		.42	(61)
Lighting		100		800		.23	(62)
Cooking	34	61	8.7×10^6 BTU	1150	.25	.20	(63)
Dishwashers		56		360		.06	(64)
Freezers		38		1200		.13	(65)
Television		160		300		.14	(66)
Other Key Appliances[6]		78		120		.03	(67)
Misc. Appliances		100		1030		.30	(68)
Potential Appliances							(69)
				Total	9.97	3.48	

[1] Heat loss is in 10^5 BTU/Year. The first number is the heat loss for old dwelling units, and the second number is for new dwelling units (i.e., those built after 1972).

[2] Energy Totals in Quads = 10^{15} BTU.

[3] Energy/Unit (Electricity) is in kilowat-hours=3412 BTU.

[4] Mobile Homes

[5] Clothes Dryer

[6] Clothes Washer

NATIONAL ENERGY DEMAND PROJECTIONS

Country USA

Year 1985

WAES Case Case E

Transportation

	(A)	(B)	(C)	(D)	(E)	(F)	(G)	
	Population/Auto	Distance/Auto/Yr.[1]	Load Factor	Fraction Urban or Rural	Energy/Distance	Fossil Fuel[2]	Electricity[2]	
Urban Auto		1,449.3 x 10^9			6417 BTU/Mi	9.30		(1)
Rural Auto					(19.5 m.p.g.)			(2)
	Activity Level[3]	% Electrified	Load Factor[4]	Fossil Fuel/ Activity Level[5]	Electricity/Activity			
Urban Bus	38.2		17.9	1570		.06		(3)
Intercity Bus	33.9		20.1	1180		.04		(4)
Rail (Passenger)	11.1			2700		.03	.004	(5)
Air (Passenger, Domestic)	271.6			7880		2.14		(6)
Air (Passenger, International)	144.9			4070		.59		(7)
Ship (Passenger)								(8)
Rail (Freight)	1034.0			630		.65		(9)
Air (Freight, Domestic)	14.3			25900		.37		(10)
Air (Freight, International)				22400		.18		(11)
Truck (Local)	114.2			6920		.79		(12)
Truck (Intercity)	648.3			2240		1.45		(13)
Ship (Freight, Domestic)	802.3			490		.39		(14)
Ship (Freight, International)						.43		(15)
Miscellaneous						4.76	.02	(16)
					Total	21.18	.024	

[1] Total Vehicle Miles

[2] Energy Totals in Quads = 10^{15} BTU

[3] 10^9 Passenger Miles or 10^9 Ton Miles

[4] $\frac{\text{Passenger Miles}}{\text{Vehicle Miles}}$ or $\frac{\text{Ton Miles}}{\text{Vehicle Miles}}$

[5] BTU/Passenger Mile or BTU/Ton Mile

NATIONAL ENERGY DEMAND PROJECTIONS

Country USA

Year 1985

WAES Case Case E

Agriculture, Mining, and Construction

Agriculture, Forestry, and Fishing (ISIC Major Division 1)	(A) Value Added	(B) Energy/Value Added	(C) Area Harvested	(D) Energy/Area	(E) Physical Output	(F) Energy/Output	(G) Fossil Fuel [1]	(H) Electricity [1]	
Farm Machinery			345×10^6 Acres	1.91×10^6 BTU/Acres			.66		(17)
Greenhouses									(18)
Other Agricultural Uses	41.3×10^9\$	18,000 BTU/\$	345×10^6 Acres	2.2×10^6 BTU/Acres	96.63×10^9\$ [2]		.63	.13	(19)
Forestry (ISIC 12)					$27.01 \times 10^9 Ft^3$				(20)
Fishing Boats (ISIC 13)									(21)
Other Fishing Use (ISIC 13)					10.43×10^9 lbs				(22)
Total Fertilizer (tons)		N ____	P ____	K ____		Total	1.29	.13	

Mining	(A) Value Added	(B) Energy/Value Added	(C) Physical Output	(D) Energy/Physical Output	(E) Fossil Fuel	(F) Electricity	
Handling and Internal Transportation							(23)
Other Energy Use	6.60×10^9		9.27×10^9 \$ [2]		.28	.06	(24)
				Total			

Construction	(A) Value Added	(B) Energy/Value Added	(C) Physical Output	(D) Energy/Physical Output	(E) Fossil Fuel	(F) Electricity	
Handling and Internal Transportation	92.2×10^9		\$$167.6 \times 10^9$ [2]		.93		(25)
Other Energy Use							(26)
				Total			
				Agriculture, Mining, and Construction Total	2.50	.19	

[1] Energy Totals in Quads = 10^{15} BTU.

[2] Dollar Value of Shipments.

NATIONAL ENERGY DEMAND PROJECTIONS

Country USA

Year 1985

WAES Case Case E

Commercial and Public [1]

	(A)	(B)	(C)	(D)	(E)	
Commercial	Floor Area [2]	Heat Gain or Loss/Area [3]	System Efficiency	Fossil Fuel [4]	Electricity [4]	
Space Heat (Fossil Fuel)	(10^9 Sq Ft) 31.94	7.10 to 6.46 x 10^4 BTU/Sq Ft	.70 to .75	3.03		(27)
Space Heat (Electricity)	4.3	6.46 x 10^4 BTU/Sq Ft	1.0		.28	(28)
Air Conditioning (Fossil Fuel)	1.27	2.430 x 10^4 BTU/Sq Ft	1.8 (COP)	.02		(29)
Air Conditioning (Electricity)	22.58	2.430 x 10^4 BTU/Sq Ft	2.0 (COP)		.27	(30)
		Energy/Area	Electricity/Area			
Lighting	36.24		1.77 x 10^4 BTU/Sq Ft		.64	(31)
Miscellaneous				.84	1.56	(32)
Presently Unknown						(33)
			Total	3.89	2.75	

Public	Floor Area	Heat Gain or Loss/Area	System Efficiency	Fossil Fuel	Electricity	
Space Heat (Fossil Fuel)						(34)
Space Heat (Electricity)						(35)
Air Conditioning (Fossil Fuel)						(36)
Air Conditioning (Electricity)						(37)
		Energy/Area	Electricity/Area			
Lighting						(38)
Miscellaneous						(39)
Presently Unknown						(40)
			Total			
			Commercial and Public Total			

[1] Public is included in the commercial data.

[2] Total Floor Space = 36.24 x 10^9 Sq. Ft.

[3] The first number is heat gain or loss/square foot for old commercial units, and the second number is for new commercial units (i.e., buildings built after 1972).

[4] Energy Totals in Quads = 10^{15} BTU.

NATIONAL ENERGY DEMAND PROJECTIONS

Country USA

Year 1985

WAES Case Case E

Industrial

	(A) Value Added	(B) Energy/Value Added	(C) Physical Output	(D) Energy/Output	(E) % Exported	(F) Fossil Fuel[1]	(G) Electricity[1]	(H) Feedstocks	
	(10^9 $)	(MBTU/$)							
Iron and Steel (ISIC 371)	30.92	202.1				5.89	.36		(41)
Nonferrous Metals (ISIC 372)	15.89	98.2				1.06	.50		(42)
Paper and Allied (ISIC 34)	19.70	104.0				1.88	.17		(43)
Chemicals and Allied (ISIC 35 minus ISIC 353)						5.19	.72	6.52	(44)
Mineral Products (ISIC 36)	19.77	103.7				1.92	.13		(45)
Food and Related (ISIC 31)	49.15	26.4				1.15	.15		(46)
Transport Eqpt. (ISIC 384)	61.39	10.8				.50	.16		(47)
Key Industry (N.E.C.) [2]								1.70	(48)
Key Industry (N.E.C.)									(49)
Misc. Manufacturing	310.51	12.2				2.95	.84		(50)
					Total	20.54	3.03	8.22	

N.E.C. = Not Elsewhere Classified

[1] Energy Totals in Quads = 10^{15} BTU.

[2] Asphalt and Road Tar Feedstocks.

NATIONAL ENERGY DEMAND PROJECTIONS

Country USA

Year 1985

WAES Case Case E

Residential

Population 234 x 10^6 Persons/Dwelling 2.71 Dwellings 86.53 x 10^6

		(A)	(B)	(C)	(D)	(E)	(F)	
Central Space Heating		Dwellings	Heat Loss [1]	System Efficiency		Fossil Fuel [2]	Electricity [2]	
Single Family	Fossil Fuels	(10^6 Units) 43.82	874 813	.63 (old units) .67 (new units)		5.95		(51)
	Electricity	8.64	734 705	1.0			.62	(52)
Town or Row Houses[4]	Fossil Fuels	4.71	495	.63 (old units) .67 (new units)		.36		(53)
	Electricity	2.63	495	1.0			.13	(54)
Apartments	Fossil Fuels	21.01	578 509	.63 (old units) .67 (new units)		1.83		(55)
	Electricity	5.72	485 466	1.0			.27	(56)
		% Saturation (Fossil Fuels)	% Saturation (Electricity)	Energy/Unit (Fossil Fuels)	Energy/Unit [3] (Electricity)			
Room Heating [5]		12	60	8.62 x 10^6 BTU	920	.09	.16	(57)
Air Conditioning (Window)			33		1256		.12	(58)
Air Conditioning (Central)			52		3162		.49	(59)
Water Heating		66	34	29.87 x 10^6 BTU	5514	1.71	.55	(60)
Refrigerators			117		1316		.45	(61)
Lighting			100		1000		.30	(62)
Cooking		32	67	9.14 x 10^6 BTU	1170	.25	.23	(63)
Dishwashers			67		367		.05	(64)
Freezers			42		1248		.15	(65)
Television			170		300		.15	(66)
Other Key Appliances [6]			80		120		.03	(67)
Misc. Appliances			100		1138		.34	(68)
Potential Appliances								(69)
					Total	10.19	4.04	

[1] Heat loss is in 10^5 BTU/Year. The first number is the heat loss for old dwelling units, and the second number is for new dwelling units (i.e., those built after 1972).

[2] Energy Totals in Quads = 10^{15} BTU.

[3] Energy/Unit (Electricity) is in kilowat-hours= 3412 BTU

[4] Mobile Homes

[5] Clothes Dryer

[6] Clothes Washer

NATIONAL ENERGY DEMAND PROJECTIONS

Country USA

Year 1985

WAES Case Case C'

Transportation

	(A)	(B)	(C)	(D)	(E)	(F)	(G)	
	Population/Auto	Distance/Auto/Yr.[1]	Load Factor	Fraction Urban or Rural	Energy/Distance	Fossil Fuel[2]	Electricity[2]	
Urban Auto		1449.3			6415 BTU/Mi	9.27		(1)
Rural Auto					(19.5 m.p.g.)			(2)
	Activity Level[3]	% Electrified	Load Factor[4]	Fossil Fuel/ Activity Level[5]	Electricity/Activity			
Urban Bus	37.8		17.9	1590		.06		(3)
Intercity Bus	34.4		20.0	1160		.04		(4)
Rail (Passenger)	11.0			2730		.03	.003	(5)
Air (Passenger, Domestic)	268.0			7540		2.02		(6)
Air (Passenger, International)	142.0			3800		.54		(7)
Ship (Passenger)								(8)
Rail (Freight)	1033.2			630		.65		(9)
Air (Freight, Domestic)	11.3			26500		.30		(10)
Air (Freight, International)						.18		(11)
Truck (Local)	114.2			6740		.77		(12)
Truck (Intercity)	651.0			2150		1.40		(13)
Ship (Freight, Domestic)	737.0			490		.36		(14)
Ship (Freight, International)						.43		(15)
Miscellaneous						4.68	.02	(16)
					Total	20.73	.023	

[1] Total Vehicle Miles

[2] Energy Totals in Quads = 10^{15} BTU

[3] 10^9 Passenger Miles or 10^9 Ton Miles

[4] $\frac{\text{Passenger Miles}}{\text{Vehicle Miles}}$ or $\frac{\text{Ton Miles}}{\text{Vehicle Miles}}$

[5] BTU/Passenger Mile or BTU/Ton Mile

NATIONAL ENERGY DEMAND PROJECTIONS

Country USA

Year 1985

WAES Case Case C'

Agriculture, Mining, and Construction

	(A)	(B)	(C)	(D)	(E)	(F)	(G)	(H)	
Agriculture, Forestry, and Fishing (ISIC Major Division 1)	Value Added	Energy/Value Added	Area Harvested	Energy/Area	Physical Output	Energy/Output	Fossil Fuel [1]	Electricity [1]	
Farm Machinery			345×10^6 Acres	1.91×10^6 BTU/Acre			.66		(17)
Greenhouses									(18)
Other Agricultural Uses	41.7×10^9 \$	18,000 BTU/\$	345×10^6 Acres	2.2×10^6 Acres	97.7×10^9 \$ [2]		.64	.13	(19)
Forestry (ISIC 12)					$18. \times 10^9$ Ft^3				(20)
Fishing Boats (ISIC 13)									(21)
Other Fishing Use (ISIC 13)					7.3×10^9 lbs				(22)
Total Fertilizer (tons)		N ____	P ____	K ____		Total	1.30	.13	

	(A)	(B)	(C)	(D)	(E)	(F)	
Mining	Value Added	Energy/Value Added	Physical Output	Energy/Physical Output	Fossil Fuel	Electricity	
Handling and Internal Transportation							(23)
Other Energy Use	6.09×10^9 \$		8.55×10^9 \$ [2]		.26	.05	(24)
				Total			

	(A)	(B)	(C)	(D)	(E)	(F)	
Construction	Value Added	Energy/Value Added	Physical Output	Energy/Physical Output	Fossil Fuel	Electricity	
Handling and Internal Transportation	93.0×10^9 \$		169.1×10^9 \$ [2]		.94		(25)
Other Energy Use							(26)
				Total			
				Agriculture, Mining, and Construction Total	2.50	.18	

[1] Energy Totals in Quads = 10^{15} BTU.

[2] Dollar Value of Shipments.

NATIONAL ENERGY DEMAND PROJECTIONS

Country USA

Year 1985

WAES Case Case C'

Commercial and Public [1]

	(A)	(B)	(C)	(D)	(E)	
Commercial	Floor Area [2]	Heat Gain or Loss/Area [3]	System Efficiency	Fossil Fuel [4]	Electricity [4]	
Space Heat (Fossil Fuel)	(10^9 Sq Ft) 31.86	7.10 to 6.18 x 10^4 BTU/Sq Ft	.70 to .75	2.97		(27)
Space Heat (Electricity)	4.25	6.18 x 10^4 BTU/Sq Ft	1.0		.26	(28)
Air Conditioning (Fossil Fuel)	1.26	2.379 x 10^4 BTU/Sq Ft	2.0 (COP)	.02		(29)
Air Conditioning (Electricity)	22.56	2.379 x 10^4 BTU/Sq Ft	2.2 (COP)		.24	(30)
		Energy/Area	Electricity/Area			
Lighting	36.11		1.63 x 10^4 BTU/Sq Ft		.59	(31)
Miscellaneous				.83	1.52	(32)
Presently Unknown						(33)
			Total	3.82	2.61	

Public	Floor Area	Heat Gain or Loss/Area	System Efficiency	Fossil Fuel	Electricity	
Space Heat (Fossil Fuel)						(34)
Space Heat (Electricity)						(35)
Air Conditioning (Fossil Fuel)						(36)
Air Conditioning (Electricity)						(37)
		Energy/Area	Electricity/Area			
Lighting						(38)
Miscellaneous						(39
Presently Unknown						(40)
			Total			
			Commercial and Public Total			

[1] Public is included in the commercial data.

[2] Total Floor Space = 36.11 x 10^9 Sq. Ft.

[3] The first number is heat gain or loss/square foot for old commercial units, and the second number is for new commercial untis (i.e., buildings built after 1972).

[4] Energy Totals in Quads = 10^{15} BTU.

NATIONAL ENERGY DEMAND PROJECTIONS

Country USA

Year 1985

WAES Case Case C'

Industrial

	(A)	(B)	(C)	(D)	(E)	(F)	(G)	(H)	
	Value Added	Energy/Value Added	Physical Output	Energy/Output	% Exported	Fossil Fuel [1]	Electricity[1]	Feedstocks	
	$(10^9\ \$)$	(MBTU/$)							
Iron and Steel (ISIC 371)	31.23	194.0				5.71	.35		(41)
Nonferrous Metals (ISIC 372)	15.89	94.4				1.02	.48		(42)
Paper and Allied (ISIC 34)	19.70	100.0				1.81	.16		(43)
Chemicals and Allied (ISIC 35 minus ISIC 353)						4.84	.67	6.49	(44)
Mineral Products (ISIC 36)	19.94	99.8				1.86	.13		(45)
Food and Related (ISIC 31)	49.46	25.9				1.13	.15		(46)
Transport Eqpt. (ISIC 384)	61.66	9.9				.46	.15		(47)
Key Industry (N.E.C.) [2]								1.71	(48)
Key Industry (N.E.C.)									(49)
Misc. Manufacturing	311.42	11.3				2.74	.78		(50)
N.E.C. = Not Elsewhere Classified					**Total**	19.57	2.87	8.20	

[1] Energy Totals in Quads = 10^{15} BTU.

[2] Asphalt and Road Tar Feedstocks.

NATIONAL ENERGY DEMAND PROJECTIONS

Country USA

Year 1985

WAES Case Case C'

Residential

Population 234×10^6 Persons/Dwelling 2.70 Dwellings 86.6×10^6

		(A)	(B)	(C)	(D)	(E)	(F)	
Central Space Heating		Dwellings	Heat Loss[1]	System Efficiency		Fossil Fuel[2]	Electricity[2]	
Single Family	Fossil Fuels	(10^6 Units) 43.82	874 743	.63 (old units) .7 (new units)		5.84		(51)
	Electricity	8.64	700 658	1.0			.58	(52)
Town or Row Houses[4]	Fossil Fuels	4.71	495	.63 (old units) .7 (new units)		.35		(53)
	Electricity	2.63	495	1.0			.13	(54)
Apartments	Fossil Fuels	21.05	578 491	.63 (old units) .7 (new units)		1.80		(55)
	Electricity	5.75	463 435	1.0			.25	(56)
		% Saturation (Fossil Fuels)	% Saturation (Electricity)	Energy/Unit (Fossil Fuels)	Energy/Unit[3] (Electricity)			
Room Heating[5]		12	57	8.1×10^6 BTU	900	.08	.15	(57)
Air Conditioning (Window)			33		1100		.11	(58)
Air Conditioning (Central)			50		2200		.33	(59)
Water Heating		66	34	28.5×10^6 BTU	5450	1.63	.55	(60)
Refrigerators			115		1290		.44	(61)
Lighting			100		800		.24	(62)
Cooking		30	65	8.7×10^6 BTU	1150	.23	.22	(63)
Dishwashers			60		360		.06	(64)
Freezers			38		1200		.13	(65)
Television			165		300		.15	(66)
Other Key Appliances[6]			78		120		.03	(67)
Misc. Appliances			100		1070		.32	(68)
Potential Appliances								(69)
					Total	9.93	3.69	

[1] Heat loss is in 10^5 BTU/Year. The first number is the heat loss for old dwelling units, and the second number is for new dwelling units (i.e., those built after 1972).

[2] Energy Totals in Quads = 10^{15} BTU.

[3] Energy/Unit (Electricity) is in kilowat-hours=3412 BTU

[4] Mobile Homes

[5] Clothes Dryer

[6] Clothes Washer